Lecture Notes in Computer Science 15015

Founding Editors

Gerhard Goos
Juris Hartmanis

AF167412

The series Lecture Notes in Computer Science (LNCS), including its subseries Lecture Notes in Artificial Intelligence (LNAI) and Lecture Notes in Bioinformatics (LNBI), has established itself as a medium for the publication of new developments in computer science and information technology research, teaching, and education.

LNCS enjoys close cooperation with the computer science R & D community, the series counts many renowned academics among its volume editors and paper authors, and collaborates with prestigious societies. Its mission is to serve this international community by providing an invaluable service, mainly focused on the publication of conference and workshop proceedings and postproceedings. LNCS commenced publication in 1973.

Szilárd Zsolt Fazekas
Editor

Implementation and Application of Automata

28th International Conference, CIAA 2024
Akita, Japan, September 3–6, 2024
Proceedings

 Springer

Editor
Szilárd Zsolt Fazekas 🆔
Akita University
Akita, Japan

ISSN 0302-9743   ISSN 1611-3349 (electronic)
Lecture Notes in Computer Science
ISBN 978-3-031-71111-4   ISBN 978-3-031-71112-1 (eBook)
https://doi.org/10.1007/978-3-031-71112-1

Preface

The 28th International Conference on Implementation and Application of Automata (CIAA 2024) was organized by members of the Mathematical Science Course of Akita University. The conference took place from September 3–6, 2024, in Akita, Japan. This event was part of the CIAA conference series, a major international venue that brings together researchers in the field of automata theory and implementation. This year marked the first time that CIAA was organized in Japan. The previous editions of the conference were held in various locations over five continents: Famagusta (2023), Rouen (2022), Bremen (2021), Košice (2019), Charlottetown (2018), Marne-la-Vallée (2017), Seoul (2016), Umeå (2015), Giessen (2014), Halifax (2013), Porto (2012), Blois (2011), Winnipeg (2010), Sydney (2009), San Francisco (2008), Prague (2007), Taipei (2006), Nice (2005), Kingston (2004), Santa Barbara (2003), Tours (2002), Pretoria (2001), London, Ontario (2000), Potsdam (WIA 1999), Rouen (WIA 1998), and London, Ontario (WIA 1997 and WIA 1996). Due to the CoVid-19 pandemic, CIAA 2020, planned to be held in Loughborough, was canceled.

This volume of Lecture Notes in Computer Science contains the scientific papers presented at CIAA 2024, and the abstracts and papers of the invited speakers. The 24 regular papers were selected from 38 submissions covering various fields in the application, implementation, and theory of automata and related structures. Each paper was reviewed by three Program Committee members with the assistance of external referees and thoroughly discussed by the Program Committee. Papers were submitted by authors from various countries: Belgium, Canada, China, Colombia, Estonia, Finland, France, Germany, India, Indonesia, Italy, Japan, Kazakhstan, Netherlands, Portugal, Russia, Serbia, Slovakia, South Africa, South Korea, Sweden, the UK, and the USA.

CIAA 2024 featured four invited talks:

- Sang-Ki Ko (University of Seoul):

  Neuro-Symbolic Approach for Efficient Regular Expression Inference

- Orna Kupferman (Hebrew University of Jerusalem):

  Playing Games on Automata

- Carl-Fredrik Nyberg-Brodda (Korean Institute for Advanced Study):

  (Semi)group Languages

- Hiroyuki Seki (Nagoya University):

  Automata and Grammars for Data Words

We wish to thank everybody who contributed to the success of this conference: the authors for submitting their carefully prepared manuscripts, the invited speakers for their excellent presentations of topics related to the theme of the conference, the Program

Committee members and external referees for evaluating the submitted manuscripts, the session chairs, and the participants who made CIAA 2024 possible. We would also like to thank the editorial team of Lecture Notes in Computer Science at Springer for the opportunity to publish the proceedings in the series and for their guidance and support during the publication process. We are grateful to the members of the Organizing Committee and the local staff members who ensured that the conference proceeded according to plan.

The administration of the reviews and discussions was handled using the EasyChair system, which made the process seamless.

CIAA 2024 was financially supported by the International Exchange Program of the National Institute of Information and Communications Technology (NICT).

July 2024   Szilárd Zsolt Fazekas

Organization

Organizing Committee

Szilárd Zsolt Fazekas	Akita University, Japan
Shinnosuke Seki	University of Electro-Communications, Japan
Ryoma Sin'ya	Akita University, Japan
Akihiro Yamamura	Akita University, Japan

Program Committee Chair

Szilárd Zsolt Fazekas	Akita University, Japan

Steering Committee

Markus Holzer (Chair)	Justus Liebig University Giessen, Germany
Oscar Ibarra	University of California, Santa Barbara, USA
Sylvain Lombardy	Université de Bordeaux, France
Nelma Moreira	Universidade do Porto, Portugal
Kai T. Salomaa (Co-chair)	Queen's University, Canada
Hsu-Chun Yen	National Taiwan University, Taiwan

Program Committee

Marie-Pierre Béal	Université Gustave Eiffel, France
Cezar Câmpeanu	University of Prince Edward Island, Canada
Pascal Caron	University of Rouen, France
Giuseppa Castiglione	Università di Palermo, Italy
Erzsébet Csuhaj-Varjú	Eötvös Loránd University, Hungary
Frank Drewes	Umeå University, Sweden
Ömer Eğecioğlu	University of California, Santa Barbara, USA
Attila Egri-Nagy	Akita International University, Japan
Szilárd Zsolt Fazekas (Chair)	Akita University, Japan
Yo-Sub Han	Yonsei University, South Korea
Markus Holzer	Justus Liebig University Giessen, Germany
Galina Jirásková	Slovak Academy of Sciences, Slovakia

Jarkko Kari	University of Turku, Finland
Manfred Kufleitner	University of Stuttgart, Germany
Martin Kutrib	Justus Liebig University Giessen, Germany
Markus Lohrey	University of Siegen, Germany
Sylvain Lombardy	LaBRI - CNRS - Institut Polytechnique de Bordeaux, France
Andreas Malcher	Justus Liebig University Giessen, Germany
Andreas Maletti	Universität Leipzig, Germany
Florin Manea	University of Göttingen, Germany
Sebastian Maneth	Universität Bremen, Germany
Ian McQuillan	University of Saskatchewan, Canada
Roland Meyer	TU Braunschweig, Germany
Victor Mitrana	Universidad Politécnica de Madrid, Spain
Nelma Moreira	University of Porto, Portugal
František Mráz	Charles University, Czech Republic
Benedek Nagy	Eastern Mediterranean University, North Cyprus
Friedrich Otto	University of Kassel, Germany
Giovanni Pighizzini	University of Milan, Italy
Luca Prigioniero	Loughborough University, UK
Rogério Reis	University of Porto, Portugal
Michel Rigo	University of Liège, Belgium
Kai Salomaa	Queen's University, Canada
Shinnosuke Seki	University of Electro-Communications, Japan
Ryoma Sin'ya	Akita University, Japan
Nicholas Tran	Santa Clara University, USA
György Vaszil	University of Debrecen, Hungary
Mikhail Volkov	Ural Federal University, Russia
Hsu-Chun Yen	National Taiwan University, Taiwan

Additional Reviewers

Sabine Broda	Thomas Haas
Olivier Carton	Eren Keskin
Emilie Charlier	Sungmin Kim
Yu-Fang Chen	Mitja Kulczynski
Jürgen Dassow	Julien Leroy
Joel Day	Didier Lime
Guilherme Duarte	Brennan Lockinger
Herman Goulet-Ouellet	Kevin Lotz
Jan Grünke	Jean-Gabriel Luque
Jens Oliver Gutsfeld	Timothy Ng
Stefan Göller	Julie Parreaux

Bruno Patrou

Erik Paul

Narad Rampersad

Priscilla Raucci

Christian Rauch

Arseny Shur

Manon Stipulanti

Bianca Truthe

Matthias Wendlandt

Di-De Yen

Invited Talks
Abstracts

Neuro-Symbolic Approach for Efficient Regular Expression Inference

Sang-Ki Ko

University of Seoul, Seoul, South Korea

Abstract. Due to the practical importance of regular expressions (regexes, for short), much research has been done on automatically generating regexes from positive and negative string examples. We tackle the problem of learning to generate regexes faster and more precisely from positive and negative examples by relying on a novel approach called 'neural example splitting.' Using a neural network trained to group similar substrings from positive strings, our approach splits each positive example string into multiple parts. We propose an effective regex synthesis framework called 'SplitRegex' that synthesizes subregexes from 'split' positive substrings and produces the final regex by concatenating the synthesized subregexes. To verify the correctness of generated independent subregexes during the subregex synthesis process, we exploit negative strings by matching against the generated subregexes. As a result, we guarantee that the final regex rejects all negative strings. SplitRegex is a divide-and-conquer framework for learning target regexes by the following process: split (=divide) positive strings, infer partial regexes for multiple parts, which is much more accurate than the whole regex inference, and concatenate (=conquer) inferred regexes. We empirically demonstrate that the proposed SplitRegex framework improves the previous regex synthesis approaches over two practical datasets and one synthesized dataset.

Playing Games on Automata

Orna Kupferman

Hebrew University of Jerusalem, Jerusalem, Israel

Abstract. The interaction between a system and its environment corresponds to a game in which the system and the environment generate a word over the alphabet of assignments to the input and output signals. The system player wins if the generated word satisfies a desired specification. A correct reactive system has to satisfy its specification in all environments, and thus corresponds to a strategy for the system player in the above game. This view is the key to the game-based approach for reactive synthesis. It reduces synthesis to the problem of generating winning strategies in two-player games whose winning conditions are on-going behaviours, which can be specified by automata on infinite words. The talk presents the game-based approach, focusing on ways to cope with the fact that the game cannot be played over nondeterministic automata.

(Semi)group Languages

Carl-Fredrik Nyberg-Brodda

Korean Institute for Advanced Study, Seoul, South Korea

Abstract. Methods from formal language theory as applied to algebra date back many decades, and give a number of exciting new ways to approach problems in group and semigroup theory. I will give an overview of some of the core questions, their history, their connections with rewriting systems, and some of the (un)decidability results that one can encounter along the way. I will also discuss recent work on undecidability results for subsets defined by monoid automata joint with I. Foniqi and R. D. Gray (East Anglia), as well as recent work with A. Carvalho (NOVA Lisbon) on defining subsets of groups by way of automata and other language-theoretic constructions.

Contents

Invited Talk

Automata and Grammars for Data Words

Hiroyuki Seki$^{(\boxtimes)}$

Nagoya University, Nagoya, Japan
seki@i.nagoya-u.ac.jp

Abstract. Register automaton (RA) and register context-free grammar (RCFG) are extensions of finite automaton and context-free grammar by adding the ability of data manipulation in a restricted way. This paper reviews definitions and basic properties of RA and RCFG. As a related topic, logics on data words, namely, linear-time temporal logic (LTL) with freeze quantifier and two-variable first-order logic with data equality are explained. Finally, nominal automaton, which can be regarded as a group-theoretic generalization of RA, is briefly described.

Keywords: register automaton · register context-free grammar · LTL with freeze quantifier · first-order logic with data equality · nominal automaton

1 Introduction

Finite automaton (abbreviated as FA) and context-free grammar (abbreviated as CFG) have been extended in various ways. Among them are extensions of automaton and grammar by adding the ability of manipulating data values. However, such an extension easily brings a Turing machine-equivalent model such as two-counter machine. Register automaton (abbreviated as RA) [8,13,17,23], pebble automaton [19,21] and data automaton [5] are extensions of FA that can manipulate data values in restricted ways. Attention has been paid to these extensions of FA as formal models of query languages for structured documents such as XML because a structured document can be modeled as a tree or a graph where data values are associated with nodes and a query on a document can be specified as a combination of a regular pattern and a condition on data values [16,18].

For query processing and optimization, the decidability of basic properties of queries is desirable. The membership problem that asks for a given query q and an element e in a document whether e is in the answer set of q is the most basic problem. The combined complexity of the membership problem is the complexity in terms of both the size of an automaton and that of an input word and the data complexity is the complexity on the size of an input word only. In application, the size of a query (an automaton) is usually smaller than that of a data (an input data word). It is desirable that the data complexity is small while the combined complexity is rather a criterion of the expressive succinctness

© The Author(s), under exclusive license to Springer Nature Switzerland AG 2024
S. Z. Fazekas (Ed.): CIAA 2024, LNCS 15015, pp. 3–16, 2024.
https://doi.org/10.1007/978-3-031-71112-1_1

of the query language. In [18], it is argued that RA is an appropriate model of query language because RA is the only model that has polynomial-time data complexity for the membership problem among the above extensions of FA. The satisfiability or (non)emptiness problem asking whether the answer set of a given query is nonempty is also important for query optimization.

RCFG was introduced in [7], which is a natural extension of both RA and CFG and shares good properties with these two models. In application, while RA has the power of expressing regular patterns on paths of a tree or a graph, it cannot represent tree patterns or patterns over branching paths. The latter kind of patterns can be represented by RCFG.

There are computational logic dealing with data words, namely, linear-time temporal logic with freeze quantifier ($\mathrm{LTL}^{\downarrow}$) [8,9] and two-variable first-order logic with data equality ($\mathrm{FO}^2(\sim, <, +1)$) [2]. A freeze quantifier in $\mathrm{LTL}^{\downarrow}$ corresponds to an update operation on registers in RA. In $\mathrm{FO}^2(\sim, <, +1)$, the equality test on data values is added to the classical (two-variable) first-order logic on words.

In RA and RCFG, it is crucial to consider instead of individual data values, the equivalence relation on the contents of registers for finitely representing a computation. The nominal computation model [3] can be seen as a group-theoretic abstraction of register models in this sense.

In this paper, we first review RA and its basic properties. Then, we proceed to RCFG and its properties, together with register type as a finite representation of computation. Next, we review the above mentioned two logics for data words. Finally, nominal automaton is briefly reviewed.

2   Register Automaton

Register automaton was introduced in [13] as finite automaton over infinite alphabet. In this paper, we follow [8,17], using the notion based on data words.

For a non-negative integer k, let $[k] = \{1, \ldots, k\}$. Assume we are given a finite alphabet Σ and a countable set D of data values containing a special element $\bot \in D$ representing the initial value. An RA has a finite control and a finite number of registers, each of which stores a data value. An input to an RA, say M, is a finite string $(a_1, d_1) \cdots (a_n, d_n)$ where $a_i \in \Sigma$ and $d_i \in D$ with $i \in [n]$ for some $n \geq 0$, which is called a *data word*. Each time M reads a pair (a_i, d_i) of a symbol and a data value, M tests whether the contents of a specified register and d_i are equal, and depending on the result of the test, M updates the contents of specified registers to d_i and also updates its finite control. Note that M can perform no operation (e.g., addition) on the contents of registers.

Let x_i be a variable representing the contents of the i-th register and let $X_k = \{x_1, \ldots, x_k\}$ be a finite set of k different variables. Let F_k be the set of *guard expressions* over k registers defined by:

$$\varphi := x_1^= \mid \ldots \mid x_k^= \mid \neg\varphi \mid \varphi \vee \varphi.$$

Intuitively, $x_i^=$ holds iff the current input data value is equal to the contents of the i-th register. We write $x_i^{\neq}$ to mean $\neg(x_i^=)$.

Definition 1 (RA). *For a non-negative integer k, a register automaton with k registers (abbreviated as k-RA) over Σ and D is a quadruple $M = (Q, Q_0, T, Q_f)$ where*

Q *is a finite set of* states,
$Q_0, Q_f \subseteq Q$ *are sets of* initial *and* final states, *and*
T *is a finite set of* transitions *having the form:*

$$q \to^a_{\varphi, \Lambda} q'$$

where $q, q' \in Q$, $a \in \Sigma$, $\varphi \in F_k$ and $\Lambda \subseteq X_k$.   □

A transition $q \to^a_{\varphi, \Lambda} q'$ means that when M is in state q and an input is (a, d), if the contents of registers and d satisfies the guard expression φ, then the contents of the i-th register such that $x_i \in \Lambda$ are updated to d and the state becomes q' after the transition.

A mapping $\theta : X_k \to D$ is called an *assignment* (of data values to k registers). Since an assignment can be considered as a k-tuple of data values, we write D^k to denote the class of all the assignments to k registers. For example, for assignment $\theta : X_k \to D$ such that $\theta(x_1) = 5$ and $\theta(x_2) = 3$, we write $\theta = [5, 3]$. For $\theta \in D^k$, $\Lambda \subseteq X_k$ and $d \in D$, define the assignment $\theta[\Lambda \leftarrow d] \in D^k$ as $\theta[\Lambda \leftarrow d](x_j) = d$ for $x_j \in \Lambda$ and $\theta[\Lambda \leftarrow d](x_j) = \theta(x_j)$ for $x_j \notin \Lambda$.

For $\theta \in D^k$, $d \in D$ and $\varphi \in F_k$, we define the satisfaction relation $\theta, d \models \varphi$ as follows: for $x_i \in X_k$, $\theta, d \models x_i^=$ iff $\theta(i) = d$ and $\theta, d \models \varphi$ is defined in the usual way for the other cases.

For $q \in Q$ and $\theta \in D^k$, (q, θ) is called a *configuration* of M. For configurations (q, θ) and (q', θ'), a transition $t = (q \to^a_{\varphi, \Lambda} q') \in T$ and a data value $d \in D$, we write $(q, \theta) \vdash_{t,d} (q', \theta')$, called a *switch*, if $\theta, d \models \varphi$ and $\theta' = \theta[\Lambda \leftarrow d]$.

A *run* of a data word $w = (a_1, d_1) \cdots (a_n, d_n)$ in M is a finite sequence of switches $c_0 \vdash_{t_1, d_1} c_1 \vdash \cdots \vdash_{t_n, d_n} c_n$ where $c_0 \in Q_0 \times \perp^k$ (an *initial configuration*), $c_n \in Q_f \times D^k$ (an *accepting configuration*) and a_i appears in the transition t_i for $i \in [n]$. The *data language recognized by M* is

$$L(M) = \{w \mid \text{there is a run of } w \text{ in } M\}.$$

Example 1. Let $\Sigma = \{a\}$ and $D = \mathbb{N} \cup \{\perp\}$ where $\mathbb{N}$ is the set of natural numbers. Let $M_1 = (\{q_0, \ldots, q_k, q_f\}, \{q_0\}, T, \{q_f\})$ be k-RA where T consists of:

$$q_{i-1} \to^a_{x_1^{\neq} \wedge \cdots \wedge x_{i-1}^{\neq}, \{x_i\}} q_i \text{ for } i \in [k], \quad q_k \to^a_{x_1^{\neq} \wedge \cdots \wedge x_k^{\neq}, \{\ \}} q_f.$$

Then, we have $L(M_1) =$

$$\{(a, d_1) \cdots (a, d_{k+1}) \mid d_i \neq d_j \text{ for any } i, j \in [k+1] \text{ such that } i \neq j\}. \quad □$$

Does the number of registers affect the language recognizing power of RA? The answer is yes [26, Corollary 1].

Theorem 1. *For each $k \geq 0$, the class of languages recognized by $(k+1)$-RA strictly includes that of k-RA.*   □

In fact, $L(M_1)$ cannot be recognized by any $(k-1)$-RA [26].

RA inherits from FA some decidability and closure properties.

Theorem 2 ([13]). *The class of data languages recognized by RA is closed under union, intersection, concatenation and Kleene-star while the class is not closed under complement.*   □

For example, the data language

$$L = \{(a_1, d_1) \cdots (a_n, d_n) \mid$$
$$n \geq 2 \text{ and } d_i \neq d_j \text{ for some } i, j \in [n] \text{ such that } i \neq j\}$$

can be recognized by an RA but $\overline{L}$ cannot. If we restrict RA to be deterministic[1] , the corresponding class is known to be closed under complement. Hence, nondeterminism increases the language recognizing power of RA.

The following theorems state the decidability and complexity of basic problems for RA.

Theorem 3 ([23]). *The combined complexity of the membership problem for RA is NP-complete while the data complexity of the problem is in P.*   □

Theorem 4. *The emptiness problem is PSPACE-complete for RA [8] and NP-complete for RA such that all register values are different [23]. The universality problem for RA is undecidable [21].*   □

In [17], regular expression (RE) is extended to regular expression with memory (REM), which has the same expressive power as RA. In [17], other extensions of RE are also proposed. Among them, for regular expression with equality (REWE), the membership (combined complexity) and emptiness problems are in P.

3   Register Context-Free Grammar

Register context-free grammar (abbreviated as RCFG) was introduced in [7] as an extension of CFG in a similar way to extending FA to RA. Though RCFG is defined as CFG generating strings over infinite alphabets in [7], we define RCFG as CFG generating data words. Just as a configuration of an RA is a pair (q, θ) of a finite state q and an assignment θ of data values to registers, an assignment is associated with each nonterminal symbol in a derivation of RCFG.

Fix a finite alphabet Σ and a countable set D of data values.

Definition 2 (RCFG). *For $k \geq 0$, a register context-free grammar with k registers (abbreviated as k-RCFG) over Σ and D is a triple $G = (V, R, S)$ where*

---

[1] An RA $M = (Q, Q_0, T, Q_f)$ is *deterministic* if for different transitions $q \to_{\varphi_1, \Lambda_1}^a q'$ and $q \to_{\varphi_2, \Lambda_2}^a q''$ in T, $\varphi_1 \wedge \varphi_2$ is unsatisfiable.

V *is a finite set of* nonterminal symbols *(abbreviated as* nonterminals*)*,
$S \in V$ *is the* start nonterminal, *and*
R *is a finite set of* production rules *(abbreviated as* rules*) of the form*

$$A \to_{\psi, \Lambda} \alpha$$

where $A \in V$, $\psi \in F_k$, $\Lambda \subseteq X_k$ *and* $\alpha \in (V \cup (\Sigma \times X_k))^*$.

$\square$

A rule $A \to_{\psi, \Lambda} \alpha$ means that when a nonterminal A associated with an assignment θ appears in a derivation and d is an input data value, if θ and d satisfy the guard expression ψ, then the contents of the i-th register such that $x_i \in \Lambda$ are updated to d, and this occurrence of A is replaced with α' in the derivation where α' is the string obtained from the right-hand side α of this rule by replacing every occurrence of $x_j \in X_k$ with the updated contents of the j-th register associated with this A.

We formally define $\Rightarrow_G$ as the smallest relation containing the instantiations of rules in R and closed under the context as follows. For $A \in V$, $\theta \in D^k$ and $Y \in ((V \times D^k) \cup (\Sigma \times D))^*$, we say (A, θ) *directly derives* Y, written as $(A, \theta) \Rightarrow_G Y$ if there exist $d \in D$ and $A \to_{\psi, \Lambda} c_1 \cdots c_n \in R$ such that

$$\theta, d \models \psi, \quad Y = c_1' \cdots c_n', \quad \theta' = \theta[\Lambda \leftarrow d]$$

where

$$c_j' = \begin{cases} (B, \theta') & \text{if } c_j = B \in V \\ (b, \theta'(x_\ell)) & \text{if } c_j = (b, x_\ell) \in \Sigma \times X_k. \end{cases}$$

The data value d in the above definition is called the *input data value* (or the *input*) of this direct derivation. If we want to emphasize the rule r and the input data value d, we write $(A, \theta) \Rightarrow_d^r Y$ instead of $(A, \theta) \Rightarrow_G Y$.

Let $\overset{*}{\Rightarrow}_G$ and $\overset{+}{\Rightarrow}_G$ be the reflexive transitive closure and the transitive closure of $\Rightarrow_G$, respectively. We abbreviate $\Rightarrow_G$, $\overset{*}{\Rightarrow}_G$ and $\overset{+}{\Rightarrow}_G$ as $\Rightarrow$, $\overset{*}{\Rightarrow}$ and $\overset{+}{\Rightarrow}$, respectively, if G is clear from the context.

Define the *data language generated by* G as

$$L(G) = \{w \in (\Sigma \times D)^* \mid (S, \bot^k) \overset{+}{\Rightarrow} w\}.$$

Example 2. Let $G = (\{S, A\}, R, S)$ be a 2-RCFG where R consists of the following three rules r_1, r_2, r_3:

$$S \to_{tt, \{x_1\}} \begin{pmatrix} a \\ x_1 \end{pmatrix} A \begin{pmatrix} a \\ x_1 \end{pmatrix}, \quad A \to_{x_1^{\neq}, \{x_2\}} \begin{pmatrix} b \\ x_2 \end{pmatrix} A \begin{pmatrix} b \\ x_2 \end{pmatrix}, \quad A \to_{x_1^{=}, \{\ \}} \begin{pmatrix} a \\ x_1 \end{pmatrix}$$

where tt represents the tautology. For example, we have the following derivation in G:

$$(S, [\bot, \bot]) \Rightarrow_1^{r_1} \begin{pmatrix} a \\ 1 \end{pmatrix} (A, [1, \bot]) \begin{pmatrix} a \\ 1 \end{pmatrix} \Rightarrow_3^{r_2} \begin{pmatrix} a \\ 1 \end{pmatrix} \begin{pmatrix} b \\ 3 \end{pmatrix} (A, [1, 3]) \begin{pmatrix} b \\ 3 \end{pmatrix} \begin{pmatrix} a \\ 1 \end{pmatrix}$$

$$\Rightarrow_1^{r_3} \begin{pmatrix} a \\ 1 \end{pmatrix} \begin{pmatrix} b \\ 3 \end{pmatrix} \begin{pmatrix} a \\ 1 \end{pmatrix} \begin{pmatrix} b \\ 3 \end{pmatrix} \begin{pmatrix} a \\ 1 \end{pmatrix}.$$

The language generated by G is $L(G) =$

$$\left\{ \begin{pmatrix} a \\ d_1 \end{pmatrix} \begin{pmatrix} b \\ d_2 \end{pmatrix} \cdots \begin{pmatrix} b \\ d_n \end{pmatrix} \begin{pmatrix} a \\ d_1 \end{pmatrix} \begin{pmatrix} b \\ d_n \end{pmatrix} \cdots \begin{pmatrix} b \\ d_2 \end{pmatrix} \begin{pmatrix} a \\ d_1 \end{pmatrix} \mid n \geq 1, d_1 \neq d_i \text{ for } i \neq 1 \right\}.$$

□

Theorem 5 ([26]). *For each $k \geq 0$, the class of languages generated by $(k+1)$-RCFG strictly includes that of k-RCFG.* □

RCFG shares some properties with CFG as RA does with FA. An *ε-rule* is a rule whose right-hand side is the empty string ε. A *unit rule* is a rule whose right-hand side is a single nonterminal symbol.

Theorem 6 ([26]). *For RCFG, we can remove both of ε-rules and unit rules without changing the generated language.* □

It is open whether unit rule removal is possible for RCFG such that an input data value can be loaded to at most one register in every rule.

Theorem 7 ([7]). *The class of data languages generated by RCFG is closed under union, concatenation, Kleene-star while the class is not closed under intersection and complement.* □

The membership and emptiness problems are already shown to be decidable in [7]. The complexities of these problems are studied in [26] for some subclasses of RCFG. Let $G = (V, R, S)$ be an RCFG. If R contains no ε-rule, G is called *ε-rule free*. If R contains neither ε-rule nor unit rule, G is called *growing*.

Theorem 8 ([26]). *The membership problem is EXPTIME-complete for general RCFG, PSPACE-complete for ε-rule free RCFG and NP-complete for growing RCFG.* □

Theorem 9 ([26]). *The emptiness problem is EXPTIME-complete for general RCFG, ε-rule free RCFG and growing RCFG.* □

Table 1. Complexity results on RCFG [26]

	general RCFG	ε-rule free RCFG	growing RCFG
Membership	EXPTIME-c	PSPACE-c	NP-c
Emptiness	EXPTIME-c		

X-c denotes 'complete for the class X.'

In both of RA and RCFG, only the equality check is possible when a transition or a rule is being applied. What happens if we allow comparisons other than the equality? Before going into details, we explain a common technique for proving the decidability of problems for RA and RCFG. In the following, we concentrate on RCFG, but a similar technique can be applied to RA and other models with registers. Note that in RA and RCFG, the essential information is not an individual data value but the equivalence among the contents of registers. From this point of view, we introduce a register type (Table 1).

A *register type* for k-RCFG (or k-register type or type) is a decomposition of X_k into non-overlapping and non-empty subsets such as $\gamma_1 = \{\{x_1, x_3\}, \{x_2\}\}$ for $k = 3$, which intuitively means that the contents of the first and third registers are the same and the contents of the second register is different from the others.

Each subset determined by the decomposition is called a partition. For a type γ, let $[x_i]_\gamma$ denote the partition containing x_i. In the above example, $[x_3]_{\gamma_1} = \{x_1, x_3\}$ and $[x_2]_{\gamma_1} = \{x_2\}$. Let Γ_k be the collection of all k-register types. For an assignment $\theta \in D^k$ and a type $\gamma \in \Gamma_k$, we say θ has type γ and write $\theta : \gamma$ when $\theta(x_i) = \theta(x_j)$ iff $[x_i]_\gamma = [x_j]_\gamma$. By definition, every assignment has a unique type. For assignments $\theta, \theta' \in D^k$, we write $\theta \sim \theta'$ if θ and θ' has a same type.

We can show that every RCFG has the following simulation property.

Definition 3 (Simulation property). *Let $G = (V, R, S)$ be a k-RCFG. We say that G has the* simulation property *if every rule $r = A \to_{\psi, \Lambda} \alpha \in R$ satisfies the following condition: For any $\theta, \theta' \in D^k$, if $(A, \theta) \Rightarrow_d^r Y$ and $\theta \sim \theta'$, there are $d' \in D$ and Y' such that*

$$(A, \theta') \Rightarrow_{d'}^r Y' \text{ and } \theta[\Lambda \leftarrow d] \sim \theta'[\Lambda \leftarrow d'].$$

□

Lemma 1 ([25]). *Every RCFG has the simulation property.*   □

Thanks to the simulation property, we can transform a given RCFG G into a CFG G' that preserves some properties of G such as the emptiness by associating each nonterminal with a type.

Now we return to a generalization of RCFG. Assume $D = \mathbb{Q}$, the set of rational numbers with comparison operators $\mathrm{OP}_\mathbb{Q} = \{=, \neq, <, \leq, >, \geq\}$. We allow atomic propositions $x_i^\bowtie$ for $\bowtie \in \mathrm{OP}_\mathbb{Q}$ and $x_i \in X_k$ in a guard expression. The semantics of $x_i^\bowtie$ is defined by

$$\theta, d \models x_i^\bowtie \text{ iff } \theta(i) \bowtie d.$$

Let $\mathrm{RCFG}(<_\mathbb{Q})$ denote RCFG with this extension while we write $\mathrm{RCFG}(=)$ to mean ordinary RCFG with data equality.

According to this extension, we extend the definition of a register type as the decomposition of X_k with $<$. That is, a register type is a sequence of partitions arranged in the ascending order. For example, we have a register type: $\{x_1, x_3\} < \{x_2, x_5\} < \{x_4\}$.

Theorem 10 ([25]). *For $RCFG(<_\mathbb{Q})$, the simulation property holds, ε-rules and unit rules can be removed and the membership and emptiness problems are EXPTIME-complete.*   □

RCFG has been extended in this way to generalized RCFG (abbreviated as GRCFG) [25]. For GRCFG, the simulation property does not always hold. For example, assume $D = \mathbb{Z}$, the set of integers with equality and inequality. Consider a register type $\{x_2\} < \{x_1, x_3\}$ and a rule $r = A \to_{x_1^> \wedge x_2^<, \{x_2\}} B$. We have $(A, [7, 4, 7]) \Rightarrow_5^r (B, [7, 5, 7])$ and $[7, 4, 7] \sim [6, 5, 6]$. However, the rule r cannot be applied to $(A, [6, 5, 6])$ with any integer as an input because there is no integer between 5 and 6. In other words, register type does not have enough information on data values with respect to the equality and inequality on integers. For GRCFG, ε-rule removal and unit rule removal are possible and the membership and emptiness are decidable if the simulation property together with some technical conditions hold (see [25] for details).

4   Logics for Data Words

Linear-time temporal logic (LTL) was extended to LTL with freeze quantifier $(\text{LTL}^\downarrow)$ [8,9]. A data value is bound to a variable in a formula and is referred to later in the scope of a freeze quantifier of that variable. The relationship among subclasses of $\text{LTL}^\downarrow$ and RA as well as the decidability and complexity of the satisfiability (non-emptiness) problems are investigated in [8,9].

$\text{FO}^2(<, +1)$ is a two-variable first-order logic where $<$ is the ancestor-descendant relation and $+1$ is the parent-child relation. In application, $\text{FO}^2(<, +1)$ corresponds to Core XPath. The logic was extended to $\text{FO}^2(\sim, <, +1)$ with data equality $\sim$ in [2].

4.1   LTL with Freeze Quantifier

Definition 4. *Let Σ be an alphabet. For $k \geq 0$, a formula in $\text{LTL}_k^\downarrow(\mathcal{O})$ over Σ where $\mathcal{O} \subseteq \{X, X^{-1}, F, F^{-1}, U, U^{-1}\}$ is defined by*

$$\phi := a \mid \neg\phi \mid \phi \vee \phi \mid O(\phi, \dots, \phi) \mid \downarrow_r \phi \mid \uparrow_r$$

where $a \in \Sigma$, $O \in \mathcal{O}$ and $r \in [k]$. □

Let D be a countable set of data values. The semantics is defined by the satisfaction relation $w, i, \theta \models \phi$ for a data word $w \in (\Sigma \times D)^*$, $i \in [|w|]$, an assignment $\theta \in D^k$ and a formula ϕ in $\text{LTL}_k^\downarrow(\mathcal{O})$.

Let $w = (a_1, d_1)(a_2, d_2) \cdots (a_m, d_m)$ with $a_j \in \Sigma$ and $d_j \in D$ for $j \in [m]$ be a data word. Define

$w, i, \theta \models a$ iff $a = a_i$,
$w, i, \theta \models \downarrow_r \phi$ iff $w, i, \theta[\{x_r\} \leftarrow d_i] \models \phi$,
$w, i, \theta \models \uparrow_r$ iff $\theta(x_r) = d_i$.

For $O \in \{X, F, U\}$, the semantics of O is defined in the usual way. O^{-1} is called a past operator, whose semantics are the dual of O.

We say that ϕ is *satisfiable* if there is a data word w such that $w, 1, \perp^k \models \phi$.

Example 3 ([8]). Let $G\phi$ denote $\neg F\neg\phi$. The following $\text{LTL}_1^\downarrow(X, F)$ formula

$$\phi_1 = G[a \Rightarrow \downarrow_1 X\{G(a \Rightarrow \neg \uparrow_1) \wedge F(b \wedge \uparrow_1)\}]$$

means that no different a-positions have a same data value and every a-position is followed by a b-position that has the same data value as the a-position has. $\square$

The expressive power of $\text{LTL}^\downarrow$ is strong so that the satisfiability is highly intractable or undecidable for important subclasses.

Theorem 11. *The satisfiability problem for $\text{LTL}_1^\downarrow(X, F)$ and $\text{LTL}_1^\downarrow(X, U)$ are decidable but not primitive recursive while the problem is undecidable for $\text{LTL}_1^\downarrow(X, F, F^{-1})$.*

The satisfiability problem for $\text{LTL}_2^\downarrow(X, F)$ and $\text{LTL}_2^\downarrow(X, U)$ is undecidable. $\square$

One of the reasons why the expressive power of $\text{LTL}^\downarrow$ exceeds that of RA lies in the fact that a conjunction $\phi_1 \wedge \phi_2$ cannot be simulated by RA. On the other hand, we need a mechanism such as modal μ-operator for representing a cycle in a state transition graph of RA [12]. In [27], a subclass of μ-calculus with freeze quantifier whose language expressive power is equivalent to RA is proposed.

4.2   Two-Variable First-Order Logic with Data Equality

In this subsection, two-variable first-order logic with data equality is defined and is briefly compared with $\text{LTL}^\downarrow$.

Definition 5. *A formula in $\text{FO}^n(\sim, <, +1)$ over a finite alphabet Σ is defined by*

$$\psi := a(x) \mid x \sim y \mid x < y \mid x = y + 1 \mid \neg\psi \mid \psi \vee \psi \mid \exists x(\psi)$$

where $a \in \Sigma$, $x, y \in X_n$. $\square$

Note that a variable is used for representing a position in a data word, not a register number. Intuitively, $x \sim y$ means that the data values of positions x and y are equal. Let D be a countable set of data values. The semantics are defined by the satisfaction relation $w, \sigma \models \psi$ for a data word $w \in (\Sigma \times D)^*$, $\sigma : X_n \to [[w]]$ and a formula ψ in $\text{FO}^n(\sim, <, +1)$. Let $w = (a_1, d_1)(a_2, d_2)\cdots(a_m, d_m)$ with $a_j \in \Sigma$ and $d_j \in D$ for $j \in [m]$ be a data word. Define

$w, \sigma \models a(x)$ iff $a = a_{\sigma(x)}$,
$w, \sigma \models x \sim y$ iff $d_{\sigma(x)} = d_{\sigma(y)}$,
$w, \sigma \models x < y$ iff $\sigma(x) < \sigma(y)$,
$w, \sigma \models x = y + 1$ iff $\sigma(x) = \sigma(y) + 1$, and
$w, \sigma \models \psi$ is defined in the usual way for the other constructs.

For a closed formula ψ, we write $w \models \psi$ to mean $w, \sigma \models \psi$ for any σ. We say that a closed formula ψ is *satisfiable* if there is a data word w such that $w \models \psi$.

Example 4 ([8]). $\text{LTL}_1^{\downarrow}(X, F)$ formula ϕ_1 in Example 3 can be expressed by the following $\text{FO}^2(\sim, <, +1)$ formula:

$$\psi_2 = \forall y[a(y) \Rightarrow \forall x\{(y < x) \wedge a(x) \Rightarrow x \not\sim y\} \wedge \exists x\{(y < x) \wedge b(x) \wedge x \sim y\}].$$

In ψ_2, the role of x is for scanning a given data word and the role of y is for keeping a data value, which corresponds to $\downarrow_1$ in ϕ_1.   □

The expressive powers of $\text{FO}^2(\sim, <, +1)$ and $\text{LTL}_1^{\downarrow}(X, U)$ are incomparable (see [2,8]).

Example 5 ([2]). Let $\Sigma = \{a, b\}$. Consider the following $\text{FO}^2(\sim, <, +1)$ formulas:

$\psi_a = \forall x \forall y (x \neq y \wedge a(x) \wedge a(y) \Rightarrow x \not\sim y$ (data values in different a-positions are different),
$\psi_{a,b} = \forall x \exists y (a(x) \Rightarrow (b(y) \wedge x \sim y))$ (for every a-position, there is a b-position that have the same data value).

Let $\psi_1 = \psi_a \wedge \psi_b \wedge \psi_{a,b} \wedge \psi_{b,a}$. If $w, 1 \models \psi_1$, the numbers of a and b in w is the same, and the set of data values in a-positions is the same as that of b-positions.
    $\text{LTL}_1^{\downarrow}(X, U)$ cannot express ψ_1. The intuitive reason is that the latter does not have a past operator F^{-1} or U^{-1}.   □

Theorem 12 ([2]). *The satisfiability problem for $\text{FO}^2(\sim, <, +1)$ is decidable and at least as hard as Petri net reachability.*   □

It is shown in [8] that the expressive powers of $\text{LTL}_1^{\downarrow}(X, X^{-1}, X^2 F, X^{-2} F^{-1})$ and $\text{FO}^2(\sim, <, +1)$ are the same.

5   Nominal Automaton

A register type in RCFG (resp. RA) is a useful tool for finitely representing the relation among data values stored in registers that appear in a derivation (resp. transition) with respect to comparison operators appearing in guard expressions. This approach can be formalized by nominal set in group theory in a more abstract way. This section is devoted to a short introduction of nominal automaton. All the definitions and theorems appear in the seminal paper [3] and we omit the respective reference to [3] in this section.

Definition 6. *Let G be a group with neutral element e. X is a G-set if it has a function $\cdot : X \times G \to X$, called a* right action, *such that*

$$x \cdot e = x \text{ and } x \cdot (\pi \sigma) = (x \cdot \pi) \cdot \sigma \text{ for every } x \in X \text{ and } \pi, \sigma \in G.$$

A data symmetry *(D, G) is a set D of data values with a subgroup $G \leq \text{Sym}(D)$ of the symmetric group (i.e., the group of permutations or bijections) of D.*   □

For example, $(\mathbb{N}, \mathrm{Sym}(\mathbb{N}))$ is a data symmetry, called the *equality symmetry* where $\mathbb{N}$ is the set of natural numbers. The *total order symmetry* is $(\mathbb{Q}, G)$ where G is the group of monotone bijections on $\mathbb{Q}$.

Note that a group action of the equality symmetry preserves the register type of an assignment for RCFG(=). For example, let $\gamma = \{\{x_1, x_3\}, \{x_2\}\}$ be a register type of 3-RCFG(=) and π be a permutation such that $\pi(1) = 5$, $\pi(3) = 1$. For assignment $\theta = [3, 1, 3]$, we have $\theta \cdot \pi = [1, 5, 1]$ and the register types of θ and $\theta \cdot \pi$ are both γ. Also, a group action of the total order symmetry preserves the register type of an assignment for RCFG($<_\mathbb{Q}$). Let $\gamma' = \{x_2\} < \{x_1, x_3\}$ and σ be a monotone bijection such that $\sigma(1.5) = 0$ and $\sigma(3) = 4.2$. Let $\theta' = [3, 1.5, 3]$, then $\theta' \cdot \sigma = [4.2, 0, 4.2]$ and θ' and $\theta' \cdot \sigma$ have the same register type γ'.

Definition 7. *Let X be a G-set. For $x \in X$, the* orbit *of x is the set $x \cdot G = \{x \cdot \pi \mid \pi \in G\} \subseteq X$.* □

Example 6. In the equality symmetry, a G-set D^2 has two orbits: $\{(d, d) \mid d \in D\}$ and $\{(d, e) \mid d, e \in D, d \neq e\}$. In the total order symmetry, D^2 has three orbits: $\{(d, d) \mid d \in D\}$, $\{(d, e) \mid d, e \in D, d < e\}$ and $\{(d, e) \mid d, e \in D, d > e\}$. □

Any G-set is partitioned into orbits in a unique way. A G-set having finite orbits is called an *orbit-finite* set. Then, an *alphabet* is any orbit-finite G-set. A finite set Σ, a set D of data values and their cartesian product $\Sigma \times D$ are examples of alphabets.

Definition 8. *A subset Y of a G-set X is* equivariant *if $Y \cdot \pi = Y$ holds for every $\pi \in G$. A G-language* over an alphabet A *is any equivariant subset $L \subseteq A^*$.*

The notion of equivariance extends to relations in a pointwise way: $R \subseteq X \times Y$ is an equivariant relation *if $(x, y) \in R$ implies $(x \cdot \pi, y \cdot \pi) \in R$ for every $\pi \in G$. Specifically, $f : X \to Y$ is an* equivariant function *if $f(x \cdot \pi) = f(x) \cdot \pi$ for every $x \in X$ and $\pi \in G$.* □

Definition 9. *Fix a data symmetry (D, G). A nondeterministic G-automaton is $M = (A, Q, Q_0, T, Q_f)$ where*

A *is an input alphabet, which is an orbit-finite G-set,*
Q *is a G-set of states,*
$Q_I, Q_f \subseteq Q$ *are the initial and final states and*
$T \subseteq Q \times A \times Q$ *is an equivariant transition relation.* □

The G-language recognized by M is defined as:

$$L(M) = \{w \in A^* \mid (q_0, w, q_f) \in T^*, q_0 \in Q_0, q_f \in Q_f\}.$$

A G-automaton is deterministic if T is a function and I is a singleton set.

A set $C \subseteq D$ *supports* $x \in X$ if $x \cdot \pi = x$ for all $\pi \in G$ that acts as identity on C, i.e., $\forall \pi \in G.(\forall c \in C.c \cdot \pi = c) \Rightarrow x \cdot \pi = x$.

Definition 10. *A G-set X is nominal in (D, G) if every $x \in X$ has a finite support.* $\square$

Intuitively, if a set C supports $x \in X$, C has all the information of x with respect to G actions. Therefore, a finite support C is regarded as a generalization of (a finite set of) registers.

Definition 11. *Fix a countable set D and a group $G \leq$ Sym(D). A G-automaton is nominal if both the input alphabet and the state set are nominal.* $\square$

We abbreviate a nondeterministic orbit-finite nominal G-automaton as G-NFA. We list some properties of G-NFA.

Theorem 13. *The class of G-languages recognized by G-NFA is closed under union and concatenation while it is not closed under intersection or complement.* $\square$

Theorem 14. *Determinization is not always possible for G-NFA.* $\square$

Theorem 15. *The expressive power of G-NFA is not changed even if ε-moves are allowed.* $\square$

The next theorem states that G-NFA is an extension of RA. Fix the equality symmetry (D, G).

Theorem 16. *Consider an alphabet $A = A_{fin} \times D$ where A_{fin} is a finite set. For every G-language $L \subseteq A^*$, L is recognized by an RA iff L is recognized by a G-NFA.* $\square$

An interesting application of RA other than querying data structures with data values is automated language learning. Angluin's L^* algorithm [1] learns the minimum FA for an unknown regular language U by constructing an observation table. The L^* algorithm has been extended for register automata (e.g. [4,6]). In these learning algorithms, an entry of an observation table is not just 0/1 but more complex information that represents the guard condition of a transition in RA, which makes the algorithm complicated. Moerman et al. proposed an L^*-style algorithm for nominal automata with equality symmetry [20]. Their algorithm recovers the simplicity of the original L^* algorithm by the abstract feature of nominal automaton, which is independent of a concrete representation of an automaton. In [22], the algorithm is extended for nominal tree automata with any data symmetry.

6   Conclusion

In this paper, we first overview extensions of classical finite automaton and context-free grammar to those with registers, which inherit some good properties of classical models. Then, we also explain and compare two logics for data words, namely, LTL with freeze quantifier and two-variable first-order logic with data equality. Lastly, we take a look at nominal automaton, which is a group-theoretic generalization of register models.

Though omitted in this paper, tree automaton has also been extended to register models [14,24,26]. RA and related models are extensively used in program verification (e.g., [11]) and in rational synthesis of reactive programs [10,15].

Acknowledgments. The author would like to thank Ryoma Senda and Yoshiaki Takata for their collaboration.

References

1. Angluin, D.: Learning regular sets from queries and counterexamples. Inf. Comput. **75**, 87–106 (1987)
2. Bojańczyk, M., David, C., Muscholl, A., Schwentick, T., Segoufin, L.: Two-variable logic on data words. ACM Trans. Comput. Logic **12**(4:27), 1–26 (2011)
3. Bojańczyk, M., Klin, B., Lasota, S.: Automata theory in nominal sets. Logical Methods Comput. Sci. **10**(3:4), 1–44 (2014). Earlier conference version: Automata with group actions, 26th Annual IEEE Symposium on Logic in Computer Science (LICS 2011), pp. 355–364
4. Bollig, B., Habermehl, P., Leucker, M., Monmege, B.: A fresh approach to learning register automata. In: Béal, M.-P., Carton, O. (eds.) DLT 2013. LNCS, vol. 7907, pp. 118–130. Springer, Heidelberg (2013). https://doi.org/10.1007/978-3-642-38771-5_12
5. Bouyer, P.: A logical characterization of data languages. Inf. Process. Lett. **84**(2), 75–85 (2002)
6. Cassel, S., Howar, F., Jonsson, B., Steffen, B.: Active learning for extended finite state machines. Formal Aspects Comput. **28**, 233–263 (2016)
7. Cheng, E.Y.C., Kaminsky, M.: Context-free languages over infinite alphabets. Acta Informatica **35**, 245–267 (1998)
8. Demri, S., Lazić, R.: LTL with freeze quantifier and register automata. ACM Trans. Comput. Logic **10**(3) (2009)
9. Demri, S., Lazić, R., Nowak, D.: On the freeze quantifier in constraint LTL: decidability and complexity. Inf. Comput. **205**(1), 2–24 (2007)
10. Exibard, L., Filiot, E., Reynier, P.-A.: Synthesis of data word transducers. In: 30th International Conference on Concurrency (CONCUR 2019). LIPIcs, vol. 140, pp. 24:1–24:15 (2019). Extended version: Logical Methods in Computer Science **17**(1), 22:1–22:25 (2021)
11. Grigore, R., Distefano, D., Petersen, R.L., Tzevelekos, N.: Runtime verification based on register automata. In: Piterman, N., Smolka, S.A. (eds.) TACAS 2013. LNCS, vol. 7795, pp. 260–276. Springer, Heidelberg (2013). https://doi.org/10.1007/978-3-642-36742-7_19
12. Jurdziński, M., Lazić, R.: Alternation-free modal mu-calculus for data trees. In: 22nd Annual IEEE Symposium on Logic in Computer Science (LICS 2007), pp. 131–140 (2007)
13. Kaminski, M., Francez, N.: Finite-memory automata. Theor. Comput. Sci. **134**, 322–363 (1994)
14. Kaminski, M., Tan, T.: Tree automata over infinite alphabets. In: Avron, A., Dershowitz, N., Rabinovich, A. (eds.) Pillars of Computer Science. LNCS, vol. 4800, pp. 386–423. Springer, Heidelberg (2008). https://doi.org/10.1007/978-3-540-78127-1_21

15. Khalimov, A., Kupferman, O.: Register-bounded synthesis. In: 30th International Conference on Concurrency (CONCUR 2019). LIPIcs, vol. 140, pp. 25:1–25:16 (2019)
16. Libkin, L., Martens, W., Vrgoč, D.: Querying graphs with data. J. ACM **63**(2), 14:1–14:53 (2016)
17. Libkin, L., Tan, T., Vrgoč, D.: Regular expressions for data words. J. Comput. Syst. Sci. **81**(7), 1278–1297 (2015)
18. Libkin, L., Vrgoč, D.: Regular path queries on graphs with data. In: 15th International Conference on Database Theory (ICDT 2012), pp. 74–85 (2012)
19. Milo, T., Suciu, D., Vianu, V.: Type checking for XML transformers. In: 19th ACM Symposium on Principles of Database Systems (PODS 2000), pp. 11–22 (2000)
20. Moerman, J., Sammartino, M., Silva, A., Klin, B., Szynwelski, M.: Learning nominal automata. In: 44th ACM SIGPLAN Symposium on Principles of Programming Languages (POPL 2017), pp. 613–625 (2017)
21. Neven, F., Schwentick, T., Vianu, V.: Finite state machines for strings over infinite alphabets. ACM Trans. Comput. Logic **5**(3), 403–435 (2004)
22. Nakanishi, R., Takata, Y., Seki, H.: Active learning for deterministic bottom-up nominal tree automata. In: Seidl, H., Liu, Z., Pasareanu, C.S. (eds.) ICTAC 2022. LNCS, vol. 13572, pp. 342–359. Springer, Cham (2022). https://doi.org/10.1007/978-3-031-17715-6_22
23. Sakamoto, H., Ikeda, D.: Intractability of decision problems for finite-memory automata. Theor. Comput. Sci. **231**, 297–308 (2000)
24. Segoufin, L.: Automata and logics for words and trees over an infinite alphabet. In: 15th EACSL Annual Conference on Computer Science Logic (CSL 2006), pp. 41–57 (2006)
25. Senda, R., Takata, Y., Seki, H.: Generalized register context-free grammars. In: Martín-Vide, C., Okhotin, A., Shapira, D. (eds.) LATA 2019. LNCS, vol. 11417, pp. 259–271. Springer, Cham (2019). https://doi.org/10.1007/978-3-030-13435-8_19
26. Senda, R., Takata, Y., Seki, H.: Complexity results on register context-free grammars and related formalisms. Theor. Comput. Sci. **923**, 99–125 (2022). Earlier conference version: Complexity results on register context-free grammars and register tree automata. In: International Colloquium on Theoretical Aspects of Computing (ICTAC 2018). LNCS, vol. 11187, pp. 415–434 (2022)
27. Takata, Y., Onishi, A., Senda, R., Seki, H.: A subclass of mu-calculus with the freeze quantifier equivalent to register automata. IEICE Trans. Inf. Syst. **E106-D**(3), 294–302 (2023)

Contributed Papers

Computing the Bandwidth of Meager Timed Automata

Eugene Asarin[1]([⊠])(iD), Aldric Degorre[1]([⊠])(iD), Cătălin Dima[2]([⊠])(iD), and Bernardo Jacobo Inclán[1]([⊠])(iD)

[1] Université Paris Cité, CNRS, IRIF, Paris, France
{asarin,adegorre,jacoboinclan}@irif.fr
[2] LACL, Université Paris-Est Créteil, Créteil, France
dima@u-pec.fr

Abstract. The bandwidth of timed automata characterizes the quantity of information produced/transmitted per time unit. We previously delimited 3 classes of TA according to the nature of their asymptotic bandwidth: meager, normal, and obese. In this paper, we propose a method, based on a finite-state simply-timed abstraction, to compute the actual value of the bandwidth of meager automata. The states of this abstraction correspond to barycenters of the faces of the simplices in the region automaton. Then the bandwidth is $\log 1/|z_0|$ where z_0 is the smallest root (in modulus) of the characteristic polynomial of this finite-state abstraction.

Keywords: Timed automata · Information · Bandwidth · Entropy

1 Introduction

We study timed automata [1] from the information-theoretic standpoint. An important characteristic of a timed automaton is its bandwidth [14], a notion generalizing the growth rate of formal languages [9,10,12,13]. In the case of timed automata, bandwidth describes the quantity of information that it can produce/transmit per time unit; however, differently from the case of discrete formal languages, information that timed automata may produce must be related to a degree of precision of an observer of the timed words generated by the timed automaton.

In the preceding work [5] we made a first step in this direction by identifying three classes of TA according to the nature of their asymptotic bandwidth: meager, normal, and obese. This analysis relies on a refinement of orbit graphs [3,19] associated with cycles in the timed automata, which themselves refine the region construction for timed automata. The three types of timed languages behave very differently. First, meager automata are almost discrete automata, or

This work was supported by the ANR project MAVeriQ ANR-CE25-0012 and by the ANR-JST project CyPhAI.

S. Z. Fazekas (Ed.): CIAA 2024, LNCS 15015, pp. 19–34, 2024.
https://doi.org/10.1007/978-3-031-71112-1_2

automata in which real-time delays are very constrained, leading to quasi-Zeno behaviors (some quantities decrease while remaining positive, as the distance between Achilles and the Tortoise, see [2, VI:9] or the more recent paper [7]). Their bandwidth is bounded by a constant, as is the case for languages of finite automata, hence being independent of the precision of an observer. Second, in normal timed languages, information can be encoded every few time units and their bandwidth is $O(\log(1/\varepsilon))$, where ε is the degree of precision by which an observer may distinguish timed words generated by the automaton. Thirdly, obese timed languages allow encoding information with a very high frequency, leading to $O(1/\varepsilon)$ bandwidth.

In this paper, we make the first step towards computing the bandwidth of a timed automaton, by attacking the case of meager automata. Our solution is based on another finite-state abstraction of orbit graphs and region automata, in which each state is associated with a barycenter of one of the faces of the regions. The bandwidth can then be computed as $\log 1/|z_0|$ where z_0 is the root of the smallest modulus of the characteristic polynomial of the adjacency λ-matrix [16] associated with the finite-state abstraction. The correctness of our construction heavily depends on investigations in [5, full version], which rely on Simon's factorization theorem [20] and Puri's reachability theory [19].

The paper is structured as follows: in Sect. 2 we recall the main notions and results concerning the bandwidth of timed automata; in Sect. 3 we show how to compute the bandwidth for the restricted class of simply-timed graphs; in Sect. 4 we present our main contribution: a method for computing the bandwidth of meager timed automata (by reduction to simply-timed ones). Most of the proofs can be found in the arXiv version [6] of this paper.

2 Preliminaries

2.1 Timed Words, Languages and Automata

Timed automata have been introduced in [1] for modeling and verification of real-time systems. We will use the main definitions in the following form:

Definition 1. *Given Σ, a finite alphabet of discrete events, a* timed word *over Σ is an element from $(\Sigma \times \mathbb{R}_+)^*$ of the form $w = (a_1, t_1) \ldots (a_n, t_n)$, with $0 \leq t_1 \leq t_2 \cdots \leq t_n$. We denote $\mathrm{dur}(w) = t_n$. A* timed language *over Σ is a set of timed words over the same alphabet.*

For a set of variables Ξ, let G_Ξ be the set of finite conjunctions of constraints of the form $\xi \sim b$ with $\xi \in \Xi$, $\sim \in \{<, \leq, >, \geq\}$ and $b \in \mathbb{N}$.

Definition 2. *A* timed automaton *is a tuple $(Q, X, \Sigma, \Delta, S, I, F)$, where*

- *Q is the finite set of discrete locations;*
- *X is the finite set of clocks;*
- *Σ is a finite alphabet;*
- *$S, I, F : Q \to G_X$ define respectively the starting, initial, and final clock values for each location;*

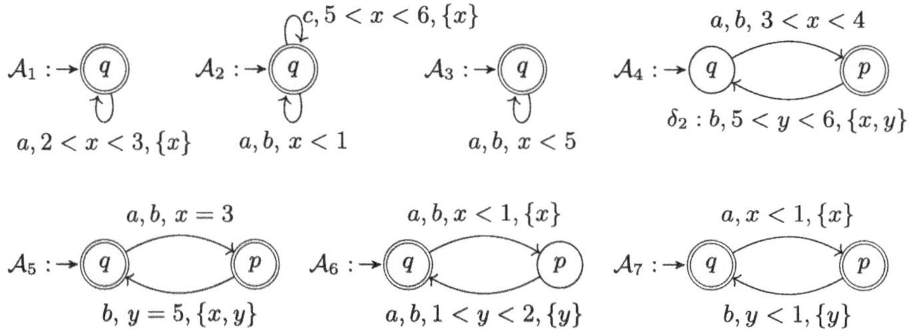

$\mathcal{A}_1 : \to \boxed{q} \quad \mathcal{A}_2 : \to \boxed{q} \quad \mathcal{A}_3 : \to \boxed{q} \quad$ with loops

$$\mathcal{A}_1 : \to (q) \quad a, 2 < x < 3, \{x\}$$

$$c, 5 < x < 6, \{x\}$$

$$a, b, x < 1$$

$$a, b, x < 5$$

$$a, b, 3 < x < 4$$
$$\mathcal{A}_4 : \to (q) \rightleftarrows (p)$$
$$\delta_2 : b, 5 < y < 6, \{x, y\}$$

$$a, b, x = 3$$
$$\mathcal{A}_5 : \to (q) \rightleftarrows (p)$$
$$b, y = 5, \{x, y\}$$

$$a, b, x < 1, \{x\}$$
$$\mathcal{A}_6 : \to (q) \rightleftarrows (p)$$
$$a, b, 1 < y < 2, \{y\}$$

$$a, x < 1, \{x\}$$
$$\mathcal{A}_7 : \to (q) \rightleftarrows (p)$$
$$b, y < 1, \{y\}$$

Fig. 1. 7 running examples of timed automata, $I(q) = \{0\}$, $F(\cdot) = \mathbf{true}$ for marked locations. An arrow labeled with a, b is a shorthand for two transitions with the same guards and resets.

– $\Delta \subseteq Q \times Q \times \Sigma \times G_X \times 2^X$ *is the transition relation, whose elements are called* edges.

A *timed automaton is* deterministic *(referred to as* DTA*) if* $\{(q, x) | x \models I(q)\}$ *is a singleton and for any two edges* $(q, q_1, a, \mathfrak{g}_1, \mathfrak{r}_1)$ *and* $(q, q_2, a, \mathfrak{g}_2, \mathfrak{r}_2)$ *with* $q_1 \neq q_2$, *the constraint* $\mathfrak{g}_1 \wedge \mathfrak{g}_2$ *is non-satisfiable.*

A small particularity of our definition is the starting clock constraint $S(q)$ defining with which clock values the run can enter each state q.

The semantics of a timed automaton is given by a *timed transition system* whose states are tuples (q, x) composed of a location $q \in Q$ and a clock valuation (vector) $x \in [0, \infty)^X$. Each edge $\delta = (q, q', a, \mathfrak{g}, \mathfrak{r}) \in \Delta$ generates many timed transitions, which are tuples $(q, x) \xrightarrow{\delta, d} (q', x')$ where $x \models S(q)$, $x + d \models \mathfrak{g}$ and $x'_c = 0$ whenever $c \in \mathfrak{r}$ and $x'_c = x_c + d$ otherwise, provided that $x' \models S(q')$. We also denote $x[\mathfrak{r}]$ the operation of resetting the clocks in $\mathfrak{r}$, hence $x' = (x + d)[\mathfrak{r}]$.

Paths are sequences of edges that agree on intermediary locations. At the semantic level, they generate *runs* of the form

$$\rho = (q_0, x_0) \xrightarrow{(\delta_1, d_1)} (q_1, x_1) \ldots \xrightarrow{(\delta_n, d_n)} (q_n, x_n),$$

which are sequences of timed transitions that agree on intermediary configurations. An *accepting run* is a run in which the first state satisfies $x_0 \models I(q_0)$ and the last one $x_n \models F(q_n)$.

Given a run ρ as above, the path associated with ρ is defined as $path(\rho) \triangleq \delta_1 \ldots \delta_n$. Furthermore, if $\delta_i = (q_i, q_{i+1}, a_i, \mathfrak{g}_i, \mathfrak{r}_i)$, then the timed word associated with ρ is defined as $word(\rho) \triangleq (a_1, d_1)(a_2, d_1 + d_2) \ldots (a_n, \sum_{i \leq n} d_i)$. The *duration* of ρ is $dur(\rho) = dur(word(\rho))$. The language of a timed automaton is the set of timed words associated with some accepting run and is denoted $L(\mathcal{A})$. We will also need *time-restricted languages*: for any timed language L and nonnegative real $T \in [0, \infty)$, $L_T = \{w \in L \mid dur(w) \leq T\}$.

For a small example consider a 3-edge path $q \to p \to q$ in the timed automaton $\mathcal{A}_6$ in Fig. 1. One of its runs is

$$(q,0,0) \xrightarrow{(a,0.8)} (p,0,0.8) \xrightarrow{(b,0.7)} (q,0.7,0) \xrightarrow{(b,0.2)} (p,0,0.2)$$

and the corresponding timed word is $(a,0.8)(b,1.5)(b,1.7)$.

Region-Split Form. Many algorithms related to timed automata utilize the finitary abstraction of the timed transition system called the *region construction* [1] that we briefly recall here. Let us denote by M the maximal constant appearing in the constraints used in the automaton. Also, for $\xi \in \mathbb{R}$, we denote $\lfloor \xi \rfloor$ its integral part, $\{\xi\}$ its fractional part and, for a clock vector $x \in [0,\infty)^X$, $x_{\text{int}} \triangleq \{c \in X | x_c \in [0,M] \cap \mathbb{N}\}$, and $x_{\text{frac}} \triangleq \{c \in X | x_c \in [0,M] \setminus \mathbb{N}\}$. Then regions are defined as equivalence classes of the region equivalence described hereafter. Two clock vectors x and y are region-equivalent iff :

- $x_{\text{int}} = y_{\text{int}}$ and $x_{\text{frac}} = y_{\text{frac}}$;
- and for any clock $c \in x_{\text{int}} \cup x_{\text{frac}}$, $\lfloor x_c \rfloor = \lfloor y_c \rfloor$;
- and for any two clocks $c_1, c_2 \in x_{\text{int}} \cup x_{\text{frac}}$, $\{x_{c_1}\} \leq \{x_{c_2}\}$ iff $\{y_{c_1}\} \leq \{y_{c_2}\}$.

Note that a region R is a simplex of some dimension $d \leq \#X$.

As in [3,5] we utilize here timed automata in which the guards defining starting clock values for each location actually define regions:

Definition 3. *A region-split TA (or RsTA for short) is a deterministic timed automaton $(Q, X, \Sigma, \Delta, S, I, F)$, such that, for any location $q \in Q$:*

- *$S(q)$ defines a non-empty bounded region, called the* starting region *of q;*
- *all states in $\{q\} \times S(q)$ are reachable from the initial state and co-reachable to a final state;*
- *either $I(q) = S(q)$ is a singleton[1] or $I(q) = \emptyset$;*
- *for any edge $(q, q', a, \mathfrak{g}, \mathfrak{r}) \in \Delta$, $(\{S(q) + d \mid d \in \mathbb{R}_+\} \cap \mathfrak{g})[\mathfrak{r}] = S(q')$, where we utilized the $(\cdot)[\mathfrak{r}]$ operator lifted to sets of clock valuations.*

A folklore result says that any DTA can be brought into a region-split form which may contain an exponentially larger set of locations. Note that this construction is similar to that of the region automaton introduced in [1], except that the outcome is slightly coarser and is typed as a timed automaton. As an example, we present on Fig. 4 the region-split form $\mathcal{A}'_6$ of the automaton $\mathcal{A}_6$ (from Fig. 1) with explicit starting regions $S(\cdot)$ for each state. In particular, state q of $\mathcal{A}_6$ is split into two states: the initial s and q corresponding to zero and non-zero clock vectors, respectively.

We also introduce the notation $\bar{A}$ for the closed version of any RsTA $\mathcal{A}$: i.e. the same automaton where all strict inequalities are changed to non-strict ones. Thanks to region-splitting, the closure retains the same set of symbolic behaviors as the original RsTA, this allows us to safely reason on runs that visit vertices and faces of regions.

---

[1] by definition of DTA this is possible for a unique location q.

2.2   Pseudo-distance on Timed Words

We are using the proximity measure on timed words introduced in [4].

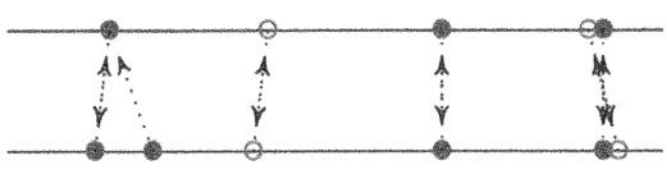

$$u = (a, .7), (b, 1.8), (a, 3), (b, 4), (a, 4.1)$$

$$v = (a, .6), (a, 1), (b, 1.7), (a, 3), (a, 4.1), (b, 4.2)$$

Fig. 2. Pseudo-distance between two timed words (dotted lines with arrows represent directed matches between letters). $\overrightarrow{d}(u, v) = 0.2$; $\overrightarrow{d}(v, u) = 0.3$, thus $d(u, v) = 0.3$.

Definition 4. *The* pseudo-distance d *between timed words*

$$w = (a_1, t_1) \ldots (a_n, t_n) \text{ and } v = (b_1, s_1) \ldots (b_m, s_m)$$

is defined as follows (with the convention $\min \emptyset = \infty$*):*

$$\overrightarrow{d}(w, v) \triangleq \max_{i \in \{1..n\}} \min_{j \in \{1..m\}} \{|t_i - s_j| : a_i = b_j\}; d(w, v) \triangleq \max(\overrightarrow{d}(w, v), \overrightarrow{d}(v, w)).$$

This pseudo-distance allows a meaningful comparison of timed words with a different number of events. It is illustrated on Fig. 2. Intuitively, two words are close to each other when they cannot be distinguished by an observer that reads the discrete letters of the word exactly (they can determine whether or not a letter has occurred) but with some imprecision w.r.t. time. So the observer cannot determine when two letters are very close to one another, which one came before the other, and not even how many times a letter was repeated within a short interval.

It is possible that d fails to distinguish timed words, that is $d(w_1, w_2) = 0$ but $w_1 \neq w_2$; this could happen when w_1 and w_2 only differ by order and quantity of simultaneous letters, e.g. for $w = (a, 1)(b, 1)$ and $v = (b, 1)(b, 1)(a, 1)$, that is why d is only a pseudo-distance.

We will sometimes use the following trick to "transform" d into a distance. We call a timed word $w = (a_1, t_1) \ldots (a_n, t_n)$ *0-free* whenever $0 < t_1 < t_2 < \cdots < t_n$. Note that d satisfies the axiom $w_1 = w_2 \Leftrightarrow d(w_1, w_2) = 0$ whenever w_1, w_2 are 0-free, hence d becomes a distance on 0-free words.

Definition 5 (0-elimination, words). *Given a timed word w over an alphabet Σ, we define its 0-elimination as the 0-free timed word $W = \nu(w)$ over the alphabet $2^\Sigma \setminus \{\emptyset\}$ as follows:*

- *let $0 < t_1 < t_2 < \cdots < t_n$ be all the distinct non-zero event times in w;*
- *let $A_i \in 2^\Sigma$ be the set of all the events in w occurring at time t_i;*
- *finally let $W = (A_1, t_1), (A_2, t_2), \ldots, (A_n, t_n)$.*

For example $\nu((c, 0)(a, 5)(b, 5)(a, 5)(c, 7)) = (\{a, b\}, 5)(\{c\}, 7)$, the events at time 0 are ignored. We summarize the properties of the operation ν:

Proposition 1. *Operation ν maps timed words on Σ to 0-free ones on $2^{\Sigma} \setminus \{\emptyset\}$. $\nu(w_1) = \nu(w_2)$ if and only if the words $w_i, i = 1, 2$ admit decompositions $w_i = u_i v_i$ with $\mathrm{dur}(u_i) = 0$ and $d(v_1, v_2) = 0$.*

In words, 0-eliminations of two timed words coincide if and only if the words are at distance 0 except for the initial time instant 0.

Fig. 3. A run of $\mathcal{A}_6$ on $(a, .9)(b, 1.2)(a, 1.8)(b, 2.4)(a, 2.7)(b, 3.5)(a, 3.6)(b, 4.52)(a, 4.59)$ $(b, 5.53)(a, 5.58)(b, 6.535)(a, 6.575)$.

2.3 Bandwidth of Timed Languages

The definition of bandwidth is based on Kolmogorov&Tikhomirov's work [15] formalizing the quantity of information in an element of a compact metric space observed with a given precision. We recall the adaptation of their definitions from [4] to the context of timed languages.

Definition 6. *Let L be a timed language, then:*

- $\mathcal{M} \subseteq L$ *is an ε-separated subset of L if $\forall x \neq y \in \mathcal{M}, d(x, y) > \varepsilon$;*
- $\mathcal{N}$ *is an ε-net of L if $\forall y \in L, \exists s \in \mathcal{N}$ s.t. $d(y, s) \leq \varepsilon$;*
- *ε-capacity of L is defined[2] as*

$$\mathcal{C}_{\varepsilon}(L) \triangleq \log \max\{\#\mathcal{M} \mid \mathcal{M}\varepsilon\text{-separated set of } L\};$$

- *ε-entropy of L is defined as*

$$\mathcal{H}_{\varepsilon}(L) \triangleq \log \min\{\#\mathcal{N} \mid \mathcal{N}\varepsilon\text{-net of } L\}.$$

As shown in [4], every time-bounded timed language is compact with respect to our distance d and thus has finite ε-capacity and ε-entropy. [14] introduced the notion of bandwidth of a timed language as the entropy or capacity per time unit and related this notion to bounded delay codes.

Definition 7 (bandwidth, [14]). *The ε-entropic and ε-capacitive bandwidths of a timed language L are*

$$\mathcal{BH}_{\varepsilon}(L) \triangleq \limsup_{T \to \infty} \frac{\mathcal{H}_{\varepsilon}(L_T)}{T}; \qquad \mathcal{BC}_{\varepsilon}(L) \triangleq \limsup_{T \to \infty} \frac{\mathcal{C}_{\varepsilon}(L_T)}{T}.$$

Definition 8 (3 classes, [5]). *A timed language L is*

- *meager whenever $\mathcal{BH}_{\varepsilon}(L) = O(1)$;*

---

[2] all logarithms in this article are base 2.

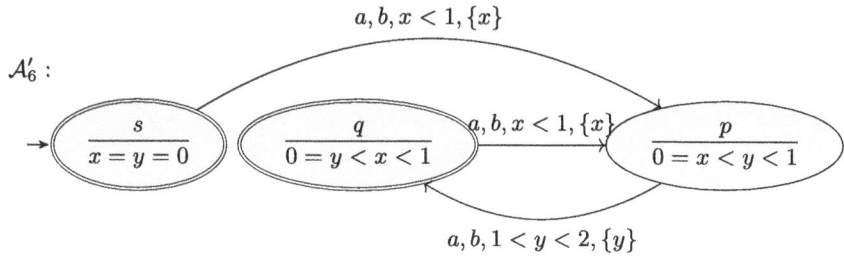

Fig. 4. The region-split form of $\mathcal{A}_6$, constraints within states correspond to start conditions $S(\cdot)$.

- normal *whenever* $\mathcal{BH}_\varepsilon(L) = \Theta(\log\frac{1}{\varepsilon})$;
- obese *whenever* $\mathcal{BH}_\varepsilon(L) = \Theta(\frac{1}{\varepsilon})$, *as* $\varepsilon \to 0$.

As the main result of [5] we proved that every timed regular language belongs to one of these classes, and provided a classification algorithm for timed automata.

Figure 1 gives a few examples of automata in the three classes: $\mathcal{A}_3$, $\mathcal{A}_5$ and $\mathcal{A}_6$ are meager, $\mathcal{A}_1$ and $\mathcal{A}_4$ are normal and $\mathcal{A}_2$ and $\mathcal{A}_7$ are obese. From now on, we concentrate on meager timed automata and compute their bandwidth. $\mathcal{A}_5$ has only discrete choices since its transitions happen at discrete dates only. $\mathcal{A}_3$ can produce a huge amount of information ($2/\varepsilon$ bits every second), but only during the first five seconds of its life. Its bandwidth is $\lim_{T\to\infty} 2/\varepsilon T = 0$. The automaton $\mathcal{A}_6$ (serving as running example below) is much less evident. It involves two flows of events, with events of the former being < 1 second apart and of the latter on the contrary > 1 second apart, as shown on Fig. 3. Such interleaving is possible for any duration T, but real-valued delays between symbols become more and more constrained.

3   Simply-Timed Graphs and Their Bandwidth

Following [11, 18], we define a class of graphs that will serve as the main abstraction for computing the bandwidth of meager timed automata.

Definition 9 (Simply-timed graphs). *A simply-timed graph (STG) is a tuple $\mathcal{G} = (Q, \Sigma, \Delta)$ with Q a finite set of states, Σ a finite alphabet, and $\Delta \subset Q \times \mathbb{Q}_+ \times \Sigma \times Q$ a finite transition relation. For a transition $(p, d, a, q) \in \Delta$, the state p is referred to as origin, q destination, a label and d delay.*

An STG is 0-free if all its delays are strictly positive. It is deterministic if for any two transitions (p, d, a, q) and (p, d, a, q') necessarily $q = q'$.

The timed language of such a graph is defined in a natural way.

Definition 10 (Semantics of STG). *A run of an STG (Q, Σ, Δ) on a timed word $w = (a_1, t_1)\ldots(a_n, t_n)$ is a sequence $q^0 \xrightarrow{d_0, a_1} q^1 \xrightarrow{d_1, a_2} \cdots q^n$, such that $t_i - t_{i-1} = d_i$, and $(q^{i-1}, d_i, a_i, q^i) \in \Delta$. The language $L(\mathcal{G})$ of an STG $\mathcal{G}$ is the set of all timed words having a run.*

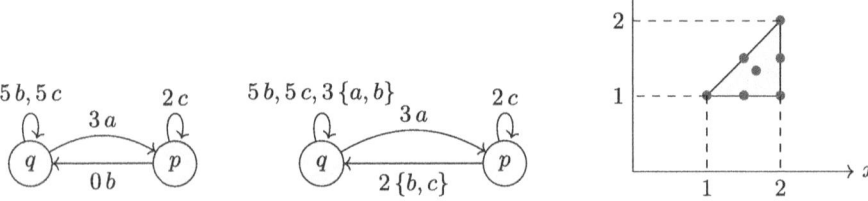

Fig. 5. A simply-timed graph (left), and its 0-free form (middle). Right: a 2-dimensional region, its 7 faces, and their barycenters.

The notions of ε-bandwidth are simpler for STG.

Definition 11. *For a set S of timed words, we call its* size *and denote by $\Upsilon(S)$ the cardinality of the largest 0-separated set in S.*

Equivalently, this is the cardinality of the smallest 0-net, and the number of equivalence classes for the binary relation $d(u,v) = 0$ over S. We remark that whenever S is 0-free, $\Upsilon(S) = \#S$.

Definition 12. *The* growth rate *of an STG $\mathcal{G}$ is defined as*

$$\beta(\mathcal{G}) = \lim_{T \to \infty} \frac{\log \Upsilon(L_T(\mathcal{G}))}{T},$$

whenever this limit exists and is finite.

Proposition 2. *Given an STG $\mathcal{G}$ with non-empty language, the growth rate $\beta(\mathcal{G})$ always exists. Moreover, taking N the common denominator of all delays, for $\varepsilon < 1/2N$ the growth rate coincides with both bandwidths:*

$$\mathcal{BC}_\varepsilon(L(\mathcal{G})) = \mathcal{BH}_\varepsilon(L(\mathcal{G})) = \beta(\mathcal{G}).$$

We present the method for computing $\beta(\mathcal{G})$, which consists of several transformations. First, we relate the notions of size and growth rate to 0-elimination.

Proposition 3. *0-elimination preserves the growth rate: $\beta(L) = \beta(\nu(L))$.*

Thanks to this proposition, instead of counting equivalence classes in L_T we will count timed words in $\nu(L)_T$. To that aim, we will transform the STG for L to that for $\nu(L)$.

Construction 1 (0-elimination in STG). *Given an STG $\mathcal{G} = (Q, \Sigma, \Delta)$, we define $\nu(\mathcal{G}) = (Q, 2^\Sigma \setminus \{\emptyset\}, \Delta')$ combining every timed transition with each subsequent instant transition: $(p, d, A, q) \in \Delta'$ whenever $d > 0$ and there exist $(p, d, a_0, q_0), (q_0, 0, a_1, q_1), \ldots, (q_{k-1}, 0, a_k, q_k) \in \Delta$ with $q_k = q$, $A = \{a_0, \ldots, a_k\}$.*

We remark that $\nu(\mathcal{G})$ can be easily computed by saturation.

Proposition 4. *Whenever an STG $\mathcal{G}$ recognizes L, the 0-eliminated graph $\nu(\mathcal{G})$ recognizes the language $\nu(L)$.*

Given an STG $\mathcal{G}$, we compute its determinization by the usual subset construction, see e.g. [17, Thm 3.3.2]. Determinization preserves the timed language and thus the growth rate.

Given a deterministic 0-free STG $\mathcal{G}$, we can compute its growth rate using the technique of generating functions.

Construction 2. *Given an STG $\mathcal{G} = (Q, \Sigma, \Delta)$ with $Q = \{q_1, \ldots, q_n\}$, we define its $n \times n$ adjacency matrix as*

$$M_{\mathcal{G}}(z) = (m_{ij}(z)) \text{ with } m_{ij}(z) = \sum_{(q_i, d, a, q_j) \in \Delta} z^d,$$

and its characteristic quasi-polynomial as $\psi_{\mathcal{G}}(z) = \det(I - M_{\mathcal{G}}(z))$.

Proposition 5. *Let $\mathcal{G}$ be a deterministic 0-free STG, and let z_0 be the smallest root (in modulus) of its characteristic equation $\psi_{\mathcal{G}}(z) = 0$. Then $\beta(\mathcal{G}) = -\log|z_0|$.*

Theorem 1. *Given an STG $\mathcal{G}$, for ε small enough, $\mathcal{BC}_\varepsilon(\mathcal{G}) = \mathcal{BH}_\varepsilon(\mathcal{G}) = \beta(\mathcal{G})$. The growth rate $\beta(\mathcal{G})$ is a logarithm of an algebraic number computable as a function of $\mathcal{G}$.*

Proof. The algorithm for computing $\beta(\mathcal{G})$ is as follows: eliminate 0, determinize, write the adjacency matrix, and solve the characteristic equation. Its correctness follows from the above propositions. □

As an example consider the STG on Fig. 5 (left). Its 0-free form is represented in the middle, it is deterministic. The characteristic matrix and equation are

$$M(z) = \begin{pmatrix} z^3 + 2z^5 & z^3 \\ z^2 & z^2 \end{pmatrix}; \quad \det(I - M(z)) = 2z^7 - 2z^5 - z^3 - z^2 + 1 = 0,$$

with the smallest root $z_0 \approx 0.698776$, the growth rate is $-\log_2 z_0 \approx 0.517098$.

4   Computing the Bandwidth of a Meager RsTA

Consider a closure of a $(c-1)$-dimensional clock region $\bar{r}$ in the clock space of $\mathcal{A}$. It is a simplex with c vertices. The convex hull of any subset of $k > 0$ vertices of a region will be referred to as a $(k-1)$-dimensional *face*. Note that each vertex and the whole closed region are considered as faces. To each such face f with vertices $v_1, \ldots, v_k$ we associate its *barycenter* $b = \frac{1}{k} \sum_{i=1}^{k} v_k$, as illustrated on Fig. 5, right. We denote the set of all such face barycenters $\alpha(r)$.

Given a clock vector $x \in \bar{r}$, we consider the smallest face $f(x)$ containing x and denote $\alpha(x)$ the barycenter of $f(x)$. Abusively, we also say that x and $\alpha(x)$ have the dimension $k - 1$.

Barycentric Abstraction. Our method for the computation of the bandwidth of an RsTA involves abstracting the RsTA by an STG whose states correspond to barycenters of the faces of each start region in the RsTA. The transitions of the STG are exactly the timed transitions of the original RsTA that allow a unique time delay and do not change face dimension. Formally:

Construction 3. *Given an* RsTA $\mathcal{A} = (Q, X, \Sigma, \Delta, S, I, F)$, *its* barycentric abstraction *is an* STG $\mathcal{G} = \alpha(\mathcal{A}) = (Q', \Sigma, \Delta')$ *obtained as follows:*

- *the state space is constituted by face barycenters of all the starting regions:*

$$Q' = \bigcup_{q \in Q} \{q\} \times \alpha(\overline{S(q)});$$

- *for each edge* $\delta = (p, p', a, \mathfrak{g}, \mathfrak{r}) \in \Delta$, *and for each couple of barycenters* $x \in \alpha(S(p))$ *and* $x' \in \alpha(S(p'))$ *of the same dimension,*
  if $L_{\bar{\delta}}((p, x), (p', x')) = \{ta\}$ *for some unique* $t \in \mathbb{Q}$, *then* Δ' *contains a transition* $((p, x), t, a, (p', x'))$. Δ' *contains no other transitions.*

Note that the abstraction leaves some edges of the RsTA without a counterpart in the built STG. In fact, these removed edges can only appear a bounded number of times within a run of a meager automaton, and thus are actually not needed to produce the full amount of bandwidth.

For our running example, the region-split automaton $\mathcal{A}'_6$ on Fig. 4 the starting region is 0-dimensional for s, just $(0, 0)$. The two other starting regions are 1-dimensional: $(0, 1) \times \{0\}$ for q and $\{0\} \times (0, 1)$ for p, in each of those there are two vertices and one 1-dimensional barycenter in the middle. The barycentric abstraction is presented on Fig. 6, top; it splits into two connected components: one on vertices (left subfigure) and one on region centers $(1/2, 0)$ for q and $(0, 1/2)$ for p (right).

The theorem below states that the bandwidth of the apparently smaller language $L(\mathcal{G})$ is indeed the same as that of $L(\bar{\mathcal{A}})$, implying that the barycentric abstraction is sufficient for computing the bandwidth of meager automata.

Theorem 2. *For any meager* RsTA $\mathcal{A}$ *and its barycentric abstraction* $\mathcal{G} = \alpha(\mathcal{A})$, *and for* ε *small enough it holds that*

$$\mathcal{BC}_\varepsilon(\mathcal{A}) = \mathcal{BH}_\varepsilon(\mathcal{A}) = \beta(L(\mathcal{G})).$$

Let us sketch the main ideas of the proof.

The **lower bound** is not difficult. By the pigeonhole principle, there exists a node $n = (p, x)$ of $\mathcal{G}$, such $L_\mathcal{G}(n, n)$ has growth rate $\beta(L(\mathcal{G}))$. Let now u a timed word leading in $\bar{\mathcal{A}}$ from I to (p, x) and v leading from (p, x) to F. Then it is easy to see (directly from Construction 3) that $u \cdot L_\mathcal{G}(n, n) \cdot v \subset L(\bar{\mathcal{A}})$ and hence $\beta(L(\mathcal{G})) = \mathcal{BH}_0(u \cdot L_\mathcal{G}(n, n) \cdot v) \leq \mathcal{BH}_0(u \cdot L_\mathcal{G}(n, n) \cdot v) \leq \mathcal{BH}_0(L(\bar{\mathcal{A}})) = \mathcal{BH}_0(L(\mathcal{A}))$.[3]

---

[3] The latter equality is a consequence of [5], full version, Lem. 35.

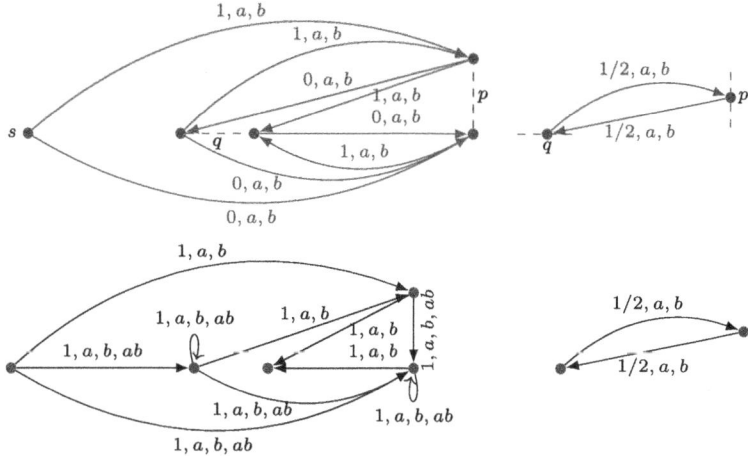

Fig. 6. Top: barycentric abstraction $\mathcal{G} = \alpha(\mathcal{A}_6')$ split in components of dimension 0 and 1. Bottom: the same graph after 0-elimination. Multiple labels on an edge mean that there is an edge for each label, all with the same timing.

For the **upper bound**, as established in [5], the totality of the bandwidth of meager automata is produced by cycles running from a unique state (q, x) (in the closure of the automaton) to itself. Moreover, each path from (q, x) to itself admits a unique timing. We prove that this bunch of cycles can be approximated by a similar bunch on the barycenter $(q, \alpha(x))$. We count elements of the latter using the STG $\mathcal{G}$.

As a comment, while our reasoning could possibly still be valid for other choices of face representatives than barycenters (any interior point of a face can represent it), our choice is not arbitrary because it allows for a terser abstraction, based on a single, canonical, representative per face. Indeed, when an edge in a cycle of a RsTA sends a face onto another face, the barycenter of the destination face is always a successor of the barycenter of the origin face.

Our main result is immediate from Theorems 1 and 2.

Theorem 3. *The bandwidth of a meager DTA $\mathcal{A}$, for ε small enough, satisfies $\mathcal{BC}_\varepsilon(\mathcal{A}) = \mathcal{BH}_\varepsilon(\mathcal{A}) = \beta$ with β a logarithm of an algebraic number computable as a function of $\mathcal{A}$.*

We remark that the overall complexity of our algorithm is doubly exponential: one exponent is due to region-splitting and taking barycenters of all faces, the other to determinizing the STG. This is to compare to PSPACE-completeness of recognizing meager DTA[5].

For the running example $\mathcal{A}_6$, we proceed with 0-elimination for the STG on Fig. 6 (top), the result is given on Fig. 6 (bottom). The graph obtained is deterministic. The adjacency matrix of the 0-free STG obtained splits into two diagonal blocks (corresponding to connected components of barycenters of dimensions 0 and 1):

$$\begin{pmatrix} 0 & 3z & 0 & 3z & 2z \\ 0 & 3z & 0 & 3z & 2z \\ 0 & 0 & 0 & 0 & 0 \\ 0 & 0 & 2z & 3z & 0 \\ 0 & 0 & 2z & 3z & 0 \end{pmatrix}; \quad \begin{pmatrix} 0 & 2z^{1/2} \\ 2z^{1/2} & 0 \end{pmatrix}.$$

The characteristic equation also splits:

$$(1 - 3z)^2(1 - 4z) = 0.$$

The smallest root (corresponding to dimension 1 barycenters) is $1/4$, the bandwidth is $-\log 1/4 = 2$.

5   Conclusions

We have solved the problem of computing the bandwidth for simply-timed graphs and based on that, for all meager automata. Our main tool is the barycentric abstraction, a new construction that can be seen as a refinement of the corner point automaton from [8] (which, in turn, refines the region automaton [1]).

The next step is to build appropriate abstractions for computing the bandwidth of normal and obese automata, using the tools from [5]. Synthesis of timed codes and their applications would be the outcome of this research program.

A   Appendix: Towards the Proof of Theorem 2

We have proved on page 10 that

$$\mathcal{BC}_\varepsilon(\mathcal{A}) \geq \beta(\alpha(\mathcal{A})); \quad \mathcal{BH}_\varepsilon(\mathcal{A}) \geq \beta(\alpha(\mathcal{A})). \tag{1}$$

The proof of the opposite inequality is based on the intermediary results presented in this appendix. The detailed proof can be read in [6].

A.1   Puri's Reachability Method

We recall some basics on reachability in timed automata, mostly due to [19] and slightly reformulated in [3,5]. We use a specific coordinate system on clock regions.

Construction 4 (Canonical barycentric coordinates). *Given a $k-1$-dimensional region r we compare its vertices as follows: $v < v'$ whenever $v \cdot \mathbf{1} < v' \cdot \mathbf{1}$ (it is easy to see that this is a total order). We enumerate all the vertices in the ascending order: $v_1 < v_2 < \cdots < v_k$. For any clock vector $x \in \bar{r}$ we define its barycentric coordinates $\boldsymbol{\lambda}(x) = (\lambda_1, \ldots, \lambda_k)$ by conditions*

$$\sum_{i=1}^{k} \lambda_i v_i = x; \quad \sum_{i=1}^{k} \lambda_i = 1; \quad 0 \leq \lambda_i \leq 1.$$

Given a path $q \xrightarrow{\pi} q'$ in a RsTA $\mathcal{A}$, and faces r and r' of $S(q)$ and $S(q')$, we aim to characterize when $x \in r$ can go to $x' \in r'$ through $\bar{\pi}$ (path corresponding to π in $\bar{\mathcal{A}}$). That is, there exists a run ρ associated with $\bar{\pi}$. The main tool of Puri's method is the *orbit graph* summarizing reachability between region vertices.

Definition 13. *The* orbit graph $\gamma(\pi)$ *is a bipartite graph from vertices of* r *to those of* r'. *It has an edge* $v \to v'$ *whenever* v *can go to* v' *through* $\bar{\pi}$. *The* orbit matrix $M(\pi)$ *is the adjacency matrix of* $\gamma(\pi)$.

Theorem 4 ([19]). *Given a path* $q \xrightarrow{\pi} q'$, *a state* (q', x') *(with* x' *in an* n'-*dimensional region face) is reachable from a state* (q, x) *(with* x *in an* n-*dimensional region face) through* $\bar{\pi}$ *iff there exists an* $n \times n'$ *stochastic matrix*[4] P *satisfying* $\boldsymbol{\lambda}(x)P = \boldsymbol{\lambda}(x')$ *and* $P \leq M(\pi)$.

We call such a P a reachability matrix from x to x' along $\bar{\pi}$. We will enrich P with duration information as follows. If for all i and j, such that $P_{ij} > 0$, there is at most a single possible run ρ_{ij} from v_i to v'_j through $\bar{\pi}$, then we define the timing matrix D^P, such that

$$D^P_{ij} = \begin{cases} \mathrm{dur}(\rho_{ij}) & \text{if } P_{ij} > 0; \\ 0 & \text{otherwise.} \end{cases}$$

Proposition 6. *Let* P *be a reachability matrix from* x *to* x' *along* $\bar{\pi}$, *with a well-defined timing matrix* D^P. *Then* x *can go to* x' *in time* $d = \sum_{i,j} \lambda_i(x)P_{ij}D^P_{ij}$.

A.2   Corollary from [5, full version]

Definition 14. *A number* $b \in \mathbb{R}$ *is called a* cycle growth bound *of an RsTA* $\mathcal{A}$ *whenever there exists a constant* C *such that for any* $q \in Q, x \in \bar{S}(q), T > 0$, *the language* $L^{\bar{\mathcal{A}}}_T((q, x), (q, x))$ *admits a 0-net of cardinality* $C2^{bT}$.

Proposition 7. *If* $b \in \mathbb{R}$ *is a cycle growth bound of a meager RsTA* $\mathcal{A}$, *then (for any* $\varepsilon > 0$)

$$\mathcal{BH}_\varepsilon(\mathcal{A}) \leq \mathcal{BC}_\varepsilon(\mathcal{A}) \leq b.$$

The quite technical (and based on Simon factorization) proof of this proposition follows verbatim that of Lemmas 68–77 and Prop. 18 from [5, full version].

A.3   On Cycles in Meager Timed Automata

Through the next two sections, we consider a cyclic path $q \xrightarrow{\pi} q$ in a meager RsTA, such that it exists a cyclic run along $\bar{\pi}$, on some clock vector x in $\bar{S}(q)$:

$$(q, x) = (q_0, x_0) \to (q_1, x_1) \to \cdots \to (q_k, x_k) = (q, x)$$

Let $r_i = \mathfrak{f}(x_i)$ be the smallest face containing x_i and P_i the reachability matrix from x_{i-1} to x_i. After establishing that all such r_i have the same dimension k and that consequently all the P_i are square, we prove the following result about reachability along such a path:

---

[4] That is, $P \cdot \mathbf{1}_n = \mathbf{1}_{n'}$.

Lemma 1. *All the matrices P_i, and the corresponding D^P, which can now be defined, have one of the following forms (for some $l, d \in \mathbb{N}$ with $l < k$):*

$$P^{kl} = \left(\begin{array}{c|c} 0_{l \times (k-l)} & I_l \\ \hline I_{k-l} & 0_{(k-l) \times l} \end{array} \right) ; D^{kld} = \left(\begin{array}{c|c} 0_{l \times (k-l)} & (d+1) \cdot I_l \\ \hline d \cdot I_{k-l} & 0_{l \times (k-l)} \end{array} \right).$$

As a side note, it also implies the following remarkable property:

Corollary 1. *Let ρ and ρ' be two cyclic runs along π, then $\mathrm{dur}(word(\rho)) = \mathrm{dur}(word(\rho'))$.*

A.4   Relating Cycles with Barycenters and Bounding the Cycle Growth

It remains to prove that the growth rate of $\mathcal{G} = \alpha(\mathcal{A})$ is a cycle growth bound for $\mathcal{A}$. To that aim, we concentrate on cycles in $\mathcal{A}$.

Lemma 2. *Let ρ be a cyclic run of path π on x, then there is a cyclic run of the same path on barycenters $\alpha(x)$ with the same reachability matrices.*

Given a timed word w accepted by a (unique) cyclic run ρ from (q, x) to (q, x), let us consider its abstraction $\alpha(\rho)$. In turn, it accepts some $w' \in L^{\mathcal{G}}((q, \alpha(x)), (q, \alpha(x)))$. We denote $w' = \alpha_{qx}(w)$.

Consider now delays in the barycentric cycle.

Lemma 3. *In a barycentric cycle, whenever the timing matrix for transition i has the form D^{kld} from Lemma 1, the transition $\alpha(x_{i-1}) \to \alpha(x_i)$ lasts $d + l/k$.*

We will use the previous lemma in the reverse direction: knowing that the barycenter takes a transition in $d + l/k$ time, we will deduce l, k, d, then matrices P^{kl} and D^{kld}, and finally the successors and timings in the concrete run of x_i.

Lemma 4. *Let $w^1, w^2 \in L^{\bar{\mathcal{A}}}_T((q, x), (q, x))$ be two timed words such that $d(w^1, w^2) > 0$ (necessarily they are recognized along different cyclic paths), and $w^{i'} = \alpha_{qx}(w^i), i = 1, 2$. Then $d(w^{1'}, w^{2'}) > 0$.*

Lemma 5. *$\beta(\mathcal{G})$ is a cycle growth bound for $\mathcal{A}$.*

Proof. Let $\mathcal{N}$ be a maximal 0-separated set in $L^{\bar{\mathcal{A}}}_T((q, x), (q, x))$ (it is also a minimal 0-net). Its image $\mathcal{N}' = \alpha_{qx}(\mathcal{N})$ by virtue of Lemma 4 is 0-separated and of the same cardinality. By construction $\mathcal{N}' \subset L^{\mathcal{G}}_T((q, \alpha(x)), (q, \alpha(x)))$. Hence $\#\mathcal{N} = \#\mathcal{N}' \leq C2^{\beta T}$, thus $\beta(\mathcal{G})$ is a cycle growth bound for $\mathcal{A}$.   □

Theorem 2 follows immediately from Proposition 7 and Lemma 5.

References

1. Alur, R., Dill, D.L.: A theory of timed automata. Theor. Comput. Sci. **126**, 183–235 (1994). https://doi.org/10.1016/0304-3975(94)90010-8
2. Aristotle: Physics (350 BCE). https://classics.mit.edu/Aristotle/physics.html
3. Asarin, E., Basset, N., Degorre, A.: Entropy of regular timed languages. Inf. Comput. **241**, 142–176 (2015). https://doi.org/10.1016/j.ic.2015.03.003
4. Asarin, E., Basset, N., Degorre, A.: Distance on timed words and applications. In: Jansen, D.N., Prabhakar, P. (eds.) FORMATS 2018. LNCS, vol. 11022, pp. 199–214. Springer, Cham (2018). https://doi.org/10.1007/978-3-030-00151-3_12
5. Asarin, E., Degorre, A., Dima, C., Jacobo Inclán, B.: Bandwidth of timed automata: 3 classes. In: Proceedings of the FSTTCS. LIPIcs, vol 284, pp. 10:1–10:17 (2023). https://doi.org/10.4230/LIPICS.FSTTCS.2023.10. Full version https://doi.org/10.48550/arXiv.2310.01941
6. Asarin, E., Degorre, A., Dima, C., Jacobo Inclán, B.: Computing the bandwidth of meager timed automata, long version (2024). https://doi.org/10.48550/arXiv.2406.12694
7. Bérard, B., Petit, A., Diekert, V., Gastin, P.: Characterization of the expressive power of silent transitions in timed automata. Fund. Inform. **36**(2–3), 145–182 (1998). https://doi.org/10.3233/FI-1998-36233
8. Bouyer, P., Brinksma, E., Larsen, K.G.: Staying alive as cheaply as possible. In: Alur, R., Pappas, G.J. (eds.) HSCC 2004. LNCS, vol. 2993, pp. 203–218. Springer, Heidelberg (2004). https://doi.org/10.1007/978-3-540-24743-2_14
9. Bridson, M.R., Gilman, R.H.: Context-free languages of sub-exponential growth. J. Comput. Syst. Sci. **64**(2), 308–310 (2002). https://doi.org/10.1006/jcss.2001.1804
10. Chomsky, N., Miller, G.A.: Finite state languages. Inf. Control **1**(2), 91–112 (1958). https://doi.org/10.1016/S0019-9958(58)90082-2
11. Dima, C.: Real-time automata. J. Automata Lang. Comb. **6**(1), 3–23 (2001). https://doi.org/10.25596/JALC-2001-003
12. Gawrychowski, P., Krieger, D., Rampersad, N., Shallit, J.: Finding the growth rate of a regular of context-free language in polynomial time. In: Ito, M., Toyama, M. (eds.) DLT 2008. LNCS, vol. 5257, pp. 339–358. Springer, Heidelberg (2008). https://doi.org/10.1007/978-3-540-85780-8_27
13. Grigorchuk, R., Machí, A.: An example of an indexed language of intermediate growth. Theor. Comput. Sci. **215**(1), 325–327 (1999). https://doi.org/10.1016/S0304-3975(98)00161-3
14. Jacobo Inclán, B., Degorre, A., Asarin, E.: Bounded delay timed channel coding. In: Bogomolov, S., Parker, D. (eds.) FORMATS 2022. LNCS, vol. 13465, pp. 65–79. Springer, Cham (2022). https://doi.org/10.1007/978-3-031-15839-1_4
15. Kolmogorov, A., Tikhomirov, V.: ε-entropy and ε-capacity of sets in function spaces. Uspekhi Matematicheskikh Nauk **14**(2), 3–86 (1959). https://doi.org/10.1007/978-94-017-2973-4_7
16. Lancaster, P.: Lambda-Matrices and Vibrating Systems. Pergamon Press (1966)
17. Lind, D., Marcus, B.: An Introduction to Symbolic Dynamics and Coding. Cambridge University Press, Cambridge (1995)
18. Markey, N., Schnoebelen, P.: Symbolic model checking for simply-timed systems. In: Lakhnech, Y., Yovine, S. (eds.) FORMATS/FTRTFT -2004. LNCS, vol. 3253, pp. 102–117. Springer, Heidelberg (2004). https://doi.org/10.1007/978-3-540-30206-3_9

19. Puri, A.: Dynamical properties of timed automata. Discret. Event Dyn. Syst. **10**(1–2), 87–113 (2000). https://doi.org/10.1023/A:1008387132377
20. Simon, I.: Factorization forests of finite height. Theor. Comput. Sci. **72**(1), 65–94 (1990). https://doi.org/10.1016/0304-3975(90)90047-L

Using Finite Automata to Compute the Base-b Representation of the Golden Ratio and Other Quadratic Irrationals

Aaron Barnoff[1]([✉]), Curtis Bright[1][iD], and Jeffrey Shallit[2][iD]

[1] School of Computer Science, University of Windsor,
Windsor, ON N9B 3P4, Canada
{barnoffa,cbright}@uwindsor.ca
[2] School of Computer Science, University of Waterloo,
Waterloo, ON N2L 3G1, Canada
shallit@uwaterloo.ca

Abstract. We show that the nth digit of the base-b representation of the golden ratio is a finite-state function of the Zeckendorf representation of b^n, and hence can be computed by a finite automaton. Similar results can be proven for any quadratic irrational. We use a satisfiability (SAT) solver to prove, in some cases, that the automata we construct are minimal.

1 Introduction

The base-b digits of famous irrational numbers, where $b \geq 2$ is an integer, have been of interest for hundreds of years. For example, William Shanks computed 707 decimal digits of π in 1873 (but only the first 528 were correct) [19]. As a high school student, the third author used a computer in 1976 to determine the first 10,000 digits of the decimal representation of $\varphi = (\sqrt{5} + 1)/2$, the golden ratio, using the computer language APL [18].

The celebrated results of Bailey, Borwein, and Plouffe [2] demonstrated that one can compute the nth bit of certain famous constants, such as π, in $O(n)$ time and $o(n)$ space.[1]

Can finite automata generate the base-b digits of irrational algebraic numbers, such as φ? This fundamental question was raised by Cobham in the late 1960's (a re-interpretation of a related question due to Hartmanis and Stearns [9]). Though Cobham believed for a time that he had proved they cannot be so generated [7], his proof was flawed, and it was not until 2007 that Adamczewski and Bugeaud [1] succeeded in proving that there is no deterministic finite automaton with output that, on input n expressed in base b, returns the nth base-b digit of an irrational real algebraic number α.

Even so, in this paper we show that, using finite automata, one *can* compute the nth digit in the base-b representation of the golden ratio φ! At first glance

---

[1] Sometimes this result is described as "computing the nth digit without having to compute the previous $n - 1$ digits". But this is not really a meaningful assertion, since the phrase "computing x without computing y" is not so well-defined.

S. Z. Fazekas (Ed.): CIAA 2024, LNCS 15015, pp. 35–50, 2024.
https://doi.org/10.1007/978-3-031-71112-1_3

this might seem to contradict the Adamczewski–Bugeaud result. But it does not, since for our theorem the input is not n expressed in base b, but rather b^n in an entirely different numeration system, the Zeckendorf representation. As we will see below, analogous results exist for any quadratic irrational.

Our result does not give a particularly efficient way to compute the base-b digits of quadratic irrationals, but it is nevertheless somewhat surprising. Using a SAT solver, in some cases (such as for the binary digits of φ) we can prove that the automaton we construct is minimal and unique. Interestingly, in other cases (such as for the ternary digits of φ) we were able to prove the minimality of our automaton, but we discovered several distinct automata with the same number of states computing the same quadratic irrational, at least up to a high precision. It is conceivable that the automata produced by our method are indeed always minimal and unique, but we leave this as an open question.

2   Number Representations and Automata

A DFAO (deterministic finite automaton with output) A consists of a finite number of states along with labeled transitions connecting them. The automaton processes an input string x by starting in the distinguished start state q_0, and then following the transitions from state to state, according to each successive symbol of x. Each state q has an output $\tau(q)$ associated with it, and the function f_A computed by the DFAO maps the input x to the output associated with the last state reached. For an example of a DFAO, see Fig. 2.

A DFA (deterministic finite automaton) is quite similar to a DFAO. The only difference is that there are exactly two possible outputs associated with each state, either 0 or 1. States with an output of 1 are called "accepting" or "final". If an input results in an output of 1, it is said to be accepted by the DFA. A *synchronized* DFA [6] is a particular type of DFA that takes two inputs in parallel; this is accomplished by making the input alphabet a set of pairs of alphabet symbols. A synchronized automaton computes a synchronized sequence $(f(n))_{n \geq 0}$; it does this by accepting exactly the inputs where the first components spell out a representation of n, and the second components spell out a representation for $f(n)$, where leading zeros may be required to make the inputs the same length; thus n and $f(n)$ are read in parallel. For more about synchronized sequences, see [16]. An example of a synchronized DFA appears in Fig. 1. Throughout the paper, integer inputs are processed starting with the most significant digit.

Let x be a non-negative real number, let $b \geq 2$ be an integer, and write the base-b representation of x in the form $x = \sum_{-\infty < i \leq t} a_i b^i = a_t a_{t-1} \cdots a_0 . a_{-1} a_{-2} \cdots$, where $a_i \in \{0, 1, \ldots, b-1\}$. For $n \geq 0$, we call a_{-n-1} the nth digit to the right of the radix point. This choice of associating n with a_{-n-1} is perhaps a little unusual, but it seems to decrease the size of the automata produced.

2.1   Zeckendorf Representation

The Fibonacci numbers are defined, as usual, by $F_0 = 0$, $F_1 = 1$, and $F_n = F_{n-1} + F_{n-2}$ for $n \geq 2$. The Zeckendorf representation [11,21] of a natural number n is the unique way of writing n as a sum of Fibonacci numbers F_i, $i \geq 2$, subject to the condition that no two consecutive Fibonacci numbers are used. We may write the Zeckendorf representation as a binary string $(n)_F = a_1 \cdots a_t$, where $n = \sum_{1 \leq i \leq t} a_i F_{t+2-i}$. For example, $(43)_F = 34+8+1 = F_9 + F_6 + F_2$ has representation 10010001. The substring 11 cannot occur due to the rule that two consecutive Fibonacci numbers cannot be used. In what follows, leading zeros in strings are typically ignored without comment. We also denote the inverse of $(\cdot)_F$ by $[\cdot]_F$; i.e., $[10010001]_F = 43$.

3   Automata and the Base-b Representation of φ

Our main result is Theorem 1 below.

Theorem 1. *For all integers $b \geq 2$, there exists a DFAO $\mathcal{A}_b$ that, on input the Zeckendorf representation of b^n, computes the nth digit to the right of the point in the base-b representation of φ.*

Proof. It is known that there exists a 7-state synchronized DFA A_1 accepting, in parallel, the Zeckendorf representations of q and $\lfloor q\varphi \rfloor$ for all $q \geq 0$ [17, Thm. 10.11.1 (a)]. Its transition diagram is depicted in Fig. 1, where accepting states are denoted by double circles, and the initial state is 0, labeled by a headless arrow entering.

The DFA A_1 is constructed using the fact that $\lfloor q\varphi \rfloor = [(q-1)_F 0]_F + 1$, where $(q-1)_F 0$ is the left shift of the string $(q-1)_F$. For example, $\lfloor 11\varphi \rfloor = 17$, and we find $(10)_F = 10010$, left-shift that to get $100100 = (16)_F$, and add 1 to get 17.

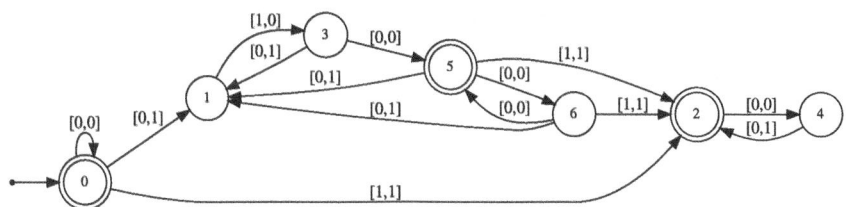

Fig. 1. Synchronized automaton for $\lfloor q\varphi \rfloor$. The inputs are the Zeckendorf representation of q and x, in parallel; it accepts iff $x = \lfloor q\varphi \rfloor$.

To understand how to use this automaton, observe that $(11)_F = 10100$ and $(\lfloor 11\varphi \rfloor)_F = (17)_F = 100101$. Since these two numbers have representations of different lengths, we need to pad the former with a leading 0. Then if $x = [0,1][1,0][0,0][1,1][0,0][0,1]$, the first components concatenated spell out 010100 and the second components spell out 100101. When we input this, starting at state 0 we visit, successively, states $1, 3, 5, 2, 4, 2$, and so we accept.

Let x be a positive real number, with base-b representation $y.a_0 a_1 a_2 \cdots$, where the period (or radix point) is the analogue of the decimal point for base b, and y is an arbitrary finite block of digits. Now $b^{n+1}x$ has base-b representation $ya_0a_1\cdots a_{n-1}a_n.a_{n+1}\cdots$ and $\lfloor b^{n+1}x\rfloor$ has base-b representation $ya_0a_1\cdots a_{n-2}a_{n-1}a_n$. Similarly, $b\lfloor b^n x\rfloor$ has base-b representation $ya_0a_1\cdots a_{n-1}0$. Hence $\lfloor b^{n+1}x\rfloor - b\lfloor b^n x\rfloor = a_n$. In the particular case where $x = \varphi$, we get a formula for the nth digit to the right of the radix point of φ, namely

$$D_b(n) := \lfloor b^{n+1}\varphi\rfloor - b\lfloor b^n\varphi\rfloor.$$

From the DFA A_1 computing $\lfloor q\varphi\rfloor$, it is possible to create another DFA A_2 accepting, in parallel, the Zeckendorf representations of q and $\lfloor bq\varphi\rfloor - b\lfloor q\varphi\rfloor$. This is based on the fact that there is an algorithm to compile a first-order logic statement involving the usual logical operations (AND, OR, NOT, etc.), the integer operations of addition, subtraction, multiplication by constants, and the universal and existential quantifiers, into an automaton that accepts the Zeckendorf representation of those integers making the statement true [13].

From the DFA A_2, we can compute b individual DFAs $A_{b,i}$ accepting the Zeckendorf representation of those q for which $\lfloor bq\varphi\rfloor - b\lfloor q\varphi\rfloor = i$, for $0 \le i < b$. Finally, we combine all the $A_{b,i}$ together into a single DFAO A_3 (using a product construction for automata) computing the difference $\lfloor bq\varphi\rfloor - b\lfloor q\varphi\rfloor$.

By substituting $q = b^n$, we see that this automaton A_3 is the desired one, computing $D_b(n)$ on input the Zeckendorf representation of b^n. $\square$

We now use `Walnut`, which is free software for compiling first-order logical expressions into automata, to explicitly compute the automata for the representation of φ in base 2 and base 3. For base 2, we need the following `Walnut` commands (further explanation follows below):

```
reg shift {0,1} {0,1} "([0,0]|[0,1][1,1]*[1,0])*":
def phin "?msd_fib (s=0 & q=0) | Ex $shift(q-1,x) & s=x+1":
def phid2 "?msd_fib Ex,y $phin(2*q,x) & $phin(q,y) & x=2*y+1":
combine FD2 phid2:
```

These produce the DFAO in Fig. 2.

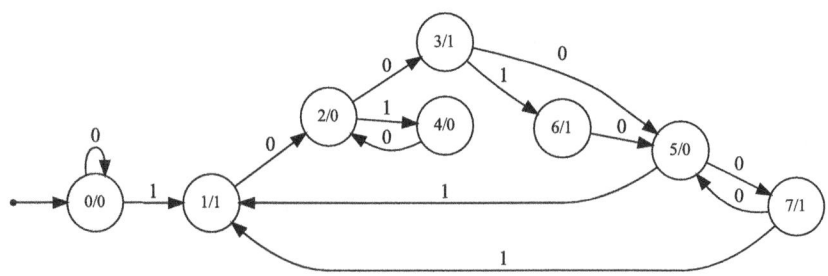

Fig. 2. Automaton $\mathcal{A}_2$ for the nth bit (base 2 digit) to the right of the binary point of φ. States are labeled in the form a/c, where a is the state number and c is the output. The input is the Zeckendorf representation of 2^n, and the output is c when the last state reached is labeled a/c.

For example, in base 2, we have $\varphi = 1.1001111000110111\cdots$. To compute the 4th digit to the right of the binary point we write $2^4 = 16$ in Zeckendorf representation, namely 100100, and feed it into the automaton, starting at state 0 and reaching states $1, 2, 3, 6, 5, 7$ successively, with output 1 at the end.

We now explain the `Walnut` commands above that generate the DFAO in Fig. 2. The first line creates a DFA called `shift`, using a regular expression; it takes two base-2 inputs and accepts only if the second is the left shift of the first. Next is the DFA `phin`, which is shown in Fig. 1 and uses `shift` to check that its two inputs have the relationship $(n)_F$ and $[(n-1)_F 0]_F + 1$, which computes the function $n \rightarrow \lfloor n\varphi \rfloor$ in a synchronized fashion. Next, the DFA `phid2`, when given the representation of q as input, accepts if $\lfloor 2q\varphi \rfloor - 2\lfloor q\varphi \rfloor = 1$, and rejects otherwise. Lastly, `combine` converts `phid2` into the DFAO of Fig. 2 by replacing the accepting and rejecting states of `phid2` with output values 1 and 0, respectively.

The automaton for base 3 (see Fig. 3) can be constructed similarly with the following `Walnut` commands:

```
reg shift {0,1} {0,1} "([0,0]|[0,1][1,1]*[1,0])*":
def phin "?msd_fib (s=0 & n=0) | Ex $shift(n-1,x) & s=x+1":
def phid31 "?msd_fib Ex,y $phin(3*n,x) & $phin(n,y) & x=3*y+1":
def phid32 "?msd_fib Ex,y $phin(3*n,x) & $phin(n,y) & x=3*y+2":
combine FD3 phid31=1 phid32=2:
```

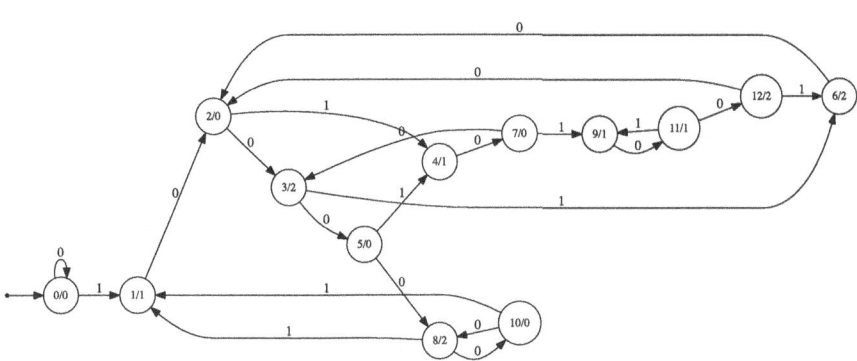

Fig. 3. Automaton for the nth digit to the right of the point of φ in base 3, with inputs as in Fig. 2.

In base 3, $\varphi = 1.1212001122021210\cdots$. To compute the 3rd digit to the right of the point we write $3^3 = 27$ in Zeckendorf representation as 1001001 and pass it to the automaton in Fig. 3, which, starting at state 0, traverses states $1, 2, 3, 6, 2, 3, 6$ successively, giving an output of 2.

There is no conceptual barrier to carrying out similar computations for any base $b \geq 2$. For base 10, for example, `Walnut` computes a finite automaton with 97 states that, on input $(10^n)_F$, returns the nth digit to the right of the decimal point in the decimal expansion of φ.

4   Other Quadratic Irrationals

There is nothing special about φ, and the same ideas can be used for any quadratic irrational. What makes quadratic irrationals special in this context is Lagrange's theorem: these numbers, and only these, have a continued fraction expansion that is ultimately periodic. This is crucial, because if this property does not hold, then the sequence of continued fraction convergents cannot satisfy a linear recurrence [12]. But a linear recurrence is needed in order to construct a numeration system with good decidability properties.

4.1   Handling $\sqrt{2}$

Another representation for the natural numbers is based on the Pell numbers, defined by $P_0 = 0$, $P_1 = 1$, and $P_n = 2P_{n-1} + P_{n-2}$ for $n \geq 2$. We can then write every natural number $n = \sum_{1 \leq i \leq t} a_i P_{t+1-i}$ where $a_i \in \{0, 1, 2\}$. To get uniqueness of the representation, we have to impose two conditions. First, we must have that $a_t \neq 2$. Second, if $a_i = 2$, then $a_{i+1} = 0$. See [4] for more details. The unique representation, over the alphabet $\{0,1,2\}$, is denoted $(n)_P$.

The Pell numeration system in `Walnut` can be used to construct automata computing the base-b digits of $\sqrt{2}$, just as we did for φ. This results in a 6-state DFAO for base 2 (see Fig. 4), and a 14-state DFAO for base 3. The `Walnut` commands for base 2 are:

```
reg pshift {0,1,2} {0,1,2}
   "([0,0]|([0,1][1,1]*([1,0]|[1,2][2,0]))|[0,2][2,0])*":
def sqrt2n "?msd_pell (s=0 & n=0) | Ex $pshift(n-1,x) & s=x+2":
def sqrt2d2 "?msd_pell Ex,y $sqrt2n(2*n,x) & $sqrt2n(n,y)
   & x=2*y+1":
combine SD2 sqrt2d2:
```

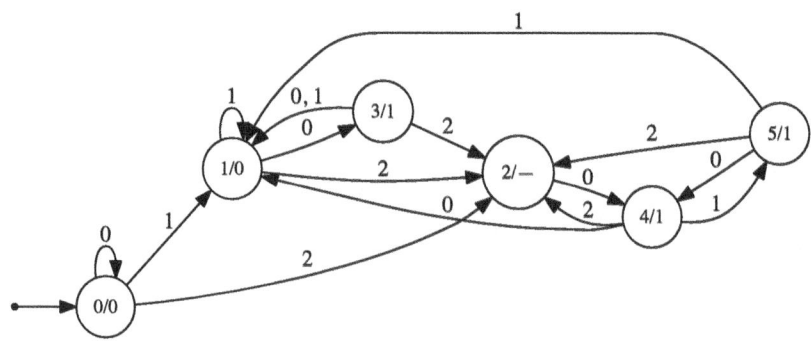

Fig. 4. Automaton for the nth bit to the right of the binary point of $\sqrt{2}$. Input is 2^n in Pell representation.

The alert reader will observe that no output is associated with state 2. This is because inputs that lead to this state, such as 12, are not valid Pell representations. However, the state cannot be removed, because 120 *is* a valid Pell representation.

4.2   Ostrowski Representation

Of course, what makes our results work is that the numeration systems are "tuned" to the particular quadratic irrational we want to compute. For φ, the numeration system is based on the Fibonacci numbers; for $\sqrt{2}$, the Pell numbers. We need to find an appropriate numeration system that is similarly "tuned" to any quadratic irrational. It turns out that the proper system is the Ostrowski numeration system [3, 14].

Every irrational real number α can be expressed uniquely as an infinite simple continued fraction $\alpha = [d_0, d_1, d_2, \ldots]$. Furthermore, q_n is called the nth denominator of a convergent for α if $q_{-2} = 1$, $q_{-1} = 0$, and $q_n = d_n q_{n-1} + q_{n-2}$ for $n \geq 0$. For example, the continued fraction for π is $[3, 7, 15, 1, \ldots]$, corresponding to the sequence $(q_n)_{n \geq 0} = 1, 7, 106, 113, \ldots$ (OEIS A002486).

The Ostrowski α-numeration system uses the sequence $(q_n)_{n \geq 0}$ of the denominators of the convergents for α to construct a unique representation for a non-negative integer N expressed as

$$N = [a_{n-1} a_{n-2} \cdots a_0]_\alpha = \sum_{0 \leq i < n} a_i q_i,$$

where the a_i have to obey the Ostrowski rules

$$0 \leq a_0 < d_1; \tag{1}$$
$$0 \leq a_i \leq d_{i+1} \text{ for } i \geq 1; \text{ and} \tag{2}$$
$$\text{for } i \geq 1, \text{ if } a_i = d_{i+1} \text{ then } a_{i-1} = 0. \tag{3}$$

The Ostrowski α-representation for $N = [a_{n-1} a_{n-2} \cdots a_0]_\alpha$ is then determined with a greedy algorithm, starting at the most significant term and choosing the largest multiple a_{n-1} for q_{n-1} that is less than N, and then applying the same algorithm recursively to $N - a_{n-1} q_{n-1}$. For example, for $\alpha = \sqrt{3} + 1 = [2, \overline{1, 2}]$, the denominators of the continued fraction convergents form the sequence $(q_n)_{n \geq 0} = 1, 1, 3, 4, 11, 15, \ldots$ (OEIS A002530). Rule 1 for the construction forces $a_0 = 0$ because $d_1 = 1$, while rule 2 requires that $a_1 \leq d_2 = 2$, $a_2 \leq d_3 = 1$, and so on. Rule 3 ensures uniqueness by enforcing the constraint that if $a_1 = d_2 = 2$, then $a_2 = 0$, and if $a_2 = d_3 = 1$, then $a_3 = 0$, and so on. Then, for example, the α-representation of 37 is $2 \cdot 15 + 4 + 3 = 2q_5 + q_3 + q_2 = [20110]_\alpha$.

In order to construct a DFAO $\mathcal{A}_b$ that, given the input of the Ostrowski α-representation of b^n, computes the nth digit to the right of the point in the base-b representation of α, we require an Ostrowski α-synchronized function $n \to \lfloor n\alpha \rfloor$. Consider a quadratic irrational $0 < \beta < 1$ with a purely periodic continued fraction $[0, \overline{d_1, d_2, \ldots, d_m}]$; here the straight bar or vinculum denotes the periodic part. Then Schaeffer et al. [15] showed that the sequence $(\lfloor n\beta \rfloor)_{n \geq 1}$ is Ostrowski β-synchronized, via the relation

$$[(n-1)_\beta 0^m]_\beta = q_m(n-1) + q_{m-1} \cdot \lfloor n\beta \rfloor, \tag{4}$$

where q_i is the denominator of the ith convergent to β, and $(n-1)_\beta 0^m$ is the β-representation of $n-1$, left-shifted m times.

Furthermore, it was shown that if $\alpha > 0$ belongs to $\mathbb{Q}(\beta)$, then $(\lfloor n\alpha \rfloor)_{n\geq 1}$ is synchronized in terms of the Ostrowski β-representation through the relation $\alpha = (a + d\beta)/c$, where $d, c \geq 1$, and

$$\lfloor n\alpha \rfloor = \left\lfloor \frac{\lfloor dn\beta \rfloor + an}{c} \right\rfloor. \tag{5}$$

This is notable because when constructing an Ostrowski α-representation with Walnut, it is assumed that $0 < \alpha < \frac{1}{2}$, which corresponds to a continued fraction with terms $d_0 = 0$ and $d_1 > 1$. If $\alpha \geq \frac{1}{2}$, then we can set $d_0 = 0$ and rotate the period until $d_1 > 1$, giving a quadratic irrational $0 < \beta < \frac{1}{2}$ corresponding to the periodic part of α. Then an Ostrowski representation for β can be constructed, and Eq. (4) is used to find an automaton for $\lfloor n\beta \rfloor$, followed by Eq. (5) to find an automaton for $\lfloor n\alpha \rfloor$. Therefore, $(\lfloor n\alpha \rfloor)_{n\geq 1}$ is synchronized in terms of the Ostrowski β-representation.

For example, for $\alpha = \sqrt{3} + 1 = [2, \overline{1,2}]$, we have $\alpha \geq \frac{1}{2}$. Since we only care about the digits after the radix point, we set $d_0 = 0$ and then rotate the period to get $\beta = [0, \overline{2,1}] = (\sqrt{3} - 1)/2 < 1/2$. This gives the sequence of denominator convergents $1, 2, 3, 8, 11, 30, \ldots$, where $m = 2$, $q_m = 3$, and $q_{m-1} = 2$, and so Eq. (4) gives $[(n-1)_\beta 00]_\beta = 3(n-1) + 2\lfloor n\beta \rfloor$. This results in a DFA for $\lfloor n\beta \rfloor$ that has 23 states. Then, we find $\alpha = (2 + 2\beta)/1$, with $a = 2$, $b = 2$, and $c = 1$, and Eq. (5) gives a DFA with 20 states, shown in Fig. 5.

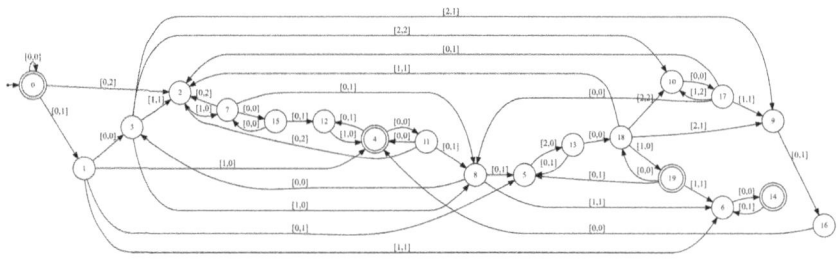

Fig. 5. Synchronized automaton for $\lfloor n\alpha \rfloor$ for $\alpha = \sqrt{3} + 1$.

Then, for example $(5)_\beta = 110$ and $(\lfloor 5\alpha \rfloor)_\beta = (13)_\beta = 10010$. When we input $[0,1][0,0][1,0][1,1][0,0]$ into the automaton, we visit states $1, 3, 8, 6, 14$ in succession, and so we accept. From here, the same general process that is outlined in Theorem 1 can be used to construct a DFA accepting in parallel the Ostrowski α-representations of q and $\lfloor bq\alpha \rfloor - b\lfloor q\alpha \rfloor$, and ultimately the DFAO $\mathcal{A}_b$ as desired.

4.3   Walnut Implementation

Constructing the DFAOs for other quadratic irrationals with Walnut requires the ost command to create custom Ostrowski representations. As explained above, `Walnut` requires that $0 < \beta < \frac{1}{2}$ to create the corresponding Ostrowski representation, and it is possible to create a DFAO for $\alpha \geq \frac{1}{2}$ by synchronizing it in terms of the Ostrowski representation for β. Presented below are the general steps for constructing a DFAO for the digits of the base-2 representation of a quadratic irrational α with `Walnut`, using the process explained above with Eqs. (4) and (5).

First, we construct the continued fraction of $\beta < \frac{1}{2}$ from α by setting $d_0 = 0$ and rotating the period until $d_1 > 1$, if necessary. Next, we determine the denominators $j = q_m$ and $k = q_{m-1}$ of the continued fraction convergent to β, where m is the number of elements in the period. Lastly, we find a, b, and c from the relation $\alpha = (a + b\beta)/c$, where $b, c > 1$. With these, we can use the following `Walnut` commands:

```
# Construct Ostrowski representation for Beta
ost ostBeta [0] [d1 d2 ... dm];
# Create a DFA of z = floor(n*Beta) using j and k
def betan "?msd_ostBeta Eu,v n=u+1 & $shift(u,v) & v=k*z+j*u":
# Create a DFA of z = floor(n*Alpha) synchronized
def alphan "?msd_ostBeta Eu $betan(b*n,u) & z=(u+a*n)/c":
# Create a DFAO for Alpha in base 2
def alphan_d2 "?msd_ostBeta Ex,y $alphan(2*n,x) & $alphan(n,y) & x!=2*y":
combine AD2 alphan_d2:
```

The `shift` DFA can be constructed from a regular expression as done above for φ, and is based on the specific representation and continued fraction sequence. If multiple left-shifts are required, it may be simpler to create a `shift` DFA that left-shifts only one position at a time, and chain its use together multiple times. For example, three left-shifts could be achieved using a 1-shift DFA by:

```
def betan "?msd_ostBeta Eu,v,w,x n=u+1 & $shift(u,v) & $shift(v,w)
      & $shift(w,x) & x=k*z+j*u":
```

Using this process, we created the DFAOs for other quadratic irrationals including the "bronze ratio" $(\sqrt{13} + 3)/2 = [3, \overline{3}]$ and several Pisot numbers. We give the `Walnut` code below.

The bronze ratio $(\sqrt{13} + 3)/2$ in bases 2 and 3:

```
# In this case m = 1, q_m = 3, and q_(m-1) = 1.
ost bt [0] [3];
reg bts {0,1,2,3} {0,1,2,3}
  "([0,0]|[0,2][2,2]*[2,0]|([0,2][2,2]*[2,3]|[0,3])
  [3,0]|([0,1]|[0,2][2,2]*[2,1])([1,1]|[1,2][2,2]*[2,1])*
  (([1,2][2,2]*[2,3]|[1,3])[3,0]|[1,2][2,2]*[2,0]|[1,0]))*":
def btbn "?msd_bt Eu,v n=u+1 & $bts(u,v) & v=1*z+3*u":
def btan "?msd_bt Eu $btbn(1*n,u) & z=(u+3*n)/1":
```

DFAO for the bronze ratio in base 2 (7 states):

```
def btn_d2 "?msd_bt Ex,y $btan(2*n,x) & $btan(n,y) & x!=2*y":
combine BTND2 btn_d2:
```

DFAO for the bronze ratio in base 3 (8 states):

```
def btn_d31 "?msd_bt Ex,y $btan(3*n,x) & $btan(n,y) & x=3*y+1":
def btn_d32 "?msd_bt Ex,y $btan(3*n,x) & $btan(n,y) & x=3*y+2":
combine BTND3 btn_d31 btn_d32:
```

Pisot number $\sqrt{3}+1$ and $(\sqrt{3}-1)/2$ in base 2:

```
# In this case m = 2, q_m = 3, and q_(m-1) = 2.
ost pv1 [0] [2 1];
reg pv1s {0,1,2} {0,1,2} "([0,0]|([0,1][1,1][1,0]|[0,1][1,0])|
    [0,2][2,0])*":
def pv1bn "?msd_pv1 Et,u,v n=t+1 & $pv1s(t,u) & $pv1s(u,v)
    & v=2*z+3*t":
```

DFAO for $(\sqrt{3}-1)/2 = [0,\overline{2,1}]$ in base 2 (see Fig. 7):

```
def pv1bn_d2 "?msd_pv1 Ex,y $pv1bn(2*n,x) & $pv1bn(n,y) & x!=2*y":
combine PV1B2 pv1bn_d2:
```

DFAO for $\sqrt{3}+1 = [2,\overline{1,2}]$ in base 2 (27 states):

```
def pv1an "?msd_pv1 Eu $pv1bn(2*n,u) & z=(u+2*n)/1":
def pv1n_d2 "?msd_pv1 Ex,y $pv1an(2*n,x) & $pv1an(n,y) & x!=2*y":
combine PV12 pv1n_d2:
```

Pisot number $(\sqrt{17}+3)/2$ and $(\sqrt{17}-3)/4$ in base 2:

```
# In this case m = 3, q_m = 7, and q_(m-1) = 4.
ost pv2 [0] [3 1 1];
reg pv2s {0,1,2,3} {0,1,2,3}
    "([0,0]|[0,1][1,0]|[0,1][1,1][1,0]|[0,2][2,0]|
    [0,2][2,1][1,0]|[0,3][3,0])*":
def pv2bn "?msd_pv2 Es,t,u,v n=s+1 & $pv2s(s,t)
    & $pv2s(t,u) & $pv2s(u,v) & v=4*z+7*s":
```

DFAO for $(\sqrt{17}-3)/4 = [0,\overline{3,1,1}]$ in base 2 (see Fig. 6):

```
def pv2bn_d2 "?msd_pv2 Ex,y $pv2bn(2*n,x) & $pv2bn(n,y) & x!=2*y":
combine PV2B2 pv2bn_d2:
```

DFAO for $(\sqrt{17}+3)/2 = [3,\overline{1,1,3}]$ in base 2 (27 states):

```
def pv2an "?msd_pv2 Eu $pv2bn(2*n,u) & z=(u+3*n)/1":
def pv2n_d2 "?msd_pv2 Ex,y $pv2an(2*n,x) & $pv2an(n,y) & x!=2*y":
combine PV22 pv2n_d2:
```

5   Are the Automata Minimal?

The automata that Walnut constructs for computing $\lfloor bq\varphi \rfloor - b\lfloor q\varphi \rfloor$ on input $q \geq 0$ are guaranteed to be minimal. However, in this paper, with our application to computing the base-b digits of φ, we are only interested in running these automata in the special case when $q = b^n$, the powers of b. Could it be that there are even smaller automata that answer correctly on inputs of the form b^n (but might give a different answer for other inputs)? After all, for each t, we are only concerned with behavior of the automaton on linearly many inputs of

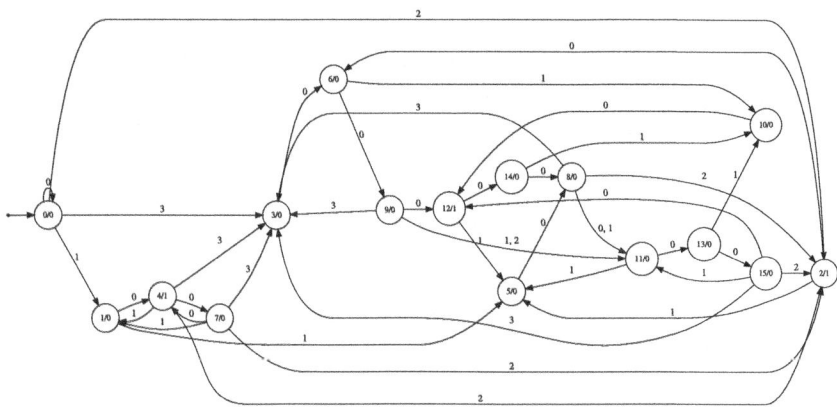

Fig. 6. Automaton for the nth bit to the right of the binary point of $(\sqrt{17} - 3)/4$ in base 2. Input is n in the Ostrowski representation corresponding to the real number $[0, 3, 1, 1, 3, 1, 1, 3, 1, 1, \ldots]$.

length t, as opposed to the exponentially large set of valid length-t Zeckendorf representations. Thus, the automaton is not very constrained.

We do not know the answer to this question, in general. The question is likely difficult; in terms of computational complexity, it is a special case of a problem known to be NP-hard, namely, the problem of inferring a minimal DFAO from incomplete data [8]. However, this problem can sometimes be solved in practice using satisfiability (SAT) solving [20].

We are able to show that some of our automata are indeed minimal, among all automata giving the correct answers on inputs of the form $q = b^n$, and satisfying two conventions: first, that leading zeroes in the input cannot affect the result, and second, that the automata obey the Ostrowski rules (1)–(3) for the particular numeration system. Our method of proving minimality, and in some cases uniqueness, uses SAT solving.

We use a modified version of a MinDFA solver called DFA-Inductor [20] to generate SAT encodings for minimal automata, which are then passed to the CaDiCaL SAT solver [5] to determine whether they have a satisfying solution. DFA-Inductor uses the *compact encoding* method given by Heule and Verwer [10], which defines eight constraints—four mandatory and four redundant— to translate DFA identification into a graph coloring problem, and then encodes those constraints into a SAT instance.

DFA-Inductor only supports DFAs (and hence only accepting or rejecting states), however, and additional output status labels were added for bases larger than 2. DFA-Inductor does not explicitly encode a "dead state" rejecting invalid strings, but a transition to a dead state can be implied by a lack of an outgoing transition on a given state. Another redundant constraint of the compact encoding method forces each state to have an outgoing transition on every symbol, which must be amended to exclude whichever symbols must transition to the implied dead state.

Our automata follow the convention that the start state consumes leading 0s in the input string. In terms of the compact encoding variables, $y_{\ell,p,q}$ indicates that state p has a transition to state q on label ℓ. This constraint is then implemented by enforcing state 0 to have a self-loop on the symbol 0 using the unit clause $y_{0,0,0}$, and the dictionary given to DFA-Inductor states that the string 0 produces output 0.

In order for the SAT solver to construct automata that obey the rules of a given Ostrowski representation, we encode the Ostrowski rules (2)–(3) as a set of constraints. Rule 1 is satisfied simply by only including strings in the dictionary that are valid in the given representation. Without these constraints, the solver may find a smaller DFAO by allowing rule-breaking transitions—such as allowing consecutive 1s for φ in the Zeckendorf representation.

5.1   Ostrowski Encoding for Purely Periodic Quadratic Irrationals

Each Ostrowski α-representation is a language made up from the set of valid strings that can be constructed using the Ostrowski rules (1)–(3). This language is recognized by a canonical DFA, and serves as the base that informs the valid structure of the final DFAO. Constructing a DFAO using only the states in the Ostrowski base DFA guarantees that rules 2 and 3 of the Ostrowski construction are never violated. Conveniently, Walnut automatically generates a DFA of the Ostrowski base during the process of constructing the representation.

Since each state in the base DFA has a unique transition set, we can refer to the ith state in the base DFA as the ith *base state*. For example, Fig. 7 shows for $\alpha = (\sqrt{3} - 1)/2 = [0, \overline{2, 1}]$ how each base state in the Ostrowski base DFA (bottom), labelled B0 to B5, correspond exactly to a state in the DFAO for returning the ith digit of α in base 2 (top).

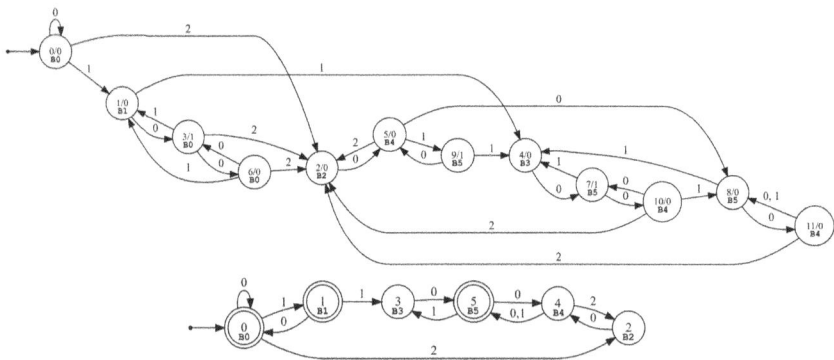

Fig. 7. Relationship between the Ostrowski base states and DFAO states for $\alpha = (\sqrt{3} - 1)/2$.

The Ostrowski rules (2)–(3) are encoded through the states in the Ostrowski base DFA by constraining each state in the DFAO to match a certain base state. Therefore, to encode the base states, we create a new variable $b_{p,t}$, which says state p in the DFAO is related to base state t in the Ostrowski base DFA. We then relate the b variable to the transition variable $y_{\ell,p,q}$, which constrains the set of valid transitions between p and q according to which base states they are associated with. The encoding is presented in Table 1.

The last constraint in the table is the only one that needs to be manually determined for each Ostrowski base DFA. For example, for $\alpha = (\sqrt{3} - 1)/2$ in Fig. 7, base state B4 is encoded as follows, where Q denotes the set of states in the DFAO and B denotes the set of states in the Ostrowski base DFA:

$$\bigwedge_{\substack{i,j \in Q \\ i \neq j}} \left((b_{i,4} \wedge b_{j,2} \rightarrow \neg y_{0,i,j}) \wedge (b_{i,4} \wedge b_{j,2} \rightarrow \neg y_{1,i,j}) \wedge (b_{i,4} \wedge b_{j,5} \rightarrow \neg y_{2,i,j}) \right)$$

$$\bigwedge_{\substack{i,j \in Q \\ i \neq j}} \bigwedge_{k \in B \setminus \{2,5\}} \bigwedge_{\ell \in \{0,1,2\}} (b_{i,4} \wedge b_{j,k} \rightarrow \neg y_{\ell,i,j})$$

Table 1. SAT encoding of Ostrowski constraints for purely periodic quadratic irrationals.

Constraints	Range	Meaning
$\neg y_{k,0,0}$	$1 \leq k \leq c$	The start state can only have a self-loop on 0.
$\neg y_{k,i,i}$	$i \in Q; i \neq 0;$ $0 \leq k \leq c$	No states other than the start state can have a self-loop on any label.
$b_{0,0}$		The start state is related to base state 0.
$b_{i,s} \rightarrow \neg b_{i,t}$	$i \in Q; s,t \in B$ $s \neq t$	Each state in the DFAO must be related to at most one base type.
$b_{i,1} \vee b_{i,2} \vee \cdots \vee b_{i,|B|}$	$i \in Q$	Each state in the DFAO must be related to at least one base type.
$(b_{i,s} \wedge b_{j,t}) \rightarrow \neg y_{k,i,j}$	$i,j \in Q; s,t \in B;$ $k \in \Sigma; \delta(s,k) \neq t$	Suppose DFAO state i is related to base state s, and state j is related to base state t. If state s in the base DFA does not have a transition to state t on label k, then i cannot have a transition to j on label k in the DFAO.

Q = set of states in DFAO; B = set of states in Ostrowski base DFA; δ is the transition function of the DFAO; Σ = alphabet; $c = \max(\Sigma)$

5.2   Results

Table 2 gives our results of DFA minimization by SAT on a few quadratic irrationals. In each of the cases, the Walnut solution was confirmed to be minimal by proving that there are no satisfying assignments of the SAT encoding with a smaller number of states than in the Walnut-produced automaton.

The dictionary containing the Ostrowski representation of the first i digits is referred to as the ith digit set. The solver is run on the SAT encoding of each digit set for a given number of states. The state count was increased every time the solver returned UNSAT, and the digit set was increased every time a satisfying assignment was found. Once the state count given by the `Walnut`-produced solution was reached, the solver was run exhaustively to find all satisfying assignments of the SAT formula and therefore *all* candidates for the minimal automata computing the quadratic irrational. However, most satisfying assignments encoded automata that only computed the given digit set correctly and did *not* correctly compute the digits of the quadratic irrational to a higher precision than what was provided in the given digit set.

Table 2. Results for computing minimal automata for various quadratic irrationals.

Quadratic Irrational	φ base 2 8 states	φ base 3 13 states	$\sqrt{2}$ base 2 6 states	$\frac{\sqrt{13}+3}{2}$ base 2 7 states	$\frac{\sqrt{13}+3}{2}$ base 3 8 states	$\frac{\sqrt{3}-1}{2}$ base 2 12 states	$\frac{\sqrt{17}-3}{4}$ base 2 16 states
Digit set size	54	197	29	64	64	27	57
SAT time (sec)	0.50	28,425.5	0.08	142.81	44.68	0.14	68.11
UNSAT time (sec)	0.18	12,123.0	0.02	0.52	24.76	0.08	2.59
Number of candidates	1	3	1	3	7	1	9

The digit set size given in Table 2 is the smallest dictionary required for the SAT solver to find the n-state `Walnut` solution. The SAT time is the time required by the solver to find the `Walnut` automaton. The UNSAT time is the time required to determine that no automata exists using $n - 1$ states. Since no candidate solutions are found at $n - 1$ states, we conclude that the n-state `Walnut` solution is minimal.

In some cases, we found multiple distinct candidates that correctly compute at least 250,000 digits of the quadratic irrational (see the last row of Table 2). For all except $(\sqrt{17} - 3)/4$, these candidate solutions differ from the `Walnut` solution only by their outgoing transitions on the start state. The candidates for φ (base 3) and $(\sqrt{13} + 3)/2$ (base 2) have differing transitions on label 1, while the candidates for $(\sqrt{13}+3)/2$ (base 3) differ on label 2. All of the candidates for $(\sqrt{17} - 3)/4$ have the same start state, but differ in their transitions on label 2. Given how similar the candidate solutions are to the `Walnut` solution and that they are correct up to a high precision, it is possible that the `Walnut` solution is not unique, though we leave this as an open problem.

Minimization of the DFAOs in some other examples presented a challenge for the SAT solver. For φ in base 4, it took over 25 h for the 78th digit set to be declared UNSAT at 13 states. For $\sqrt{2}$ in base 3, it took over 55 h for the 258th digit set to be declared SAT at 11 states, but the satisfying assignment found by the solver corresponded to an automaton that incorrectly computed the ternary digits of $\sqrt{2}$ starting at the 321st digit.

Acknowledgments. We thank the referees for several useful suggestions.

References

1. Adamczewski, B., Bugeaud, Y.: On the complexity of algebraic numbers I. Expansions in integer bases. Ann. Math. **165**, 547–565 (2007)
2. Bailey, D., Borwein, P., Plouffe, S.: On the rapid computation of various polylogarithmic constants. Math. Comput. **66**, 903–913 (1997)
3. Baranwal, A.R., Schaeffer, L., Shallit, J.: Ostrowski-automatic sequences: theory and applications. Theor. Comput. Sci. **858**, 122–142 (2021)
4. Baranwal, A.R., Shallit, J.: Critical exponent of infinite balanced words via the Pell number system. In: Mercaş, R., Reidenbach, D. (eds.) WORDS 2019. LNCS, vol. 11682, pp. 80–92. Springer, Cham (2019). https://doi.org/10.1007/978-3-030-28796-2_6
5. Biere, A., Fazekas, K., Fleury, M., Heisinger, M.: CaDiCaL, Kissat, Paracooba, Plingeling and Treengeling entering the SAT Competition 2020. In: Proceedings of the of SAT Competition 2020 – Solver and Benchmark Descriptions, Department of Computer Science Report Series B, vol. B-2020-1, pp. 51–53. University of Helsinki (2020)
6. Carpi, A., Maggi, C.: On synchronized sequences and their separators. RAIRO Inform. Théor. App. **35**, 513–524 (2001)
7. Cobham, A.: On the Hartmanis-Stearns problem for a class of tag machines. In: IEEE Conference Record of 1968 Ninth Annual Symposium on Switching and Automata Theory, pp. 51–60 (1968). Also appeared as IBM Research Technical Report RC-2178, 23 August 1968
8. Gold, M.E.: Complexity of automaton identification from given data. Inform. Control **37**, 302–320 (1978)
9. Hartmanis, J., Stearns, R.E.: On the computational complexity of algorithms. Trans. Am. Math. Soc. **117**, 285–306 (1965)
10. Heule, M.J.H., Verwer, S.: Exact DFA identification using SAT solvers. In: Sempere, J.M., García, P. (eds.) ICGI 2010. LNCS (LNAI), vol. 6339, pp. 66–79. Springer, Heidelberg (2010). https://doi.org/10.1007/978-3-642-15488-1_7
11. Lekkerkerker, C.G.: Voorstelling van natuurlijke getallen door een som van getallen van Fibonacci. Simon Stevin **29**, 190–195 (1952)
12. Lenstra, H.W., Jr., Shallit, J.O.: Continued fractions and linear recurrences. Math. Comput. **61**, 351–354 (1993)
13. Mousavi, H., Schaeffer, L., Shallit, J.: Decision algorithms for Fibonacci-automatic words, I: basic results. RAIRO Inform. Théor. App. **50**, 39–66 (2016)
14. Ostrowski, A.: Bemerkungen zur Theorie der Diophantischen Approximationen. Abh. Math. Sem. Hamburg **1**, 77–98, 250–251 (1922). Reprinted in Collected Mathematical Papers, vol. 3, pp. 57–80
15. Schaeffer, L., Shallit, J., Zorcic, S.: Beatty sequences for a quadratic irrational: decidability and applications (2024). arxiv preprint arXiv:2402.08331 [math.NT]
16. Shallit, J.: Synchronized sequences. In: Lecroq, T., Puzynina, S. (eds.) WORDS 2021. LNCS, vol. 12847, pp. 1–19. Springer, Cham (2021). https://doi.org/10.1007/978-3-030-85088-3_1
17. Shallit, J.: The Logical Approach To Automatic Sequences: Exploring Combinatorics on Words with *Walnut*, London Mathematical Society Lecture Note Series, vol. 482. Cambridge University Press (2023)
18. Shallit, J.: Calculation of $\sqrt{5}$ and ϕ (the golden ratio) to 10,000 decimal places (1976). Reviewed in Math. Comput. **30**, 377 (1976)

19. Shanks, W.: On the extension of the numerical value of π. Proc. R. Soc. Lond. **21**, 318–319 (1873)
20. Zakirzyanov, I., Shalyto, A., Ulyantsev, V.: Finding all minimum-size DFA consistent with given examples: SAT-based approach. In: Cerone, A., Roveri, M. (eds.) SEFM 2017. LNCS, vol. 10729, pp. 117–131. Springer, Cham (2018). https://doi.org/10.1007/978-3-319-74781-1_9
21. Zeckendorf, E.: Représentation des nombres naturels par une somme de nombres de Fibonacci ou de nombres de Lucas. Bull. Soc. Roy. Liège **41**, 179–182 (1972)

PDFA Distillation with Error Bound Guarantees

Robert Baumgartner[✉] and Sicco Verwer

Technical University of Delft, Delft, Netherlands
{r.baumgartner-1,s.e.verwer}@tudelft.nl

Abstract. Active learning algorithms to infer probabilistic finite automata (PFA) have gained interest recently, due to their ability to provide surrogate models for some types of neural networks. However, recent approaches either cannot guarantee determinism, which makes the automaton harder to understand and compute, or they rely on techniques that bound errors on individual transitions. In this work we propose a derivative of the recent $L^\#$ algorithm to learn deterministic PFA (PDFA) from systems returning a distribution over a set of tokens given an input string. Along with determinism, we can give error bounds on probabilities assigned to whole strings with an easy to understand approach. We show formal correctness of our algorithm and test it on neural networks trained to model three datasets from computer- and network-systems respectively. We show that the algorithm can learn the network's behaviour closely, and provide an example application of how the model can be used to interpret the network. We note that our approach is in theory applicable in general to learn deterministic weighted finite automata. We provide the source code of our algorithm and relevant scripts on our public repository.

Keywords: Active Automata Learning · PDFA distillation · Explainable AI

1 Introduction

Active learning of automata has had its advent in the work of Dana Angluin, who introduced the L* algorithm [1] to learn deterministic finite automata (DFA) from an unknown target system. Multiple derivatives of the L* have come out since then, optimizing one or more parts of the algorithm (for more details, see e.g. [7]). Starting with the work of Weiss et al. [20] these algorithms have gained interest again by distilling DFA from a neural network (NN) trained to recognize an unknown target language. Follow up work constitutes e.g. of the work of Mayr and Yovine [11], who introduce a derivative of the L* they call bounded L* and show its properties to be PAC-bounded. Muškardin et al. [14] investigate the effect of the counterexample search strategy on the resulting automata.

While training neural networks to recognize unknown (regular) languages provides an interesting theoretical basis for deep learning, these types of networks

S. Z. Fazekas (Ed.): CIAA 2024, LNCS 15015, pp. 51–65, 2024.
https://doi.org/10.1007/978-3-031-71112-1_4

are rarely found outside academic environments. More interesting networks are language models: Given a set of possible tokens Σ, the network returns conditional probabilities of the form $P(a|x)$, where $a \in \Sigma$ and x is a string in Σ^*, the set of all possible strings over Σ. In words, $P(a|x)$ models the probability of a token a to occur after having seen the substring x. Weiss et al. [19] proposed an adaptation of the L* algorithm to learn PDFA from such networks. A similar approach is taken by Eyraud and Ayache [4], which use a spectral approach from [2] to extract weighted finite automata (WFA) from the real-valued matrix. Okudono et al. [15] focus their work on the counterexample search. While powerful, a drawback of the spectral approach is that the resulting automata are not deterministic. A slightly new approach is taken in [13] and [12]. Unlike the previous approaches these algorithms use an observation tree and minimize it via merging states. To this end they quantize the distributions of the states and define congruence over strings, which leads them to different notions of similarity of states.

In our work we build on the recent $L^\#$ algorithm (Vandraager et al., [18]), an algorithm designed to learn Mealy-automata, similar to L*. Unlike L* it builds a tree of observations and identifies states by searching for distinguishing criteria. We build an observation tree modeling probability distributions and introduce a simple notion of state similarity to generalize the model. Unlike $L^\#$ we do not have to distinguish a state from the rest, but can use the numerical output of PDFAs and choose states that minimize induced errors, thus relaxing the strictness of our merge-requirements. Our similarity measure guarantees that, given a parameter $\mu > 0$ as input, for each string x that has been seen by the algorithm that $|P(x) - \pi(x)| \leq \mu$, where $P(x)$ is the assigned probability by the network, and $\pi(x)$ is the probability of x assigned by our inferred PDFA. We further support this notion of proximity between the network's output and the PDFA's output via our proposed equivalence test, giving us PDFAs that mimic the network with intuitive requirements. We show the capability of our algorithm on three datasets, namely the CTU-13 dataset, the BGL dataset, and the HDFS dataset. We train a recurrent neural network (RNN) on each of these respectively, and then distill PDFA varying μ to investigate the effect on the distilled models, and show an example application of the surrogate model. For reproducibility and verification purposes we provide the source code of our method on our public repository[1].

2   Background

2.1   PDFA

A PDFA is an automaton defined over a tuple $\mathcal{A} = \{q_0, Q, \Sigma, \tau, \pi\}$. In this tuple, Q denotes a set of states, and q_0 is a special initial state. The alphabet Σ is a finite set of tokens that the PDFA can accept as input. We denote a as an individual token in Σ, and x is an arbitrary string from the set of all possible

---

[1] https://github.com/tudelft-cda-lab/FlexFringe.

strings over Σ, denoted by Σ^*. We write in short $|x|$ for the length of string x, and λ is the special empty string with length $|\lambda| = 0$.

An input string $x = a_0 a_1 ... a_n$ traverses $\mathcal{A}$ via the transition function τ : $Q \times \Sigma \to Q$ recursively: $\tau(q, \lambda) = q$ and $\tau(q, ax) = \tau(\tau(q, a), x)$.... Note that we introduced shorthand notation for $\tau(q, x)$ as the recursive traversal through $\mathcal{A}$ with a string x. A state q' is reachable from state q iff $\exists x : q' = \tau(q, x)$. We write shorthand $Q_{\tau(q)}$ to denote the set of all states that are reachable from state q, and we denote by $X_{\tau(q)}$ the set of strings needed to reach them starting in state q. Traversing the automaton with a string x results in a sequence of states $q^0 q^1 ... q^{n-1}$ being visited. We call this sequence of states the trace of x and write $T(x)$.

Lastly, the mapping $\pi : Q \times \Sigma \to [0, 1]$, $\pi : Q \to [0, 1]$ maps state-symbol pairs and states to probabilities. A PDFA requires $\forall q \in Q$: $\sum_{a \in \Sigma} \pi(q, a) + \pi(q) = 1$. We call $\pi(q)$ the stopping probability of state q, modeling the probability of a string to reach state q and end there. Given an input string $x = a_0 a_1 ... a_{m-1}$ and its associated trace of states $T(x) = q^0 q^1 ... q^{m-1} q^m$ we can compute the probability of string x to occur via $\pi(x) = \prod_{i=0}^{m-1} \pi(q^i, a_i) \cdot \pi(q^m)$. Figure 1 depicts an observation tree and its minimal PDFA.

2.2 Observation Tree, Closed PDFA, and State Merging

We say that a PDFA is an observation tree $\mathcal{OT}$ if for each state $q \in \mathcal{OT}$ there exists a unique string x_q^A s.t. $q = \tau(x_q^A)$, and q is only reachable from q_0 by this string. We call x_q^A the access string of q, and $x_{q_0}^A = \lambda$. Contrary to the observation tree, we call a PDFA $\mathcal{A}$ consisting of state set Q closed iff $\forall q \in Q$, $\forall a \in \Sigma$: $\tau(q, a) \in Q$.

In this work we build observation trees and minimize them via state merging techniques [7]. Building deterministic models furthermore requires a determinization process whenever two states are merged, subsequently merging state pairs such that $\tau(q, a)$ is uniquely identified $\forall q \in Q, \forall a \in \Sigma$. State merging induces a mapping in between the observation tree and the PDFA $\mathcal{A}$. In Fig. 1 we show an observation tree along with a related closed PDFA obtained via 1. Merge q_1 into q_0, which merges q_2 into q_3 as part of the determinization procedure. 2. Merge q_4 and q_5 into q_3, in arbitrary order.

3 Learning Algorithm

Our algorithm leans heavily on the works of Vandraager et al. [18]. We define a teacher, which is an abstraction of the system-under-learning (SUL), and a learner, an abstraction of the learning algorithm. The learner can ask the teacher two question: Firstly a membership query $\mathcal{MQ}(xa)$ for string xa, to which the teacher replies with a conditional probability $P(a|x), a \in \Sigma_\xi$. Here, we enhanced the set Σ with a unique stopping symbol ξ indicating the end of a sequence[2].

---

[2] The presence of such a symbol is a reasonable assumption. E.g. in natural language processing this is commonly referred to as the <EOS> (end-of-sentence) symbol.

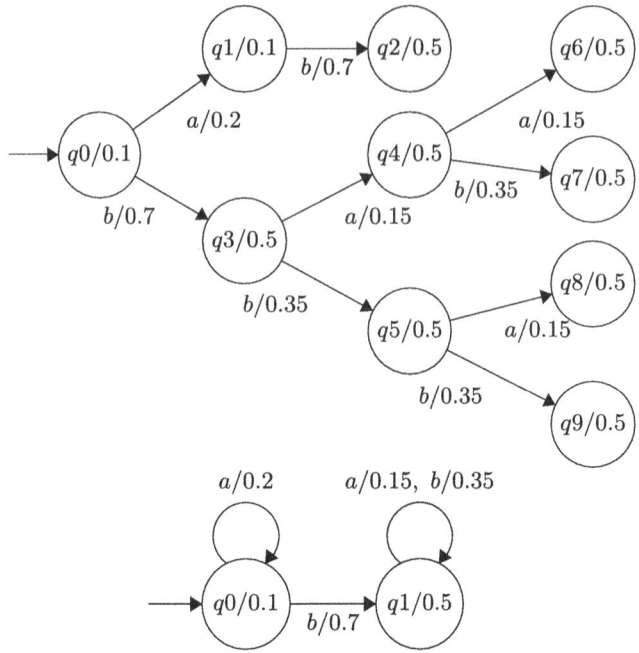

Fig. 1. An observation tree (above) and a corresponding (closed) PDFA.

Secondly, the learner can ask an equivalence query $\mathcal{EQ}(\mathcal{T}, \mathcal{H})$ between the target SUL $\mathcal{T}$ and the current hypothesis $\mathcal{H}$. The teacher will respond with either a 'yes', meaning the systems are behaving similar enough, or a 'no' along with a counterexample string x_{cex} which violates the error bound.

The learner has two main routines: In the first routine it tries to find a hypothesis candidate for target $\mathcal{T}$. Whenever it finds a valid hypothesis candidate $\mathcal{H}$ the second routine starts, either validating $\mathcal{H}$ and terminating or processing the counterexample returned by the teacher. The learner repeats these two routines until it has found a hypothesis that can pass $\mathcal{EQ}(\mathcal{T}, \mathcal{H})$. We describe the two routines in the following.[3]

3.1   Finding a Hypothesis

In order to find a hypothesis we employ the red-blue-framework [9], which distinguishes a core of red states, which are considered states of the final automaton, blue states and white states. Blue states are states q' who have a transition $q' = \tau(q, a)$ s.t. q' is not red but q is. States that are neither red nor blue are

---

[3] Because we describe our algorithm at all stages, it was not always clear on how we should refer to the individual states and nodes of the automaton. For simplicity we decided to stick to the word 'state' meaning both state and node from here on, as they are mostly interchangeable for us.

Algorithm 1. Extend fringe

Input: State q, alphabet Σ
   for all $a \in \Sigma$ **do**
      Create state q' satisfying $q' = \tau(q, a)$
      $InitializeState(q')$ ▷ Algorithm 2
      $\mathcal{F}_n \leftarrow \mathcal{F}_n \cup \{q'\}$
   end for

Algorithm 2. Initialize state

Input: State q, access to target $\mathcal{T}$ via teacher
   Save $P(x_q^A)$ in state
   $\pi(q) \leftarrow \mathcal{MQ}(x_q^A \xi)$
   for all $a \in \Sigma$ **do**
      $\pi(q, a) \leftarrow \mathcal{MQ}(x_q^A a)$
   end for

---

white. Initially, the observation tree $\mathcal{OT}$ consists only of one red state q_0. The learner initializes q_0 via Algorithm 2 and creates a set of blue states around q_0 via Algorithm 1. Then the learner attempts to minimize $\mathcal{OT}$ via state merging.

We call a pair of a red state q_r and a blue state q_b consistent under $\mu \in [0, 1)$ iff it holds that

$$d(q_r, q_b) = \left| \frac{\pi(x_{q_b}^A)}{\pi(q_b)} \cdot \pi(q_r) - P(x_{q_b}^A) \right| \leq \mu. \tag{1}$$

Note that we introduced a distance between q_r and q_b here, written $d(q_r, q_b)$ in short. The term $\frac{\pi(x_{q_b}^A)}{\pi(q_b)} \cdot \pi(q_r)$ represents $\pi(x_{q_b}^A)$ if the merge were to happen. We call q_r and q_b mergeable iff $\forall q_b' \in Q_{\tau(q_b)}$ and their respective set of strings $X_{\tau(q_b)}$ it holds that either $\tau(q_r, x')$, $x' \in X_{\tau(q_b)}$ is not defined, or $q_r' = \tau(q_r, x')$ and $d(q_r', q_b') \leq \mu$. In words: When we want to merge two states the inequality (1) needs to hold for all states that are merged through determinization as well. However, with these merge requirements we are able to guarantee that $|P(x) - \pi(x)| \leq \mu$ for each string x for which $\tau(x)$ is defined on the observation tree.

In order to find a hypothesis the learner continues growing the hypothesis until it has found a set of red nodes that form a closed PDFA. In every step the learner fixes $\mathcal{C}$, the current set of red states, and $\mathcal{F}$, the current set of blue states, and compares each of the blue states with each of the red states. If a blue state is consistent with a red state, the blue state will be merged into the red state. If a blue state q_b can be merged with multiple red states, it will be merged into the red state q_r satisfying $q_r = argmin_{q_r' \in \mathcal{C}} \{d(q_r', q_b)\}$. A blue state that cannot be merged with any red state will turn red at the end of the iteration, and will be extended via Algorithm 1. If during an iteration all blue states were able to merge into a red state, then the learner has found a closed PDFA. The entire subroutine is depicted in Algorithm 3.

Algorithm 3. Find hypothesis

Input: Unmerged observation tree with root state q_0, alphabet Σ, threshold μ
Output: Hypothesis $\mathcal{H}$

  Initialize $\mathcal{H}$ with root state q_0
  while Hypothesis not found **do**
    Fix red states in set $\mathcal{C}$, blue states in set $\mathcal{F}$
    $\mathcal{R} \leftarrow \emptyset$
    for all Blue states $q_b \in \mathcal{F}$ **do**
      $s_{min} \leftarrow 1,\ m \leftarrow$ NULL $\triangleright$ m is placeholder for best merge
      for all Red states $q_r \in \mathcal{C}$ **do**
        if $Consistent(q_r, q_b)$ and $Score(q_r, q_b) < s_{min}$ **then** $\triangleright$ See Section 3.1
          $s_{min} \leftarrow Score(q_r, q_b)$
          $m \leftarrow (q_r, q_b)$
        end if
      end for
      if m not NULL **then**
        Merge q_b into q_r
      else $\triangleright$ No merge found for q_b
        $\mathcal{R} \leftarrow \mathcal{R} \cup \{q_b\}$
      end if
    end for
    if $\mathcal{R}$ is not empty **then**
      for all $q_b \in \mathcal{R}$ **do**
        Mark q_b red
        $ExtendFringe(q_b)$ $\triangleright$ Algorithm 1
      end for
    else
      Return $\mathcal{H}$
    end if
  end while

3.2 Counterexample Search and Processing

Once the learner found a valid hypothesis, it asks the teacher for an equivalence query $\mathcal{EQ}(\mathcal{T}, \mathcal{H})$. The query can have two possible outcomes: Either the teacher deems $\mathcal{T}$ and $\mathcal{H}$ sufficiently close, or it rejects $\mathcal{H}$ and provides a counterexample string x_{cex} for which the following requirement does not hold. In accordance with our merging procedure we require closeness for $\mathcal{T}$, modeling $P(x)$, and $\mathcal{H}$, modeling $\pi(x)$, via $|P(x) - \pi(x)| \leq \mu$.

Finding a counterexample can mean one of two possible cases: In the first case the string x_{cex} has been seen before. In this case a merge between a red state q_r and a blue state q_b has been performed, and after the merge one or more states q' have been added s.t. q' was reachable by q at the time q' was created. In this case the merge of q_r and q_b has been wrong. In the second scenario x_{cex} has not been seen yet. Thus $\mathcal{H}$ does not have sufficient information, yet.

We opted for a simple strategy that can deal with both scenarios: Our learner remembers an inverse mapping from $\mathcal{H}$ to $\mathcal{OT}$. To deal with counterexamples we first reset $\mathcal{H}$ to $\mathcal{OT}$, and then parse x_{cex} via $\tau(x_{cex})$. Whenever transitions are not defined on τ the learner creates the missing states, and initializes any new state q' through subroutines 1 and 2. Once the counterexample is processed the algorithm returns to finding a valid hypothesis. We show the counterexample processing subroutine in Algorithm 4, and the main loop of our algorithm in Algorithm 5.

Algorithm 4. Process Counterexample

Input: Observation tree starting in root node q_0, counterexample string x_{cex}

  $q' \leftarrow q_0$
  $m \leftarrow |x_{cex}| - 1$
  for i in $0...m$ **do** $\triangleright$ $x_{cex} = a^0 a^1 ... a^m$
    if $\tau(q', a^i)$ not defined **then**
      $ExtendFringe(q')$
    end if
    $q' \leftarrow \tau(q', a^i)$
  end for

Algorithm 5. Main routine

Input: Access to target $\mathcal{T}$ via teacher, alphabet Σ, error bound μ
Output: Hypothesis $\mathcal{H}$

  $Initialize(q_0)$
  $ExtendFringe(q_0)$ $\triangleright$ Algorithm 1
  while Hypothesis not found **do**
    $\mathcal{H} \leftarrow FindHypothesis(q_0, \Sigma, \mu)$
    Perform $\mathcal{EQ}(\mathcal{T}, \mathcal{H})$
    if Counterexample x_{cex} returned by $\mathcal{EQ}(\mathcal{T}, \mathcal{H})$ **then**
      Reset $\mathcal{H}$ $\triangleright$ q_0 holds observation tree again
      $ProcessCounterexample(q_0, x_{cex})$
    else
      Return $\mathcal{H}$
    end if
  end while

3.3   Practival Considerations

Equivalence Oracle. While practical in theory, in reality it is impossible to check $\mathcal{H}$ for each possible string x. Multiple search strategies exist to find counterexamples [14]. Here we opted for a simple solution, generating random strings over Σ^* assuming a uniform distribution over one and the maximum string length, and a uniform distribution over Σ for each token a of x. If a maximum number of strings has been suggested without finding a counterexample the oracle deems $\mathcal{H}$ and $\mathcal{T}$ equivalent.

Early Stopping. Depending on the complexity of the underlying problem the hypothesis $\mathcal{H}$ can grow to a very large model. In these cases it is desirable to have early stopping criteria. We set a limit n_{max} on the number of red states.

The first time the number of red states reaches n_{max} we *force-merge* the set of remaining blue states into the set of current red states: For each of the blue states q_b, find a red state q_r that minimizes $q_r = argmin_{q'_r \in C} d(q'_r, q_b)$ and merge q_b into q_r.

4 Correctness

Lemma 1. *Every iteration of FindHypothesis (Algorithm 3) over the current set of blue states, comparing them with the set of red states, either results in a complete basis $\mathcal{B}$, or it identifies a new red state, growing $\mathcal{H}$ by at least one new state.*

The proof of this Lemma is by design of the algorithm. It is important to note that every new red state q_r accepts at least its access string $x^A_{q_r}$ s.t. $P(x^A_{q_r}) = \pi(x^A_{q_r})$. Therefore, each identified red state increases the number of strings satisfying $|P(x) - \pi(x)| \leq \mu$ by at least one and therefore the quality of the result.

Lemma 2. *Every identified red state in the observation tree q directly corresponds to a state q' in target $\mathcal{T}$ s.t. $\pi(q, a) = \pi(q', a)$, $\forall a \in \Sigma$ and $\pi(q) = \pi(q')$.*

Proof. We prove by contradiction: Assume the learner identified a red state q' that is not part of $\mathcal{T}$. We further assume that no states of $\mathcal{OT}$ have been merged yet, i.e. $\mathcal{H}$ is an observation tree still. Then, $\exists a \in \Sigma_\varepsilon$: $P(x^A_{q'} a) \neq \pi(x^A_{q'} a)$. By design however on $\mathcal{OT}$ models probabilities precisely, therefore this event cannot happen. The case for $\mathcal{H}$ with performed merges follows from the fact that our merge routine holds error bounds $\forall \mu \in [0, 1)$.

We have to note that assuming the target $\mathcal{T}$ to be a PDFA provides value by helping us in showing correctness of our algorithm. Many real-world applications however have underlying systems of higher complexity. In those cases the number of states to model the target system cannot easily be bound by a fixed number of states, but we know that the probability assigned to strings decreases monotonically in length: $P(x) \geq P(xa)$. Therefore it does make sense to define an upper bound on the length of the strings we want $\mathcal{T}$ to model, and then have the algorithm cover those cases. We now focus on the *ProcessCounterexample* routine.

Lemma 3. *Assuming the teacher rejected $\mathcal{H}$ along with counterexample string x_{cex}, there are only two possible reasons:*

1. *x_{cex} has been seen before by the learner, in which case the learner merged two inconsistent states.*
2. *x_{cex} or a substring of it have not yet been seen by the learner. In this case there exist states in $\mathcal{T}$ that are not yet part of $\mathcal{H}$ and its observation tree.*

As explained in Sect. 3.1, our merge routine ensures that $\forall x \in \Sigma^*$ s.t. $\tau(x)$ is well defined on the unmerged observation tree $\mathcal{OT}$ it is $|P(x) - \pi(x)| \leq \mu$ at the time of a merge. Therefore, case 1 of Lemma 3 can only happen on states the learner added to $\mathcal{H}$ after a merge has happened. Resetting $\mathcal{H}$ to the observation tree will solve this problem, since the merge check will ensure that this merge will not happen again. In case 2 of Lemma 3 resetting to the observation tree and adding x_{cex} will create the unknown states necessary to accept x_{cex}. In this case we either add new states to the automaton, or we remove one or more wrongly performed merges on $\mathcal{H}$. Either case is going to increase the number of accepted traces x by the automaton by a minimum of one, namely x_{cex} will be accepted from there on.

Theorem 1. *Assuming $\mathcal{T}$ is a PDFA with n states, and the membership queries $\mathcal{MQ}(x)$ be noise free. Then the algorithm will terminate after a finite number of iterations and output a hypothesis $\mathcal{H}$ with $n' \leq n$ states s.t. $\forall x \in \Sigma^* : |P(x) - \pi(x)| \leq \mu$.*

Proof. The proof follows from the previous lemmas. The idea is that in each turn, one of the two main routines will increase the number of identified states by at least one. Since the target has a finite number of states the algorithm must terminate.

5   Experiments and Results

To test our algorithm we applied it to three datasets, namely the CTU-13 dataset, the HDFS dataset, and the BGL dataset. We extracted sliding windows and trained a language model of the form $P(a|\sigma), a \in \Sigma_\xi$ for each of them. We then distilled PDFA from the networks.

To test how well the distilled models approximate the underlying neural networks we did the following: We set the threshold μ to the values of $1e^{-3}$, $1e^{-5}$, $1e^{-7}$, and $1e^{-10}$ respectively. We report for each tested value of μ the values $min_x(|P(x) - \pi(x)|)$, $max_x(|P(x) - \pi(x)|)$, $MSE = \sqrt{\sum_x (P(x) - \pi(x))^2}$, as well as the number of states. In all these instances, x is over the corresponding test-set of the respective dataset. Our test sets were all limited to a size of $20,000$, due to the slow prediction of the neural networks.

Additionally, we made sure that the neural networks learned the underlying datasets correctly in the following manner: All selected datasets consist of normal and anomalous data. We consider an extracted sliding window malign if it contains one or more malign tokens, else it is benign. After training the models, we can assign labels: Given a sequence of length $|x| = m$ and $x = a^0 a^1 ... a^{m-1}$, we obtain predicted probabilities $P_1 = P(a^0|\lambda)$, $P_2(a^1|a^0)$, ..., $P_m = P(a^{m-1}|a^0...a^{m-2})$ from the network. We assign a score of $1 - min_{P_i, i \in \{0...m\}}$ to each sequence, over which we compute an receiver operating characteristic (ROC) curve and report the area-under-curve (AUC). The closer the AUC to a value of 1, the better the classification works. Because the inferred PDFAs behave similar to the underlying neural network in their input-output-behavior, we can use the same method to assign anomaly scores as we did with the networks.

5.1 Datasets

CTU-13. The CTU-13 dataset [5] consists of captured network traffic grouped together in Netflow format [8] called flows, and each flow is labeled as either background or malicious. The dataset comes in 13 individual scenarios. We picked scenario 10 and used the algorithm of Pellegrino et al. [17] to encode the netflows into an alphabet of size $|\Sigma| = 92$ with 10 even percentiles. To create sequences we sorted via time stamp and grouped by connection, and then extracting sliding windows of size 10 over the connections. Connections with fewer flows resulted in a single window.

We considered a sequence as malign if it contained a malicious data packet, else benign. We counted both the number of benign n_{benign} and malign sequences n_{malign}, and computed $f = \frac{n_{malign}}{n_{benign}}$. Malign sequences were automatically assigned to the test-set. Benign sequences were randomly assigned to the test-set with a uniform probability of f, and else to the train-set. This way, the number of benign and malign sequences in the test-set was roughly even, and the train-set consisted only of benign sequences, making sure the network learned only the normal behavior of the network. Because the train-set was too large for our machine, we randomly sampled $400,000$ sequences.

HDFS. Another dataset that we tested our algorithm on is the HDFS dataset [21], which represents logs generated from the Hadoop File System run on Amazon's Elastic Compute cloud. Because the system logs come semi-structured, they first have to be tokenized. We used the already preprocessed dataset as made available by [3].

BGL. The last dataset we tested on was the BGL dataset [16]. The dataset consists of log-data collected on the Blue Gene/L supercomputer as of 2006, which has been labelled by alerts or normal activity. To obtain tokens we applied the DRAIN algorithm [6] implemented by Zhu et al. [22], giving us an alphabet of size $|\Sigma| = 321$. We then grouped the templates by node and rolled a sliding window of size 10 the obtained tokens, and then split the resulting sequences into train- and a test-set. We split the dataset into train- and test-set the same way we did on the CTU-13 dataset. Due to the large size of the test-set, we further randomly sampled $100,000$ sequences from the test-set for evaluation.

5.2 Results and Discussion

Approximation Results. The initial results of the experiments can be taken from Table 1. We omitted the entries for $\mu = 1e^{-5}$ on the HDFS dataset, because

Table 1. Results of the experiments. We show the approximation errors on the respective test-sets, as well as the resulting number of states that the PDFA has.

Dataset	Metric	$\mu = 1e^{-3}$	$\mu = 1e^{-5}$	$\mu = 1e^{-7}$	$\mu = 1e^{-10}$
CTU-13	$min_x(|P(x) - \pi(x)|)$	1.27e{−78}	1.27e{−78}	1.27e{−78}	1.27e{−78}
	$max_x(|P(x) - \pi(x)|)$	4.08e{−4}	9.69e{−5}	7.20e{−5}	1.40e{−4}
	MSE	1.71e{−10}	4.56e{−12}	2.43e{−12}	5.27e{−12}
	# States	14	121	1493	3644
HDFS	$min_x(|P(x) - \pi(x)|)$	5.16e{−256}	−	5.16e{−256}	5.16e{−256}
	$max_x(|P(x) - \pi(x)|)$	1.15e{−6}	−	1.28e{−7}	7.44e{−8}
	MSE	2.00e{−13}	−	1.55e{−15}	7.65e{−16}
	# States	1	1	14	798
BGL	$min_x(|P(x) - \pi(x)|)$	6.70e{−177}	6.59e{−190}	3.62e{−193}	3.91e{−186}
	$max_x(|P(x) - \pi(x)|)$	3.02e{−3}	3.02e{−3}	6.47e{−4}	3.02e{−3}
	MSE	8.90e{−8}	8.90e{−8}	7.09e{−9}	8.90{e − 08}
	# States	16	165	918	3500

the distilled model is the same as for $\mu = 1e^{-3}$. Because of the increased complexity of the BGL dataset we had to use our early-stop criterion at 3500 states, as the machine became too large for our hardware with 16GB RAM memory.

Obviously, the HDFS dataset was the easiest to learn. We can see from the results that the model could be represented with a single state and self loops even for a value of $\mu = 1e^{-5}$, and only 14 states for $\mu = 1e^{-7}$. Compared with the HDFS dataset the CTU-13 and the BGL datasets learned larger models. This is likely due to their larger alphabet sizes, which increases the number of possible search paths exponentially.

Decreasing the μ parameter greatly increased the number of states of the distilled models, and in the end many more states have to be added for less gain. Interestingly, on the CTU-13 and BGL datasets the performance in terms of maximum error and MSE decreased slightly in the model with the largest number of states compared with a smaller one. We again attribute this to the fact that larger models are harder to check. An example: In the example of the BGL dataset, we identified the sequence consisting of template 314 repeated ten times. The probabilities are predicted precise until after seven applications of τ the distilled model suddenly underestimates the real probability by several orders of magnitude, caused by a wrong merge. Due to the size of the model this wrong merge is much harder to detect by the counterexample search and remains in the final model. Due to the fact that this string appears a lot in the test-set the MSE of the model becomes larger again. The model with 918 states does not make this wrong merge and predicts the same sequence better. This is a problem common to all active learning algorithms, but can be improved with better search strategies. For a more elaborate discussion of this we refer the reader to the work of Marzouk and de la Higuera [10].

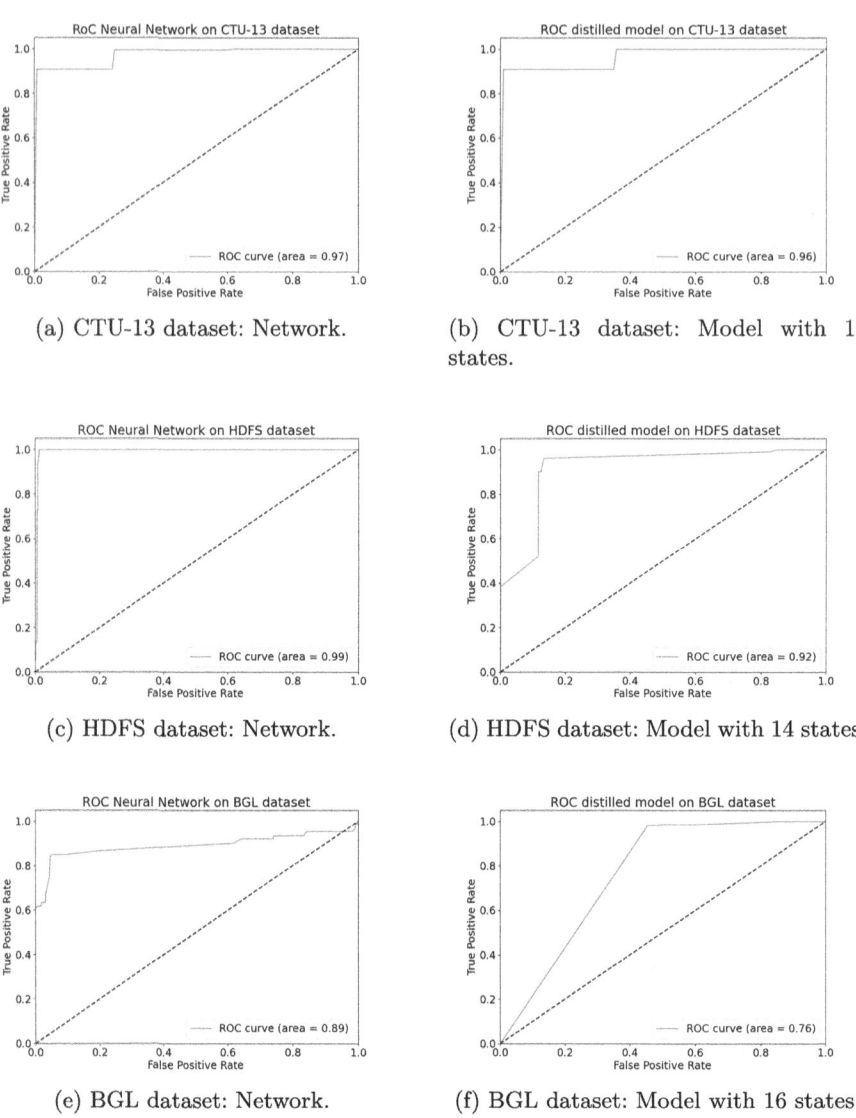

Fig. 2. ROC curves obtained on the datasets. On the left are the ROC curves of the neural network, contrasted with a respective distilled PDFA to the right.

Anomaly Detection Results. We show some results for the anomaly detection, a proxy for how well the networks approximated the dataset, as well as an example for a distilled in Fig. 2. We can see that the networks performed very strong on the CTU-13 dataset and the HDFS dataset, with an AUC of 0.97 and 0.99 respectively. The BGL dataset proved to be more challenging with an AUC of 0.89 for the network. This has to do with the underlying dataset, which is harder to learn than the other two, as well as with how we chose to extract tokens from it. However, the network is still capable of detecting anomalies significantly above random guesses, showing that it learned the data it was given.

Not very surprising the distilled models did not match the underlying network's performances perfectly: The model with 14 states on the CTU-13 dataset obtained an AUC of 0.96, just very slightly below the neural network. On the HDFS dataset however the model with 14 states has an AUC of just 0.92, 0.07 below the neural network, and the distilled model with 15 states on the BGL dataset has an AUC of just 0.76, 0.13 below the neural network. Interestingly, on most of the datasets the smaller models were the better ones, and in most cases the detection performance decreased with larger models. We attribute this to the following: Larger models do have more paths to cover. Therefore, ensuring consistency with the error bound becomes much harder with smaller μ. An exception to this was posed by the HDFS dataset, where the model with 1 state had an AUC of 0.4 only, indicating that an error of $1e^{-5}$ is not sufficient for any anomaly detection on this problem.

Use Case Example. In this subsection we want to demonstrate how the models can be used. We take the distilled model with 14 states we extracted from the on the HDFS data trained network. Figure 3 shows the model. For simplicity we omitted all transitions with a probability less than $\pi(q, a) \leq 1e^{-4}$, marked all transitions with probabilities larger than 0.01 blue, and all with probabilities larger than 0.1 green. The following is a malign string from our test-set: *17, 26, 26, 28, 16*. In our model we can see that the first three transitions were highly

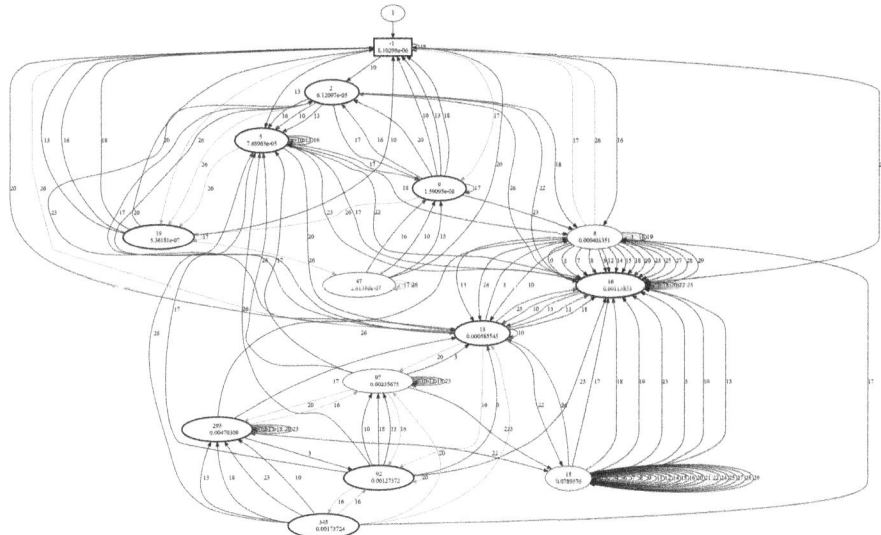

Fig. 3. Visual representation of the 14-state PDFA extracted from the NN trained on the HDFS dataset.

probable. However, the transition after the second *26* is very low according to our model. In fact the substring *17, 26, 26, 28* does not appear in our training set at all. We can thus use our distilled model to get a clean visual representation of our network, helping us to predict low probability strings.

6   Conclusion and Future Work

In this work we introduced a new algorithm that actively learns PDFAs and demonstrated its capabilities learning simple models that can still achieve good detection results in anomaly detection examples. We also showed a simple use case example of how the distilled PDFA can be used as a global explanation for an underlying neural network. An advantage of our method is that bounds are guaranteed to hold on every tested string. Other methods use merge criteria such as the variation distance, which is a much less intuitive metric when parsing whole strings. Compared with spectral methods and in accordance with some other works we are also able to infer deterministic models.

Just as in the case of the original $L^\#$ algorithm, the run-time of our algorithm is a drawback. Even smaller machines can lead to a large observation tree, leading to many asked queries. We could pinpoint the bulk of the run-time to these queries. Tackling this could for example include pruning the tree when predicted probabilities become too small to violate the error bounds. Another possible optimization of the algorithm is obviously a better counterexample search. We also consider the further investigation of applications such as the anomaly detection an interesting research direction.

Acknowledgments. This work is supported by NWO TTW VIDI project 17541 - Learning state machines from infrequent software traces (LIMIT).

References

1. Angluin, D.: Learning regular sets from queries and counterexamples. Inf. Comput. **75**(2), 87–106 (1987)
2. Balle, B., Carreras, X., Luque, F.M., Quattoni, A.: Spectral learning of weighted automata: a forward-backward perspective. Mach. Learn. 33–63 (2014)
3. Du, M., Li, F., Zheng, G., Srikumar, V.: Deeplog: anomaly detection and diagnosis from system logs through deep learning. In: Proceedings of the 2017 ACM SIGSAC Conference on Computer and Communications Security, CCS 2017, pp. 1285–1298. Association for Computing Machinery, New York (2017)
4. Eyraud, R., Ayache, S.: Distillation of weighted automata from recurrent neural networks using a spectral approach. Mach. Learn. (2021)
5. García, S., Grill, M., Stiborek, J., Zunino, A.: An empirical comparison of botnet detection methods. Comput. Secur. **45**, 100–123 (2014)
6. He, P., Zhu, J., Zheng, Z., Lyu, M.R.: Drain: an online log parsing approach with fixed depth tree. In: 2017 IEEE International Conference on Web Services (ICWS), pp. 33–40 (2017)
7. de la Higuera, C.: Grammatical Inference: Learning Automata and Grammars. Cambridge University Press, Cambridge (2010)

8. Hofstede, R., et al.: Flow monitoring explained: from packet capture to data analysis with NetFlow and IPFIX. IEEE Commun. Surv. Tutor. **16**(4), 2037–2064 (2014)

9. Lang, K.J., Pearlmutter, B.A., Price, R.A.: Results of the Abbadingo one DFA learning competition and a new evidence-driven state merging algorithm. In: Honavar, V., Slutzki, G. (eds.) ICGI 1998. LNCS, vol. 1433, pp. 1–12. Springer, Heidelberg (1998). https://doi.org/10.1007/BFb0054059

10. Marzouk, R., de la Higuera, C.: Distance and equivalence between finite state machines and recurrent neural networks: computational results. CoRR (2020). https://arxiv.org/abs/2004.00478

11. Mayr, F., Yovine, S.: Regular inference on artificial neural networks. In: Holzinger, A., Kieseberg, P., Tjoa, A.M., Weippl, E. (eds.) CD-MAKE 2018. LNCS, vol. 11015, pp. 350–369. Springer, Cham (2018). https://doi.org/10.1007/978-3-319-99740-7_25

12. Mayr, F., Yovine, S., Carrasco, M., Pan, F., Vilensky, F.: A congruence-based approach to active automata learning from neural language models. In: Coste, F., Ouardi, F., Rabusseau, G. (eds.) Proceedings of 16th edition of the International Conference on Grammatical Inference. Proceedings of Machine Learning Research, vol. 217, pp. 250–264. PMLR (2023)

13. Mayr, F., Yovine, S., Pan, F., Basset, N., Dang, T.: Towards efficient active learning of PDFA (2022)

14. Muškardin, E., Aichernig, B.K., Pill, I., Tappler, M.: Learning finite state models from recurrent neural networks. In: ter Beek, M.H., Monahan, R. (eds.) Integrated Formal Methods, pp. 229–248. Springer, Cham (2022). https://doi.org/10.1007/978-3-031-07727-2_13

15. Okudono, T., Waga, M., Sekiyama, T., Hasuo, I.: Weighted automata extraction from recurrent neural networks via regression on state spaces (2019)

16. Oliner, A., Stearley, J.: What supercomputers say: a study of five system logs. In: 37th Annual IEEE/IFIP International Conference on Dependable Systems and Networks (DSN 2007), pp. 575–584 (2007)

17. Pellegrino, G., Lin, Q., Hammerschmidt, C., Verwer, S.: Learning behavioral fingerprints from netflows using timed automata. In: 2017 IFIP/IEEE Symposium on Integrated Network and Service Management (IM), pp. 308–316 (2017)

18. Vaandrager, F., Garhewal, B., Rot, J., Wißmann, T.: A new approach for active automata learning based on apartness. In: Fisman, D., Rosu, G. (eds.) TACAS 2022. LNCS, vol. 13243, pp. 223–243. Springer, Cham (2022). https://doi.org/10.1007/978-3-030-99524-9_12

19. Weiss, G., Goldberg, Y., Yahav, E.: Learning deterministic weighted automata with queries and counterexamples. In: Wallach, H., Larochelle, H., Beygelzimer, A., d' Alché-Buc, F., Fox, E., Garnett, R. (eds.) Advances in Neural Information Processing Systems, vol. 32. Curran Associates, Inc. (2019)

20. Weiss, G., Goldberg, Y., Yahav, E.: Extracting automata from recurrent neural networks using queries and counterexamples. Mach. Learn. (2022)

21. Xu, W., Huang, L., Fox, A., Patterson, D., Jordan, M.I.: Detecting large-scale system problems by mining console logs. In: Proceedings of the ACM SIGOPS 22nd Symposium on Operating Systems Principles, SOSP 2009, pp. 117–132. Association for Computing Machinery, New York (2009)

22. Zhu, J., et al.: Tools and benchmarks for automated log parsing. In: Proceedings of the 41st International Conference on Software Engineering: Software Engineering in Practice, ICSE-SEIP 2019, pp. 121–130. IEEE Press (2019)

Constructing a BPE Tokenization DFA

Martin Berglund[1]([✉]), Willeke Martens[1], and Brink van der Merwe[2,3]

[1] Department of Computing Science, Umeå University, Umeå, Sweden
{mbe,wms}@cs.umu.se
[2] Department of Computer Science, Stellenbosch University,
Stellenbosch, South Africa
abvdm@cs.sun.ac.za
[3] National Institute for Theoretical and Computational Sciences,
Stellenbosch, South Africa

Abstract. Many natural language processing systems operate over *tokenizations* of text to address the open-vocabulary problem. In this paper, we give and analyze an algorithm for the efficient construction of deterministic finite automata (DFA) designed to operate directly on tokenizations produced by the popular byte pair encoding (BPE) technique. This makes it possible to apply many existing techniques and algorithms to the tokenized case, such as pattern matching, equivalence checking of tokenization dictionaries, and composing tokenized languages in various ways.

1 Introduction

Many modern natural language processing systems, such as large language models, operate over *tokenizations* of strings. In very notable examples, the OpenAI GPT models [6], which underpin ChatGPT [5] (and many others, e.g., Meta's Llama series of models), use byte pair encoding (BPE) tokenizers. BPE [7] is a technique based on data compression, successively merging pairs of adjacent tokens based on a dictionary of rules, with the dictionary built based on how common pairs are in some training set. See [1] for a thorough formalization of the formal languages implications of this approach.

For example, the GPT-2 tokenization of this string CIA A ⎵2024 ⎵in ⎵Ak ita , ⎵Japan is indicated by the boxes. That is, the tokenization makes '2024', 'in' and 'Japan' single tokens (being very common), but both 'CIAA' and 'Akita' are too uncommon to fit in the 50,000 rules the GPT-2 BPE dictionary features, and get constructed from more common parts (such as the common acronym 'CIA').

However, the way this string is built from tokens is *not unique*, the tokens ⎵CI, AA, ⎵Aki, and ta are also in the dictionary, so the string 'CIAA' can be formed by ⎵CI AA, and 'Akita' by ⎵Aki ta. Using this alternative tokenization is however *not correct*, since the (token) dictionary used specifies rules in order of priority, and the first example above is the correct result of using the BPE algorithm given later in Algorithm 1. Real-world implementations of (variants of) this algorithm use a priority queue of possible rule applications to produce

S. Z. Fazekas (Ed.): CIAA 2024, LNCS 15015, pp. 66–78, 2024.
https://doi.org/10.1007/978-3-031-71112-1_5

the tokenization, which makes it non-obvious how to combine it with automata-based string processing techniques.

That is, string processing is traditionally concerned with strings over some fixed alphabet of symbols, where every string is denoted by precisely one sequence of symbols (i.e. it is based on the free monoid), but here many distinct sequences of tokens may concatenate into the same string. Only one of those token sequences is considered *correct*, however. Conversely, a particular substring may be tokenized differently based on its context, complicating pattern matching. Luckily, the language of correct tokenizations (i.e. the string language where token boundaries are marked) is regular [1], and here we consider practical algorithms for constructing finite automata which operate on correct tokenizations.

Such automata are of great practical importance. Token sequences are already common in communication with language services, and if such services keep getting more popular, more and more text may be encoded in such a way, at least when being transmitted in that context. Specific use-cases may include pattern matching for filtering (e.g., on the network or for safety systems), validating the correctness of tokenizations (incorrect tokenizations can severely confuse models, similarly to occurrences of extremely rare tokens [4]), and with further development, to *perform* tokenizations or rewrite tokenized text directly.

2   Notation and Basic Definitions

An *alphabet* Σ is a finite set of symbols. As we also consider alphabets of strings (or 'tokens'), we consistently refer to a normal alphabet consisting of 'indivisible' symbols as a *base alphabet*. Let Σ^* denote the set of all strings (including the empty string ε) over the base alphabet Σ and let $\Sigma^+ = \Sigma^* \smallsetminus \{\varepsilon\}$.

A *token alphabet* Γ over the base alphabet Σ is a finite subset $\Gamma \subset \Sigma^+$ where $\Sigma \subseteq \Gamma$ and for all $w \in \Gamma$ with $|w| > 1$ there are some $u, v \in \Gamma$ such that $w = uv$. We refer to the elements of such Γ as *tokens*. A sequence of tokens $u_1, \ldots, u_n \in \Gamma$ is denoted $u_1 \wr \cdots \wr u_n$ and called a *tokenization*. The set of all sequences of tokens over Γ is denoted $\Gamma^\wr$. For any arbitrary token alphabet Γ and corresponding base alphabet Σ, we define $\pi : \Gamma^\wr \to \Sigma^*$ as $\pi(u_1 \wr \cdots \wr u_n) = u_1 \cdots u_n$, i.e. the concatenation of the tokens. For a tokenization τ, let $|\tau|$ denote the number of tokens in τ, i.e. $|u_1 \wr \cdots \wr u_n| = n$ for $u_1, \ldots, u_n \in \Gamma$.

Observe that while both tokenizations and strings are sequences, they are different *kinds* of sequences even when they contain the same elements. E.g., the tokenization $\alpha_1 \wr \cdots \wr \alpha_n \in \Gamma^\wr$ is distinct from the string $\alpha_1 \cdots \alpha_n$ despite $\alpha_1, \ldots, \alpha_n \in \Sigma$. That is, they are both obtained by using the same sequence of elements from Σ, but are combined with the concatenation and tokenization operator respectively. The tokenization $\alpha_1 \wr \cdots \wr \alpha_n \in \Gamma^\wr$ with $\alpha_i \in \Sigma \subseteq \Gamma$, is called the *base tokenization* of the string $\alpha_1 \cdots \alpha_n$.

To keep our exposition clear, we adopt some conventions. We let Σ denote some base alphabet, and Γ a token alphabet over Σ. When giving examples, we use $\Sigma = \{\alpha, \beta, \gamma, \ldots\}$. Furthermore, we let α, β, γ be variables denoting symbols (so elements from the base alphabet), let u, v, w be strings or tokens, and τ, ϕ tokenizations. In each case we also reserve all sub-/super-scripted variants of these symbols for those same purposes. As such we have, for example, $\alpha_3 \in \Sigma$ and $\hat{\tau} \in \Gamma^{\wr}$. When writing $\phi = u \wr \tau \wr v$, we have that ϕ is a tokenization where the first token is u, the last token is v, and the intervening tokens form the tokenization τ, so $|\phi| = |\tau| + 2$.

Definition 1. *A token deterministic finite automaton (DFA) is a tuple $A = (Q, \Gamma, q_0, \delta, F)$ where:*

- *Q is the finite set of* states,
- *Γ is the* token alphabet,
- *$q_0 \in Q$ is the* initial state,
- *$\delta : Q \times \Gamma \to Q$ is the* transition function, *which may be partial,*
- *$F \subseteq Q$ is the set of* final states.

A run of A is a sequence $q_1, u_1, q_2, u_2, q_3, \ldots, u_{n-1}, q_n$ for some $n \geq 0$, $q_1, \ldots, q_n \in Q$, $u_1, \ldots, u_{n-1} \in \Gamma$, such that $\delta(q_i, u_i) = q_{i+1}$ for all $1 \leq i < n$. We denote such a run as $q_1 \xrightarrow{u_1} q_2 \xrightarrow{u_2} \cdots \xrightarrow{u_{n-1}} q_n$, or with the shorthand $q_1 \xrightarrow{u_1 \wr \cdots \wr u_{n-1}} q_n$ if the intermediate states are not of interest. We say that such a run reads the string $\pi(u_1 \wr \cdots \wr u_{n-1})$. If not otherwise specified, a run has $q_1 = q_0$ (i.e. it begins in the initial state).

The run is accepting iff $q_1 = q_0$ and $q_n \in F$. The language accepted by A, denoted $\mathcal{L}(A) \subseteq \Gamma^{\wr}$ is the set of all tokenizations τ for which there exists an accepting run $q_0 \xrightarrow{\tau} q_n$.

Token DFAs give rise to some additional questions about determinism. Consider the automaton for the (finite) language $\{a \wr bc, a \wr b \wr c\}$, as shown in Fig. 1.

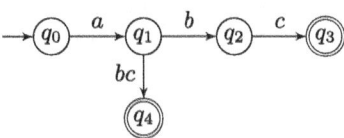

Fig. 1. A token DFA for the language $\{a \wr b \wr c, a \wr bc\}$ which indicates the existence of two distinct tokenizations of the string abc.

This automaton is deterministic in the classical sense, but there are two distinct accepting runs which read abc (i.e. $\pi(a \wr b \wr c) = \pi(a \wr bc)$). As we aim to use token DFAs to represent BPE tokenizations of strings, which are well-defined, we introduce a stronger condition which precludes such situations.

Definition 2. *A token DFA $A = (Q, \Gamma, q_0, \delta, F)$ is* context-invariant *if for runs $q_1 \xrightarrow{\varphi} q_n$ and $q'_1 \xrightarrow{\varphi'} q'_m$ in A (with q_1 or q'_1 not necessarily being equal to q_0) for which $\pi(\varphi) = \pi(\varphi')$, we have $\varphi = \varphi'$.*

Remark 1. Context-invariance ensures that every string has a unique tokeniza-tion, e.g., the token DFA in Fig. 1 is not context-invariant, as its two runs $q_0 \xrightarrow{a \wr b \wr c} q_3$ and $q_0 \xrightarrow{a \wr bc} q_4$ violate Definition 2. Context-invariance is a stronger property than simply requiring the uniqueness of tokenizations, consider the automaton in Fig. 2. Uniqueness is the special case of Definition 2 where we only consider runs where $q_1 = q_1'$ is the initial state, and both runs are accepting. However, as we will see BPE tokenization does fulfill context-invariance, so we demonstrate this stronger property for the automata our algorithm produces.

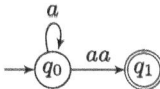

Fig. 2. A token DFA which has unique tokenizations $\{aa, a \wr aa, a \wr a \wr a \wr aa, \ldots\}$ but is not context-invariant, with runs such as $q_0 \xrightarrow{a \wr aa} q_1$ and $q_0 \xrightarrow{a \wr a \wr a} q_0$ violating the property.

Next, we define BPE tokenization, as in [1].

Definition 3. *A* byte pair dictionary D over Σ *is a sequence of pairs of tokens over* Σ, *denoted* $D = [u_1 \wr v_1, \ldots, u_n \wr v_n]$. *We call each* $u_i \wr v_i$ *a* rule, *and say that* $u_i \wr v_i$ *has* higher priority *than* $u_j \wr v_j$ *if* $i < j$.
  A dictionary is proper *if for each* j *with* $|u_j| > 1$ *there exists some* $i < j$ *such that* $u_j = u_i v_i$, *and, symmetrically, for each* j *with* $|v_j| > 1$ *there exists some* $i < j$ *such that* $v_j = u_i v_i$.
  Observe that when D *is proper this makes* $\Sigma \cup \{u_1 v_1, \ldots, u_n v_n\}$ *a valid token alphabet, which we call the* token alphabet *of* D.

When not otherwise stated, we assume that all dictionaries are proper. We adopt the convention that D and all its sub-/superscripted variants, always refer to some (proper) dictionary. In [1] both SentencePiece [3] and HuggingFace [2] BPE tokenization semantics are described, but the semantics are demonstrated to coincide for proper dictionaries. As such we arbitrarily choose to restate the HuggingFace semantics in Algorithm 1.

Algorithm 1. BPE tokenization

1: **Input:** a string $\alpha_1 \cdots \alpha_n \in \Sigma^*$, a proper dictionary $D = [u_1 \wr v_1, \ldots, u_m \wr v_m]$
2: initialize $\tau = \alpha_1 \wr \cdots \wr \alpha_n$
3: **for** each rule $u_i \wr v_i$ in D in priority order: **do**
4:      **while** there are ϕ and ϕ' such that $\phi \wr u_i \wr v_i \wr \phi' = \tau$ **do**
5:          choose such ϕ and ϕ' minimizing $|\phi|$
6:          update $\tau = \phi \wr u_i v_i \wr \phi'$
7:      **end while**
8: **end for**
9: **Output:** the correct tokenization τ.

We denote the BPE tokenization of a string w according to dictionary D as $\mathbb{T}^D(w)$. Note that $\mathbb{T}^\varnothing(\alpha_1 \cdots \alpha_n)$ yields $\alpha_1 \wr \cdots \wr \alpha_n$, the so-called *base tokenization* of $\alpha_1 \cdots \alpha_n$. The definition of $\mathbb{T}^D$ is naturally extended to sets so that $\mathbb{T}^D(L) = \{\mathbb{T}^D(w) \mid w \in L\}$ for a language $L \subseteq \Sigma^*$. In particular, $\mathbb{T}^D(\Sigma^+)$ is the set of *all* BPE tokenizations over Σ^+ according to dictionary D.

Example 1. Let $D = [a \wr a, a \wr b, b \wr c, ab \wr c]$ and $w = aaacbcabc$. Then $\mathbb{T}^D(w) = aa \wr a \wr c \wr bc \wr abc$.

3   Construction Procedure

We now give a procedure to construct a token DFA A for a given regular language L and dictionary D over Σ such that $\mathcal{L}(A) = \mathbb{T}^D(L)$.

For $D = [u_1 \wr v_1, \ldots, u_n \wr v_n]$ and $0 \le i \le n$, define the dictionary D_i as $[u_1 \wr v_1, \ldots, u_i \wr v_i]$ so that $D_0 = [\,]$ and $D_n = D$. We inductively construct a sequence of DFAs A_i satisfying $\mathcal{L}(A_i) = \mathbb{T}^{D_i}(L)$, for a regular language L. The base token DFA A_0 (accepting $\mathbb{T}^\varnothing(L)$) is trivially obtained from the string DFA for L by changing all symbols in Σ into their corresponding tokens in Γ. For $1 \le i \le n$ we construct A_i by merging the rule $u_i \wr v_i$ into A_{i-1} as described in Algorithm 2.

Algorithm 2 merges the rule $u \wr v$ into $A = (Q, \Gamma, q_0, \delta, F)$ by considering each $(s_1, s_2, s_3) \in Q^3$ where $\delta(s_1, u) = s_2$ and $\delta(s_2, v) = s_3$. These runs reading $u \wr v$ are replaced with ones reading uv instead, by adding the transition $\delta(s_1, uv) = s_3$ (line 5) and removing the transition $\delta(s_1, u) = s_2$ (line 20). However, this might inadvertently eliminate runs not containing the sequence $u \wr v$, i.e. runs that read a u to transition to s_2, but then take a different transition from s_2 than the transition on v. This is remedied by adding a new state $fresh(s_2)$ in line 10, which inherits all outgoing transitions from s_2 except the ones on v and potentially uv in lines 13–17. In particular, if $u = v$, a transition on uv from s_2 should *not* be copied to $fresh(s_2)$, as this would create the run $s_1 \xrightarrow{u} fresh(s_2) \xrightarrow{uu} q$, which should never be part of a run in the DFA produced by Algorithm 2. This can most easily be seen by inspecting Algorithm 1, the only way a token uu can be created is by the rule $u \wr u$ being applied, and it is then applied first on the left, producing $uu \wr u$. Formal correctness proofs follow in Sect. 4. Observe that the transition on uu at s_2 can not come from A (the input DFA), otherwise this implies A has runs on both uu and $u \wr u$, and this contradicts the context-invariant assumption on A. Rather that transition would have just been added to s_2 by step 5 in Algorithm 2. Example 2 illustrates both the case where $u = v$ and the transition is *not* copied from s_2 to $fresh(s_2)$ (the step from A_0 to A_1), and the case where $u \ne v$ and the transition *is* copied (the step from A_1 to A_2, observe that the token ba occurs on two transitions in A_2).

Algorithm 2 will at times produce useless states, which might simply be trimmed afterwards.

Example 2. For byte pair dictionary $D = [a \wr a, b \wr a]$ over $\Sigma = \{a, b\}$, constructing the token DFA A that accepts $\mathbb{T}^D(\Sigma^*)$ requires two iterations. Starting from the base token DFA A_0 in Fig. 3, Algorithm 2 is first applied with the merge $a \wr a$,

Algorithm 2. Applying a Merge

```
 1: Input: a context-invariant DFA A = (Q, Γ, q₀, δ, F), and a rule u ≀ v
 2: let S = {(s₁, s₂, s₃) ∈ Q³ | δ(s₁, u) = s₂, δ(s₂, v) = s₃}
 3: let S₂ = {s₂ | (s₁, s₂, s₃) ∈ S}
 4: for (s₁, s₂, s₃) ∈ S do
 5:     add new transition by defining δ(s₁, uv) = s₃
 6: end for
 7: add uv to Γ
 8: for s₂ ∈ S₂ do
 9:     create a fresh state from s₂, denote it fresh(s₂)
10:     add fresh(s₂) to Q, and if s₂ ∈ F add fresh(s₂) to F as well
11: end for
12: for every s₂ ∈ S₂ do
13:     if u ≠ v then
14:         add new transitions by defining δ(fresh(s₂), α) = δ(s₂, α) for all α ∈ Γ ∖ {v}
15:     else
16:         add new trans. by defining δ(fresh(s₂), α) = δ(s₂, α) for all α ∈ Γ ∖ {v, uv}
17:     end if
18: end for
19: for q ∈ Q with δ(q, u) ∈ S₂ do
20:     replace the transition by defining δ(q, u) = fresh(δ(q, u))
21: end for
22: output the resulting DFA
```

yielding the depicted token DFA A_1. Next, Algorithm 2 is applied to A_1 with the merge $b \wr a$, generating the token DFA A_2 which accepts $\mathbb{T}^D(\Sigma^*)$.

Remark 2. It is interesting to note when and how we obtain loops at $fresh(s_2)$ when applying Algorithm 2. This happens when we have a loop on u (in the case where $u \neq v$). This loop is replaced by two transitions on u, one going from s_2 to $fresh(s_2)$ and the other being a loop on $fresh(s_2)$. This happens by first in line 12 adding a transition from $fresh(s_2)$ to s_2 (on u), and then in line 19, changing the target of this (newly added) transition to be $fresh(s_2)$ rather than s_2, but

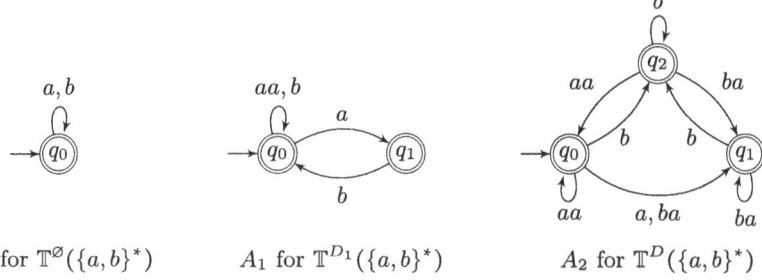

A_0 for $\mathbb{T}^\emptyset(\{a,b\}^*)$ A_1 for $\mathbb{T}^{D_1}(\{a,b\}^*)$ A_2 for $\mathbb{T}^D(\{a,b\}^*)$

Fig. 3. The three token DFA constructed in Example 2, A_0 is the initial universal base token DFA, A_1 is A_0 with the rule $a \wr a$ merged, and A_2 is A_1 with the rule $b \wr a$ merged.

also adding a transition from s_2 to $\mathit{fresh}(s_2)$ on u (in line 19). An example of this can be seen in Fig. 3 A_1 to A_2, with the loop on b at q_0 in A_1 being replaced by a transition from q_0 to q_2 and a loop at state q_2 (both on b) in A_2. Contrast this to with what happens when $u = v$, by considering the loop on a at q_0 in A_0, and $\mathit{fresh}(q_0)$ being q_1 in A_1.

4   Proof of Correctness

Observe that whenever a new transition is created by Algorithm 2, we update the (partial) transition function of the input token DFA, thus yielding a new token DFA. The remainder of this section sets out to demonstrate that the procedure in Sect. 3 yields the token DFA for the BPE tokenizations of the given language L and dictionary D.

Example 3. If we apply Algorithm 2 on the non-context-invariant token DFA in Fig. 1 with rule $b \wr c$, we obtain a nondeterministic automaton (state q_1 gaining a second transition on the token bc), but the result stated in the next lemma still holds otherwise.

Further, when considering the non-context-invariant token DFA A in Fig. 2, and the rule $a \wr a$, Algorithm 2 again produces a nondeterministic automaton A', in which there is for example no $\phi \in \mathcal{L}(A')$, with $\pi(\phi) = \pi(a \wr aa)$, even though $a \wr aa \in \mathcal{L}(A)$.

These examples illustrate the necessity of assuming context-invariance in the next lemma and in this section in general.

It turns out that Algorithm 2 preserves context-invariance. However, before proving this, we first consider the following useful lemma that gets us most of the way in terms of correctness of Algorithm 2.

Lemma 1. *For any rule $u \wr v$ and context-invariant DFA $A = (Q, \Gamma, q_0, \delta, F)$ applying Algorithm 2 produces a DFA $A' = (Q', \Gamma', q'_0, \delta', F')$ such that there is a run $q'_1 \xrightarrow{\tau'} q'_m$ in A' if and only if there is a run $q_1 \xrightarrow{\tau} q_n$ in A (with q'_1 and q_1 any state in Q' and Q respectively) where $\pi(\tau') = \pi(\tau)$, and both runs end at either accepting or rejecting states. Thus, if both runs start at q'_0 and q_0 respectively, then both are either accepting or rejecting.*

Proof. The lemma trivially holds when in A we have no run of the form $s_1 \xrightarrow{u} s_2 \xrightarrow{v} s_3$, since then $S = \varnothing$ in Algorithm 2, producing an unchanged $A' = A$. Observe that the lemma therefore especially holds when A contains a transition on uv due to its presupposed context-invariance.

To relate the runs of the automata, we relate their states. Observe that Q' contains either one or two copies of each state from Q. Define $\gamma : Q' \to Q$ to recover the state in A from which a state in A' was created, so $\gamma(q) = q$ for $q \in Q$, and $\gamma(\mathit{fresh}(q)) = q$ for all $\mathit{fresh}(q) \in Q' \setminus Q$ produced by line 10.

To find a run in A corresponding to the run $q'_1 \xrightarrow{u_1 \wr \cdots \wr u_{m-1}} q'_m$ in A', note that the run $\gamma(q'_1) \xrightarrow{u_1 \wr \cdots \wr u_{m-1}} \gamma(q'_m)$ in A is well-defined unless the original run in A' contains a step of the form $q \xrightarrow{uv} q'$ since this, by Algorithm 2, is the only transition not directly inherited from $\gamma(q)$. If such a step of the form $q \xrightarrow{uv} q'$ exists, then $(\gamma(q), s, \gamma(q')) \in S$ for some $s \in Q$, and can be replaced by the two steps $\gamma(q) \xrightarrow{u} s \xrightarrow{v} (\gamma(q'))$ instead. Repeating this procedure for all steps of the form $q \xrightarrow{uv} q'$ produces a run in A. Given that $\gamma(q'_0) = q_0$ and that $q' \in F'$ if and only if $\gamma(q') \in F$, it follows trivially that both runs must be either accepting or rejecting.

In the other direction, we proceed by induction on the length of runs to demonstrate that for all runs $q_1 \xrightarrow{\tau} q_n$ in A, there exists a run $q'_1 \xrightarrow{\tau'} q'_m$ such that $q_n = \gamma(q'_m)$ and $\pi(\tau) = \pi(\tau')$. The base case is trivial since a zero-length run has $\pi(\tau) = \pi(\tau') = \varepsilon$ and $q_1 = \gamma(q'_1)$ for some $q'_1 \in Q'$. As the inductive step, extend an arbitrary run r of length $n-1$ in A with a transition $q_n \xrightarrow{w} q_{n+1}$. This changes the corresponding run of r in A' according to the following three cases depending on how w relates to the rule $u \wr v$ being merged.

1. If $w \neq v$: the corresponding run is extended with a transition on w to $\delta'(q'_m, w)$, which respects the required relationship between the two runs given that $\delta(\gamma(q'), w) = \gamma(\delta'(q', w))$ for all $q' \in Q$ when $w \neq v$.
2. If $w = v$ and q'_m is a fresh state: by construction $\delta'(q'_m, v)$ is undefined. However, observe that then the *previous* step in the corresponding run must be $\delta'(q'_{m-1}, u) = q'_m$ (the only way to reach a fresh state). Hence, $(\gamma(q'_{m-1}), \gamma(q'_m), q) \in S$ in Algorithm 2 for some $q \in Q$. Therefore, replace that previous step with $q'_{m-1} \xrightarrow{uv} q$. Note that when $u = v$ fresh states do not get transitions on uv (see line 16), but in such a case q'_{m-1} cannot be fresh, as fresh states never have outgoing transitions on v.
3. If $w = v$ and q'_m is *not* a fresh state: just like in case 1 then $\delta'(q'_m, w)$ is defined and can be used.

By the same argument as above, straightforward inspection of γ establishes that either both runs are accepting or both are rejecting.   □

Now we have the tools to prove that Algorithm 2 preserves context-invariance.

Lemma 2. *For any rule $u \wr v$ and context-invariant DFA A, applying Algorithm 2 produces a context-invariant DFA A'.*

Proof. By contradiction assume that A' is *not* context-invariant. Then A' has runs on τ and φ so that $\tau \neq \varphi$ but $\pi(\tau) = \pi(\varphi)$.

Let $\tau = w_1 \wr \cdots \wr w_n$ and $\varphi = w'_1 \wr \cdots \wr w'_m$ where i is the smallest index such that $w_i \neq w'_i$. The mapping defined in the proof of Lemma 1 instructs to replace all occurrences of uv with $u \wr v$ in both τ and φ to yield respective corresponding runs τ' and φ' in A that read $\pi(\tau)$. Since A is context-invariant, it must be that $\tau' = \varphi'$, in turn implying $\{w_i, w'_i\} = \{u, uv\}$. Without loss of generality let $w_i = u$ and $w'_i = uv$. Then consider the token w_{i+1}, which by the defined mapping must have v as a prefix, so either

- $w_{i+1} = v$, but this implies a run $q \xrightarrow{u} q' \xrightarrow{v} q''$ in A', impossible by Algorithm 2, or,

– $u = v$ and then $w_{i+1} = vv$, but line 16 prevents the introduction of a transition on uv in the case where $u = v$ if u has just been read.

No other alternatives for w_{i+1} are possible since any other token with v as a prefix is unaffected by the mapping (in which case $\tau' \neq \varphi'$). As such, we reach a contradiction, i.e. no such τ and φ exist.   □

Next we show the relationship between tokenizations of two corresponding accepting runs as shown to exist in Lemma 1.

Lemma 3. *Let $A = (Q, \Gamma, q_0, \delta, F)$ be a context-invariant token DFA and $A' = (Q', \Gamma', q_0', \delta', F')$ the resulting token DFA of applying Algorithm 2 with rule $u \wr v$. Suppose that for a given accepting run $q_1 \xrightarrow{\tau} q_n$ in A, the corresponding accepting run in A' is $q_1' \xrightarrow{\tau'} q_m'$ where $\pi(\tau) = \pi(\tau') = w$. Then, if there exists a proper dictionary D such that $\mathbb{T}^D(w) = \tau$, and the dictionary D' obtained by concatenating the rule $u \wr v$ to D is also proper, then $\mathbb{T}^{D'}(w) = \tau'$.*

Proof. From Lemma 2, it follows that A' is context-invariant. Hence, we can use the mapping from the proof of Lemma 1, to deduce that τ' can be obtained from τ by only altering the subtokenizations $u \wr v$:

– In case $u \neq v$, each such subtokenization is simply replaced by uv.
– In case $u = v$, A' cannot read the token u followed by a token uu, subsequently $u \wr u$ is replaced by uu starting from the subtokenization's left-most occurrence.

Now, let D be a proper dictionary such that $\mathbb{T}^D(w) = \varphi$ and suppose D' obtained by concatenating the rule $u \wr v$ to D also is proper. Since Algorithm 1 applies each rule $u_i \wr v_i$ in order of priority to w, $\mathbb{T}^{D'}(w)$ can be directly obtained by applying the rule $u \wr v$ to τ. When applying $u \wr v$ to τ, we repeatedly decompose τ into $\phi \wr u \wr v \wr \phi'$ so that $|\phi|$ is minimal and replace $u \wr v$ with uv until no such decomposition can be found. Clearly, this yields that $\mathbb{T}^{D'}(w) = \tau'$.   □

Observe that Lemma 3 does not hold without the requirement that the involved dictionaries have to be proper. For starters, if a dictionary D is not proper, recall that the SentencePiece and HuggingFace semantics do not necessarily yield the same BPE tokenization, so $\mathbb{T}^D$ is not well-defined. Further, we can demonstrate by example that $\mathcal{L}(A)$ in such case not necessarily corresponds to either the SentencePiece or HuggingFace tokenizations.

Example 4. Take the dictionary $D = [ab \wr a, a \wr b]$ (which is not proper) and let $L = \{a, b\}^*$. The generated token DFA A is depicted in Fig. 4. (Merging the rule $ab \wr a$ into the base token DFA just results in the base token DFA as no runs on $ab \wr a$ are present.) Now, the SentencePiece tokenization of the string $w = abababababa$ is $aba \wr b \wr aba \wr b \wr a$ (see [1]) and the HuggingFace tokenization (refer to Algorithm 1) is $ab \wr ab \wr ab \wr ab \wr a$, which the token DFA A obviously accepts. Strictly speaking, for non-proper dictionaries, HuggingFace is not forced to select rules in decreasing order of priority (where for proper, it will by default), but only select the highest priority applicable rule, and keep on applying it left to right (see [1]). Thus, with this formulation, the HuggingFace tokenization will be $ab \wr ab \wr ab \wr aba$, and thus, token DFA A accepts neither HuggingFace nor SentencePiece tokenization, since $\delta(\cdot, aba)$ is undefined for any given state in A.

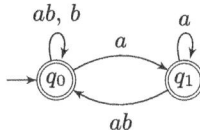

Fig. 4. Automaton A generated for dictionary $D = [ab \wr a, a \wr b]$ by applying the merge $ab \wr a$ followed by the the merge $a \wr b$ to the base token DFA.

Finally, we are ready to prove that if we use Algorithm 2 iteratively to produce each time the token DFA A_i, using A_{i-1} and the rule $u_i \wr v_i$ from the proper dictionary $D = [u_1 \wr v_1, \ldots, u_n \wr v_n]$ as input (with A_0 being the base token DFA of a string DFA A), then the language of A_i is the the tokenization of $\mathcal{L}(A)$ using $D_i = [u_1 \wr v_1, \ldots u_i \wr v_i]$.

Theorem 1. *Let A be a string DFA over base alphabet Σ. Let D be the proper dictionary $D = [u_1 \wr v_1, \ldots, u_n \wr v_n]$ over Σ. Further, let $A_0, \ldots, A_n$ be the sequence of token DFAs where A_0 is the base token DFA for A and for each $i > 0$, A_i is produced by applying the merge $u_i \wr v_i$ to A_{i-1}. Then, $\mathcal{L}(A_i) = \mathbb{T}^{D_i}(\mathcal{L}(A))$ for all $i \in \{0, \ldots, n\}$.*

Proof. We proceed by induction. Let $L = \mathcal{L}(A)$. It is given that $\mathcal{L}(A_0) = \mathbb{T}^{\varnothing}(L)$. Assume that $\mathcal{L}(A_i) = \mathbb{T}^{D_i}(L)$. Observe that since A is a string DFA, A_0 is trivially context-invariant. Hence, each of the token DFAs in the sequence $A_1, \ldots, A_n$ must also be context-invariant by Lemma 2. It further follows from Lemma 1 that for every $\varphi \in \mathcal{L}(A_i)$ there is some $\varphi' \in \mathcal{L}(A_{i+1})$ such that $\pi(\varphi) = \pi(\varphi')$. Moreover, since $\mathbb{T}^{D_i}(\pi(\varphi)) = \varphi$, then $\mathbb{T}^{D_{i+1}}(\pi(\varphi)) = \varphi'$ by Lemma 3. So, for all $\varphi \in \mathbb{T}^{D_i}(L)$, there exists some $\varphi' \in \mathcal{L}(A_{i+1})$ such that $\varphi' \in \mathbb{T}^{D_{i+1}}(L)$. It only remains to show that all $\psi \in \mathcal{L}(A_{i+1})$ are in $\mathbb{T}^{D_{i+1}}(L)$. If $\psi \in \mathcal{L}(A_{i+1})$ then by Lemma 1 $\psi' \in \mathcal{L}(A_i)$ so that $\pi(\psi) = \pi(\psi')$, meaning $\psi' \in \mathbb{T}^{D_i}(L)$. We derive $\psi \in \mathbb{T}^{D_{i+1}}(L)$ from Lemma 3, hence $\mathcal{L}(A_{i+1}) = \mathbb{T}^{D_{i+1}}(L)$. This completes the induction step. $\square$

The case where A is a universal string DFA is of particular interest, as the produced token DFA then accepts the language of all correct tokenizations.

5   Complexity Bounds

A cursory reading of Algorithm 2 might suggest that *each* application of the algorithm could double the size of the token DFA (i.e. we may copy *all* states if $Q = \{s_2 \mid (s_1, s_2, s_3) \in S\}$), which would increase the number of states by a factor $2^{|D|}$ when applying Theorem 1. However, this turns out to be too pessimistic.

Definition 4. *For a DFA* $A = (Q, \Sigma, q_0, \delta, F)$ *and* $\alpha \in \Sigma$ *the* target count *of* α *in* A, *denoted* $targetc(A, \alpha)$, *is defined as* $targetc(A, \alpha) = |\{q' \mid \exists q \in Q, \delta(q, \alpha) = q'\}|$. *If all* $\alpha \in \Sigma$ *have target count at most one we say that* A *is* single target.

We then relate the number of states added by a merge to the values targetc in the original automaton.

Lemma 4. *Assume Algorithm 2 is applied to the context-invariant token DFA* $A = (Q, \Gamma, q_0, \delta, F)$ *with rule* $u \wr v$ *and produces* $A' = (Q', \Gamma \cup \{uv\}, q_0, \delta, F)$ *as output with* $uv \notin \Gamma$ *(if* A *had a transition on* uv *we would have* $A = A'$). *Then:*

1. $|Q'| \le |Q| + targetc(A, u)$,
2. $targetc(A', uv) \le targetc(A, v)$, *and*,
3. $targetc(A', \beta) = targetc(A, \beta)$ *for all* $\beta \in \Gamma$.

Proof. Observe that $targetc(A, u)$ forms an upper bound on how many distinct states s_2 can be found in Algorithm 2: the only new state ever introduced is $fresh(s_2)$, and s_2 can only be states with incoming transitions on u by construction. Since the number of states in A' will be the number of states in A plus one copy of each such state s_2 the bound in (1) is obtained.

Similarly, for (2), $targetc(A, v)$ forms an upper bound on how many states gain an incoming transition labelled uv, as Algorithm 2 adds transitions labelled uv only going to a state s_3 (lines 5 and 14), and those must by definition have an incoming transition labelled v.

The equality in (3) is a matter of inspecting each step of Algorithm 2. For $\beta \notin \{u, v\}$ (observe that $\beta = uv$ is also excluded in (3)), only step 14 of Algorithm 2 adds such transitions, where if $\delta(s_2, \beta) = s'$ we add a transition $\delta'(fresh(s_2), \beta) = s'$. However, as s' is then by construction already a target for a transition labelled β this does not increase $targetc(A', \beta)$. Similarly, for $\beta = u$, line 14 might add the transition $\delta'(fresh(s_2), u) = \delta(s_2, u)$ without increasing $targetc(A', u)$, since in such case $\delta(s_2, u)$ already must have had such an incoming edge. While line 20 removes every incoming edge labeled u from s_2, it adds a corresponding incoming edge labeled u to $fresh(s_2)$, implying that $targetc(A', u) = targetc(A, u)$. Finally, trivially $targetc(A', v) = targetc(A, v)$ as Algorithm 2 never adds or removes a transition labelled v. □

Theorem 2. *Let* A *be a context-invariant token DFA over* Γ *with* n *states and* $k = \max\{targetc(A, \alpha) \mid \alpha \in \Gamma\}$. *Let* D *a proper dictionary. Then iteratively merging all rules from* D *into* A *(as in Theorem 1) produces a DFA with at most* $n + k|D|$ *states. There exists* A *and* D *which achieve this bound.*

Proof. This result follows from Lemma 4. For $|D| = 0$ the automaton is unchanged and the result trivially holds. Otherwise, inductively merge the highest priority rule of D into A to produce A', and observe that the lemma dictates that A' has at most k more states than A by condition (1), and the maximum targetc (i.e. the value k) for A' is the same as that for A. As such we can iterate this argument on A' and the next highest priority rule in D, which after $|D|$ steps produces the indicated bound.

To reach the bound consider the single-state DFA for a^* and the dictionary $D = [a \wr a, aa \wr aa, aaaa \wr aaaa, \dots, a^{2^{|D|-1}} \wr a^{2^{|D|-1}}]$. Observe that the resulting token DFA accepts a token sequence if and only if it begins some number of instances of the token $a^{2^{|D|}}$, followed by zero or one instances of the token $a^{2^{|D|-1}}$, followed by zero or one instances of the token $a^{2^{|D|-2}}$, and so on, ending with zero or one instances of the token a^{2^0} (i.e. the token 'a'). This clearly requires $|D| + 1$ states to represent, and as we have $|Q| = 1$ and $k = 1$ this equals the bound.   □

Theorem 3. *Algorithm 2 can be applied to a suitably encoded context-invariant token DFA $A=(Q, \Sigma, q_0, \delta, F)$ in time $\mathcal{O}(|Q||\Sigma|)$. Applying Algorithm 2 iteratively as in Theorem 1 with a proper dictionary D can be done in time $\mathcal{O}(|Q||\Gamma||D|^2)$, where $\Gamma = \Sigma \cup \{uv \mid u \wr v \in D\}$.*

Proof. We assume looking up $\delta(q, \alpha)$ and adding or removing transitions can be can be done time $\mathcal{O}(1)$, and that a state can be added in time $\mathcal{O}(|\Sigma|)$. The time complexity of Algorithm 2 is dominated by the loop over S_2 from lines 11 to 17, running $\mathcal{O}(|Q|)$ times, with each iteration running in time $\mathcal{O}(|\Sigma|)$. Furthermore, we can construct S in time $\mathcal{O}(|Q|)$ by for each $q \in Q$ looking up $\delta(q, u)$ and $\delta(\delta(q, u), v)$. If both are defined these states form a triple in S. Iterating the procedure, Theorem 2 dictates that no automaton in the process will have more than $|Q| + k|D|$ states (where $k = \max\{\text{targetc}(A, \alpha) \mid \alpha \in \Sigma\}$), with the token alphabet increasing from Σ to Γ. Putting it together, we apply Algorithm 2 $|D|$ times, each with a running time in $\mathcal{O}((|Q| + k|D|)|\Gamma|)$, giving a time complexity of $\mathcal{O}(|Q||\Gamma||D| + k|\Gamma||D|^2)$, which is $\mathcal{O}(|Q||\Gamma||D|^2)$, since $k \leq |Q|$.   □

6   Conclusions and Future Work

To summarize, we have given an algorithm for constructing finite automata which recognize correct BPE tokenizations, proven it correct, and demonstrated it efficient. In immediate applications this enables efficient validation of tokenization correctness, efficient pattern matching on tokenized text, and equivalence checking of proper BPE dictionaries. In general, it makes the many algorithms and closure properties of regular languages available in the tokenized setting.

As this is a practical and very popular area, there is a lot of potential future work to consider. Section 5 demonstrates that the state complexity of the automata constructed is quite modest, the base alphabet will in practice be Unicode code points, which means that each state may have a very large number of transitions. As such the encoding of the automata should be done with some care, potentially investigating a link to symbolic automata. Further, beyond automata one should consider the transducer generalization, where the input is a normal string and the output its tokenization. This is very straightforward to do, but the state- and computational complexity should be investigated with care.

Acknowledgments. This work was partially supported by the Wallenberg AI, Autonomous Systems and Software Program (WASP) funded by the Knut and Alice Wallenberg Foundation.

References

1. Berglund, M., van der Merwe, B.: Formalizing BPE tokenization. In: 13th International Workshop on Non-classical Models of Automata and Applications, NCMA 2023, Famagusta, Cyprus, 18–19 September 2023, pp. 16–27. Open Publishing Association (2023)
2. Hugging Face: Transformers (2023). https://github.com/huggingface/transformers/blob/v4.28.1/src/transformers/models/gpt2/tokenization_gpt2.py
3. Kudo, T., Richardson, J.: SentencePiece: a simple and language independent subword tokenizer and detokenizer for neural text processing. CoRR abs/1808.06226 (2018)
4. Li, Y., et al.: Glitch tokens in large language models: categorization taxonomy and effective detection. arXiv preprint arXiv:2404.09894 (2024)
5. OpenAI: ChatGPT: optimizing language models for dialogue (2022). https://openai.com/blog/chatgpt/
6. Radford, A., Wu, J., Child, R., Luan, D., Amodei, D., Sutskever, I., et al.: Language models are unsupervised multitask learners. OpenAI Blog **1**(8), 9 (2019)
7. Sennrich, R., Haddow, B., Birch, A.: Neural machine translation of rare words with subword units. In: Proceedings of the 54th Annual Meeting of the Association for Computational Linguistics (Volume 1: Long Papers), Berlin, Germany, pp. 1715–1725. Association for Computational Linguistics (2016). https://doi.org/10.18653/v1/P16-1162

SAT-Based Automated Completion for Reachability Analysis

Yohan Boichut[1]([✉]), Vincent Hugot[1,2], and Adrien Boiret[1,2]

[1] LIFO, UR 4022, Université d'Orléans, Orléans, France
yohan.boichut@univ-orleans.fr
[2] INSA Centre Val de Loire, Bourges, France
{vincent.hugot,adrien.boiret}@insa-cvl.fr

Abstract. Reachability analysis in rewriting has served as a verification technique in recent decades, despite the underlying issue being undecidable. Regular tree model-checking has found application in verifying security protocols, Java programs, and concurrent systems. The premise in these approaches is to represent the targeted system as a state system and encode its transitions using a term rewriting system or a tree transducer. The crucial aspect lies in calculating a fixed point that represents the set of configurations or states that can be reached. While this is generally uncomputable, it is sufficient to compute an overapproximation for the purpose of verifying safety properties.

Let A, B, and $\mathcal{R}$ represent, respectively, an initial set of terms, a set of forbidden ("bad") terms, and a term rewriting system. The question is whether there exists a regular approximation $A^\star$ of the set of reachable terms such that $A^\star \supseteq \mathcal{R}^\star(A)$ and $A^\star \cap B = \varnothing$.

Finding suitable approximations requires, in practice, the use of heuristics, steered towards an anticipated conclusive fixed point by the intervention of human domain experts. The parameters upon which they act may take the form of term equations, normalizing rules, predicate abstractions, etc., but in all cases boil down to carefully choosing states to merge during the fixpoint computation, forcing convergence while avoiding overshooting the approximation into B.

We propose a practical, scalable automated method offloading that expert work to a SAT solver.

1 Introduction

Over the past few decades, the verification of critical systems has been the subject of intense study. Various formal methods, such as model-checking, symbolic model-checking, and theorem proving, have been applied extensively.

Theorem proving techniques often rely on human interaction to prove safety properties. For finite state systems, model-checking has been successfully employed, with its main limitation being the size of the state space. However, verifying infinite state systems using symbolic and regular model-checking poses challenges for automatic verification. The overarching issue of system verification

S. Z. Fazekas (Ed.): CIAA 2024, LNCS 15015, pp. 79–93, 2024.
https://doi.org/10.1007/978-3-031-71112-1_6

is undecidable. In *regular model-checking* [18], the problem of system verification is transformed into a reachability problem within term rewriting or term transduction.

Given two tree automata (TA) A and B, representing respectively the initial and the forbidden languages, and a TRS $\mathcal{R}$, the *tree automata completion algorithm* strives to generate a TA A^* of language closed under $\mathcal{R}$ and not containing any forbidden term. It does this by adding new rules and states to A. There is no guarantee of termination, however.

Consider the TRS $\mathcal{R}$ composed of a single rule $f(x) \rightarrow f(a(x))$ and a TA A that accepts only the ground term $f(\bot)$. Then A^* must accept all terms of the form $f(a^*(\bot))$. While this language is regular, without expert intervention to guide the reuse or merging of new states during the computation of A^*, it would diverge, adding a new state to recognize each term $f(a^n(\bot))$ (cf. Example 18[p9]).

The key parameter which the expert must manipulate can be conceptualized as an equational theory. In practice, it may take various forms, including priority transitions & approximation rules [15], approximation functions [8], sets of term equations, e.g. $a(x) = x$ [9, 11], etc. These approaches have shown effectiveness in verifying security protocols, Java programs or functional programs [4, 5, 12, 13].

Automating this technique has been a challenging pursuit that has motivated us for several years. While some solutions have been explored in [1], they are specific to certain application fields, such as security protocols. Other solutions presented in [12] focus on verifying safety properties of higher-order functions. One significant limitation of [12] is that the term rewriting system must be terminating.

The workshop paper [2] made an initial attempt to automate the reachability problem in its broadest sense: given $A, B, \mathcal{R}$ as above, can mergings of states be computed for A such that the resulting language both avoids forbidden terms and is closed under $\mathcal{R}$? This was automated by converting the question into a WS1S formula and feeding it to the MONA [19] solver. The key limitations of [2] are that the language of B must be finite, and that the approach did not scale at all: the cutoff point was ~20 variables.

This paper expands on [2]: we greatly simplify the formalisation—correcting various errors in the process—and remove the key limitation. On the practical side, we target classical propositional logic to leverage the power of SAT solvers, which are highly optimised. As we shall see, this pays off in the experimental results.

Section 2 fixes definitions and notations regarding tree automata and rewriting. Section 3 presents a characterisation of conclusive approximations in propositional logic. In Sect. 4 integrates the method in the completion process. Section 5 analyses complexity and experimental results.

2   Notations and Reminders About Tree Automata

Given a set S and an equivalence relation $\approx$ on S, we write $S/_\approx$ the quotient set (set of equivalence classes), and $[x]_\approx$ or $[x]$ the equivalence class of $x \in S$ in S wrt. $\approx$.

Let $\mathbb{X}$ be a finite set of variables and Σ a finite ranked alphabet, where the arity of each symbol $f \in \Sigma$ is denoted by $\mathrm{ar}(f)$. The set of **terms** built from Σ and $\mathbb{X}$ is denoted by $\mathcal{T}(\Sigma, \mathbb{X})$. The set of **ground terms** (without variables) is written $\mathcal{T}(\Sigma)$. The notions of **position**, and **substitution** are defined as usual.

For a term $t \in \mathcal{T}(\Sigma, \mathbb{X})$, $\mathbb{X}(t)$ is the set of variables appearing in t, $\mathcal{P}(t)$ is the set of positions of t, as words of $[\![0, n]\!]^*$. Thus, a term is seen as a function $t \in (\mathcal{P}(t) \to \Sigma \cup \mathbb{X})$. It is **linear** if each variable of $\mathbb{X}(t)$ appears at exactly one position.

A **rewrite rule** is a pair of terms, written $l \to r$, such that $\mathbb{X}(l) \supseteq \mathbb{X}(r)$ and $l \notin \mathbb{X}$. It is **left-** (resp. **right-**)**linear** if l (resp. r) is linear, and just **linear** if it is both left- and right-linear. A **term rewriting system (TRS)** $\mathcal{R}$ is a finite set of rewrite rules. It is **left-** (resp. **right-**)**linear** if all its rules are. It induces a rewrite relation $\to_\mathcal{R}$ on $\mathcal{T}(\Sigma)$, whose reflexive and transitive closure is written $\to_\mathcal{R}^*$. We let $\mathcal{R}^*(L) = \{\, v \mid u \in L \wedge u \to_\mathcal{R}^* v \,\}$.

Let $\mathbb{Q}$ be a countably infinite set of elements called **states**, all of arity 0, and such that $\mathbb{X}$, $\mathbb{Q}$, and Σ are all pairwise disjoint. A **configuration** is a (ground) term $c \in \mathcal{T}(\Sigma \cup \mathbb{Q})$. A **transition** is a rewrite rule $c \to q \in \mathcal{T}(\Sigma \cup \mathbb{Q}) \times \mathbb{Q}$; it is **normal** if either $c \in \mathbb{Q}$ (in which case it is an **ε-transition**) or $c = f(q_1, \ldots, q_n)$, with $q_k \in \mathbb{Q}$, $\forall k$.

Definition 1. A **nondeterministic bottom-up tree automaton (TA)** A is a tuple $\langle \Sigma, Q, F, \Delta \rangle$, where Σ is a ranked alphabet, $Q \subseteq \mathbb{Q}$ is a set of states, $F \subseteq Q$ is a set of final states, and $\Delta \subseteq \mathcal{T}(\Sigma, Q) \times Q$ is a set of normal transitions. It is **ε-free** if it has no ε-transition, which we shall safely assume whenever convenient [6, p23].

Given a state $q \in Q$, its **language in A** is defined as $[\![q]\!]_A = \{\, t \in \mathcal{T}(\Sigma) \mid t \to_\Delta^* q \,\}$. We write $[\![q]\!]$ when A is clear from context. The **language of A** is $[\![A]\!] = \bigcup_{q \in F} [\![q]\!]$. We often write $\to_A$ for $\to_\Delta$. The **synchronised product of two TA, $A \times B$**, is defined as usual [6, p30].

By convention, any TA A will be assumed to be defined as above. We employ object-like notations: e.g. $B.Q$ is the set of states of B, and $B = \langle\, \Sigma := f(\Sigma),\ Q := \ldots \,\rangle$ defines a new automaton whose alphabet is a function of Σ (meaning $A.\Sigma$ in context, given the definition of A above), etc. We can define a new automaton from another, keeping most attributes implicitly unaltered, but automatically updating them if needed; for instance, $B = A \langle \Delta := \Delta \cup \{\, f(p) \to q \,\} \rangle$ defines B to be the same as A, but with the addition of rule $f(p) \to q$. It is thus shorthand for $B = \langle \Sigma := \Sigma, Q := Q \cup \{\, p, q \,\},\ F := F,\ \Delta := \Delta \cup \{\, f(p) \to q \,\} \rangle$.

Given a TA A and a TRS $\mathcal{R}$, a **critical pair** is a tuple $(\sigma, l, r, q)^1$ where $\sigma : \mathbb{X} \to Q$ is a substitution and $l \to r \in \mathcal{R}$ is a rewrite rule, such that there is a state q with $\sigma l \to_A^* q$ and $\sigma r \not\to_A^* q$. We denote by $\mathcal{C}_{\mathcal{R},A}^{\mathrm{rit}}$ the set of critical pairs:

$$\mathcal{C}_{\mathcal{R},A}^{\mathrm{rit}} = \{\, (\sigma, l, r, q) \mid l \to r \in \mathcal{R},\ \sigma l \to_A^* q,\ \sigma r \not\to_A^* q \,\} . \qquad (\mathcal{C}_{\mathcal{R},A}^{\mathrm{rit}})$$

---

[1] This is usually presented as the pair $(\sigma l, \sigma r)$, hence the name "critical pair".

A is said to be **acritical** (wrt. $\mathcal{R}$) if there are no critical pairs. If so, A is also **complete**, that is to say $\mathcal{R}^*(\llbracket A \rrbracket) \subseteq \llbracket A \rrbracket$. The converse is not true.

A **completion step** consists in solving all critical pairs by adding transitions – in dotted line – such that

$$\begin{array}{ccc} \sigma\,l & \xrightarrow{\ \mathcal{R}\ } & \sigma\,r \\ A\downarrow{\scriptstyle *} & {\scriptstyle *}\diagup & \\ q & \xleftarrow{\ \ \ } & A \end{array} \quad.$$

3   SAT Solving for Overapproximations

Let A and B be TA and $\mathcal{R}$ a TRS. Here A can be the initial language, or may already be extended by one or several steps of completion. In any case, the pivotal issue is testing $\mathcal{R}^*(\llbracket A \rrbracket) \cap \llbracket B \rrbracket = \varnothing$. While this question is undecidable, a well-chosen over-approximation of $\mathcal{R}^*(\llbracket A \rrbracket)$ can suffice to conclude. In this section, we give a logical characterisation of whether such an approximation can be obtained purely by merging states of A—and, if so, how.

Definition 2. Given a TA A and an equivalence relation $\approx$ on Q, which we call simply an **approximation**, the **quotient automaton** $A/_{\approx}$ is defined as usual:

$$A/_{\approx} \;=\; \left\langle \begin{array}{ll} \Sigma := \Sigma \quad Q := Q/_{\approx} \quad F := \big\{\, [q]_{\approx} \mid q \in F \,\big\} \\ \Delta := \Big\{\, f(\ldots [q_i]_{\approx} \ldots) \to [q]_{\approx} \;\Big|\; f(\ldots q_i \ldots) \to q \in \Delta \,\Big\} \end{array} \right\rangle \tag{1}$$

Note that we have obviously $\llbracket A \rrbracket \subseteq \llbracket A/_{\approx} \rrbracket$, for any A and $\approx$.

Definition 3. Given TA A, B and a TRS $\mathcal{R}$, an approximation $\approx$ is **acritical** if $A/_{\approx}$ has no critical pair (which implies that it is also **complete**: $\mathcal{R}^*(\llbracket A \rrbracket) \subseteq \llbracket A/_{\approx} \rrbracket$) and **suitable** if $\llbracket A/_{\approx} \rrbracket \cap \llbracket B \rrbracket = \varnothing$. It is **conclusive** if it is both acritical and suitable.

Problem 4 (States Merging). *Given TA A, B and a left-linear[2] TRS $\mathcal{R}$, find (if it exists) a conclusive approximation $\approx$, or prove its non-existence.*

In all that follows, A, B, and $\mathcal{R}$ are defined as in Problem 4.

We reduce that problem to SAT by building a formula of propositional logic whose valuation describes $\approx$. The **relevant variables** of that formula are $\mathbb{V} = \{\, p \approx q \mid p, q, \in Q \,\}$. We call such a formula φ an **approximation formula**.

The semantics of approximation formulæ are defined as usual:

$$\llbracket \varphi \rrbracket \;=\; \{\, V \mid V \models \varphi \,\} \,. \tag{2}$$

---

[2] This the usual restriction for TA completion. See [14, Sec. 4.4.1] for a discussion.

A valuation V defines a relation $\operatorname{rel} V$ (V is an **underlying valuation** of $\operatorname{rel} V$):

$$\operatorname{rel} V = \{ (p,q) \mid V(p \approx q) = \top \} . \tag{3}$$

Thus, a formula φ defines a set of relations:

$$\{\!|\varphi|\!\} = \{ \operatorname{rel} V \mid V \in [\![\varphi]\!] \} . \tag{4}$$

Now we need to ensure that those relations are **(1)** approximations that are **(2)** suitable and **(3)** acritical.

To ensure that those atoms of the form "$p \approx q$" do indeed describe an equivalence relation, and therefore an approximation, we start by enforcing reflexivity, symmetry, and transitivity.

Definition 5 (φ^{e}). We denote by φ^{e}_A, or simply φ^{e}, the **equivalence formula** defined as:

$$\varphi^{\mathrm{e}}_A = \bigwedge_{p \in Q} p \approx p \ \wedge \bigwedge_{\substack{p,q \in Q \\ p<q}} p \approx q \Leftrightarrow q \approx p \ \wedge \bigwedge_{\substack{p,q,r \in Q \\ p \neq q \neq r \\ p<r}} (p \approx q \wedge q \approx r) \Rightarrow p \approx r .$$
$$\tag{5}$$

The conditions $p < q$ and $p \neq q \neq r \wedge p < r$ in φ^{e} are trivial simplifications that can be made without loss of generality given any arbitrary order on Q to generate smaller formulæ. In the implementation, atoms $p \approx p$ are never actually generated in any formula, but directly replaced by $\top$.

Proposition 6 (Equivalence). $\{\!|\varphi^{\mathrm{e}}|\!\}$ *is the set of all approximations.*

To enforce suitability, we translate, rather straightforwardly, the non-existence of any accepting run for $A \times B$.

Definition 7 (φ^{s}). Let $P = A \times B$. We denote by $\varphi^{\mathrm{s}}_{\mathcal{R},A,B}$, or simply φ^{s}, the **suitability formula** on variables $P.Q \cup \mathbb{V}$, defined as:

$$\varphi^{\mathrm{s}}_{\mathcal{R},A,B} = \overbrace{\left[\neg \bigvee_{X \in P.F} X \right]}^{\varphi^{\mathrm{s}}_l} \wedge \overbrace{\bigwedge_{(p,q) \in P.Q} \left[(p,q) \iff \bigvee_{\substack{f(\ldots,(p_i,q_i),\ldots) \to (p',q) \\ \in P.\Delta}} p \approx p' \ \wedge \ \bigwedge_i (p_i,q_i) \right]}^{\varphi^{\mathrm{s}}_r} \tag{6}$$

The (p,q) variables are not *relevant* in the sense defined above, and we shall ignore their valuation when extracting the approximation, but they serve as "scaffolding" for the formula. Without them, this would be longer and much more difficult. They are used only in φ^{s}.

Proposition 8 (Suitability). $\{\!|\varphi^{\mathrm{e}} \wedge \varphi^{\mathrm{s}}|\!\}$ *is the set of all suitable approximations.*

Proof. ⊆ **suitable.**

Let $\approx \in \{\!|\varphi^{\mathrm{e}} \wedge \varphi^{\mathrm{s}}|\!\} \subseteq \{\!|\varphi^{\mathrm{e}}|\!\}$. By Proposition 6, it is an approximation. We show it is suitable.

Let us start by showing that, if the state $([p]_{\approx}, q)$ is accessible in $^A\!/_{\approx} \times B$ then the variable (p, q) is true in any underlying valuation; this is done by structural induction on trees of $[\![([p]_{\approx}, q)]\!]$.

◇ *Base case: $t = a$:*

If $a \in [\![([p]_{\approx}, q)]\!]$ then $a \to [p]_{\approx} \in {}^A\!/_{\approx}.\Delta$ and $a \to q \in B.\Delta$, thus there exists p' such that $p' \approx p$ and $a \to p' \in A.\Delta$. Therefore $a \to (p', q) \in P.\Delta$, and that rule satisfies the disjunction on the rules of $P.\Delta$ of φ^{s}_r, which means that (p, q) is true.

◇ *Inductive case: $t = f(t_1, \ldots, t_n)$:*

If $t \in [\![([p]_{\approx}, q)]\!]$ then $\forall i \in [\![1, n]\!]$ we have p_i, q_i such that:
(1) $f([p_1]_{\approx}, \ldots, [p_n]_{\approx}) \to [p]_{\approx} \in {}^A\!/_{\approx}.\Delta$, **(2)** $f(q_1, \ldots, q_n) \to q \in B.\Delta$, **(3)** $t_i \in [\![([p_i]_{\approx}, q_i)]\!]$. By the first point there must exist $p', p'_1, \ldots, p'_n$ such that $p' \approx p$, all $p'_i \approx p_i$, and $f(p'_1, \ldots, p'_n) \to p' \in A.\Delta$. Combined with $f(q_1, \ldots, q_n) \to q \in B.\Delta$, this gives $f((p'_1, q_1), \ldots, (p'_n, q_n)) \to (p', q) \in P.\Delta$. By $t_i \in [\![([p_i]_{\approx}, q_i)]\!]$ and $p'_i \approx p_i$ we also have that $t_i \in [\![([p'_i]_{\approx}, q_i)]\!]$, which by induction means that we can assume (p'_i, q_i) to be true. The rule $f((p'_1, q_1), \ldots, (p'_n, q_n)) \to (p', q)$ satisfies the disjunction on rules of $P.\Delta$ and the truth of (p'_i, q_i) satisfies the conjunction on i of φ^{s}_r, which means that (p, q) is true.

Now, by φ^{s}_l, no $(p, q) \in P.F$ is true, so no $([p]_{\approx}, q) \in ({}^A\!/_{\approx} \times B).F$ can be accessible, thus $[\![{}^A\!/_{\approx}]\!] \cap [\![B]\!] = \varnothing$, and $\approx$ is suitable.

⊇ **suitable.**

Let $\approx$ be a suitable approximation, let us show that $\approx \in \{\!|\varphi^{\mathrm{s}}|\!\}$. We'll complete the underlying valuation of $\approx$ so that (p, q) is true iff $[\![([p]_{\approx}, q)]\!] \neq \varnothing$, and show this satisfies φ^{s}.

First, φ^{s}_l. Since $\approx$ is suitable, we have $[\![{}^A\!/_{\approx}]\!] \cap [\![B]\!] = \varnothing$, thus there can be no $(p, q) \in (A \times B).F$ such that $[\![([p]_{\approx}, q)]\!] \neq \varnothing$. Hence, all states of $P.F$ are set to false, and φ^{s} is satisfied.

Second, φ^{s}_r. By our definition (p, q) is true iff there exists a term $t = f(t_1, \ldots, t_n) \in [\![([p]_{\approx}, q)]\!]$. This is to say that t is evaluated by rules of the form $f(\ldots, [p_i]_{\approx}, \ldots) \to [p]_{\approx} \in {}^A\!/_{\approx}.\Delta$ and $f(\ldots, q_i, \ldots) \to q \in B.\Delta$. Equivalently, **(1)** there is a rule $f(\ldots, (p_i, q_i), \ldots) \to (p', q) \in P.\Delta$, **(2)** for all i, $t_i \in [\![([p_i]_{\approx}, q_i)]\!]$, which means all (p_i, q_i) are set to true, and **(3)** $p' \approx p$. Thus (p, q) is true iff the matching disjunctive clause is true; φ^{s}_r is satisfied. □

There remains to enforce acriticality. We are going to translate "under the approximation, there are no critical pairs".

To do so, we need to compute under which approximations a term can be evaluated (i.e. rewritten) into a given state; this will enable us to enforce, for all rules $l \to r \in \mathcal{R}$, that if a term t matching l can be evaluated in q under some approximation, then there exists a compatible approximation under which the rewritten term can also be evaluated in q, thus enforcing acriticality.

For this, we shall implement a notion of *unifiers*, U_q^t, giving us suitable formulæ and substitutions under which t evaluates in q.

Definition 9 (Unifier). Let $t \in \mathcal{T}(\Sigma, Q \cup \mathbb{X})$ be a term linear in $\mathbb{X}$, and let $q \in Q$. The set of **unifiers** of t and q, written U_q^t, is defined as:

$$
U_q^t =
\begin{cases}
\{(\top, \{t \mapsto q\})\} & \text{if } t \in \mathbb{X} \\
\{(t \approx q, \varnothing)\} & \text{if } t \in Q \\
\displaystyle\bigcup_{f(q_1,\dots,q_n) \to p \in \Delta} \{(p \approx q, \varnothing)\} \otimes \bigotimes_{k=1}^{n} U_{q_k}^{t_k} & \text{if } t = f(t_1,\dots,t_n)
\end{cases}
, \quad (7)
$$

where $(\varphi, \sigma) \otimes (\psi, \rho) = (\varphi \wedge \psi,\ \sigma \cup \rho)$, neutral element $(\top, \varnothing)$, and

$$
\{\dots, (\varphi_i, \sigma_i), \dots\} \otimes \{\dots, (\psi_j, \rho_j), \dots\} = \{\dots, (\varphi_i, \sigma_i) \otimes (\psi_j, \rho_j), \dots\} . \quad (8)
$$

Note that left-linearity of $\mathcal{R}$ justifies the functionality of $\sigma \cup \rho$.

It may seem a bit unprincipled to have U_q^t operate "modulo $\approx$" in all cases but $t \in \mathbb{X}$. A more natural definition would certainly be $\{(p \approx q, \{t \mapsto p\}) \mid p \in Q\}$. We generally opted, for this paper, to favour simplicity and directness in the formulæ and proofs over performance, and to reserve discussion of optimisations to further works. This is the exception. The practical performance impact of that single choice would be colossal, reducing the scope of the method by several orders of magnitude in most cases. The cost of choosing the computationally cheaper $\{(\top, \{t \mapsto q\})\}$ is paid in the remaining proofs: in many instances we need to explicitly reason "modulo $\approx$", because the unifier won't do it for us. To do that, given $\sigma, \sigma' : \mathbb{X} \to \mathbb{Q}$, we write $\sigma \approx \sigma'$ if $\forall x,\ \sigma(x) \approx \sigma'(x)$.

We consider $A/_\approx$ for some given $\approx$. We extend the definition of $[\cdot]_\approx$ from states of Q to configurations $\mathcal{T}(\Sigma \cup Q)$ inductively: $[f(t_1, \dots, t_n)]_\approx = f([t_1]_\approx, \dots, [t_n]_\approx)$. Put another way, $[\cdot]_\approx$ lifts configurations of A to corresponding configurations of $A/_\approx$. We write $[\cdot]_\approx$ simply as $[\cdot]$, since $\approx$ is the same throughout our proofs.

We now offer a characterisation of unifiers.

Lemma 10 (Unifier characterisation). *Let $A/_\approx$ be an ε-free quotient automaton, t a term of $\mathcal{T}(\Sigma \cup Q, \mathbb{X})$ linear in $\mathbb{X}$, and $\sigma : \mathbb{X}(t) \to Q$ a substitution. Then $[\sigma t] \to_{A/_\approx}^* [q]$ iff there exists $(\varphi, \sigma') \in U_q^t$ such that $\approx \in \{\!|\varphi|\!\}$ and $\sigma' \approx \sigma$.*

Proof. This can be proven by induction on t.

   Case $t = p \in Q$: We have $\sigma = \varnothing$ and $[p] \to_{A/_\approx}^* [q]$ iff $[p] = [q]$ iff $p \approx q$. Since $U_q^p = \{(t \approx q, \varnothing)\}$, the equivalence holds.

   Case $t = x \in \mathbb{X}$: We have $\sigma = \{x \mapsto p\}$ for some state p, and by definition $U_q^x = \{(\top, \sigma' := \{x \mapsto q\})\}$. Then $[\sigma t] = [p] \to_{A/_\approx}^* [q]$ iff $p \approx q$ iff $\sigma' \approx \sigma$.

   Case $t = f(t_1, \dots, t_n)$: There are σ_i such that $\sigma t = f(\dots, \sigma_i t_i, \dots)$ and $\sigma = \bigcup \sigma_i$—the functionality of σ is guaranteed by the linearity of t. We have $[\sigma t] = f(\dots, [\sigma_i t_i], \dots) \to_{A/_\approx}^* [q]$ iff there exists $q_1, \dots, q_n$ such that **(1)** for all

i, $[\sigma_i t_i] \to^*_{A/\approx} [q_i]$ and **(2)** $f(\ldots, [q_i]_\approx, \ldots) \to [q]_\approx \in A/\approx . \Delta$ The latter condition can, by definition of $A/\approx$, be replaced by **(2')** $\exists p : f(\ldots, q_i, \ldots) \to p \in \Delta \wedge p \approx q$.

By induction, $[\sigma_i t_i] \to^*_{A/\approx} [q_i]$ iff $\exists (\varphi_i, \sigma'_i) \in U^{t_i}_{q_i} : \approx \in \{\!|\varphi_i|\!\} \wedge \sigma'_i \approx \sigma_i$. Thus, by definition of $\otimes$, we can reformulate **(1)** as **(1')** $\exists (\varphi := \bigwedge \varphi_i,\ \sigma' := \bigcup \sigma'_i) \in \bigotimes^n_{i=1} U^{t_i}_{q_i} : \approx \in \{\!|\varphi|\!\} \wedge \sigma' \approx \sigma$.

In the end, $[\sigma t] \to^*_{A/\approx} [q]$ iff there exists a rule $f(\ldots, q_i, \ldots) \to p \in \Delta$, $\sigma' \approx \sigma$ and φ such that $(\varphi, \sigma') \in \bigotimes^n_{i=1} U^{t_i}_{q_i}$ and $\approx \in \{\!|p \approx q \wedge \varphi|\!\}$, that is to say iff there exists φ' such that $(\varphi', \sigma) \in U^t_q$. □

Definition 11 (φ^{a}). We denote by $\varphi^{\mathrm{a}}_{\mathcal{R},A}$, or simply φ^{a}, the **acriticality formula**:

$$\varphi^{\mathrm{a}}_{\mathcal{R},A} \;=\; \bigwedge_{\substack{l \to r \in \mathcal{R} \\ q \in Q}} \bigwedge_{(\alpha,\sigma) \in U^l_q} \left[\alpha \;\Rightarrow\; \bigvee_{(\beta,\varnothing) \in U^{r\sigma}_q} \beta \right] \tag{9}$$

Proposition 12 (Acriticality). $\{\!|\varphi^{\mathrm{e}} \wedge \varphi^{\mathrm{a}}|\!\}$ *is the set of all acritical approximations.*

Proof. Let $\approx \in \{\!|\varphi^{\mathrm{e}} \wedge \varphi^{\mathrm{a}}|\!\}$. By Proposition 6, it is an approximation, and Lemma 10 applies.

Since all configurations of $A/\approx$ have representatives in A, the acriticality of $A/\approx$ can be expressed as: for every rule $l \to r \in \mathcal{R}$, state $q \in Q$, and substitution σ, if $[\sigma l] \to^*_{A/\approx} [q]$ then $[\sigma r] \to^*_{A/\approx} [q]$.

By Lemma 10, this means that for every l, r, q, σ, if there exists $(\alpha, \sigma') \in U^l_q$ such that $\sigma \approx \sigma'$, and $\approx \in \{\!|\alpha|\!\}$, then there exists $(\beta, \varnothing) \in U^{\sigma r}_q$ and $\approx \in \{\!|\beta|\!\}$. Since $\sigma' \approx \sigma$, we have, by definition, $U^{\sigma r}_q = U^{\sigma' r}_q$. We can wlog. quantify directly on the substitutions actually occurring in the unifiers—here σ'.

By simple reformulation of the quantifiers, this is equivalent to: for every rule $l \to r \in \mathcal{R}$ and state $q \in Q$ (first conjunction of φ^{a}), for all $(\alpha, \sigma') \in U^l_q$ (second conjunction of φ^{a}), if $\approx$ satisfies α, then there exists β such that $(\beta, \varnothing) \in U^{\sigma' r}_q$, and $\approx$ satisfies β (inner disjunction of φ^{a}).

Hence, $A/\approx$ is acritical iff $\approx \in \{\!|\varphi^{\mathrm{a}}|\!\}$. □

Definition 13. We denote by $\varphi_{\mathcal{R},A,B}$, or simply φ, the **conclusiveness formula**:

$$\varphi_{\mathcal{R},A,B} \;=\; \varphi^{\mathrm{e}} \wedge \varphi^{\mathrm{s}} \wedge \varphi^{\mathrm{a}} \tag{10}$$

Theorem 14 (Conclusiveness). $\{\!|\varphi_{\mathcal{R},A,B}|\!\}$ *is the set of all conclusive approximations.*

4   The Completion Algorithm

Given a TA, we can now automatically find a suitable approximation of it by feeding φ to a SAT solver. There remains to integrate this to the completion algorithm.

Section 4.1 recalls the notions and fixes our notations for classical completion. Section 4.2 discusses how to integrate our search for suitable approximations in a completion process.

4.1   Reminders About Classical Completion

Given two TA A, B and a TRS $\mathcal{R}$, the tree automata completion algorithm introduced in [10] attempts to compute a fixpoint automaton $A^\star$ such that $\mathcal{R}^*([\![A^\star]\!]) \subseteq [\![A^\star]\!]$ and $[\![A^\star]\!] \cap [\![B]\!] = \varnothing$. The search for $A^\star$ commences from A and extends it with new states and rules to recognise more and more rewritten terms via successive completion steps—cf. Eq. $(\mathcal{C}_{\mathcal{R},A}^{\mathrm{rit}})_{[\mathrm{p}4]}$ and below. However, σr may be a term of arbitrary depth, and so the new rules $\sigma r \to q$ must be normalised.

Definition 15 (Rule Normalisation). Let $t \to q$ be a transition and $Q \subseteq \mathbb{Q}$ a set of states. The normalisation of $t \to q$ avoiding Q is written $\overline{t \to q}^Q$ and defined as:

$$\overline{f(t_1,\dots,t_n) \to q}^Q = \{ f(t_1^!,\dots,t_n^!) \to q \} \cup \bigcup_{i=1}^n \overline{t_i \to t_i^!}^Q \tag{11}$$

$$\overline{p \to q}^Q = \{ p \to q \}, \quad \forall p \neq q \in Q \tag{12}$$

$$\overline{q \to q}^Q = \varnothing, \quad \forall q \in Q, \tag{13}$$

where, for all i, $t_i^! = t_i$ if $t_i \in Q$, and otherwise $t_i^!$ is some fresh state from $\mathbb{Q} \setminus Q$. The operation is extended to sets of rules in the obvious way.

Definition 16 (Completion operation). We denote by $\mathcal{R}^\|(A)$[3] the automaton obtained from A by **one step of completion** wrt. $\mathcal{R}$. It is defined as:

$$\mathcal{R}^\|(A) = A\left\langle \Delta := \Delta \cup \overline{\left\{ \sigma r \to q \mid (\sigma, l, r, q) \in \mathcal{C}_{\mathcal{R},A}^{\mathrm{rit}} \right\}}^Q \right\rangle. \tag{14}$$

Definition 17 (Classical Completion Algorithm). Given TA A and a TRS $\mathcal{R}$, we let

$$\Gamma_{\mathcal{R}}(A) = \begin{cases} A & \text{if } \mathcal{C}_{\mathcal{R},A}^{\mathrm{rit}} = \varnothing \\ \Gamma_{\mathcal{R}}(\mathcal{R}^\|(A)) & \text{otherwise.} \end{cases} \tag{15}$$

The completion algorithm in this pure form will generally not terminate.

Example 18. Let $\mathcal{R} = \{ f(x) \to f(a(x)) \}$. Let A be a TA such that $F = \{q_f\}$ and $\Delta = \{ \bot \to q_\bot, f(q_\bot) \to q_f \}$; we have $[\![A]\!] = \{ f(\bot) \}$. A first completion step is then performed as follow:

$$\mathcal{R}^\|(A) = A\left\langle \Delta := \Delta \cup \overline{f(a(q_\bot)) \to q_f}^Q \right\rangle$$

where $\overline{f(a(q_\bot)) \to q_f}^Q = \{ a(q_\bot) \to p_0, \ f(p_0) \to q_f \}$. Consequently, performing a second completion step would lead to:

$$\mathcal{R}^\|(\mathcal{R}^\|(A)) = \mathcal{R}^\|(A)\left\langle \Delta := \Delta \cup \overline{f(a(p_0)) \to q_f}^Q \right\rangle.$$

And so on ad infinitum.

---

[3] Intuitively, a completion step enables A to recognise all terms obtainable by applying any number of rules of $\mathcal{R}$ *in parallel*—and exactly those terms if $\mathcal{R}$ is right-linear—hence the notation.

This is where human intervention is required: during each completion step the expert must notice such problems and suggest, for instance, to merge p_0 and $q_\perp$ in $\mathcal{R}^\parallel(A)$ (either directly or through more abstract tools; e.g. in [9] the expert would give the equation $a(x) = x$) making $\mathcal{R}^\parallel(A)$ acritical and yielding the expected language $f(a^*(\perp))$—which in this example happens to be $\mathcal{R}^*(\llbracket A \rrbracket)$ exactly. Of course the expert would also take care to avoid B; for instance merging q_f with the others is unsuitable if $\llbracket B \rrbracket = \{\perp\}$.

Our goal is of course to automate this manual step, replacing the human expert by a SAT solver.

4.2   Completion with Automated State Merging, and Variants

The modified algorithm is basically the same as before, but the termination criterion is not whether the automaton is acritical, but whether it can be conclusively approximated.

Definition 19 (Automated Approximate Completion Semi-Algorithm). Given TA A, B and a TRS $\mathcal{R}$, we let

$$\Gamma_{\mathcal{R},B}^{\mathrm{aut}}(A) = \begin{cases} A/_{\approx} & \text{if } \approx \in \{\varphi_{\mathcal{R},A,B}\} \\ \Gamma_{\mathcal{R},B}^{\mathrm{aut}}\left(\mathcal{R}^\parallel(A)\right) & \text{otherwise.} \end{cases} \tag{16}$$

In cases where several conclusive approximations $\approx$ are available, we select one arbitrarily. Whether $\exists \approx \in \{\varphi_{\mathcal{R},A,B}\}$ is of course tested using a SAT solver. Some obvious optimisations can be made—for instance, the "live-states" simplification: starting from the first completion, there is no use trying to merge the states of the original A directly, which shortens the formula a bit.

Let's apply this to Example 18, for $\llbracket B \rrbracket = \{\perp\}$, with live-states simplification in the unifiers. We have, for φ^{a}, once all $\top$ and $\perp$ are simplified away:

$$(p_0 \approx q_f \Rightarrow ((p_0 \approx q_f \wedge q_\perp \approx p_0) \vee p_0 \approx q_f))$$
$$\wedge\ (p_0 \approx q_f \Rightarrow ((p_0 \approx q_f \wedge q_\perp \approx p_0 \wedge q_\perp \approx p_0) \vee (p_0 \approx q_f \wedge q_\perp \approx p_0)))$$
$$\wedge\ ((q_\perp \approx p_0 \wedge q_\perp \approx p_0) \vee q_\perp \approx p_0)$$

We also have $\varphi^{\mathrm{s}} = \neg\langle q_f, p_0 \rangle \wedge \langle q_\perp, p_0 \rangle \wedge (\langle q_f, p_0 \rangle \iff q_f \approx q_\perp)$. Note that the seemingly unconditional conjuncts come from $\top\perp$-simplifications, for instance $\langle q_\perp, p_0 \rangle$ was originally $\langle q_\perp, p_0 \rangle \Leftrightarrow (\top \wedge \top)$.

This gives us two possible solutions: $q_\perp \approx p_0$ and $q_\perp \approx p_0 \approx q_f$. Of these, only the first is suitable.

The purpose of performing completion steps is the same as usual: the starting automaton may lack the structure to capture $\mathcal{R}$, regardless of approximation. In Example 18, A could not at all handle the symbol a before completion. While we think this is the most *practical* way of integrating State Merging into completion, there are other ways to do so.

Whereas the classical approach involves expert intervention at each completion step, performing mergings that carry on to the next step, (with possibility of backtracking), this algorithm performs just one merge at the end. This is a good heuristic, but is less general than the fully interactive approach in that, in some cases, even if a conclusive approximation exists, we are not guaranteed to find it. Consider

$$\mathcal{R} = \left\{ \begin{array}{l} f(x,y) \to f\left(s(x), s(y)\right),\ f\left(s(x), s(y)\right) \to f(x,y), \\ f\left(0, s(x)\right) \to a,\ f\left(s(x), 0\right) \to a, \end{array} \right\}. \qquad (17)$$

Let A be a TA such that $[\![A]\!] = \{ f(0,0) \}$. Reachability analysis searches for, and can theoretically reach [11], given the right expert choices for state mergings after each completion step, any $\mathcal{R}$-closed regular over-approximation L of $[\![A]\!]$. We know by [3] that in all cases, we must have $a \in L$. A valid example would be

$$L = \left\{ f\left(s^n(0), s^m(0)\right) \mid n, m \geqslant 0 \right\} \cup \{ a \}. \qquad (18)$$

Yet, this language cannot be found by performing just one final state merging in a fully automatic application of Definition 19, no matter how many completion steps are done beforehand.

Indeed, since $\mathcal{R}$ is linear and we do not manually merge states at each step of Definition 19, the completion by itself cannot introduce any approximation, but merely adds rules to recognise, exactly, new terms of the form $f\left(s^n(0), s^n(0)\right)$ ad infinitum, exactly as in Example 18[p9].

Thus, at no point of the process can a state merging generate L, because rules like $f\left(0, s(x)\right)$ can never be evaluated by the automaton, and thus the automaton cannot contain any transition of the form $a \to q$. Those can only be introduced by solving a critical pair *after* a merging has made unbalanced terms recognisable, which cannot occur in Definition 19.

Certainly, we could develop a semi-algorithm that alternates between rewriting steps and state merging, testing all possibilities or backtracking when needed, but this is impractical—equivalent to enumerating all TA.

An intermediate approach would be that of [2, Algo. 6.1]: not performing any actual completion, instead adding at each step, for each $l \to r \in \mathcal{R}$, normalised rules $\{ \ldots, x_i \mapsto q_i, \ldots \} r \to q$ for fresh q_i, q. This adds the necessary structure to the automaton, leaving the task of actually making the fresh states reachable to the State Merging problem. However, this can introduce a lot of redundant structure, making the approach costly compared to the heuristic of Definition 19.

5   Complexity, Implementation and Tests

How scalable can this approach be in practice? Both φ^e and φ^s are clearly polynomial (cubic and quadratic, respectively), so the critical question is the behaviour of φ^a.

Proposition 20 (Worst-Case). $\left|\varphi^a_{\mathcal{R},A,B}\right| = O\left(|Q| \times |\mathcal{R}| \times N \times |\Delta|^N\right)$, where $N = \max_{r \in \mathcal{R}} |r|$.

Proof. The size of a rule is defined as $|l \to r| = |l| + |r|$, and the size of an element of unifier is defined a $|(\alpha, \sigma)| = |\alpha| + |\sigma|$.

The size of each formula of U_q^t is exactly $|t|$, and $\left|U_q^t\right| \leqslant |\Delta|^{|t|}$.

It is easy to prove by induction: for $t \in \mathbb{X}$ and $t \in q$, U_q^t has one element of size 1. For $t = f(t_1, \ldots, t_n)$, we say that for all rules $f(q_1, \ldots, q_n) \to p \in \Delta$, U_q^t contains the combinations by $\otimes$ of $(p \approx q, \varnothing)$ and of one element of each $U_{q_i}^{t_i}$. The size of each of these combinations is the sum of the size of its parts, that is to say $1 + |t_1| + \cdots + |t_n| = |t|$. The number of these combinations is the product of the sizes $\left|U_{q_i}^{t_i}\right| \leqslant |\Delta|^{|t_i|}$. In total, each rule $f(q_1, \ldots, q_n) \to p \in \Delta$ can add no more than $|\Delta|^{|t_1| + \cdots + |t_n|}$ elements. Thus, the total number of elements in U_q^t is no greater than $|\Delta| \times |\Delta|^{|t_1| + \cdots + |t_n|} = |\Delta|^{|t|}$.

From these bounds on U_q^t, we give an upper bound for the size of φ^{a} (cf. Definition 11[p8]):

$\diamond$ The disjunction on β is of size $O\left(|r| \times |\Delta|^{|r|}\right)$

$\diamond$ The conjunction on α, σ is of size $O\left(|l \to r| \times |\Delta|^{|l \to r|}\right)$

$\diamond$ The size of the conjunction on q and $l \to r$ is the sum of the above for each rule of $\mathcal{R}$ and state of Q. If let N be the size of the largest rule, we have $O(|Q| \times |\mathcal{R}| \times N \times |\Delta|^N)$. $\qquad\square$

To evaluate how SAT solvers deal with those inputs, we conducted various experiments on a proof-of-concept (PoC) implementation of a State-Merging solver[4]. Our PoC is implemented in Python 3.11 and uses PySAT [17] to interface with the CNF-SAT–solver backend, Minisat 2.2 [7]. All experiments were run on the same machine: Arch Linux, 12-core AMD Ryzen 9 5900X, 32G RAM, Python 3.11.8. All tests were single-threaded.

Qualitatively, the PoC does, instantly, produce the right answers on our usual small examples, such as $f(x) \to f(a(a(x)))$, $f(x, y) \to f(a(x), b(y))$, etc. Quantitatively, State Merging solving scales quite differently depending on which characteristic of the input is considered, some having a linear impact and some an exponential one, as predicted by Proposition 20. To observe that, we experiment on simple values of $A, B, \mathcal{R}$ and graph the solution times and sizes in Fig. 1 (mind that some scales are log and some linear).

Our test automaton is in all cases A_n, recognising lists $f(a_1, f(a_2, \ldots f(a_k, \perp) \ldots)$, $a_i \in [\![1, n]\!]$, using a separate non-deterministic state for each possible value of leaves a_i. A human being would undoubtedly write that TA in 2 states for any n, but the curves are scarcely affected by the structure of A, so this is as good an example as any. Figure 1a shows a polynomial growth across all metrics (time and size of φ), as $|A_n|$ increases. For large formulæ, about $1/15$th of the total time is spent in the SAT-solver, by Fig. 1b (including time spent reading the Python `list-of-list-of-int` structure containing the CNF clauses). The rest is spent computing φ and converting it to Conjunctive Normal Form (CNF; we use the linear Tseytin transformation). The ratios of Fig. 1b tell the same story regardless of $A, B, \mathcal{R}$. Our formulæ do not seem to be pathological cases for the solver; we are mostly limited by how fast we can generate φ.

---

[4] https://github.com/vincent-hugot/CIAA-2024-SAT-Completion.

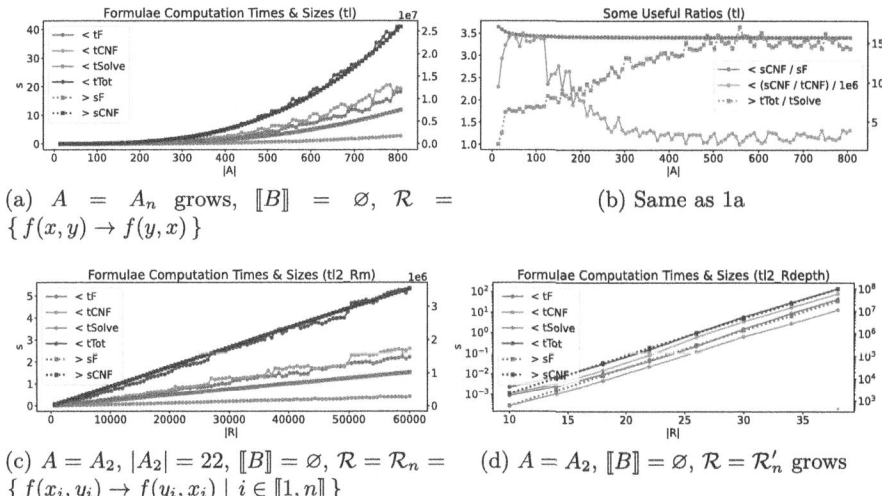

(a) $A = A_n$ grows, $\llbracket B \rrbracket = \varnothing$, $\mathcal{R} = \{ f(x,y) \to f(y,x) \}$

(b) Same as 1a

(c) $A = A_2$, $|A_2| = 22$, $\llbracket B \rrbracket = \varnothing$, $\mathcal{R} = \mathcal{R}_n = \{ f(x_i, y_i) \to f(y_i, x_i) \mid i \in \llbracket 1, n \rrbracket \}$

(d) $A = A_2$, $\llbracket B \rrbracket = \varnothing$, $\mathcal{R} = \mathcal{R}'_n$ grows

Fig. 1. Some experimental results. $<, >$ indicates whether the curve refers to the y-axis on the left or right. t = time, s = size, F = φ, CNF = φ in CNF, tSolve = time spent in the SAT solver backend, tTot = total time to solution.

No graph is provided regarding the influence of B because there is not much to discuss: $|\varphi^s|$ is linear in $|B|$, and the tests bear that out. The effect of $\mathcal{R}$ is more complicated. In terms of the number of rules, all metrics grow linearly, viz. Fig. 1c. The size of the rules, however, has an exponential impact. Consider Fig. 1d, where $\mathcal{R}'_n = \{f(0, f(0, \dots f(x, y) \dots)) \to f(\dots f(f(y, x), 0) \dots, 0)\}$ operates on lists of length n – the results are the same with variables $x_1, \dots, x_n$ instead of constants.

The practicality of the approach is therefore highly dependent upon the form of the TRS. In all cases, and even accounting for hardware differences, this is a marked improvement compared to [2], whose own experiments could not exceed $|A| = 20$.

Engineering Considerations: A well-engineered implementation could vastly outperform our PoC. Most of the time is spent doing formula manipulation via structural pattern-matching; on a synthetic test on large formulæ, OCaml outperforms Python on such tasks by a factor of 16.4. Furthermore, φ is naturally a large conjunction of independent clauses which could trivially be generated in parallel, fed to the solver incrementally, and discarded. On our 12-core CPU, expecting a 10x time speedup is not unreasonable, and this would also immensely reduce the maximum RAM necessary, decoupling it from the current $O(|\varphi|)$. Those two improvements combined would probably yield a performance increase in the vicinity of 150x on our machine, even if the SAT solver itself is still shared and single-threaded, since it is *very* far from being the bottleneck. Since SAT-solving can itself be parallelised to some extent using divide-and-conquer methods, e.g. [16], massively parallel implementations are theoretically possible.

6   Conclusion and Future Works

In order to automate the steps of reachability analysis which require expert intervention, we have introduced a SAT encoding of the *state merging problem (SMP)*: **In:** TA A, B, and TRS $\mathcal{R}$; **Out:** $\approx$ such that $A/_{\approx}$ is closed wrt. $\mathcal{R}$ and does not intersect B. We have presented various ways to integrate SAT-backed SMP solving into completion algorithms to perform reachability analysis. Our experiments and complexity analysis show that, for relatively flat TRS, the method can be applied to fairly large instances and is potentially quite scalable wrt. available computing power. Its main limitation is the exponential impact of large, deep rewrite rules.

Future Works: our main focus is of course to overcome that limitation. An interesting idea involves changing our proxy criterion for "completeness wrt. $\mathcal{R}$" from acriticality to something more permissive, involving partial orderings on states instead of partitioning, such that applying a rewriting rule always leads to a "weaker" state, rather than an equivalent one. As this constraint is strictly easier to satisfy, this would allow not only for our algorithm to find solutions for more instances, but also for solutions to potentially be found with fewer completion steps. The impact of this change is difficult to gauge a priori, but as it loads more work on the highly optimized SAT-solvers and potentially lessens the number of iterations before a solution is reached, we can hope for a performance increase.

From a different angle, any applicable TRS-flattening technique would enable us to blunt the impact of the largest rules.

References

1. Boichut, Y., Courbis, R., Héam, P.-C., Kouchnarenko, O.: Finer is better: abstraction refinement for rewriting approximations. In: Voronkov, A. (ed.) RTA 2008. LNCS, vol. 5117, pp. 48–62. Springer, Heidelberg (2008). https://doi.org/10.1007/978-3-540-70590-1_4
2. Boichut, Y., Dao, T.-B.-H., Murat, V.: Characterizing conclusive approximations by logical formulae. In: Delzanno, G., Potapov, I. (eds.) RP 2011. LNCS, vol. 6945, pp. 72–84. Springer, Heidelberg (2011). https://doi.org/10.1007/978-3-642-24288-5_8
3. Boichut, Y., Héam, P.-C.: A theoretical limit for safety verification techniques with regular fix-point computations. IPL **108**(1), 1–2 (2008)
4. Boichut, Y., Héam, P.-C., Kouchnarenko, O.: Approximation-based tree regular model-checking. Nord. J. Comput. **14**(3), 216–241 (2008)
5. Bouajjani, A., Habermehl, P., Rogalewicz, A., Vojnar, T.: Abstract regular (tree) model checking. STTT **14**(2), 167–191 (2012)
6. Comon, H., Dauchet, M., Gilleron, R., Lugiez, D., Tison, S., Tommasi, M.: Tree Automata Techniques and Applications (TATA) (2007)
7. Eén, N., Sörensson, N.: An extensible SAT-solver. In: Giunchiglia, E., Tacchella, A. (eds.) SAT 2003. LNCS, vol. 2919, pp. 502–518. Springer, Heidelberg (2004). https://doi.org/10.1007/978-3-540-24605-3_37

8. Feuillade, G., Genet, T., Tong, V.V.T.: Reachability analysis over term rewriting systems. J. Autom. Reason. **33**(3–4), 341–383 (2004)
9. Genet, T., Rusu, V.: Equational Tree Automata Completion. J. Symb. Comput. **45**, 2010 (2010)
10. Genet, T.: Decidable approximations of sets of descendants and sets of normal forms. In: Nipkow, T. (ed.) RTA 1998. LNCS, vol. 1379, pp. 151–165. Springer, Heidelberg (1998). https://doi.org/10.1007/BFb0052368
11. Genet, Th.: Completeness of tree automata completion. In: FSCD 2018. LIPIcs, vol. 108, pp. 16:1–16:20 (2018)
12. Genet, T., Haudebourg, T., Jensen, T.: Verifying higher-order functions with tree automata. In: Baier, C., Dal Lago, U. (eds.) FoSSaCS 2018. LNCS, vol. 10803, pp. 565–582. Springer, Cham (2018). https://doi.org/10.1007/978-3-319-89366-2_31
13. Genet, T., Klay, F.: Rewriting for cryptographic protocol verification. In: McAllester, D. (ed.) CADE 2000. LNCS (LNAI), vol. 1831, pp. 271–290. Springer, Heidelberg (2000). https://doi.org/10.1007/10721959_21
14. Genet, T.: Reachability analysis of rewriting for software verification. Habilitation thesis (habilitation à diriger des recherches), University of Rennes I (2009)
15. Genet, T., Tong, V.V.T.: Reachability analysis of term rewriting systems with timbuk. In: Nieuwenhuis, R., Voronkov, A. (eds.) LPAR 2001. LNCS (LNAI), vol. 2250, pp. 695–706. Springer, Heidelberg (2001). https://doi.org/10.1007/3-540-45653-8_48
16. Heule, M.J.H., Kullmann, O., Wieringa, S., Biere, A.: Cube and conquer: guiding CDCL SAT solvers by lookaheads. In: Eder, K., Lourenço, J., Shehory, O. (eds.) HVC 2011. LNCS, vol. 7261, pp. 50–65. Springer, Heidelberg (2012). https://doi.org/10.1007/978-3-642-34188-5_8
17. Ignatiev, A., Morgado, A., Marques-Silva, J.: PySAT: a Python toolkit for prototyping with SAT oracles. In: Beyersdorff, O., Wintersteiger, C.M. (eds.) SAT 2018. LNCS, vol. 10929, pp. 428–437. Springer, Cham (2018). https://doi.org/10.1007/978-3-319-94144-8_26
18. Resten, Y., Maler, O., Marcus, M., Pnueli, A., Shahar, E.: Symbolic model checking with rich assertional languages. In: Grumberg, O. (ed.) CAV 1997. LNCS, vol. 1254, pp. 424–435. Springer, Heidelberg (1997). https://doi.org/10.1007/3-540-63166-6_41
19. Klarlund, N., Møller, A.: MONA Version 1.4 User Manual, January 2001. Notes Series NS-01-1. http://www.brics.dk/mona/

Non-emptiness Test for Automata over Words Indexed by the Reals and Rationals

Bernard Boigelot[ID], Pascal Fontaine[ID], and Baptiste Vergain[(✉)][ID]

Montefiore Institute, B28, University of Liège, Liège, Belgium
{Bernard.Boigelot,Pascal.Fontaine,bvergain}@uliege.be

Abstract. Automata have been defined to recognize languages of words indexed by linear orderings, which generalize the usual notions of finite, infinite, and ordinal words. The reachability problem for these automata has already been solved for scattered linear orderings.

In this paper, we design an analogous procedure that solves reachability over the specific domains $\mathbb{R}$ and $\mathbb{Q}$. Given an automaton on linear orderings, this procedure decides in polynomial time whether this automaton accepts at least one word indexed by $\mathbb{R}$ or by $\mathbb{Q}$. We claim that this algorithm constitutes an essential step to designing effective decision procedures for the first-order monadic theory of order interpreted over $\mathbb{R}$ or $\mathbb{Q}$.

Keywords: Automata · Linear Orderings · Real Domain · Rational Domain · Emptiness Test · Reachability

1 Introduction

In [3], Bruyère and Carton introduce automata on words indexed by linear orderings, which generalize the concept of word, and notably encapsulate the usual notions of finite, infinite, and ordinal words. Although initially restricted to scattered linear orderings (i.e., orderings that do not contain a dense subordering), these automata have later been extended to deal with all linear orderings [2]. Notably, a Kleene-like theorem asserting the equivalence between languages accepted by automata on linear orderings and languages described by an extended form of rational expressions is proved in [2,3]. The strong connections between rational languages and the monadic second-order theory of linear orderings MSO($<$) is investigated in [1], namely, every language accepted by an automaton on linear orderings is definable in MSO($<$). The converse however only holds when the orderings are countable and scattered.

We are interested in designing effective decision procedures for monadic first-order theories of order MFO(D, $<$) interpreted over a fixed dense domain D, in

---

This work is partially supported by the FNRS-DFG PDR Weave (SMT-ART) grant 40019202.

S. Z. Fazekas (Ed.): CIAA 2024, LNCS 15015, pp. 94–108, 2024.
https://doi.org/10.1007/978-3-031-71112-1_7

particular $\mathbb{R}$ and $\mathbb{Q}$. The decidability of MFO($\mathbb{Q}$, $<$) derives directly from the decidability of MSO($\mathbb{Q}$, $<$) [15]. For $\mathbb{R}$ however, Shelah [19] showed that MSO($\mathbb{R}$, $<$) is undecidable. The decidability of MFO($\mathbb{R}$, $<$) has been established in [6]. Some theoretical bounds for the complexity of decision procedures for MFO($\mathbb{Q}$, $<$) and MFO($\mathbb{R}$, $<$) have been obtained in [16,17]. Our long-term goal consists in designing an automata-based decision procedure in the spirit of the one introduced by Büchi to deal with S1S [4], which roughly corresponds to MSO($\mathbb{N}$, $<$), later extended to deal with all words indexed by countable ordinals [5]. The first step would consist in building an automaton recognizing the models of a given formula without any restriction on the domain. Then, a dedicated procedure would decide the emptiness of the language over a specific linear ordering, equal to $\mathbb{R}$ or $\mathbb{Q}$ in our case. This work focuses on the latter part.

Testing for the non-emptiness of the language accepted by an automaton on linear orderings reduces to deciding whether an accepting state is reachable from an initial state. Reachability for automata on scattered linear orderings is discussed in [9]. A generalization over every linear ordering is sketched in [11]. In this paper, we investigate how to design an analogous procedure that decides reachability over the domains $\mathbb{R}$ and $\mathbb{Q}$.

The paper is structured as follows. We first recall in Sect. 2 the useful definitions relating to linear orderings, rational expressions, and automata on linear orderings. In Sect. 3, we discuss the state of the art regarding the reachability problem. In Sect. 4, we provide an algorithm for deciding emptiness over $\mathbb{R}$ in polynomial time. Finally, in Sect. 5, we show how to adapt the previous algorithm to deal with $\mathbb{Q}$.

2   Preliminaries

2.1   Linear Orderings

We first give basic definitions and results about linear orderings, and refer to [18] for further details.

A *linear ordering* J is a totally ordered set, i.e., a set equipped with a binary relation $<_J$ that is irreflexive, asymmetric, transitive, and total. Two linear orderings J and K—respectively associated with the order relations $<_J$ and $<_K$—are *order-isomorphic* if there exists an order-preserving bijection between J and K. Formally, let b be such a bijection (from J to K), then $b(j_1) <_K b(j_2)$ iff $j_1 <_J j_2$ for all $j_1, j_2 \in J$. We denote by $-J$ the *backwards* linear ordering that corresponds to J with its ordering reversed. The class of all linear orderings is denoted by $\mathcal{L}$.

The *order type* of a linear ordering J is the class of all linear orderings order-isomorphic to J. The order types of a singleton, the set composed of the N first natural numbers, $\mathbb{N}$, $\mathbb{Z}$, $\mathbb{Q}$, and $\mathbb{R}$ are respectively denoted by 1, N, ω, ζ, η, and λ. Notice that the order type of any non-empty open interval of $\mathbb{R}$ is λ, and that the order type of any non-empty open interval of $\mathbb{Q}$ is η.

The concatenation of two linear orderings J and K (with respective order relations $<_J$ and $<_K$) is denoted by $J + K$. It corresponds to the linear ordering

Fig. 1. The linear ordering $\mathbb{N} + \mathbb{Z}$.

composed of the set of pairs $\{(j,1) \mid j \in J\} \cup \{(k,2) \mid k \in K\}$, equipped with the order relation $<$ defined by $(j_1, 1) < (j_2, 1)$ if $j_1 <_J j_2$, $(k_1, 2) < (k_2, 2)$ if $k_1 <_K k_2$, and $(j, 1) < (k, 2)$ for every $j \in J$ and $k \in K$.

More generally, let K and J_k for all $k \in K$ be linear orderings. The linear ordering $\Sigma_{k \in K} J_k$ is obtained by concatenating the orderings J_k w.r.t. the ordering K. Formally, the *sum* $\Sigma_{k \in K} J_k$ is the linear ordering defined over the set of pairs (j, k) such that $j \in J_k$ and $k \in K$, with the order relation $(j_1, k_1) < (j_2, k_2)$ iff either $k_1 <_K k_2$, or $k_1 = k_2$ and $j_1 <_{J_{k_1}} j_2$. When $J_k = J$ for every $k \in K$, we may write J^K in place of $\Sigma_{k \in K} J_k$. These operators naturally extend to order types. For instance, the order type ω^ω is the class of all linear orderings order-isomorphic to the linear ordering $\mathbb{N}^\mathbb{N}$.

A *cut* of a linear ordering J is a partition of J into two sets (K, L) such that for every pair $(k, \ell) \in K \times L$, one has $k < \ell$. The set of all cuts of an ordering J is denoted by $\widehat{J}$. Notice that $\widehat{J}$ is a linear ordering as well, with $(K_1, L_1) < (K_2, L_2)$ iff $K_1 \subsetneq K_2$, and that $\widehat{J}$ always has a first element $(\emptyset, J)$ and a last element $(J, \emptyset)$. For every $j \in J$, the *consecutive* cuts $(\{k \mid k < j\}, \{\ell \mid j \leq \ell\})$ and $(\{k \mid k \leq j\}, \{\ell \mid j < \ell\})$ are (respectively) denoted by j^- and j^+. Let $c = (K, L)$ be a cut that is neither the first nor the last one of J, i.e., $K \neq \emptyset$ and $L \neq \emptyset$. If K admits a greatest element and L admits a smallest element, then c is called a *jump*. If K does not admit a greatest element and L does not admit a smallest element, then c is called a *gap*.

A linear ordering J is *(Dedekind) complete* if the set $\widehat{J}$ does not contain any gap. For instance, $\mathbb{N}$ and $\mathbb{R}$ are complete, while $\mathbb{Q}$ is incomplete. A linear ordering J is *dense* if between any two distinct elements lies another one, i.e., for every $j_1, j_2 \in J$ such that $j_1 <_J j_2$, there exists j_3 such that $j_1 <_J j_3 <_J j_2$. A linear ordering is *scattered* if it does not contain any dense sub-ordering. Given a dense linear ordering J, a sub-ordering $J' \subseteq J$ is said to be *dense in* J if between any two distinct elements of J lies an element of J'. For instance, $\mathbb{Q}$ and $\mathbb{R} \setminus \mathbb{Q}$ are both dense in $\mathbb{R}$.

Example 1. Consider the linear ordering $\mathbb{N} + \mathbb{Z}$ corresponding to the set of natural numbers followed by the set of integers. A depiction of this set and its cuts is given in Fig. 1. The elements of this ordering are represented as dots, while the cuts are represented as vertical lines. Notice that this ordering admits a single gap, hence it is incomplete. It is also scattered.

2.2   Words and Rational Expressions

We now define the notions of words and rational expressions on linear orderings, as introduced in [2,3].

A *word indexed by a linear ordering* is a totally ordered sequence of letters. Formally, given an alphabet Σ and a linear ordering J, a word $w = (\alpha_j)_{j \in J}$ indexed by J is a function $w : J \to \Sigma$. The *length* $|w|$ of w is J itself. The *empty word* ε denotes the word indexed by the empty set. Two words $w_1 = (\alpha_j)_{j \in J}$ and $w_2 = (\beta_k)_{k \in K}$ are equal if there exists an order-preserving bijection b from J to K such that for all $j \in J$, $\alpha_j = \beta_{b(j)}$. Informally, for a given word w, only the order type of its underlying linear ordering $|w|$ is relevant, not the linear ordering $|w|$ itself.

We now define a product operator on words. Let K and J_k for all $k \in K$ be linear orderings. Let $w_k = (\alpha_{j,k})_{j \in J_k}$ be a word of length J_k, for every $k \in K$. The *product* $\Pi_{k \in K} w_k$ is the word w of length $|w| = \Sigma_{k \in K} J_k$ equal to $(\alpha_{j,k})_{(j,k) \in |w|}$. The product of two words w_1 and w_2 is denoted by $w_1 \cdot w_2$.

A notion of rational expressions has been introduced in [2,3] to describe languages of words indexed by linear orderings. Recall that in this paper, we consider a fixed domain that is either $\mathbb{R}$ or $\mathbb{Q}$. For this reason, we only introduce the rational operators that are relevant to describe words indexed by $\mathbb{R}$ or $\mathbb{Q}$.

Let X and Y be two sets of words on linear orderings. We define the following operators:

- *Concatenation*: $X \cdot Y = \{x \cdot y \mid x \in X, y \in Y\}$,
- *Infinite repetition*: $X^\omega = \{\Pi_{i \in \mathbb{N}} x_i \mid \forall i \in \mathbb{N}, x_i \in X\}$,
- *Reverse infinite repetition*: $X^{-\omega} = \{\Pi_{i \in (-\mathbb{N})} x_i \mid \forall i \in -\mathbb{N}, x_i \in X\}$.

We also define a *shuffle* operator that generates density. Let $\Sigma^\diamond$ denote the set of all the words indexed by any linear ordering[1] over the alphabet Σ, i.e., $\Sigma^\diamond = \{w \mid |w| \in \mathcal{L}\}$.

Let $X_1, \ldots, X_n \subseteq \Sigma^\diamond$, for some $n \geq 1$. We denote by $\mathrm{sh}(X_1, \ldots, X_n)$ the set of words of $\Sigma^\diamond$ that can be written as $\Pi_{r \in R} x_r$ such that (1) R is a non-empty, dense, and complete linear ordering without first or last element, and (2) there exists a partition $\{R_1, \ldots, R_n\}$ of R such that every R_i is dense in R, and for every $r \in R$ and $i \in [1, n]$, if $r \in R_i$ then $x_r \in X_i$.

Some remarks are in order. Notice that the order type of the dense and complete ordering R mentioned in the shuffle definition can be different from λ. This will be further discussed in Sect. 4.1. Also notice that, as mentioned in [2], although the shuffle operator relies on a complete ordering R, if there exists $i \in [1, n]$ such that $\varepsilon \in X_i$ (but $\bigcup_{i \in [1,n]} X_i \neq \{\varepsilon\}$), then there exists $w \in \mathrm{sh}(X_1, \ldots, X_n)$ such that $|w|$ is incomplete. Intuitively, every element of R_i corresponding to ε does not belong to $|w|$, but matches a gap in $\widehat{|w|}$.

2.3   Automata

We now recall definitions initially introduced in [3].

---

[1] This definition slightly differs from what is commonly found in the literature: it implies that $\varepsilon \in \Sigma^\diamond$ holds, while the usual definition excludes the empty word from the set $\Sigma^\diamond$.

Definition 1. *An* automaton on linear orderings *is a tuple* $\mathcal{A} = (Q, \Sigma, I, F, \Delta^s, \Delta^\ell)$, *where* Q *is a finite set of states,* Σ *is a finite alphabet,* $I \subseteq Q$ *is a set of initial states,* $F \subseteq Q$ *is a set of accepting states,* $\Delta^s \subseteq Q \times \Sigma \times Q$ *is a set of* successor transitions, *and* $\Delta^\ell \subseteq (2^Q \times Q) \cup (Q \times 2^Q)$ *is a set of (*left *and* right*)* limit transitions.

The set Δ^s contains *successor transitions* of the form (q, α, q'), with $q, q' \in Q$ and $\alpha \in \Sigma$, alternatively written $q \xrightarrow{\alpha} q'$, which are similar to the transitions of finite and infinite-word automata. Notice however that automata on linear orderings do not admit ε-transitions. The set Δ^ℓ is composed of *left-limit transitions* of the form (P, q), with $P \subseteq Q$ and $q \in Q$, written $P \to q$, and *right-limit transitions* of the form (q, P), with $q \in Q$ and $P \subseteq Q$, written $q \to P$. We sometimes write sets of limit transitions in a more compact form, e.g., we write $p, q \to \{p, q\} \to p$ in place of the set $\{p \to \{p, q\}, q \to \{p, q\}, \{p, q\} \to p\}$.

We now define the notion of *path* in such an automaton. Let J be a linear ordering, Q a finite set of states, and $\pi = (q_c)_{c \in \widehat{J}}$ a word over the alphabet Q (i.e., each cut $c \in \widehat{J}$ is associated with an element of Q). The *left-limit set* and *right-limit set* of π at c are the two subsets $\lim_c^- \pi$ and $\lim_c^+ \pi$ of Q defined as follows:

- $\lim_c^- \pi = \{q \in Q \mid \forall i < c, \exists k : i < k < c \land q = q_k\}$,
- $\lim_c^+ \pi = \{q \in Q \mid \forall c < i, \exists k : c < k < i \land q = q_k\}$.

The left limit at c is therefore the set of states that can be found before c and infinitely close to c in π. If c has a predecessor, then this set is empty (intuitively it means that a *successor transition*, and not a *limit transition*, has to be taken to reach the state mapped to c). The case of a right limit is handled symmetrically. For the first cut c_{min} and the last cut c_{max} of J, we set $\lim_{c_{min}}^- \pi = \emptyset$, and $\lim_{c_{max}}^+ \pi = \emptyset$, although the definition above implies that both sets are equal to Q.

Definition 2. *Let* $\mathcal{A} = (Q, \Sigma, I, F, \Delta^s, \Delta^\ell)$ *be an automaton and* $w = (\alpha_j)_{j \in J}$ *a word of length* J. *A* path π *labeled by* w *is a sequence of states* $\pi = (q_c)_{c \in \widehat{J}}$ *of length* $\widehat{J}$ *such that*

- *For every* $j \in J$, $(q_{j-}, \alpha_j, q_{j+})$ *is a successor transition in* Δ^s.
- *For every cut* $c \in \widehat{J}$ *that is not the first cut and does not have a predecessor,* $\lim_c^- \pi \to q_c$ *is a left-limit transition in* Δ^ℓ.
- *For every cut* $c \in \widehat{J}$ *that is not the last cut and does not have a successor,* $q_c \to \lim_c^+ \pi$ *is a right-limit transition in* Δ^ℓ.

A path is *accepting* if it starts in a state $p \in I$ and ends in a state $q \in F$. A word is accepted by $\mathcal{A}$ if it labels an accepting path.

Note that a *finite-word automaton* can be seen as an automaton on linear orderings that has an empty set of limit transitions (i.e., $\Delta^\ell = \emptyset$). Infinite-word automata [4,14] and automata on ordinal words [5,10] also correspond to particular cases of automata on linear orderings.

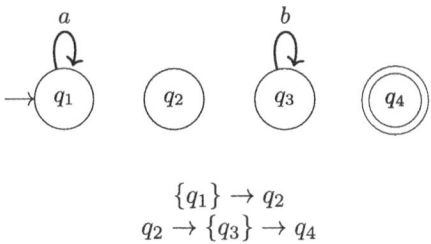

$$\{q_1\} \rightarrow q_2$$
$$q_2 \rightarrow \{q_3\} \rightarrow q_4$$

Fig. 2. An automaton accepting $a^\omega \cdot b^{-\omega} \cdot b^\omega$.

Example 2. Consider the word $w = a^\omega \cdot b^{-\omega} \cdot b^\omega$ defined over the alphabet $\{a, b\}$. Notice that w is indexed by the linear ordering considered in Example 1. The automaton depicted in Fig. 2 accepts $\{w\}$. An accepting path can be described as follows: it maps the cuts (including the first one) of the first sub-ordering of order type ω to the state q_1, the gap to the state q_2, the cuts of the sub-ordering of order type $-\omega + \omega$ (or equivalently ζ) to the state q_3, and the last cut to the state q_4.

Example 3. Consider the automata depicted in Figs. 3a and 3b. Both accept a non-empty language. The automaton in Fig. 3a accepts words indexed by $\mathbb{R}$, and more generally by any non-empty dense and complete ordering J with no first and no last element, as long as the set of indices labeled by the letter a and those labeled by b partition J, and are both dense in J. For instance, consider the word $w : \mathbb{R} \rightarrow \{a, b\}$, such that $w(x) = a$ if $x \in \mathbb{Q}$, and $w(x) = b$ if $x \in \mathbb{R} \setminus \mathbb{Q}$. Let us describe an accepting path π labeled by w. The first cut $(\emptyset, \mathbb{R})$ of $\mathbb{R}$ is abbreviated by $-\infty$, and the last cut $(\mathbb{R}, \emptyset)$ by $+\infty$. We set $\pi(-\infty) = q_1$ and $\pi(+\infty) = q_6$. Since $\mathbb{R}$ is complete, every other cut of $\mathbb{R}$ is of the form r^- or r^+ for some $r \in \mathbb{R}$. If $r \in \mathbb{Q}$, then we set $\pi(r^-) = q_2$ and $\pi(r^+) = q_3$. If $r \in \mathbb{R} \setminus \mathbb{Q}$, then we set $\pi(r^-) = q_4$ and $\pi(r^+) = q_5$. The automaton depicted in Fig. 3a does not however accept words indexed by $\mathbb{Q}$, or any incomplete ordering, since this automaton does not contain any state that can be mapped to a gap. The language accepted by this automaton is described by the rational expression $\text{sh}(a, b)$.

On the other hand, the automaton depicted in Fig. 3b accepts words indexed by $\mathbb{Q}$, but does not accept any word indexed by $\mathbb{R}$. Indeed, any accepting path necessarily visits the state q_4, which can only correspond to a gap, since it does not admit any incoming or outgoing successor transition. The language accepted by this automaton is described by the rational expression $\text{sh}(a, \varepsilon)$.

3   State of the Art

Given an automaton $\mathcal{A}$ and two states p and q, the reachability problem consists in deciding whether $\mathcal{A}$ admits a path starting at p and ending at q. In [9],

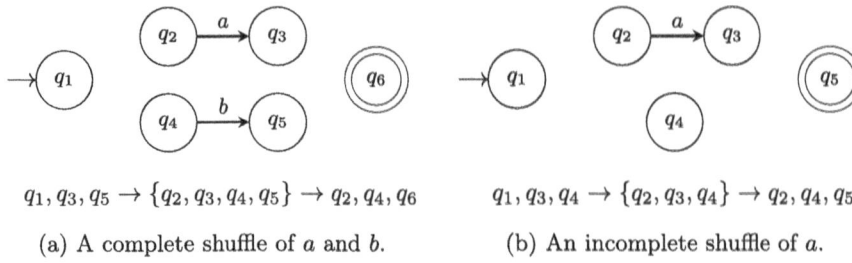

$q_1, q_3, q_5 \rightarrow \{q_2, q_3, q_4, q_5\} \rightarrow q_2, q_4, q_6$

(a) A complete shuffle of a and b.

$q_1, q_3, q_4 \rightarrow \{q_2, q_3, q_4\} \rightarrow q_2, q_4, q_5$

(b) An incomplete shuffle of a.

Fig. 3. Automata accepting words indexed by dense orderings.

Carton describes a polynomial procedure for deciding reachability on scattered linear orderings only, i.e., deciding whether there exists a path labeled by a word indexed by such an ordering. Given an input automaton on linear orderings $\mathcal{A}$, Carton's algorithm generates a finite-word automaton $\mathcal{A}_F$ sharing the same set of states as $\mathcal{A}$, that accounts for every possible path in $\mathcal{A}$ while storing the set of states visited in the labels read along the paths. More precisely, the existence of a transition (p, P, q) in $\mathcal{A}_F$ implies that there exists in $\mathcal{A}$ a path $p \rightarrow q$ labeled by a word indexed by a scattered linear ordering that visits exactly the set of states P. Reciprocally, the existence in $\mathcal{A}$ of a path $p \rightarrow q$ labeled by a word indexed by a scattered linear ordering implies the existence of a path $p \rightarrow q$ in $\mathcal{A}_F$, although it may not necessarily be composed of a single transition. By not explicitly building a transition for every possible path, Carton's solution remains polynomial. Reachability in $\mathcal{A}$ then reduces to reachability in $\mathcal{A}_F$.

A generalization of this algorithm over arbitrary linear orderings is sketched in [11], where the author introduces a specific additional rule for dealing with shuffles. To the best of our knowledge, a proof of this generalization has not been published, and it seems that the additional rule is actually only able to deal with complete shuffles. Actually, a small modification of this rule (somewhat similar to what will be done in Sect. 5) is enough to deal with both complete and incomplete shuffles altogether, so as to accurately cover any linear ordering.

4   Reachability over the Reals

4.1   Characterization

Properties like completeness, density, and not having first or last elements can be expressed by automata on linear orderings. However, there is no equivalence between accepting words indexed by complete and dense orderings without first or last elements, and accepting words indexed by $\mathbb{R}$. In particular, we have the following result.

Theorem 1. *There exists an automaton on linear orderings $\mathcal{A}_{\mathbb{R}}^{\emptyset}$ that accepts words indexed by non-empty, dense, and complete linear orderings that do not have first or last elements, such that $\mathcal{A}_{\mathbb{R}}^{\emptyset}$ does not accept words indexed by $\mathbb{R}$.*

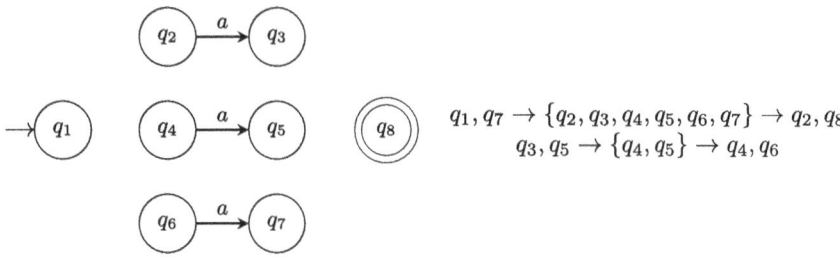

Fig. 4. The automaton $\mathcal{A}_{\mathbb{R}}^{\emptyset}$, that does not accept words indexed by $\mathbb{R}$.

The automaton $\mathcal{A}_{\mathbb{R}}^{\emptyset}$ depicted in Fig. 4 matches the description given in Theorem 1. The proof of Theorem 1 ensues from the two following lemmas.

Lemma 1. *The automaton $\mathcal{A}_{\mathbb{R}}^{\emptyset}$ does not accept any word indexed by $\mathbb{R}$.*

Proof Sketch. Let w be a word accepted by $\mathcal{A}_{\mathbb{R}}^{\emptyset}$. By studying the structure of the automaton $\mathcal{A}_{\mathbb{R}}^{\emptyset}$, one first proves that the ordering $|w|$ must be of the form $\Sigma_{k \in K} J_k$, where for all $k \in K$, J_k is an infinite, dense, and complete ordering with a first and a last element, and K is an infinite, dense, and complete ordering without first and last elements. Intuitively, each J_k is recognized by a sub-path that starts with $q_2 \xrightarrow{a} q_3$, then follows a dense mix of $q_4 \xrightarrow{a} q_5$ transitions, and finally ends with $q_6 \xrightarrow{a} q_7$.

Then, one shows that $\mathbb{R}$ cannot be expressed as such an ordering $\Sigma_{k \in K} J_k$. By contradiction, assume that $\mathbb{R}$ can be partitioned into an infinite, dense, and complete set I of pairwise disjoint closed intervals $[a_i, b_i]$ with $a_i < b_i$. On the one hand, thanks to Cantor's isomorphism theorem [8], there does not exist an infinite, dense, and complete ordering that is also countable, therefore the set I must be uncountable. On the other hand, I must be countable since each interval $[a_i, b_i] \in I$ contains a rational number that is not shared with any other, which yields a contradiction. $\qquad\square$

Lemma 2. *The automaton $\mathcal{A}_{\mathbb{R}}^{\emptyset}$ accepts a word indexed by a non-empty, dense, and complete linear ordering that does not have a first or a last element.*

Proof. Consider the linear ordering $S = [0, 1]^{\mathbb{R}}$, where $[0, 1]$ denotes the closed interval of $\mathbb{R}$ between 0 and 1. Intuitively, S corresponds to the real line in which every real number has been replaced by a copy of the closed interval $[0, 1]$. The ordering S is dense, complete, and does not admit a first or a last element. We can represent elements of S as pairs $(x, y) \in [0, 1] \times \mathbb{R}$, ordered by the relation $<_S$ defined by $(x_1, y_1) <_S (x_2, y_2)$ if either $y_1 < y_2$, or $y_1 = y_2$ and $x_1 < x_2$, where $<$ denotes the usual order relation on both $\mathbb{R}$ and $[0, 1]$. Now, consider the word w indexed by S (i.e., $|w| = S$) obtained by labeling each element of S with the symbol a. Let us describe an accepting run of $\mathcal{A}_{\mathbb{R}}^{\emptyset}$ reading w. We denote by $\widehat{s}_{min}$ (resp. $\widehat{s}_{max}$) the first (resp. last) cut of S. Consider the mapping $\pi : \widehat{S} \to Q$ defined as follows:

- $\pi(\widehat{s}_{min}) = q_1$,
- $\pi(\widehat{s}_{max}) = q_8$,
- for all $y \in \mathbb{R}$, $\pi((0,y)^-) = q_2$ and $\pi((0,y)^+) = q_3$,
- for all $y \in \mathbb{R}$, $\pi((1,y)^-) = q_6$ and $\pi((1,y)^+) = q_7$,
- for all $x \in (0,1)$ and for all $y \in \mathbb{R}$, $\pi((x,y)^-) = q_4$,
- for all $x \in (0,1)$ and for all $y \in \mathbb{R}$, $\pi((x,y)^+) = q_5$.

It is immediate to establish that π describes an accepting run. $\square$

This concludes the proof of Theorem 1.

4.2   Algorithm

We now introduce an algorithm that decides reachability over $\mathbb{R}$.

Let $\mathcal{A} = (Q, \Sigma, I, F, \Delta^s, \Delta^\ell)$ be an automaton and $q_I, q_F \in Q$ two states. We consider the problem of deciding reachability between q_I and q_F. The first step of the algorithm consists in constructing a sequence of finite-word automata $\mathcal{A}_0, \ldots, \mathcal{A}_M$, where $M = \max\{|P| \mid \exists q \in Q : q \to P \in \Delta^\ell \vee P \to q \in \Delta^\ell\}$. For every $j \in [0, M]$, the automaton $\mathcal{A}_j$ is of the form $(Q, 2^Q, I, F, \Delta_j, \emptyset)$, i.e., the successor transitions in Δ_j (which we define later) are labeled by subsets of Q. The purpose of each automaton $\mathcal{A}_j$ is to characterize every path in $\mathcal{A}$ that is labeled by a word indexed by $\mathbb{R}$, and that only involves limit transitions on sets of cardinality less than or equal to j.

We introduce the following useful definitions. An *open $\mathbb{R}$-path* $\bar{r}$ from q_1 to q'_n in $\mathcal{A}_j$ of label $P = P_1 \cup \cdots \cup P_n$ is a finite sequence of transitions $(q_1, P_1, q'_1), \ldots, (q_n, P_n, q'_n) \in \Delta_j$ (of length $n \geq 1$) such that, for every $i \in [1, n-1]$, there exists a successor transition $(q'_i, \alpha_i, q_{i+1}) \in \Delta^s$ for some $\alpha_i \in \Sigma$. Intuitively, such a path corresponds to a finite concatenation of open intervals of $\mathbb{R}$ (corresponding to transitions in $\mathcal{A}_j$), connected by single elements (corresponding to successor transitions in $\mathcal{A}$), resulting in an open interval of $\mathbb{R}$. Analogously, a *closed $\mathbb{R}$-path* $\bar{r}$ from p to p' in $\mathcal{A}_j$ of label $P \cup \{p, p'\}$ is the combination of an open $\mathbb{R}$-path from a state q_1 to a state q'_n (of label P) with two transitions $(p, \alpha, q_1), (q'_n, \alpha', p') \in \Delta^s$, for some $\alpha, \alpha' \in \Sigma$.

The transition relation Δ_j is defined for $j \in [0, M]$ by recursion over j. For $j > 0$, $\Delta_j = \Delta_{j-1} \cup S_j \cup L_j \cup R_j$, where S_j, L_j, and R_j are sets of transitions constructed from $\mathcal{A}_{j-1}$ as described below, and $\Delta_0 = \emptyset$. For every $j \in [1, M]$, the sets S_j, L_j, and R_j are initially empty. They are then filled according to the following rules.

Shuffle rule: For every pair of transitions $q \to P \in \Delta^\ell$ and $P \to q' \in \Delta^\ell$ such that $|P| = j$, the transition $(q, P \cup \{q, q'\}, q')$ is added to S_j if and only if:

- There exists a (possibly empty) set $\{\bar{r}_1, \ldots, \bar{r}_n\}$ of closed $\mathbb{R}$-paths in $\mathcal{A}_{j-1}$ with $n \geq 0$, where each $\bar{r}_i$ starts in a state q_i such that $P \to q_i \in \Delta^\ell$, and ends in a state q'_i such that $q'_i \to P \in \Delta^\ell$. The label of the $\mathbb{R}$-path $\bar{r}_i$ is denoted by Q_i.

– There exists a (necessarily non-empty) set of successor transitions $\{(s_1,\gamma_1,s_1'),$ $\dots,(s_m,\gamma_m,s_m')\} \subseteq \Delta^s$, with $m > 0$, such that, for each $i \in [1,m]$, $s_i, s_i' \in P$, $P \to s_i \in \Delta^\ell$, and $s_i' \to P \in \Delta^\ell$.
– $\bigcup_{i \in [1,m]} \{s_i, s_i'\} \cup \bigcup_{i \in [1,n]} Q_i = P$.

We refer to the case $n = 0$ as a *simple shuffle*, and to the case $n > 0$ as a *Cantor shuffle*. Intuitively, a Cantor shuffle represents a dense combination of intervals of order type $1 + \lambda + 1$ (corresponding to closed $\mathbb{R}$-paths in $\mathcal{A}_{j-1}$) and single elements (corresponding to successor transitions in $\mathcal{A}$). A simple shuffle represents a dense combination of single elements only.

Infinite repetition rule: For every transition $P \, \rangle \, q \subset \Delta^\ell$ such that $|P| = j$, and for every state $s \in P$, the transition $(s, P \cup \{q\}, q)$ is added to L_j if and only if there exist $p, p' \in P$ such that:

– $\mathcal{A}_{j-1}$ admits an open $\mathbb{R}$-path from p to p' labeled by P of the form (r_1, P_1, r_1'), $\dots, (r_n, P_n, r_n')$ such that $s = r_i$ for some $i \in [1, n]$, and
– $\mathcal{A}$ admits a successor transition of the form $(p', \alpha, p) \in \Delta^s$.

Intuitively, we look for a cycle that alternates between open intervals of order type λ (corresponding to open $\mathbb{R}$-paths in $\mathcal{A}_{j-1}$) and single elements (corresponding to successor transitions in $\mathcal{A}$) such that the set of states visited by the whole cycle is exactly P. If such a cycle exists, it can be repeated infinitely many times, therefore allowing to follow the limit transition $P \to q$. A similar principle applies to the *reverse infinite repetition rule* described below.

Reverse infinite repetition rule: For every transition $q \to P \in \Delta^\ell$ such that $|P| = j$, and for every state $s \in P$, the transition $(q, P \cup \{q\}, s)$ is added to R_j if and only if there exist $p, p' \in P$ such that:

– $\mathcal{A}_{j-1}$ admits an open $\mathbb{R}$-path from p to p' labeled by P of the form (r_1, P_1, r_1'), $\dots, (r_n, P_n, r_n')$ such that $s = r_i'$ for some $i \in [1, n]$, and
– $\mathcal{A}$ admits a successor transition of the form $(p', \alpha, p) \in \Delta^s$.

The final step of the algorithm consists in searching whether there exists an open $\mathbb{R}$-path in $\mathcal{A}_M$ that starts in q_I and ends in q_F. In the positive case, the algorithm returns *yes*, otherwise *no*.

The definitions above ensure that for every $j \in [0, M]$, if $\mathcal{A}$ admits a path from a state q to a state q' that visits a set of states P, is labeled by a word indexed by $\mathbb{R}$, and only involves limit transitions on sets of cardinality less than or equal to j, then $\mathcal{A}_j$ admits an open $\mathbb{R}$-path from q to q' labeled by P.

4.3   Correctness

We have the two following results, asserting the correctness of the algorithm of Sect. 4.2.

Theorem 2. *If there exists an open* $\mathbb{R}$*-path in* $\mathcal{A}_M$ *from* q_I *to* q_F *of label* P, *then there exists a path in* $\mathcal{A}$ *from* q_I *to* q_F, *labeled by a word indexed by* $\mathbb{R}$, *that visits exactly the states in* P.

Proof Sketch. The proof proceeds by explicitly building a word indexed by $\mathbb{R}$ or, equivalently, by any open interval (x, y) of $\mathbb{R}$, that labels a path in $\mathcal{A}$.

Let $\bar{r}$ be an open $\mathbb{R}$-path in $\mathcal{A}_M$ corresponding to the sequence of transitions $(q_1, P_1, q_1'), \ldots, (q_n, P_n, q_n') \in \Delta_M$. The proof is by induction on the sequence of rules involved in the generation of the transitions (q_i, P_i, q_i'). The base case corresponds to an open $\mathbb{R}$-path $\bar{r}$ composed of a single transition $(q, P, q') \in \Delta_M$ that stems from the *simple shuffle* rule, since this rule does not rely on preexisting $\mathbb{R}$-paths. Let $(s_1, \gamma_1, s_1'), \ldots, (s_m, \gamma_m, s_m') \in \Delta^s$ be the successor transitions that triggered the application of the simple shuffle rule. The domain $\mathbb{R}$ can be partitioned into m non-empty disjoint subsets $R_1, \ldots, R_m$ such that each R_i is dense in $\mathbb{R}$. The word $w : \mathbb{R} \to \Sigma$ indexed by $\mathbb{R}$ and defined by $w(x) = \gamma_i$ if $x \in R_i$ labels a path from q to q'.

For the inductive case, we first consider the *infinite repetition* rule. It is handled by considering an infinite sequence of words indexed by consecutive intervals of $\mathbb{R}$, e.g., $(-\infty, 0)$, $(0, 1)$, $(1, 2)$, … and concatenating them to form a word indexed by the full set $\mathbb{R}$. The case of the *reverse infinite repetition* is handled symmetrically.

To reason about the remaining rule, that is, the *Cantor shuffle*, we consider a Cantor set construction [7,12] that partitions the interval $(0, 1)$ into a countable set I of disjoint open intervals, together with an uncountable set S of isolated points. For every $n > 0$, the set I can be partitioned into n subsets $I_1, \ldots, I_n$ such that for all $k \in [1, n]$, an interval in I_k lies between any two intervals in I. Similarly, for every $m > 0$, the set S can be partitioned into m subsets $S_1, \ldots, S_m$ such that for all $j \in [1, m]$, the set S_j is dense in S. By associating each $\mathbb{R}$-path $\bar{r}_i$ involved in the Cantor shuffle with the set of intervals I_i, and each successor transition (s_j, γ_j, s_j') with the set of isolated points S_j, we obtain a word indexed by $(0, 1)$ (and equivalently $\mathbb{R}$) that labels a path of the desired form.   □

Theorem 3. *If the automaton* $\mathcal{A}$ *accepts a word indexed by* $\mathbb{R}$, *then there exists an open* $\mathbb{R}$*-path in* $\mathcal{A}_M$ *from a state* $q_I \in I$ *to a state* $q_F \in F$.

The proof is essentially based on mechanisms introduced in [13] for establishing the decidability of the first-order theory of order, and on ideas introduced in [9].

Proof Sketch. The proof consists in showing that for all $j \in [0, M]$, if $\mathcal{A}$ admits a path from a state q to a state q' that visits a set of states P, labeled by a word indexed by $\mathbb{R}$, and that only follows limit transitions on sets of cardinality less than or equal to j, then $\mathcal{A}_j$ admits an open $\mathbb{R}$-path from q to q' labeled by P.

The proof relies on the property that for all $j \in [1, M]$, the construction of $\mathcal{A}_j$ only relies on transitions generated in $\mathcal{A}_{j-1}$. This ensues from the fact that at Step j, every application of a rule generates a transition of one of the two following forms. The first case is that the generated transition is labeled by a set of states of cardinality strictly greater than j. In that case, this transition can clearly only be involved in a rule at a later Step $j + k$ for some $k \geq 1$. The other possibility is to have a transition labeled by a set of cardinality of size exactly equal to j. In that case, using this transition to apply another rule in $\mathcal{A}_j$ (this time not only based on transitions in $\mathcal{A}_{j-1}$, but also transitions generated at Step j itself) would only generate new transitions that are redundant w.r.t. the existence of open $\mathbb{R}$-paths in $\mathcal{A}_j$. □

4.4 Example

We now give a short example to illustrate our algorithm, and more specifically the construction of the automata $\mathcal{A}_0, \ldots, \mathcal{A}_M$. Consider the automaton $\mathcal{A}$ depicted in Fig. 5. The automaton $\mathcal{A}_0$ is shown in Fig. 6, and does not contain any transition. Since no limit transition involves a set of cardinality 1 or 2, we have $\mathcal{A}_2 = \mathcal{A}_1 = \mathcal{A}_0$. In the automaton $\mathcal{A}_3$ given in Fig. 7, the transition $(q_4, \{q_3, q_4, q_5\}, q_3)$ (resp. $(q_5, \{q_3, q_4, q_5\}, q_3)$) results from the application of the shuffle rule on the set $\{q_3, q_4, q_5\}$, with $q_I = q_4$ (resp. $q_I = q_5$) and $q_F = q_3$. For the same reason as before, we have $\mathcal{A}_4 = \mathcal{A}_3$. The last automaton $\mathcal{A}_5$ is illustrated in Fig. 8. The new transitions result from the application of the shuffle rule on the set $\{q_1, q_2, q_3, q_4, q_5\}$, for every $q_I \in \{q_1, q_4, q_5\}$ and $q_F \in \{q_2, q_3\}$.

4.5 Complexity

The algorithm described in Sect. 4.2 runs in polynomial time w.r.t. the size of $\mathcal{A}$, i.e. $|Q| + |\Delta^s| + |\Delta^\ell|$. More precisely, we have the following result.

Proposition 1. *The automaton $\mathcal{A}_M$ can be computed in polynomial time in the size of $\mathcal{A}$.*

Proof Sketch. A first argument is that for all $j \in [0, M]$, each rule is called only polynomially many times during the construction of the automaton $\mathcal{A}_j$. Indeed, the shuffle rule is only considered for every combination of two limit transitions in $\mathcal{A}$, and the (reverse) infinite repetition is only considered for every combination of a limit transition in $\mathcal{A}$ with one state of $\mathcal{A}$.

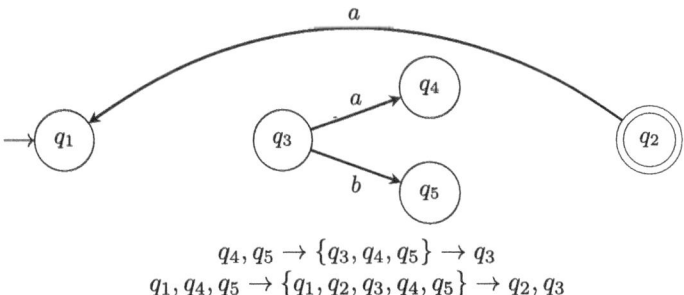

$$q_4, q_5 \rightarrow \{q_3, q_4, q_5\} \rightarrow q_3$$
$$q_1, q_4, q_5 \rightarrow \{q_1, q_2, q_3, q_4, q_5\} \rightarrow q_2, q_3$$

Fig. 5. The input automaton $\mathcal{A}$.

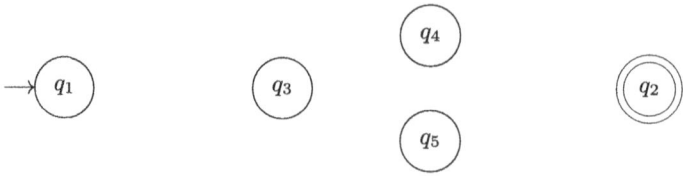

Fig. 6. The finite-word automata $\mathcal{A}_0$, $\mathcal{A}_1$, and $\mathcal{A}_2$.

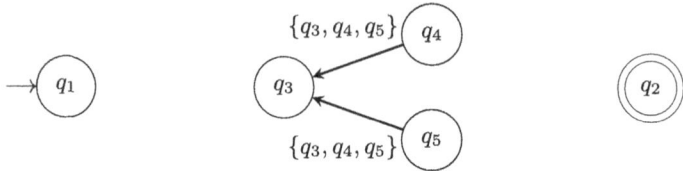

Fig. 7. The finite-word automata $\mathcal{A}_3$ and $\mathcal{A}_4$.

It remains to show that each rule can be applied in polynomial time. The case of the simple shuffle rule is immediate. The other rules require to either be able to check the existence of an open $\mathbb{R}$-path labeled by a given set of states P (to deal with the infinite and reverse infinite repetition rules), and to be able to check the existence of a set of closed $\mathbb{R}$-paths and successor transitions such that their combined labels and visited states are equal to P (to deal with the shuffle rule). To perform these two operations in polynomial time, in this specific context, we rely on a polynomial procedure that, given a set of states P, two states $q, q' \in P$, and the automaton $\mathcal{A}_j$, computes the union of the labels of every possible (open or closed) $\mathbb{R}$-path in $\mathcal{A}_j$ from q to q' labeled by a subset of P. More precisely, this procedure computes the set $S_{\text{open}}(\mathcal{A}_j, P, q, q') = \{r \in P \mid \mathcal{A}_j \text{ admits an open } \mathbb{R}\text{-path from } q \text{ to } q', \text{ of label } R \subseteq P, \text{ s.t. } r \in R\}$, or analogously the set $S_{\text{closed}}(\mathcal{A}_j, P, q, q')$ that deals with closed $\mathbb{R}$-paths. Let $\delta_{\overrightarrow{P,j,q}} \subseteq \Delta_j$ (resp. $\delta_{\overleftarrow{P,j,q'}}$) be the set of transitions occurring in (open or closed) $\mathbb{R}$-paths of $\mathcal{A}_j$ starting in q (resp. ending in q'), and labeled by a subset of P. The sets $\delta_{\overrightarrow{P,j,q}}$ and $\delta_{\overleftarrow{P,j,q'}}$ can be computed in polynomial time by performing a (reverse) traversal of the automaton $\mathcal{A}_j$ that respects the structure of $\mathbb{R}$-paths, i.e., alternating between transitions in Δ_j and transitions in Δ^s. The set $S_{\text{open}}(\mathcal{A}_j, P, q, q')$ (resp. $S_{\text{closed}}(\mathcal{A}_j, P, q, q')$) is then obtained by taking the union of the labels of the transitions in $\delta_{\overrightarrow{P,j,q}} \cap \delta_{\overleftarrow{P,j,q'}}$.

Thanks to the (reverse) infinite repetition rule, given a set of states P and two states $p, p' \in P$ such that $S_{\text{open}}(\mathcal{A}_j, P, p, p') = P$, there exists an open $\mathbb{R}$-path from p to p' in $\mathcal{A}_j$ labeled by P. Such a path can be constructed by first choosing, for each $q \in P$, an open $\mathbb{R}$-path $\overline{r}_q$ labeled by $R \subseteq P$ such that $q \in R$, and then concatenating these paths by means of transitions of the form $(p', \alpha, p) \in \Delta^s$, which must exist for the infinite repetition rule to be applicable.

Now, regarding the shuffle rule, consider two set of states P, S such that $S \subseteq P$. We define $L_P = \{p \in P \mid P \rightarrow p \in \Delta^\ell\}$, and $R_P = \{p \in P \mid p \rightarrow P \in \Delta^\ell\}$.

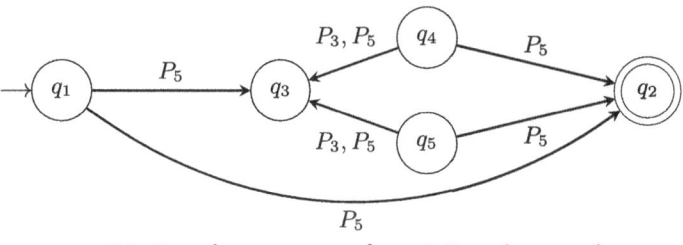

with $P_5 = \{q_1, q_2, q_3, q_4, q_5\}$, and $P_3 = \{q_3, q_4, q_5\}$.

Fig. 8. The finite-word automaton $\mathcal{A}_5$.

Using a similar argument as before, one shows that checking the existence of a set of closed $\mathbb{R}$-paths $\bar{r}_1, \ldots, \bar{r}_n$, each labeled by Q_i, starting from a state in L_P, and ending in a state in R_P, such that $(P \setminus S) \subseteq \bigcup_{i \in [1,n]} Q_i$, reduces to checking that the inclusion $(P \setminus S) \subseteq \bigcup_{(q,q') \in L_P \times R_P} S_{\text{closed}}(\mathcal{A}_j, P, q, q')$ holds. $\square$

5   Reachability over the Rationals

The algorithm introduced in Sect. 4.2 can be adapted to solve reachability over $\mathbb{Q}$. The incompleteness of $\mathbb{Q}$ requires the rules to be modified, in order to account for the presence of gaps.

We first define a notion of valid path over $\mathbb{Q}$, analogous to the notion of $\mathbb{R}$-path. A $\mathbb{Q}$-path from q_1 to q'_n in $\mathcal{A}_j$ of label $P = P_1 \cup \cdots \cup P_n$ is a finite sequence of transitions $(q_1, P_1, q'_1), \ldots, (q_n, P_n, q'_n) \in \Delta_j$ such that for each $i \in [1, n-1]$, either $q'_i = q_{i+1}$, or there exists a successor transition $(q'_i, \alpha_i, q_{i+1}) \in \Delta^s$. Intuitively, such a path corresponds to a finite concatenation of open intervals of $\mathbb{Q}$ (corresponding to transitions in $\mathcal{A}_j$) connected either by single elements (corresponding to successor transitions in $\mathcal{A}$, if the shared bound between the two consecutive interval belongs to $\mathbb{Q}$), or by gaps (if the shared bound belongs to $\mathbb{R} \setminus \mathbb{Q}$). For a given set of states P appearing in a limit transition, both the simple and Cantor shuffle rules additionally require a non-empty set of states $(g_1, \ldots, g_k)$ such that $g_i \to P \to g_i$, for all $i \in [1, k]$. Finally, the left and right infinite repetition rules are modified to allow for the required $\mathbb{Q}$-path to be cyclic, i.e., $q'_n = q_1$.

References

1. Bedon, N., Bès, A., Carton, O., Rispal, C.: Logic and rational languages of words indexed by linear orderings. Theory Comput. Syst. **46**, 737–760 (2010)
2. Bès, A., Carton, O.: A Kleene theorem for languages of words indexed by linear orderings. Int. J. Found. Comput. Sci. **17**(03), 519–541 (2006)
3. Bruyère, V., Carton, O.: Automata on linear orderings. J. Comput. Syst. Sci. **73**(1), 1–24 (2007)

4. Büchi, J.R.: On a decision method in restricted second order arithmetic. In: Proceedings of the International Congress on Logic, Methodology and Philosophy of Science, 1960 (1962)
5. Büchi, J.R.: Transfinite automata recursions and weak second order theory of ordinals. In: Proceedings of the International Congress on Logic, Methodology, and Philosophy of Science, 1965 (1965)
6. Burgess, J.P., Gurevich, Y.: The decision problem for linear temporal logic. Notre Dame J. Formal Logic **26**(2), 115–128 (1985)
7. Cantor, G.: Über unendliche, lineare Punktmannichfaltigkeiten. Math. Ann. **20**(1), 113–121 (1882)
8. Cantor, G.: Beiträge zur Begründung der transfiniten Mengenlehre. Math. Ann. **46**, 481–512 (1895)
9. Carton, O.: Accessibility in automata on scattered linear orderings. In: Diks, K., Rytter, W. (eds.) MFCS 2002. LNCS, vol. 2420, pp. 155–164. Springer, Heidelberg (2002). https://doi.org/10.1007/3-540-45687-2_12
10. Choueka, Y.: Finite automata, definable sets, and regular expressions over ω^n-tapes. J. Comput. Syst. Sci. **17**(1), 81–97 (1978)
11. Cristau, J.: Automata and temporal logic over arbitrary linear time. In: Proceedings of the International Conference on Foundations of Software Technology and Theoretical Computer Science. LIPIcs, vol. 4, pp. 133–144 (2009)
12. DiBenedetto, E.: Real Analysis. Springer, Cham (2002)
13. Läuchli, H., Leonard, J.: On the elementary theory of linear order. Fundam. Math. **59**(1), 109–116 (1966)
14. Muller, D.E.: Infinite sequences and finite machines. In: Proceedings of the Annual Symposium on Switching Circuit Theory and Logical Design, pp. 3–16. IEEE Computer Society (1963)
15. Rabin, M.O.: Decidability of second-order theories and automata on infinite trees. Trans. Am. Math. Soc. **141**, 1–35 (1969)
16. Rabinovich, A.: Temporal logics over linear time domains are in PSPACE. Inf. Comput. **210**, 40–67 (2012)
17. Reynolds, M.: The complexity of temporal logic over the reals. Ann. Pure Appl. Logic **161**(8), 1063–1096 (2010)
18. Rosenstein, J.G.: Linear Orderings. Academic Press, Cambridge (1982)
19. Shelah, S.: The monadic theory of order. Ann. Math. **102**(3), 379–419 (1975)

On Bidirectional Deterministic Finite Automata

Simon Dieck$^{(\boxtimes)}$ and Sicco Verwer

Delft University of Technology, Delft, The Netherlands
{S.Dieck,S.E.Verwer}@tudelft.nl

Abstract. Bidirectional deterministic finite automata (biDFA) are a recent innovation with many potential applications. In this paper, we present novel theoretical results for bidirectional automata. We show that there exist regular languages, where minimal biDFA models are exponentially smaller than minimal DFA models. We show this for a language that has a structure common to software logs. This makes biDFA especially interesting when inferring models from such data. However, we also prove that the problem of biDFA minimization is NP-hard. As our key contribution, we provide a Myhill-Nerode style congruence-based characterization for the languages they can recognize. Since most algorithms for learning DFAs are based on such a congruence, this characterization is an important building block for obtaining learning algorithms.

Keywords: DFA · Regular languages · Context-free languages · Linear languages

1 Introduction

Deterministic finite automata (DFA) are one of the fundamental building blocks of computer science. They have applications in many diverse fields. Among others, they can be used for detecting fingerprints in security applications [6], as a part of parsers in the programming language field [8] or as interpretable surrogate models in machine learning [15,16]. While several improvements have been introduced to DFA over the years, like probabilistic [20], nested [3], or timed [1] versions, their core capabilities remained largely the same. They recognize the regular languages and do so in linear time.

Linear time parsing of languages more complex than regular languages is typically done using deterministic push-down automata (PDA). These models can recognize a proper superset of the regular languages, that is a proper subset of the context-free languages [13]. Jirásková and Klíma recently introduced deterministic biautomata, which are models that provide deterministic linear time parsing for a proper superset of the regular languages and proper subset of the linear languages [11]. Such models are of special interest since they can recognize some languages deterministic PDAs cannot. This set was even conjectured by

© The Author(s), under exclusive license to Springer Nature Switzerland AG 2024
S. Z. Fazekas (Ed.): CIAA 2024, LNCS 15015, pp. 109–123, 2024.
https://doi.org/10.1007/978-3-031-71112-1_8

Knuth as "[...], some grammar which is translated using "both ends towards the middle"." [13].

We believe that the deterministic biautomata introduced by Jiráskova and Klíma are models with significant potential. Normally, increased expressiveness comes at a cost. For example, a PDA is more expressive than a DFA but requires a potentially arbitrarily sized stack to parse languages. It can also come at a cost in terms of parsing time like with most non-deterministic models, which need to branch on certain decisions. While being more expressive than DFA, interestingly biautomata require no extra memory and the same parsing time. In this paper, we show that not only is the minimal biautomata never larger than the minimal DFA, there even exist languages, where it is exponentially smaller.

This smaller number of states is especially interesting for applications on software traces. This is a common and useful application for DFA, where some software is used as an opaque model and the inferred DFA forms a surrogate model that can be analyzed [4,14]. There one often encounters long distance dependencies. That is paired calls to the system, one close to the beginning of the trace and the other close to the end. These normally occur due to resource allocation like opening and closing a connection, locking and unlocking a database or simply claiming and freeing memory. To resolve these correctly DFA require a significant blowup in their number of states. In this paper, we will show that biautomata can resolve long distance dependencies without this blowup.

This paper aims to provide theoretical results that allow for a better analysis of deterministic biautomata bringing them closer to actual application. For this, we adjust Jiráskova's definition of deterministic biautomata slightly to be more similar to that of DFA resulting in a model we call bidirectional deterministic finite automaton (biDFA) to also better distinguish them from the biautomaton introduced earlier by Klíma [12].

One of the nice properties of DFAs is how they relate to the free monoid. For a given language one can define a congruence relation on the free monoid such that the congruence classes directly map to the states of a minimal deterministic automaton for that language. This property, known as the Myhill-Nerode Theorem, is widely used in proofs and algorithms regarding DFA. It is also the basis for many algorithms that infer DFA from traces [2,5,21]. In this paper, we provide a similar congruence-based characterization for biDFA. We believe this is an important step towards developing inference algorithms for biDFA.

Finally, we use this congruence identity to prove the negative result that, unlike DFA, minimizing an existing biDFA is NP-hard.

1.1   Related Work

The concept of a biautomaton was first introduced by Klíma and Polák [12]. The basic underlying idea is that an automaton has two "heads" with which it can read a word from two sides moving inwards. They then imposed some additional conditions on the transition functions, which limits these models to only recognize regular languages.

Holzer and Jakobi then extended these biautomata to allow nondeterministic transitions and found that such machines could recognize all context-free linear languages [9].

Ultimately Jirásková and Klíma took the non-deterministic models from Holzer and Jakobi and restricted it to have only deterministic transitions, which they call deterministic biautomata [11]. These models do not have the conditions originally imposed by Klíma and Polák. In order to avoid confusion, we call these models biDFA in this paper. In their paper, Jiráskova and Klíma analyze the kind of languages such a model can recognize, a class they call **DB**. They identify it as a proper subset of the linear languages that is incomparable to the set of languages that can be recognized by a one-turn PDA [11]. Further, they show that the emptiness, finiteness and universality problem as well as the problem of determining equality to a regular set are NL-complete.

In this context, one should also mention $5 \to 3$ Watson-Crick automata [17]. They are a model developed for parsing DNA and thus also parse a word from both ends moving towards the center. Especially the deterministic, 1-limited and sensing variation of this model bears significant similarity to biDFA [18,19]. Whether the classes **DB**, recognized by biDFA and 2detLIN, recognized by deterministic, 1-limited and sensing $5 \to 3$ Watson-Crick automata are equivalent is an open question.

We are interested in learning biDFAs. Most heuristic algorithms for DFAs (see, e.g., [5]) start from a prefix tree data structure, which emphasizes patterns at the start of words and ignores those at the end. This limits the use of learned DFAs. Using a biDFA model, we can circumvent this limitation. Many learning algorithms rely on a congruence characterization to identify a model's states, see, e.g., [2,7,21]. In this paper, we take an important step towards biDFA learning by obtaining the first congruence characterization for biDFA.

2 Definitions

First, we will cover some basic notations and concepts used in the definitions and proofs of this paper. For a more thorough overview, we refer to the textbook [5].

Given a set Σ, which we call an alphabet with its members being called characters or symbols. We consider all possible ordered sequences of elements from Σ, which we call words, where λ represents the empty word. The set of all such sequences is called Σ^*. We introduce the string concatenation operator "$\cdot$" s.t. given two sequences $(a_1, a_2, \ldots, a_n) \cdot (b_1, b_2, \ldots, b_m) = (a_1, a_2, \ldots, a_n, b_1, b_2, \ldots, b_m)$. As a shorthand, we write these sequences normally without brackets, e.g. $a_1 a_2 \cdots a_n$ and $aa \cdot bb = aabb$. We normally call a set $L \subseteq \Sigma^*$ a language over the alphabet Σ.

Definition 1 (DFA). *A deterministic finite automaton (DFA) is a five-tuple $A = (Q, \Sigma, \delta, q_\lambda, F)$, where Q is a set of states, Σ is an alphabet, $\delta : Q \times \Sigma \to Q$ a transition function, $q_\lambda \in Q$ the initial state and $F \subseteq Q$ the set of accepting states.*

The transition function δ is often extended with a recursive definition to accept any word $w = w_1 \cdots w_n \in \Sigma^*$:

$$\delta^*(q, w) = \begin{cases} q & \text{if } w = \lambda \\ \delta(q, w) & \text{if } |w| = 1 \\ \delta^*(\delta(q, w_1), w_2 \cdots w_n) & \text{otherwise} \end{cases}$$

With this we define the language recognized by a DFA A as $L_A = \{w \in \Sigma^* | \delta^*(q_\lambda, w) \in F\}$. These definitions also directly imply a procedure that decides membership in L_A for any word $w \in \Sigma^*$ in $O(n)$, where $n = |w|$.

2.1   The Bidirectional Deterministic Finite Automaton

We can view δ^* as a procedure that in each step consumes a character on the left of the word to provide a new state. Intuitively, the core idea of the bidirectional deterministic finite automaton is that characters can also be consumed from the right.

Definition 2 (biDFA). *A bidirectional deterministic finite automaton (biDFA) is a six-tuple $B = (Q_l, Q_r, \Sigma, \delta, q_\lambda, F)$, where $Q_l \cup Q_r$, with $Q_l \cap Q_r = \emptyset$, forms a set of states, Σ is an alphabet, $\delta : Q_l \cup Q_r \times \Sigma \to Q_l \cup Q_r$ a transition function, $q_\lambda \in Q_l \cup Q_r$ the initial state and $F \subseteq Q_l \cup Q_r$ the set of accepting states.*

Note that the biDFA definition is almost identical to the DFA definition, except for splitting Q into two disjoint subsets. The bigger change is in the definition of $\delta^* : Q_l \cup Q_r \times \Sigma^* \to Q_l \cup Q_r$:

$$\delta^*(q, w) = \begin{cases} q & \text{if } w = \lambda \\ \delta(q, w) & \text{if } |w| = 1 \\ \delta^*(\delta(q, w_1), w_2 \cdots w_n) & \text{if } |w| > 1 \text{ and } q \in Q_l \\ \delta^*(\delta(q, w_n), w_1 \cdots w_{n-1}) & \text{if } |w| > 1 \text{ and } q \in Q_r \end{cases}$$

Transitions out of states in Q_l work like regular DFA transitions, but transitions out of states in Q_r look at the rightmost remaining character instead. We will call the states of Q_l and Q_r left and right states respectively. Note that we can still define the language recognized by a biDFA as $L_B = \{w \in \Sigma^* | \delta^*(q_\lambda, w) \in F\}$. Similarly deciding this membership also only takes $O(n)$ time. This is due to each step in the recursive definition of δ^* reducing the size of the remaining word by 1.

The definition we give here is equivalent to that of Jirásková and Klíma, but we encode the direction of the action on the state, where they have two different transition functions [11]. This also necessitates them to allow transitions to be empty. We do not allow this.

Another nice property of DFAs is that they can not only decide membership for the language they represent in linear time but also produce members of the language it represents. For this, we treat the DFA as a labeled digraph where Q

is the set of nodes and E the set of edges, which contains (q_u, q_v) with label σ iff $\delta(q_u, \sigma) = q_v$. Any walk on (Q, E), starting at q_λ and ending in a node that is in F represents a word in L_A. We can construct this word, by taking the labels of the edges in the order they were crossed. For example if σ_0 is the label of the first edge and σ_n the label of the last, we obtain $w = (\sigma_0, \sigma_1, \ldots, \sigma_n)$.

We can construct words with biDFAs in a similar fashion. We construct the graph (Q, E) in the same way and perform a walk that starts in q_λ and ends in a state in F. An example of such a graph representation can be found in Fig. 2. However, to obtain a word from a walk we need to consider edges that originate from left states differently from those coming from right states. We consider the sequences of those edges separately, where $w_l = (\sigma_0^l, \sigma_1^l, \ldots, \sigma_n^l)$ is the sequence of edge labels that were visited from left states and $w_r = (\sigma_0^r, \sigma_1^r, \ldots, \sigma_n^r)$ those originating from right states. We form $w = w_l \cdot w_r^R$, where w_r^R is the reversed sequence of w_r. Intuitively we build two words at the same time. One from left to right, like a regular DFA and one from right to left. In the end, we join these words together in the middle.

3   Properties

We have seen that we can use a biDFA in essentially the same way as a DFA. In the following section, we will first provide a simple Lemma comparing the size of minimal DFA to minimal biDFA on regular languages and then provide a novel congruence-based characterization of biDFA.

3.1   Size of Minimal BiDFA on Regular Languages

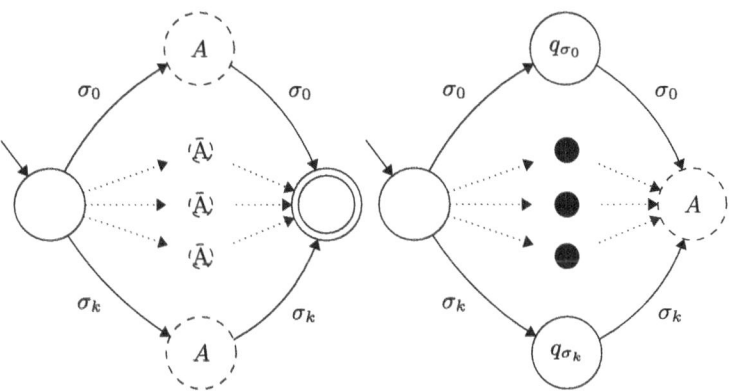

(a) Construction for the DFA.   (b) Construction for the biDFA

Fig. 1. Comparison of the constructions for the DFA and biDFA in Lemma 3. The dashed circles labeled A represent the entire automaton A.

Note that for any DFA $A = (Q, \Sigma, \delta, q_\lambda, F)$ we can construct an equivalent biDFA $B = (Q, \emptyset, \Sigma, \delta, q_\lambda, F)$. We simply ignore the possibility of right states.

Thus, given a minimal DFA, we can always construct a biDFA of equal size w.r.t. the number of states.

Lemma 3. *Given a regular language L, and a minimal DFA $A = (Q, \Sigma, \delta, q_\lambda, F)$ that recognizes L, there exists languages where for a biDFA $B = (Q_l, Q_r, \Sigma, \delta^b, q_\lambda^b, F^b)$ that also recognizes L, $|Q_l| + |Q_r| \leq (n+1) \cdot (i+1)$ and $|Q| \geq n^{(i+1)}$ for any $n, i \geq 1$.*

Proof. Consider a language L_A over the alphabet Σ whose minimal DFA A has n states. We define Σ' as an alphabet with k symbols and $\Sigma' \cap \Sigma = \emptyset$. We use this to construct $L_D = \{sws \in (\Sigma \cup \Sigma')^* | s \in \Sigma', w \in L_A\}$. In words, L_D is the language of words that have the same character of Σ' at the start and end while having a word from L_A in between.

We will now construct a minimal DFA and a biDFA for L_D. This construction is illustrated in Fig. 1. Since there exists a minimal DFA where each state corresponds to a right congruence class of L_D, we can use the number of congruence classes to obtain a lower bound for the size of the minimal DFA recognizing L_D. Consider $a, b \in \Sigma'$ and $u \in \Sigma^*$, where there exists a suffix r, s.t. $u \cdot r \in L_A$. The words $a \cdot u$ and $b \cdot u$ are not right congruent w.r.t L_D, since $a \cdot u \cdot r \cdot a \in L_D$, while $b \cdot u \cdot r \cdot a \notin L_D$. Since u could be in any of n different congruence classes and there are k different symbols in Σ', this implies that L_D has at least $n \cdot k$ different right congruence classes. Thus any DFA recognizing L_D has at least nk states.

For a biDFA, we can construct it as follows: For each symbol σ in Σ' create a state $q_\sigma \in Q_r$ s.t. $|Q_r| = k$. Let $A = (Q^a, \Sigma, \delta, q_\lambda, F)$ be a minimal automaton recognizing L_A. We define the biDFA $B = (Q^a \cup \{q_\lambda^b, q_{sink}\}, Q_r, \Sigma \cup \Sigma', \delta^b, q_\lambda^b, F)$, with:

$$\delta^b(q, v) = \begin{cases} q_\sigma & \text{if } q = q_\lambda^b \text{ and } v = \sigma, \forall \sigma \in \Sigma' \\ q_\lambda & \text{if } q = q_\sigma \text{ and } v = \sigma, \forall \sigma \in \Sigma' \\ \delta(q, v) & \text{if } q \in Q^a \text{ and } v \in \Sigma \\ q_{sink} & \text{otherwise} \end{cases}$$

This biDFA first processes the first and last character and then is left with only the word that needs to be in L_A for which it can just use the minimal DFA of L_A. As such, this biDFA needs only $n + 2 + k$ states to recognize L_D. Now if we set $k = n$ we obtain $|Q_l| + |Q_r| = (n+2) + n = 2n + 2$, while for any minimal DFA recognizing L_D $|Q| \geq n^2$. Now we can just choose L_D as our new language L_A and repeat this construction. So for i repetitions we obtain $L_{D^i} = \{d \cdot w \cdot d^R \in (\Sigma \cup \Sigma')^* | |d| = i, d \in (\Sigma')^*, w \in L_A\}$. Essentially L_{D^i} contains words in L_A which are surrounded by a palindrome of exactly length i. With the same argument as before we know for a minimal DFA $|Q| \geq n^{(i+1)}$. If we just repeat the construction of the biDFA we increase the number of states by $(n+1)$ each time, assuming we reuse the sink state. Therefore there exists a biDFA with $|Q_l| + |Q_r| = (n+1) \cdot (i+1)$. Since there might be a biDFA that recognizes L_A with less than n states we can relax this equality to obtain the inequality claimed in the Lemma. $\square$

The construction used in Lemma 3 also works for reverse DFA, that is DFA that read words backwards. This construction is also especially interesting for the application of learning DFA surrogate models from software traces. The base case before we repeat the construction mimics a long distance dependency. If there are multiple dependencies and they are properly nested we obtain a fixed length palindrome structure similar to the one used in Lemma 3. Proper nesting will occur if the calls are for example at the beginning and end of a recursive function. But even in the case where the order on one side is reversed or even shuffled the blowup will be limited to processing this pre- and suffix and no duplication of some internal logic as in Fig. 1 (a) will occur. So especially in the field of learning surrogate models from software traces, biDFA could result in more concise models.

3.2 Characterisation of Languages Recognized by BiDFA

As can be seen in Fig. 2a we can construct biDFAs that recognize non-regular languages. In this section, we will first give some definitions and then use them to give a congruence-based characterization of this class of languages, similar to the Myhill-Nerode theorem. We refer to this class of languages, which Jiráskova and Klíma call **DB**, as the *symmetric languages* [11]. While Jiráskova and Klíma have placed this class of languages in the context of better-known language classes they do not provide such a characterization [11]. A congruence-based characterization is very useful, as it provides a necessary and sufficient condition for a language to allow a biDFA that recognizes it. Further, most inference algorithms for DFA use congruence to decide which words can be represented by the same state [2,5,7,21]. Accordingly, such a congruence-based characterization is an important step toward a learning algorithm for biDFA.

Definition 4. *Given an alphabet Σ a center function $c : \Sigma^* \to \mathbb{N}$ is a function that chooses a center for each word s.t. $\forall w \in \Sigma^* : c(w) \in [1, |w| + 1]$. A center function splits any word into a left and a right by taking all characters up to the indicated position for the left part and the rest as the right part. We define these splits and their shorthand notation as follows: $l_{c,w} = w_1 w_2 \ldots w_{c(w)-1}$ and $r_{c,w} = w_{c(w)} w_{c(w)+1} \ldots w_{|w|}$. We call a center function stable, iff for all words $w \in \Sigma^*$ and all symbols $s \in \Sigma$: $c(w) \leq c(l_{c,w} s r_{c,w}) \leq c(w) + 1$.*

One can think of a stable center function as a center function that shifts forward by either 0 or 1 when a single character is inserted.

The following definition uses a center function to define an equivalence relation over all words formed from an alphabet.

Definition 5. *Given an alphabet Σ and a stable center function c, as well as a language $L \subseteq \Sigma^*$, we say that two words w and w' inter-equivalent w.r.t. L and c, denoted $w \equiv_{c,L} w'$, if and only if the following holds:*

$$\forall m \in \Sigma^* : l_{c,w} m r_{c,w} \in L \iff l_{c,w'} m r_{c,w'} \in L$$

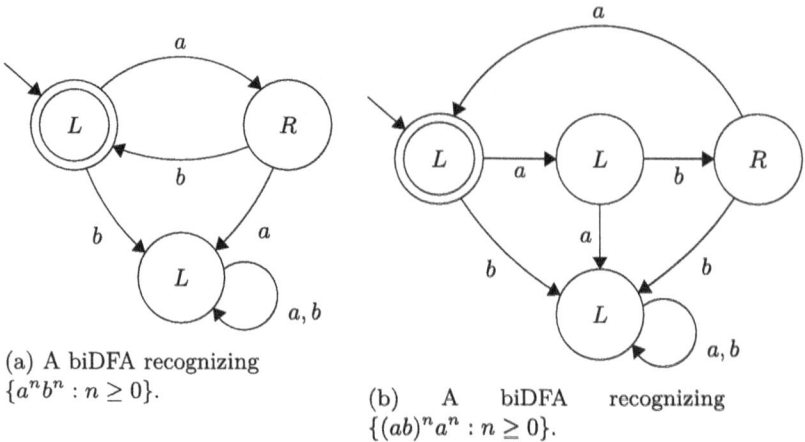

(a) A biDFA recognizing
$\{a^n b^n : n \geq 0\}$.

(b)   A   biDFA   recognizing
$\{(ab)^n a^n : n \geq 0\}$.

Fig. 2. The states labeled with "L" are in Q_l and the states labelled "R" are in Q_r. The states with a double circle are in F and the unconnected arrow indicates the starting state.

In words, Definition 5 states that two words are inter-equivalent if they behave the same with regards to membership in L if the same interfix is inserted at the center indicated by c. This equivalence can be interpreted as a generalization of the left and right congruence. If for all $w \in \Sigma^* : c(w) = 1$ this is the left congruence and conversely if for all $w \in \Sigma^* : c(w) = |w| + 1$ this is the right congruence.

Definition 6. *We call a stable center function* strictly stable *if it satisfies the following condition:*

$$\forall w, w' \in \Sigma^* : w \equiv_{c,L} w' \implies$$
$$\forall s, d \in \Sigma : c(l_{c,w} r_{c,w}) = c(l_{c,w} s r_{c,w}) + 1 \iff c(l_{c,w'} r_{c,w'}) = c(l_{c,w'} d r_{c,w'}) + 1.$$

Strict stability intuitively means that if two words are inter-equivalent the center function for them shifts in the same way for all single characters. So either both shift by 0 or both shift by 1.

Note that strict stability also holds if $s = d$ or $w = w'$. Thus, with a strictly stable center function the same word will shift its center in the same direction no matter which single character is inserted.

Note, that with this definition of a strictly stable center function c the following holds for any $w, w', u \in \Sigma^* : w \equiv_{c,L} w' \implies l_{c,w} u r_{c,w} \equiv_{c,L} l_{c,w'} u r_{c,w'}$. This is because $l_{c,w} u r_{c,w}$ and $l_{c,w'} u r_{c,w'}$, will have the center at the same position relative to u. For this to hold strict stability is necessary. It is also apparent that the center functions associated with the left and right congruence, namely $\forall w \in \Sigma^* : c(w) = 1$ and $\forall w \in \Sigma^* : c'(w) = |w| + 1$, are strictly stable.

Definition 7. *We define the* interjection function *w.r.t. to a center function c and an alphabet Σ, "$\cdot_c$": $\Sigma^* \times \Sigma^* \to \Sigma^*$ s.t. $w \cdot_c u = l_{c,w} u r_{c,w}$.*

Note that the inter equivalence with regard to c and L is a congruence relation w.r.t. "$\cdot_c$" if c is strictly stable. We denote the associated congruence classes with $[w]_{c,L} \in [\Sigma^*]_{c,L}$.

Lemma 8. *If c is strictly stable then the set Σ generates Σ^* with the function* "$\cdot_c$". *This means $\forall w \in \Sigma^*$ with $|w| \geq 1 : \exists s \in \Sigma, w' \in \Sigma^* : w' \cdot_c s = w$.*

Proof. Consider two sequences of interjections $w = \lambda \cdot_c a_1 \cdot_c a_2 \cdot_c \cdots \cdot_c a_k \cdot_c \ldots$ and $w' = \lambda \cdot_c b_1 \cdot_c b_2 \cdot_c \cdots \cdot_c b_k \cdot_c \ldots$. Assume $a_i = b_i$ for $i < k$ but $a_k \neq b_k$. We claim that $w \neq w'$. To prove this we first consider the sequence until $i = k-1$ since the sequences are the same they will produce the same word $p = \lambda \cdot_c a_1 \cdot_c a_2 \cdot_c \cdots \cdot_c a_{k-1}$. Say $c(p) = j$, then due to the strict stability of c $j \leq c(p \cdot_c a_k) = c(p \cdot_c b_k) \leq j+1$. We first handle the case of it being $j+1$. If we consider position j of w and w' we know $w_j = a_k$ and $w'_j = b_k$. This is due to the center for all future interjections being at a position greater or equal to $j+1$ because of the stability of c and position j accordingly never being updated with a different value. Accordingly, the claim holds for this case. For the other case of $c(p \cdot_c a_k) = j$ we similarly know that a_k and b_k are at position $|w| - (k-j)$. This position relative to the end of the word will also never be changed because the center can at most increase by 1 while the length of the word will increase by 1 for every interjection. Our claim is thus proven and we obtain that no two distinct sequences of interjections can produce the same word. The Lemma follows from the observation that there are Σ^n possible sequences of interjections of length n as well as possible words of length n. Thus each word is generated by a unique sequence of interjections. $\square$

Given Definitions 4, 5, 6, 7 and Lemma 8 we can now characterize the class of languages biDFA can recognize as follows:

Theorem 9. *Given an alphabet Σ, a language $L \subseteq \Sigma^*$ can be recognized by a biDFA if and only if there exists a strictly stable center function c such that the number of inter congruence classes on Σ^* w.r.t. c and L is finite.*

Proof. One direction of this Theorem is simple, as any given biDFA directly implies a strictly stable center function. You can recall that a biDFA implicitly constructs a left and right word simultaneously. At any point, we define c to place the center in between this implied left and right word. If this is done it is also immediately apparent that if two words map to the same state in this biDFA they are inter congruent w.r.t. this center function. Since the number of states is finite and all words map to a state the number of congruence classes also must be finite.

For the other direction, we will construct a "canonical" biDFA w.r.t. c and L from the inter-congruence classes. Note that Σ is given. We will first create a state for each congruence class. The strictly stable property allows us to assign a direction to each congruence class, which then forms the left and right states, Q_l and Q_r. Left if they shift by 1 and right if they shift by 0. This allows for a mapping $m : [\Sigma^*]_{c,L} \rightarrow Q_l \cup Q_r$. For δ we use the interjection function. If given a character $s \in \Sigma$ with $[w]_{c,L} \cdot_c s = [w']_{c,L}$ for some $w, w' \in \Sigma^*$ we define

$\delta(m([w]_{c,L}), s) = m([w \cdot_c s]_{c,L})$. This fully defines δ since $[\Sigma]_{c,L}$ is a generating set with "$\cdot_c$" for $[\Sigma^*]_{c,L}$ due to the strict stability of c and is possible since the inter-equivalence is a congruence w.r.t. "$\cdot_c$". We define the starting state as $q_0 = m([\lambda]_{c,L})$, the state associated with the congruence class that contains the empty word. Each congruence class that contains words that are in L will be added to F. We will now prove that the biDFA constructed this way recognizes L by induction over the length of words. We claim that $\delta(q_0, w) = m([w]_{c,L})$. This claim is equivalent to the biDFA recognizing L, since $m([w]_{c,L}) \in F$ iff $w \in L$ by our construction. Also by construction, the claim holds for $[\lambda]_{c,L}$. Now we consider any word w with $|w| = n$. For all $s \in \Sigma$ our construction defines $\delta(m([w]_{c,L}), s) = m([w \cdot_c s]_{c,L})$ but by our induction hypothesis $m([w]_{c,L}) = \delta(q_0, w)$ which implies $\delta(m([w]_{c,L}), s) = \delta(\delta(q_0, w), s) = \delta(q_0, w \cdot_c s) = m([w \cdot_c s]_{c,L})$ completing the induction step. This covers all words of length $n + 1$ since c is strictly stable and thus $[\Sigma]_{c,L}$ is a generating set.   □

Corollary 10. *For a given strictly stable center function c and a language L the number of states of the minimal biDFA defining the same center function and recognizing L is equivalent to the number inter-equivalence classes of the inter-equivalence w.r.t. c and L. There exists an isomorphism between such a minimal biDFA and the "canonical" minimal biDFA w.r.t. c and L.*

Proof. Given a minimal biDFA B that defines a center function c we can consider an equivalence relation, $w \equiv_B w' \iff \delta(q_0, w) = \delta(q_0, w')$, that considers two words equivalent if they end up in the same state. But since the biDFA recognizes L and defines c $w \equiv_B w'$ implies $\forall m : w \cdot_c m \in L \iff w' \cdot_c m \in L$. Therefore, this equivalence implies inter-equivalence $w \equiv_B w' \implies w \equiv_{c,L} w'$. Thus, each inter-equivalence class is a union of the equivalence classes of $\equiv_B$. Now assume there exist two states $u, v \in Q_l \cup Q_r$ s.t. there exists an equivalence class $[w]_{c,L}$ with $[u]_B \cup [v]_B = [w]_{c,L}$. In other words, we assume the "canonical" biDFA we construct in Theorem 9 is not minimal. Since all words arriving in u are inter-equivalent w.r.t. c and L to all words arriving in v we can merge u and v without it affecting c or L. This is a contradiction to the minimality of B. For the merging we will also need to merge all $x, y \in Q_l \cup Q_r : \exists s \Sigma \delta(u, s) = x \wedge \delta(v, s) = y$ to obtain a deterministic model. But since c is strictly stable the inter-equivalence is a congruence w.r.t. "$\cdot_c$" and accordingly for all $w, w' \in \Sigma^*$ $w \equiv_{c,L} w' \implies w \cdot_c s \equiv_{c,L} w' \cdot_c s$ so this is possible.

Thus every minimal biDFA has a number of states which is equal to the number of inter-equivalence classes w.r.t. c and L. This also provides a mapping between states and equivalence classes, which are the states of the "canonical" biDFA w.r.t. c and L. The argument provided for making deterministic merges provides a mapping for the edges. It follows that there exists an isomorphism between a minimal biDFA respecting c and L to the "canonical" biDFA w.r.t. c and L.   □

3.3   NP-Hardness of the BiDFA Minimization Problem

One important difference of Theorem 9 to the Myhill-Nerode Theorem is that it does not provide an isomorphism to the globally minimal biDFA for a given language. Corollary 10 only provides minimality with regard to a given center function. Indeed the size of the biDFA strongly depends on the choice of this center function. Given a DFA for a language, finding a minimal DFA for the same language is possible in polynomial time [10]. Since the smaller size of biDFA is one of their advantages over DFA, efficient minimization would be a nice property. However, in the following Theorem we show that, unlike the minimization problem for DFA, the minimization problem for biDFA is NP-hard. We do this by reducing the vertex cover problem to finding a minimal biDFA. We will first provide formal definitions of both problems.

Definition 11 (Minimal Vertex Cover Problem). *Given a graph $G = (V, E)$ a vertex cover $C \subseteq V$ is a subset of nodes such that each edge $e \in E$ is adjacent to at least one node $c \in C$. The minimal vertex cover problem given a graph G is the problem of finding a cover C whose size $|C|$ is minimal among all possible covers of G.*

Definition 12 (Minimal biDFA Problem). *The minimal biDFA problem, given a language L defined by a biDFA A, is to find a biDFA B recognizing L such that its number of states $|Q_l| + |Q_r|$ is minimal among all biDFA recognizing L.*

The idea for the reduction is that we construct a language where we introduce for each edge $e = (v_x, v_y)$ the two words $e_j v_x e_j$ and $e_j e_j v_y$. If a center function assigns say $c(e_j e_j) = 2$ the next character processed needs to be v_x to be accepting, for $c(e_j e_j) = 3$ it needs to be v_y. Words waiting for the same character can be sent to the same state due to their inter-congruence. So a biDFA with minimal states will choose the center function in such a way that as many words as possible are waiting for the same character. But due to our construction, a word waiting for a character to be accepting can be mapped to its corresponding edge being covered by the vertex represented by this character. Accordingly, a minimal number of states maximizes the overlap of coverage and thus corresponds to a minimal vertex cover. The construction used in Theorem 13 is illustrated in Fig. 3.

Theorem 13. *There exists a polynomial time reduction from the vertex cover problem to the minimal biDFA problem. This also holds for a biDFA that recognizes a finite language.*

Proof. We are given an instance of vertex cover, the graph $G = (V, E)$. From this, we construct an alphabet $\Sigma = E \cup V$. So for each edge and node e_j, v_i the alphabet contains a corresponding character. We now construct a language L s.t. for each edge $e_j = (v_x, v_y) \in E$ it contains the two words $e_j v_x e_j$ and $e_j e_j v_y$. Accordingly, $|L| = 2|E|$. Since this is a finite language we can construct a DFA which recognizes it. Since every DFA is also a biDFA we can use this as an input for the minimal biDFA problem. We will first show that if there exists

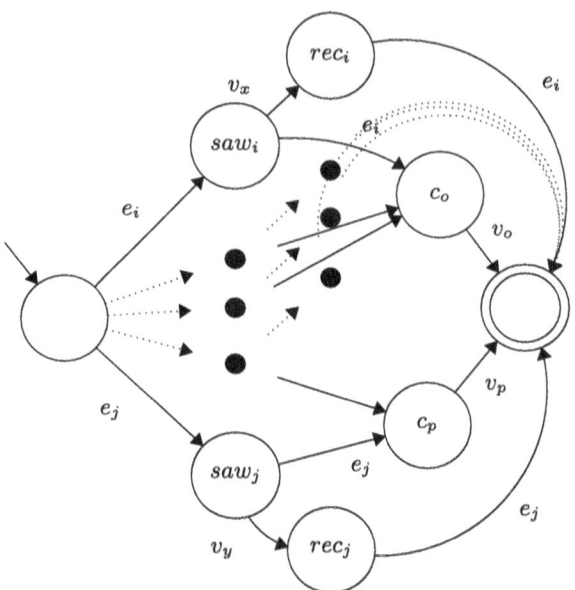

Fig. 3. Sketch of the construction of the biDFA representing a minimal vertex cover that is built in the proof of Theorem 13. In this example there exist edges $e_i = (v_x, v_o)$ and $e_j = (v_y, v_p)$, where e_i and two other edges are covered by v_o and e_j and one other edge is covered by v_p.

a vertex cover of size k on G, then there exists a biDFA with $2 \cdot |E| + k + 3$ states recognizing L. After this, we will show that there exists a vertex cover of size k if there exists a biDFA of size $2|E| + k + 3$ recognizing L. This is sufficient to prove that the minimal vertex cover problem reduces to the minimal biDFA problem.

Vertex Cover to biDFA: We assume there exists a vertex cover $C \subseteq V$ for G, with $|C| = k$. From this, we will provide a constructive proof for a biDFA with $2 \cdot |E| + k + 3$ states that recognizes L. We will construct a biDFA with a starting state q_0 an accepting state a and a sink state s, all in Q_l. Additionally, we have two states for each edge e_j saw_j and rec_j. We create a state c_i for each node v_i in C. All states rec_j and c_i will be in G_l only some of the states saw_j will we use the capabilities of the biDFA to read from the right and be in Q_r. For a state saw_j corresponding to an edge $e_j = (v_x, v_y)$ we define saw_j to be in Q_l if $v_y \in C$ and in Q_r otherwise. It is also clear, that this biDFA has the claimed $2 \cdot |E| + k + 3$ states. The transitions are constructed as follows: The sink is rejecting and always transitions to itself, while the accepting state is the only state in F and always transitions to the sink s. For the starting state we have the transitions $\delta(q_0, e_j) = saw_j$ and $\delta(q_0, v_i) = s$. For the transitions of saw_j with $e_j = (v_x, v_y)$ we have $\delta(saw_j, e_j) = c_y$ if $v_y \in C$ and $\delta(saw_j, e_j) = c_x$ otherwise. This is possible since one of v_x, v_y needs to be in the cover and thus a corresponding state exists. From c_i we transition into the accepting state with v_i and to the sink otherwise. As currently given the biDFA will process and accept for each $e_j = (v_x, v_y)$ one of the two strings $e_j v_x e_j$ and $e_j e_j v_y$. It will process and accept $e_j v_x e_j$ if only $v_x \in C$ and $e_j e_j v_y$ if both v_x and v_y are in the cover or only v_y is. For two strings corresponding to edges e_j, e_l covered by the same node v_x, $e_j e_j$ and $e_l e_l$ will have their center placed in such a way that $e_j e_j \cdot_c v_x \in L$ and $e_l e_l \cdot_c v_x \in L$. Accordingly, both can transition to the same state c_x. To process and accept the remaining of the two strings we define for an edge $e_j = (v_x, v_y)$ $\delta(saw_j, v_x) = rec_j$ if $v_y \in C$ and $\delta(saw_j, v_y) = rec_j$ otherwise. All transitions not mentioned specifically for saw_j will transition to the sink state. From rec_j we transition to the accepting state if we see e_j and to the sink otherwise. This way the biDFA is fully defined and accepts exactly L.

biDFA to Vertex Cover: We will consider a biDFA as previously constructed with $2|E| + k + 3$ states. We claim that if a biDFA with fewer states exists there also exists a corresponding vertex cover with fewer nodes. First observe that the accepting state, the starting state and the sink state must all be unique from all other states independent of the center function. Since the starting state accepts when interjecting any $w \in L$, the accepting state accepts when interjecting λ and the sink state always rejects. Since all words in L have length 3 the only loop that can exist is from and to the sink state and words which are not congruent with the sink can only be congruent with words of equal length. With the starting and accepting state we've already considered all words with lengths 0 and 3, so we only have to consider words of lengths 1 and 2. First, let us assume the starting state is indeed in Q_l, with the corresponding implications for the center function. Clearly no word e_j is congruent with a word e_i, where $e_i \neq e_j$ since even if they are adjacent to the same node v_y $e_j e_j v_y \in L$ but $e_i e_j v_k \notin L$. It follows that there are at least $|E|$ congruence classes of words of length 1. Accordingly, for all $e_i = (v_x, v_y)$ e_i maps to some unique state saw_i. First, we consider the words of length 2, that start with an edge character, followed by a vertex character, like e_i, v_x. There must be a unique non sink state rec_i, to which $e_i v_x$ maps if $e_i = (v_x, v_y)$ and $saw_i \in Q_l$ or $e_i = (v_y, v_x)$ and $saw_i \in Q_r$. Otherwise, it maps to the sink state. There must be at least $|E|$ such states, since assuming $saw_i \in Q_l$ for all $e_j \neq e_i$ and $v_p \in V$ $e_j v_p e_i \notin L$ but $e_i v_x e_i \in L$. Similarly if saw_i was in Q_r when considering $e_i v_y$ it holds for all $e_j \neq e_i$ that $e_j e_i v_p \notin L$ but $e_i e_i v_p \in L$. Accordingly, under the assumption of the starting state being in Q_l, we have shown that any biDFA will need at least $2|E| + 3$ states to handle the words we have analyzed so far. Only the words of length 2 which contain two characters corresponding to edges, e.g. $e_j e_j$, are left to analyze. Two such words are congruent iff both allow for the same final character which corresponds to a vertex. Now assume there exists an assignment of the states saw_i to Q_l and Q_r, s.t. less than k such congruence classes existed. Since each of these states only has one outgoing transition to the accepting state with a label that corresponds to a vertex, we can take this set of vertexes. This set is a vertex cover of size smaller k, which was our claim.

To finalize the proof we need to look at the case of the starting state being in Q_r. This leads to $|E| + l$ congruence classes for words of length 1, with $l \geq k$, since e_i and e_j with $e_i \neq e_j$ cannot be congruent with the same argument as above and v_x and v_y with $v_x \neq v_y$ cannot be congruent since there exists an e_j s.t. $e_j e_j v_x \in L$ but $e_j e_j v_y \notin L$ and there are at least l unique vertex characters as suffixes. The same argument leads to $|E| + l$ states for the words of length 2, thus having the starting state in Q_r will always lead to a biDFA larger than the one we constructed.   □

4   Discussion

In this paper, we have provided new theoretical results for deterministic biautomata first introduced by Jirásková and Klíma [11]. We prove that minimal biDFA can be exponentially smaller than the corresponding DFA, strongly encouraging further research into where biDFA could replace DFA. The inter-congruence relationship introduced in this paper provides a good theoretical foundation for developing algorithms that learn biDFA. It provides a test of whether two words should end up in the same state which could be used in algorithms such as L^* [2]. Further, the ability to recognize languages in linear time, that deterministic PDAs cannot, like palindrome, could be of interest to the programming language field. Here the inter-congruence relationship we define also provides a strategy to prove whether a biDFA exists for a given language.

An advantage that remains for DFA over biDFA is that DFA do not require knowledge of where a string ends. This is especially important in applications on data streams where there quite possibly is no end to a word. In such applications, it will be hard to use biDFA.

Aside from this, the main limitation of biDFA is the concept of the center, defining which states are right and left. While this center for each word gets defined by an already existing biDFA, finding a good center function for a language is not a trivial problem. Indeed, this fact was used in the proof of Theorem 13. We expect that the difficulty of this will be the main obstacle in adapting algorithms for identifying or minimizing DFA from some given data or grammar. Furthermore, Theorem 13 provides only a lower bound. The upper bound for the complexity of this problem is even an open question. It likely depends on the complexity of the language equality problem for biDFA which is also an open question.

Nevertheless, we do not exclude the existence of efficient heuristics for identifying center functions. Such heuristics are the main open problem for obtaining learning algorithms for biDFA. However, the potential of having exponentially fewer states provides significant motivation for finding such heuristics.

Acknowledgement. This work is supported by NWO TTW VIDI project 17541 - Learning state machines from infrequent software traces (LIMIT).

References

1. Alur, R., Dill, D.L.: A theory of timed automata. Theor. Comput. Sci. **126**(2), 183–235 (1994)
2. Angluin, D.: Learning regular sets from queries and counterexamples. Inf. Comput. **75**(2), 87–106 (1987)
3. Chatterjee, K., Henzinger, T.A., Otop, J.: Nested weighted automata. ACM Trans. Comput. Logic (TOCL) **18**(4), 1–44 (2017)
4. Cook, J.E., Wolf, A.L.: Discovering models of software processes from event-based data. ACM Trans. Softw. Eng. Methodol. (TOSEM) **7**(3), 215–249 (1998)
5. De la Higuera, C.: Grammatical Inference: Learning Automata and Grammars. Cambridge University Press, Cambridge (2010)
6. Ficara, D., Giordano, S., Procissi, G., Vitucci, F., Antichi, G., Di Pietro, A.: An improved DFA for fast regular expression matching. ACM SIGCOMM Comput. Commun. Rev. **38**(5), 29–40 (2008)
7. Heule, M.J.H., Verwer, S.: Software model synthesis using satisfiability solvers. Empir. Softw. Eng. **18**(4), 825–856 (2013)
8. Hoe, A.V., Sethi, R., Ullman, J.D.: Compilers-Principles, Techniques, and Tools. Pearson Addison Wesley Longman, Boston (1986)
9. Holzer, M., Jakobi, S.: Minimization and characterizations for biautomata. Fund. Inform. **136**(1–2), 113–137 (2015)
10. Hopcroft, J.E., Motwani, R., Ullman, J.D.: Introduction to automata theory, languages, and computation. ACM SIGACT News **32**(1), 60–65 (2001)
11. Jirásková, G., Klíma, O.: On linear languages recognized by deterministic biautomata. Inf. Comput. **286**, 104778 (2022)
12. Klíma, O., Polák, L.: On biautomata. RAIRO-Theor. Inform. Appl. **46**(4), 573–592 (2012)
13. Knuth, D.E.: On the translation of languages from left to right. Inf. Control **8**(6), 607–639 (1965)
14. Lee, D., Yannakakis, M.: Principles and methods of testing finite state machines-a survey. Proc. IEEE **84**(8), 1090–1123 (1996)
15. Mayr, F., Yovine, S., Pan, F., Basset, N., Dang, T.: Towards efficient active learning of PDFA. arXiv preprint arXiv:2206.09004 (2022)
16. Muškardin, E., Aichernig, B.K., Pill, I., Tappler, M.: Learning finite state models from recurrent neural networks. In: ter Beek, M.H., Monahan, R. (eds.) IFM 2022. LNCS, vol. 13274, pp. 229–248. Springer, Cham (2022). https://doi.org/10.1007/978-3-031-07727-2_13
17. Nagy, B.: On a hierarchy of $5' \rightarrow 3'$ sensing watson-crick finite automata languages. J. Log. Comput. **23**(4), 855–872 (2013)
18. Nagy, B., Kovács, Z.: On deterministic 1-limited $5' \rightarrow 3'$ sensing watson-crick finite-state transducers. RAIRO-Theor. Inform. Appl. **55**, 5 (2021)
19. Nagy, B., Parchami, S.: On deterministic sensing $5' \rightarrow 3'$ watson-crick finite automata: a full hierarchy in 2detlin. Acta Informatica **58**, 153–175 (2021)
20. Rabin, M.O.: Probabilistic automata. Inf. Control **6**(3), 230–245 (1963)
21. Vaandrager, F., Garhewal, B., Rot, J., Wißmann, T.: A new approach for active automata learning based on apartness. In: Fisman, D., Rosu, G. (eds.) TACAS 2022. LNCS, vol. 13243, pp. 223–243. Springer, Cham (2022). https://doi.org/10.1007/978-3-030-99524-9_12

Block Languages and Their Bitmap Representations

Guilherme Duarte[1] , Nelma Moreira[1]([envelope]) , Luca Prigioniero[2] ,
and Rogério Reis[1]

[1] CMUP & DCC, Faculdade de Ciências da Universidade do Porto,
Rua do Campo Alegre, 4169-007 Porto, Portugal
{guilherme.duarte,nelma.moreira,rogerio.reis}@fc.up.pt
[2] Department of Computer Science, Loughborough University, Loughborough, UK
L.Prigioniero@lboro.ac.uk

Abstract. In this paper we consider block languages, namely sets of
words having the same length, and we propose a new representation
for these languages. In particular, given an alphabet of size k and a
length ℓ, a block language can be represented by a bitmap of length k^ℓ,
where each bit indicates whether the corresponding word, according to
the lexicographical order, belongs, or not, to the language (bit equal to 1
or 0, respectively). This representation turns out to be a good tool for
the investigation of several properties of block languages, making proofs
simpler and reasoning clearer. First, we show how to convert bitmaps into
deterministic and nondeterministic finite automata. We then focus on the
size of the machines obtained from the conversion and we prove that their
size is minimal. Finally, we give an analysis of the maximum number of
states sufficient to accept every block language in the deterministic and
nondeterministic case.

1 Introduction

In the area of formal languages and automata theory, the class of regular languages is one of the most investigated. Classical recognizers for this class are finite automata, in both deterministic and nondeterministic variants. The capabilities of these machines to represent languages in a more or less succinct way have been widely studied in the area of *descriptional complexity*. In this context, the *size* of a model is measured in terms of number of symbols used to write down its description. In the specific case of finite automata, it is often considered the number of states as a measure of complexity. In this area, the minimality of finite automata has been also studied. For example, it is well known that, given a language, the deterministic finite automaton of minimal size accepting it is unique (up to isomorphisms), and there exist efficient algorithms for the minimization

This work was partially supported by CMUP, member of LASI, which is financed by
national funds through FCT – Fundação para a Ciência e a Tecnologia, I.P., under the
projects with reference UIDB/00144/2020 and UIDP/00144/2020.

S. Z. Fazekas (Ed.): CIAA 2024, LNCS 15015, pp. 124–137, 2024.
https://doi.org/10.1007/978-3-031-71112-1_9

of these machines [2]. The situation in the nondeterministic case is more challenging as minimal nondeterministic finite automata are not necessarily unique. Furthermore, given an integer n, deciding whether there is a nondeterministic finite automaton with less than n states accepting a language is a PSPACE-hard problem [17]. In this paper we consider finite languages where all words have the same length, which are called *homogeneous* or *block* languages. Their investigation is mainly motivated by their applications to several contexts such as code theory [4,9] and image processing [6,7]. A typical problem in code theory is the construction of (maximal) block languages (codes) capable of detecting and correcting errors. Several properties of block codes using automata theory have also been studied, e.g. [15]. On the other hand, an image can be represented by a set of words of a same length (pixels). Then, automata can be used to generate, compress, and manipulate images.

As a subclass of finite languages, block languages inherit some properties known for that class. For instance, the minimization of deterministic finite automata can be done in linear time in the case of finite (and hence also block) languages [13]. Due to the fact that all words have the same length, there are some gains in terms of descriptional complexity. It is known that the elimination of nondeterminism from an n-state nondeterministic finite automaton for a block language costs $2^{\Theta(\sqrt{n})}$ in size [7], which is smaller than the general case, for which the cost in size is $2^{\Theta(n)}$ [11,14]. The maximum number of states of minimal deterministic finite automata for finite and block languages were studied by Câmpeanu and Ho [3], and Hanssen and Liu determined the number of block languages that attain the maximum state complexity [8]. Minimal deterministic finite automata for finite languages were enumerated by Almeida et al. [1]. Asymptotic estimates and exact formulae for the number of n-state minimal deterministic finite automata accepting finite languages over alphabets of size k were obtained by J. Priez [12] and by Price et al. [5].

Here we propose a new representation for block languages. In particular, given an alphabet of size k and a length ℓ, each block language can be represented by a binary string of length k^ℓ, also called *bitmap*, in which each symbol (or *bit*) indicates whether the correspondent word, according to the lexicographical order, belongs to the language (bit equal to 1) or not (bit equal to 0). We use this representation as a tool to investigate several properties of block languages. More precisely, in Sects. 4 and 5 we show how to convert bitmaps into deterministic and nondeterministic finite automata, respectively. It is important to notice that the devices yielded by such conversions have minimal size. While the conversion to deterministic finite automata can be done in polynomial time in the size of the bitmap, we prove that the transformation in the nondeterministic case is NP-complete. For the deterministic case, we also refine the analysis of the state complexity of block languages given by Câmpeanu and Ho [3], and we present a family of languages that witnesses the optimality of such costs (Sect. 4.1). On the other hand, for nondeterministic finite automata, we determine the sufficient number of states to accept every block language (Sect. 5.1).

2   Preliminaries

In this section we review some basic definitions about finite automata and languages and fix notation. Given an *alphabet* Σ, a *word* w is a sequence of symbols, and a *language* $L \subseteq \Sigma^\star$ is a set of words on Σ. The empty word is represented by ε. The *(left) quotient* of a language L by a word $w \in \Sigma^\star$ refers to the set $w^{-1}L = \{w' \in \Sigma^\star \mid ww' \in L\}$. The *reversal* of a word $w = \sigma_0\sigma_1\cdots\sigma_{n-1}$ is denoted as w^R and is obtained by reversing the order of the symbols of w, that is $w^R = \sigma_{n-1}\sigma_{n-2}\cdots\sigma_0$. Given two integers i, j with $i < j$, let $[i, j]$ denote the range from i to j, including both i and j, namely $\{i, \ldots, j\}$. Moreover, we shall omit the left bound if it is equal to 0, thus $[j] = \{0, \ldots, j\}$.

A *nondeterministic finite automaton* (NFA) is a five-tuple $\mathcal{A} = \langle Q, \Sigma, \delta, I, F \rangle$ where Q is a finite set of states, Σ is a finite alphabet, $I \subseteq Q$ is the set of initial states, $F \subseteq Q$ is the set of final states, and $\delta : Q \times \Sigma \rightarrow 2^Q$ is the transition function. We consider the *size* of an NFA as its number of states. The transition function can be extended to words and sets of states in the natural way. When $I = \{q_0\}$, we use $I = q_0$. An NFA accepting a non-empty language is *trim* if every state is accessible from an initial state and every state leads to a final state. Given a state $q \in Q$, the *right language* of q is $\mathcal{L}_q q(\mathcal{A}) = \{w \in \Sigma^\star \mid \delta(q, w) \cap F \neq \emptyset\}$, and the *left language* is $\overleftarrow{\mathcal{L}}_q(\mathcal{A}) = \{w \in \Sigma^\star \mid q \in \delta(I, w)\}$. The *language accepted* by $\mathcal{A}$ is $\mathcal{L}(\mathcal{A}) = \bigcup_{q \in I} \mathcal{L}_q(\mathcal{A})$. An NFA $\mathcal{A}$ is *minimal* if it has the smallest number of states among all NFAs that accept $\mathcal{L}(\mathcal{A})$. An NFA is *deterministic* (DFA) if $|I| = 1$ and $|\delta(q, \sigma)| \leq 1$, for all $(q, \sigma) \in Q \times \Sigma$. We can convert an NFA into an equivalent DFA by using the well-known subset construction. Two states q_1, q_2 of an automaton $\mathcal{A}$ are *equivalent* (or *indistinguishable*) if $\mathcal{L}_{q_1}(\mathcal{A}) = \mathcal{L}_{q_2}(\mathcal{A})$. If a DFA is *minimal* it has no equivalent states and it is unique up to renaming of states. If $\mathcal{A}$ is the minimal DFA for L, then, for each state q, $\mathcal{L}_q(\mathcal{A}) = w_q^{-1}L$ for some $w_q \in \Sigma^\star$, and if $q \neq q'$ then $w_q^{-1}L \neq w_{q'}^{-1}L$. The *state complexity* of a language L, denoted $sc(L)$, is the size of the minimal DFA accepting L. The *nondeterministitic state complexity* of a language L, denoted $nsc(L)$, is defined analogously.

A trim NFA $\mathcal{A} = \langle Q, \Sigma, \delta, I, F \rangle$ for a non-empty finite language, whose longest word is $\ell \geq 0$, is acyclic and ranked, i.e., the set of states Q can be partitioned into $Q_0 \cup Q_1 \cup \cdots \cup Q_\ell$ such that for every state q in rank Q_r, $\mathcal{A}$ reaches a final state by words of length at most r ($Q_r = \{q \in Q \mid \forall w \in \Sigma^\star, \delta(q, w) \in F \implies |w| \leq r\}$) and all transitions from states in rank Q_r lead only to states in rank Q_s, with $s < r$. We define the *width* of a rank Q_r as the cardinality of Q_r, the *width* of $\mathcal{A}$ to be the maximal width of all ranks. In the following, for the ease of notation, we shall denote the ranks by their indices, e.g., we refer to rank Q_r as rank r.

A DFA for a finite language is also ranked but it may have a *dead state* Ω, which is the only cyclic state, is usually omitted, and has no rank. Formally, $\mathcal{A} = \langle Q \cup \{\Omega\}, \Sigma, \delta, q_0, F \rangle$, with $F \subseteq Q$, $q_0 \neq \Omega$, and, for each symbol $\sigma \in \Sigma$, $\delta(\Omega, \sigma) = \Omega$. Moreover, for each nonempty word w and each state $q \in Q$, $\delta(q, w) \neq q$. In a trim acyclic automaton, two states q and q' are equivalent if they are both in the same rank, either final or not final, and their transition functions lead to equivalent states, i.e., $\delta(q, w) \in F \iff \delta(q', w) \in F$, for each word $w \in \Sigma^*$. An acyclic DFA can be minimized by merging equivalent states and the resulting algorithm runs in linear time in the size of the automaton (Revuz algorithm, [1,13]).

In the following we shall consider sequences of Boolean values, $\mathsf{B} \in \{0,1\}^n$ that we denote by *bitmaps*. Given two bitmaps $\mathsf{B}_1, \mathsf{B}_2 \in \{0,1\}^n$, $\mathsf{B}_1 \circ \mathsf{B}_2$ represents the bitmap obtained by carrying out the bitwise operation $\circ \in \{\vee, \wedge\}$, and $\overline{\mathsf{B}}_1$ the bitwise complement of B_1.

3   Block Languages and Bitmaps

Given an alphabet $\Sigma = \{\sigma_0, \ldots, \sigma_{k-1}\}$ of size $k > 0$ and an integer $\ell \geq 0$, a *block language* $L \subseteq \Sigma^\ell$ is a set of words of length ℓ over Σ. The language L can be characterized by a word in $\{0,1\}^{k^\ell}$ that we call *bitmap* and denote as

$$\mathsf{B}(L) = b_0 \cdots b_{k^\ell - 1},$$

where $b_i = 1$ if the word w is in L, and $i \in [k^\ell - 1]$ is the index of w in the lexicographical ordered list of all the words of Σ^ℓ. We will denote the bitmap of a language as B when it is unambiguous to which language the bitmap refers to. Moreover, each bitmap $\mathsf{B} \in \{0,1\}^{k^\ell}$ represents a block language of length ℓ over a k-ary alphabet, thus one can use any alphabet of size k. Boolean bitmap bitwise operations trivially correspond to boolean set operations on block languages of the same length.

Example 1. Let $L = \{aaaa, aaba, aabb, abab, abba, abbb, babb, bbaa, bbab, bbba\}$ be a language over $\{a, b\}$ and $\ell = 4$. The bitmap of L is $\mathsf{B}(L) = 1011011100011110$.

A bitmap $\mathsf{B} \in \{0,1\}^{k^\ell}$ can be split into factors of length k^r, for $r \in [\ell]$. Let $s_i^r = b_{ik^r} \cdots b_{(i+1)k^r - 1}$ denote the i-th factor of length k^r, for $i \in [k^{\ell-r} - 1]$. Since each factor of length k^r can also be split into k factors, s_i^r is inductively defined as:

$$s_i^r = \begin{cases} b_i, & \text{if } r = 0, \\ s_{ik}^{r-1} \cdots s_{(i+1)k-1}^{r-1}, & \text{otherwise.} \end{cases}$$

The following lemma formally introduces the observation that each factor s_i^r, with $r \in [\ell]$ and $i \in [k^{\ell-r} - 1]$, represents the bitmap of a quotient of L.

Lemma 1. *Let $L \subseteq \Sigma^\ell$ be a block language, $|\Sigma| = k$, $\ell \geq 0$, and B the bitmap of L. Let $r \in [\ell]$, $i \in [k^{\ell-r} - 1]$, and $w \in \Sigma^{\ell-r}$ be the word of index i of size $\ell - r$, in lexicographic order. Then, s_i^r corresponds to the bitmap of $w^{-1}L$.*

Proof. Let us prove by induction on $r \in [\ell]$:

- *Base case $r = 0$:* by definition, $s_i^0 = b_i$. Additionally, we have that $b_i = 1$ if the word w is the i-th word in Σ^ℓ and $w \in L$. Since $|w| = \ell$ either $w^{-1}L = \{\varepsilon\}$ or $w^{-1}L = \emptyset$, according to the membership of w in L.

– *Inductive step:* since $r > 0$, we have that $s_i^r = s_{ik}^{r-1} \cdots s_{(i+1)k-1}^{r-1}$. By hypothesis, we have that s_{ik+j}^{r-1} corresponds to the bitmap of the language $w_j^{-1}L$, where w_j denotes the $(ik+j)$-th word of length $\ell-(r-1)$ over Σ, for each $j \in [k-1]$. One can observe that the words $(w_j)_{j \in [k-1]}$ are all equal on the first $\ell - r$ symbols corresponding to the i-th word of size $\ell - r$. Thus, s_i^r corresponds to the bitmap of the quotient of L by the i-th word of length $\ell - r$.   □

Example 2. Recall the Example 1, where $\mathsf{B} = 1011011100011110$, $k = 2$ and $\ell = 4$. We have that $s_0^2 = 1011$ is the bitmap of $(aa)^{-1}L = \{aa, ba, bb\}$, $s_1^3 = 00011110$ the bitmap of $b^{-1}L = \{abb, baa, bab, bba\}$, and $s_0^4 = \mathsf{B}$ the bitmap of L.

Given a bitmap $\mathsf{B} \in \{0,1\}^{k^\ell}$, let $\mathcal{B}_r$ be the set of factors of B of length k^r, for $r \in [\ell]$, in which there is at least one bit different than zero, that is,

$$\mathcal{B}_r = \{\, s \in \{0,1\}^{k^r} \mid \exists i \in [k^{\ell-r} - 1] : s = s_i^r \text{ and } s_i^r \neq 0^{k^r} \,\}.$$

Example 3. Consider the bitmap of Example 1, $\mathsf{B} = 1011011100011110$, with $k = 2$ and $\ell = 4$. We have $\mathcal{B}_0 = \{1\}$, which contains the only factor of length 1 in B different than 0, $\mathcal{B}_1 = \{01, 10, 11\}$, which contains the factors of length 2 different than 00 occurring in even positions of B, $\mathcal{B}_2 = \{0001, 0111, 1011, 1110\}$, which contains the factors of length 4 in positions multiple of 4 in B. Analogously, we have $\mathcal{B}_3 = \{00011110, 10110111\}$ and $\mathcal{B}_4 = \{\mathsf{B}\}$.

The size of $\mathcal{B}_r$ is bounded by the number of factors with size k^r. Additionally, each factor is a composition of factors from the previous set. These two conditions are formally stated in the following lemma.

Lemma 2. *Let $L \subseteq \Sigma^\ell$ be a block language of words of length $\ell \geq 0$ over an alphabet Σ of size $k > 0$, with a correspondent bitmap B. Then, the cardinality of $\mathcal{B}_r$ is bounded by:*

$$|\mathcal{B}_r| \leq \begin{cases} 1, & \text{if } r = 0, \\ \min(k^{\ell-r}, (|\mathcal{B}_{r-1}| + 1)^k - 1), & \text{otherwise.} \end{cases}$$

Proof. The case $r = 0$ is trivial, as $\mathcal{B}_0$ contains at most the factor 1. This happens when L is not empty. Since there are at most $k^{\ell-r}$ unique factors of size k^r in a bitmap of size k^ℓ, for $r \in [\ell]$, then $|\mathcal{B}_r| \leq k^{\ell-r}$. Now, let $s \in \mathcal{B}_r$, for some $r \in [1, \ell]$. By definition, $s = s_0 \cdots s_{k-1}$ (with $|s_j| = k^{r-1}$), and either $s_j \in \mathcal{B}_{r-1}$ or it is composed only by zeros, for every $j \in [k-1]$. Since $s \in \mathcal{B}_r$, it must have at least one bit equal to 1. Therefore, $|\mathcal{B}_r| \leq (|\mathcal{B}_{r-1}| + 1)^k - 1$.   □

The sets $\mathcal{B}_r$ are related to the states of the finite automata representing the block language with bitmap B, as described in the next sections. A finite automaton for a block language is, of course, also acyclic and ranked. If two states belong to the same rank, their right languages contain only words of the same length. All final states belong to Q_0 and, therefore, can be merged. Additionally, if an NFA for a block language has multiple initial states, they can also be merged.

4   Bitmaps for Block Languages and Minimal DFAs

In this section we relate the bitmap of a block language to its minimal DFA. Given a bitmap B associated with a block language $L \subseteq \Sigma^\ell$, with $|\Sigma| = k$ and $\ell \geq 0$, one can directly build a minimal DFA $\mathcal{A}$ for L. Let $Q = \bigcup_{r \in [\ell]} \mathcal{B}_r$ be the set of states of $\mathcal{A}$, and the transition function mapping the states in $\mathcal{B}_r$ with the ones in $\mathcal{B}_{r-1}$, $r \in [1, \ell]$. We will now detail this construction. We start by the final state, that is the factor $1 \in \mathcal{B}_0$, as well as the dead state corresponding to the factor 0. Then, for each rank $r = 1, \ldots, \ell$, we consider every factor $s \in \mathcal{B}_r$ as a state in rank r. As stated in Lemma 1, every factor s corresponds to the bitmap of the quotient of the language L by some word w. The transitions from s are then given by the decomposition of s into $s_0 \cdots s_{k-1}$, where $|s_i| = k^{r-1}$, for every $i \in [k-1]$. Then, we set $\delta(s, \sigma_i) = s_i$, where $s_i \in \mathcal{B}_{r-1}$. Note that, if the language L is not empty, this construction creates exactly one initial and one final state, since $|\mathcal{B}_0| = |\mathcal{B}_\ell| = 1$.

Lemma 3. *Let $L \subseteq \Sigma^\ell$ be a block language with bitmap B, where $\ell \geq 0$. Then, the DFA $\mathcal{A}$ obtained by applying the above construction to B accepts L, that is, $\mathcal{L}(\mathcal{A}) = L$.*

Proof. Let us show that both $\mathcal{L}(\mathcal{A}) \subseteq L$ and $L \subseteq \mathcal{L}(\mathcal{A})$.

- $\mathcal{L}(\mathcal{A}) \subseteq L$: Let $w \in \Sigma^\ell \setminus L$. By construction, each state of $\mathcal{A}$ is also a factor which, by Lemma 1, is the bitmap of the quotient of L by some word. Since $w \notin L$, w can be split into two words $w = w_1 w_2$ such that $w_1^{-1} L = \emptyset$. Then, since for every word x, $x^{-1}\emptyset = \emptyset$, we get $w^{-1}L = (w_1 w_2)^{-1}L = w_2^{-1}(w_1^{-1}L) = \emptyset$. The empty language corresponds to the bitmap associated with the dead state, therefore $w \notin \mathcal{L}(\mathcal{A})$.
- $L \subseteq \mathcal{L}(\mathcal{A})$: Let $w \in L$. Then, $w^{-1}L = \{\varepsilon\}$, whose bitmap is the final state, therefore $w \in \mathcal{L}(\mathcal{A})$. □

Lemma 4. *Let $L \subseteq \Sigma^\ell$ be a block language with bitmap B, where $\ell \geq 0$. Then, the DFA $\mathcal{A}$ obtained by applying the above construction to B is minimal.*

Proof. Let s_1, s_2 be two distinct states of $\mathcal{A}$. It can be noticed that if s_1 and s_2 do not belong to the same rank, they are distinguishable. Otherwise, if they belong to the same rank, by construction, $s_1 \neq s_2$, and consequently they have distinct right languages. Therefore, s_1 and s_2 are distinguishable. □

Combining the results of Lemmas 3 and 4, we obtain:

Theorem 1. *Let $L \subseteq \Sigma^\ell$ be a block language. The above construction of a DFA from the bitmap B(L) yields the minimal DFA for L.*

Example 4. Let $L \subseteq \{a, b\}^4$ be the language of Example 1 with bitmap $B(L) = 1011011100011110$. The correspondent minimal DFA is depicted in Fig. 1. The final state (in rank 0) labeled by s_0^0 represents the bitmap factor 1 and the dead state is omitted as well as all transitions from and to it. States in rank

1 correspond to 2-bit factors, in this case: $s_0^1 = 10$, $s_1^1 = 11$, and $s_2^1 = 01$. We have $\delta(s_0^1, a) = s_0^0$, $\delta(s_2^1, b) = s_0^0$, etc. States in rank 2 correspond to $2^2 = 4$-bit words: $s_0^2 = 1011$, $s_1^2 = 0111$, $s_2^2 = 0001$ and $s_3^2 = 1110$. And we have, for instance, $\delta(s_0^2, a) = s_0^1$ and $\delta(s_0^2, b) = s_1^1$. Similarly for ranks 3 and 4. The initial state corresponds to B.

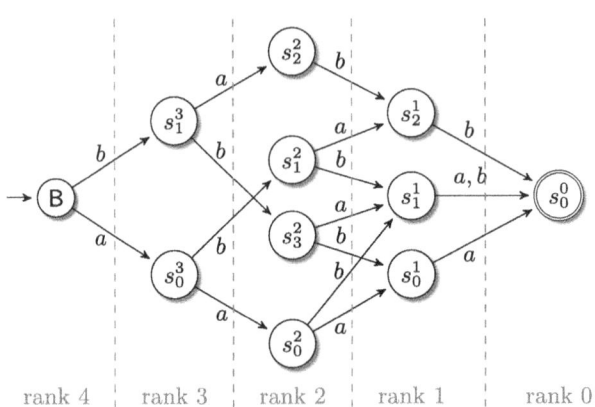

Fig. 1. Minimal DFA accepting the language of Example 1, where $s_0^0 = 1$, $s_0^1 = 10$, $s_1^1 = 11$, $s_2^1 = 01$, $s_0^2 = 1011$, $s_1^2 = 0111$, $s_2^2 = 0001$, $s_3^2 = 1110$, $s_0^3 = 10110111$, $s_1^3 = 00011110$, and B = 1011011100011110.

4.1   Maximal Size of Minimal DFAs for Block Languages

Câmpeanu and Ho [3] showed that the number of states of a DFA accepting a block language $L \subseteq \Sigma^\ell$, over an alphabet of size k and $\ell \geq 0$, is at most $\frac{k^{\ell-r}-1}{k-1} + \sum_{n=0}^{r-1}(2^{k^n} - 1) + 1$, where $r = \min\{n \in [\ell] \mid k^{\ell-n} \leq 2^{k^n} - 1\}$. In the next result we give an estimation of the value of r.

Theorem 2. *Let $\ell > 0$, $k > 1$, and $r = \min\{n \in [\ell] \mid k^{\ell-n} \leq 2^{k^n} - 1\}$. Then, $r = \lfloor \log_k \ell \rfloor + 1 + x$, for some $x \in \{-1, 0, 1\}$.*

Proof. By definition of r, we have that both the following inequalities hold: $k^{\ell-r} \leq 2^{k^r} - 1$ and $k^{\ell-(r-1)} > 2^{k^{r-1}} - 1$. From the first inequality we obtain:

$$k^{\ell-r} \leq 2^{k^r} - 1 \implies k^{\ell-r} < 2^{k^r} \implies (\ell-r) \cdot \log_2 k < k^r \implies$$

$$\implies \log_k(\ell-r) + \log_k(\log_2 k) < r \implies$$

$$\implies \log_k(\ell-r) < r \implies \log_k(\ell(1 - \frac{r}{\ell})) < r \implies$$

$$\implies \log_k \ell + \log_k(1 - \frac{r}{\ell}) < r \implies \log_k \ell < r.$$

While, from the second inequality, we get:

$$k^{\ell-r+1} > 2^{k^{r-1}} - 1 \implies k^{\ell-r+1} \geq 2^{k^{r-1}} \implies$$
$$\implies (\ell - r + 1) \cdot \log_2 k \geq k^{r-1} \implies$$
$$\implies \log_k(\ell - r + 1) + \log_k(\log_2 k) \geq r - 1 \implies$$
$$\implies \log_k(\ell - r + 1) > r - 2 \implies \log_k \ell + \log_k(1 + \frac{1-r}{\ell}) > r - 2$$
$$\implies \log_k \ell > r - 2.$$

$\square$

We now present a family of *witness languages* recognized by minimal DFAs of maximal size, according to the bounds given in Theorem 2. The bitmaps of these languages correspond to sequences of binary representations of the first m positive integers, as we will see in Lemma 5. Given an integer i, let us denote by $i_{[2]}$ its binary representation, and let $\mathsf{pad}(s,t)$ be the function that adds leading zeros to a binary string s until its length equals t. Moreover, to indicate that the j-th least significant bit of $i_{[2]}$ is 1, we will use the notation $i \wedge 2^j \neq 0$. Given a block language $L \subseteq \Sigma^\ell$ and a word $w \in L$, we denote by $\mathsf{ind}(w)$ the index of w in the lexicographical ordered list of the words of Σ^ℓ.

Let $\ell > 0$, $k = 2$, and $r = \min\{n \in [\ell] \mid 2^{\ell-n} \leq 2^{2^n} - 1\}$ as in Theorem 2. In a minimal DFA with maximal size, the rank having the largest size is either r or $r - 1$, depending on whether $2^{\ell-r} > 2^{2^{r-1}} - 1$ or not, respectively. Let r_ℓ be that rank and $m = \max(2^{\ell-r}, 2^{2^{r-1}} - 1)$ its width. Then, we consider the following family of witnesses, defined for every $\ell > 0$:

$$\mathsf{MAX}_\ell = \{w_1 w_2 \mid w_1 \in \Sigma^{\ell-r_\ell}, w_2 \in \Sigma^{r_\ell},$$
$$i = \mathsf{ind}(w_1), j = \mathsf{ind}(w_2), (i+1) \wedge 2^j \neq 0\}.$$

Example 5. For $\ell = 5$, we have $r = \min\{n \in [5] \mid 2^{5-n} \leq 2^{2^n} - 1\} = 2$. Moreover, $m = \max(2^{5-2}, 2^{2^{2-1}} - 1) = \max(8, 3) = 8$, implying that $r_\ell = r$. Then, consider for $\Sigma = \{a, b\}$,

$$\mathsf{MAX}_5 = \{aaaaa, aabab, abaaa, ababa, abbba, baaaa,$$
$$baaba, babab, babba, bbaaa, bbaab, bbaba, bbbbb\}.$$

For example, let $w_1 = baa$ where $i = \mathsf{ind}(w_1) = 4$ and $(i+1)_{[2]} = 101$. For $j = 0$ and $j = 2$, we have that $(i+1) \wedge 2^j \neq 0$, which correspond to the words aa and ba, respectively. Thus, $baaaa, baaba \in \mathsf{MAX}_5$.

Lemma 5. *Let* r, r_ℓ, *and* m *be defined as before for* $\ell > 0$ *and alphabet size* $k = 2$. *Let* $P_{m,r_\ell} = \prod_{i=1}^m \mathsf{pad}(i_{[2]}, 2^{r_\ell})^{\mathrm{R}}$. *Then, the bitmap of the language* MAX_ℓ *is given by*

$$\mathsf{B}(\mathsf{MAX}_\ell) = \begin{cases} P_{m,r_\ell}, & \text{if } m = 2^{\ell-r}, \\ P_{m,r_\ell} \cdot 0^{2^{r_\ell}}, & \text{if } m = 2^{2^{r-1}} - 1. \end{cases}$$

Proof. Let us show that the bitmap is correct, for either value of m.

1. $m = 2^{\ell-r}$:

    In this case $r_\ell = r$. Also, $|\mathsf{B}(\mathsf{MAX}_\ell)| = 2^\ell$. Let $w_1 \in \Sigma^{\ell-r_\ell}$, $i = \mathrm{ind}(w_1)$, $w_2 \in \Sigma^{r_\ell}$, and $j = \mathrm{ind}(w_2)$. We must prove that $w_2 \in w_1^{-1}\mathsf{MAX}_\ell$ if, and only if, the j-th most significant bit of $(i+1)_{[2]}^{\mathrm{R}}$ is set to 1, and vice versa. If $w_2 \in w_1^{-1}\mathsf{MAX}_\ell$ then, by definition, the j-th bit of the binary representation of $i+1$ is set to 1, that is, $(i+1) \wedge 2^j \neq 0$, thus the condition holds. In the other direction a similar argument applies.

2. $m = 2^{2^{r-1}} - 1$:

    Now, $r_\ell = r - 1$. Let us first prove that the size of $\mathsf{B}(\mathsf{MAX}_\ell)$ is 2^ℓ. It can be noticed that both $2^{2^{r-1}} - 1 > 2^{\ell-r}$ and $2^{2^{r-1}} - 1 < 2^{\ell-r+1}$ hold. These two conditions imply that $2^{r-1} = \ell - r + 1$. Then, $|\mathsf{B}(\mathsf{MAX}_\ell)| = 2^{r-1}2^{2^{r-1}} = 2^{\ell-r+1+r-1} = 2^\ell$.

    Since m is odd, we add a padding of 2^{r_ℓ} zeros to ensure that the length of the bitmap is 2^ℓ. By Lemma 1, these particular bits of the bitmap correspond to $w_1^{-1}\mathsf{MAX}_\ell$ such that $\mathrm{ind}(w_1) = 2^{2^{r-1}} - 1$. Thus, we need to prove that $w_1^{-1}\mathsf{MAX}_\ell = \emptyset$. By the definition of MAX_ℓ, for $(i+1) \wedge 2^j \neq 0$ to hold, j must be at least 2^{r-1}, but for every $w_2 \in \Sigma^{r_\ell}$ we have $\mathrm{ind}(w_2) \leq 2^{r-1} - 1$. Thus, $w_1^{-1}\mathsf{MAX}_\ell = \emptyset$. $\square$

Example 6. According to Lemma 5, the bitmap of the language MAX_5 given in Example 5 is

$$\mathsf{B}(\mathsf{MAX}_5) = \prod_{i=1}^{8} \mathsf{pad}(i_{[2]}, 4)^{\mathrm{R}} = 10000100110000101010011011100001.$$

To have a DFA of maximal size for a block language contained in Σ^ℓ, for some $\ell \geq 0$, the width of each rank $r' \in [\ell]$ must be either $2^{2^{r'}} - 1$, for $r' \in [r-1]$, or $2^{\ell-r'}$, for $r' \in [r, \ell]$, from which the result from Câmpeanu and Ho was established [3]. From this observation, it follows:

Lemma 6. *Let r, r_ℓ, and m be defined as before for $\ell > 0$ and alphabet size $k = 2$. Then, the minimal DFA accepting the language MAX_ℓ has maximal size.*

Proof. Let $Q = Q_0 \cup \ldots \cup Q_\ell$ be the set of states of the minimal DFA for MAX_ℓ, such that $q \in Q_{r'}$ is in rank $r' \in [\ell]$. Then, the DFA has maximal size if $|Q_{r'}| = 2^{2^{r'}} - 1$, for $r' \in [r-1]$, and if $|Q_{r'}| = 2^{\ell-r'}$, for $r' \in [r, \ell]$. To that aim, one analyses the cardinalities of the sets $\mathcal{B}_{r'}$, with $r' \in [\ell]$, for the possible values of m.

1. $m = 2^{\ell-r}$:

    (a) $(\forall r' \in [r-1])|\mathcal{B}_{r'}| = 2^{2^{r'}} - 1$: we have that $\mathcal{B}_{r'} = \{\mathsf{pad}(i_{[2]}, 2^{r'})^{\mathrm{R}} \mid \forall i \in [1, m]\}$. Since both $|\mathcal{B}_{r'}| \leq 2^{2^{r'}} - 1$ and $m \geq 2^{2^{r'}} - 1$, the proposition holds.

(b) $(\forall r' \in [r, \ell]) |\mathcal{B}_{r'}| = 2^{\ell - r'}$: clearly $|\mathcal{B}_r| = 2^{\ell - r}$. Let $x \in [\ell - r]$ and $r' = r + x$. The set $\mathcal{B}_{r'}$ is given by splitting $\mathsf{B}(\mathsf{MAX}_\ell)$ into factors of length $2^{r'}$. Of course, $|\mathcal{B}_{r'}| = 2^{\ell - r'}$.

2. $m = 2^{2^{r-1}} - 1$:

(a) $(\forall r' \in [r - 1]) |\mathcal{B}_{r'}| = 2^{2^{r'}} - 1$: analogous to the first case (1a).

(b) $(\forall r' \in [r, \ell]) |\mathcal{B}_{r'}| = 2^{\ell - r'}$: it suffices to show that $|\mathcal{B}_r| = 2^{\ell - r}$, as we saw on the previous case (1b). By construction, $\mathsf{B}(\mathsf{MAX}_\ell)$ is composed by m different blocks of length 2^{r_ℓ} and a single block of zeros. Since m is odd, each element of $\mathcal{B}_r$, consisting of blocks of length 2^r, will be equal either to the binary representation of two consecutive numbers or the second number represented is zero. Therefore, $|\mathcal{B}_r| = 2^{2^{r-1}-1}$ and, as mentioned in proof of Lemma 5, $2^{r-1} = \ell - r + 1$. Then, $|\mathcal{B}_r| = 2^{\ell - r}$. $\qquad\square$

5   Bitmaps for Block Languages and Minimal NFAs

We now show that the bitmap of a block language L can be used to obtain a minimal NFA for L. However, in this case the problem is NP-complete. Given a bitmap B associated with a block language $L \subseteq \Sigma^\ell$, one can build a minimal NFA similarly to the previous construction for minimal DFAs, by iteratively finding the minimal number of states required at each rank. The main difference with the deterministic case is that the quotients of the language, corresponding to factors from the bitmap, are represented by sets of states, instead of single ones.

First, let us define what a *cover* is. Let $\mathcal{C}$ be a finite set of binary words of length n, that is, $\mathcal{C} \subseteq 2^{\{0,1\}^n}$, for some $n \in \mathbb{N}$. We say that $\mathcal{C}$ is a *cover for a word* $s \in \{0,1\}^n$ (or, alternatively, s *is covered* by $\mathcal{C}$) if there is a subset of words in $\mathcal{C}$ such that the bitwise disjunction of those words equal s. Formally, $\mathcal{C}$ covers s if and only if $(\exists m \in [1, |\mathcal{C}|])(\exists \{c_1, \ldots, c_m\} \subseteq \mathcal{C})(\bigvee_{i=1}^m c_i = s)$.

We extend this definition to sets in the natural way, that is, $\mathcal{C}$ is a *cover of a finite set* of binary words $\mathcal{B}$ if every word in $\mathcal{B}$ is covered by $\mathcal{C}$. Moreover, we say that $\mathcal{C}$ is a *minimal cover* for $\mathcal{B}$ if there is no smaller set that covers $\mathcal{B}$.

Example 7. Let $\mathcal{C} = \{1100, 1010, 0001\}$ and $\mathcal{B} = \{1100, 1110, 1101, 1111\}$. Then, $\mathcal{C}$ covers $\mathcal{B}$ because $\mathcal{C}$ covers every word from $\mathcal{B}$:

- $1100 = 1100$;
- $1110 = 1100 \vee 1010$;
- $1101 = 1100 \vee 0001$;
- $1111 = 1100 \vee 1010 \vee 0001$.

One can observe that the set $\mathcal{C}$ is a minimal cover for $\mathcal{B}$, but it is not unique since $\mathcal{C}' = \{1100, 0010, 0001\}$ also cover $\mathcal{B}$ and $|\mathcal{C}| = |\mathcal{C}'|$.

Let us now show how to obtain an NFA for a non-empty block language $L \subseteq \Sigma^\ell$, for some $\ell \geq 0$ and an alphabet $\Sigma = \{\sigma_0, \ldots, \sigma_{k-1}\}$ of size k, and with bitmap representation B. The construction starts as before, where the final state 1 is added at rank 0. Additionally, we define the function $\rho : \{0, 1\}^* \to 2^{\{0,1\}^*}$ which maps a factor into the set that covers it. First, let $\rho(1) = \{1\}$. Then, for each rank $i = 1, 2, \ldots, \ell$, we look for the minimal set $\mathcal{C}_i$ that covers the set $\mathcal{B}_i$, the collection of factors of B with length k^i with at least one bit set to 1.

The rank i in the NFA will be $\mathcal{C}_i$. Subsequently, for each factor $s \in \mathcal{B}_i$ we set $\rho(s) = \{c_0, \ldots, c_{m-1}\} \subseteq \mathcal{C}_i$, such that $\rho(s)$ covers s.

The transitions from rank i to rank $i-1$ will then be determined in a similar way to the DFA construction. For each state c in rank i, we split $c = c_0 \cdots c_{k-1}$, where $|c_j| = k^{i-1}$, for every $j \in [k-1]$, and set $\delta(c, \sigma_j) = \rho(c_j)$, if $c_j \neq 0^{k^{i-1}}$.

We must also guarantee that ρ is defined in c_j, i.e., that $c_j \in \mathcal{B}_{i-1}$. For that, we need to limit the search space of the cover $\mathcal{C}_i$, so that each word in the set is a concatenation of k words from $\mathcal{B}_{i-1}$ or $0^{k^{i-1}}$. Formally, $\mathcal{C}_i \subseteq (\mathcal{B}_{i-1} \cup 0^{k^{i-1}})^k \setminus 0^{k^i}$.

Also, as we previously saw, $\mathcal{B}_\ell = \{B\}$, so the minimal cover for $\mathcal{B}_\ell$ is itself. This result implies that B will be the single initial state at rank ℓ.

Lemma 7. *Let $L \subseteq \Sigma^\ell$ be a block language, for some $\ell \geq 0$, with bitmap B. Then, the NFA $\mathcal{A}$ given by the above construction applied to B accepts L, that is, $\mathcal{L}(\mathcal{A}) = L$.*

Proof. Recall the construction of the minimal DFA from a bitmap in Lemma 3. For $r \in [\ell]$, let $s \in \mathcal{B}_r$ be a state from the DFA. By construction, $\rho(s) = \{c_0, \ldots, c_{n-1}\}$, where c_i are states in $\mathcal{A}$ that exactly cover the bitmap s. As $\mathcal{L}(s) = \bigcup_{i \in [n-1]} \mathcal{L}(c_i)$, one concludes that $\mathcal{L}(\mathcal{A}) = L$. □

Lemma 8. *Let $L \subseteq \Sigma^\ell$ be a block language, for some $\ell \geq 0$, with bitmap B. Then, the NFA $\mathcal{A}$ given by the above construction applied to B is minimal.*

Proof. Let $w \in \Sigma^{\ell-r}$, for some $r \in [\ell]$, and P be the set of states reachable from the initial state q_0 of $\mathcal{A}$ after consuming w, that is, $P = \delta(q_0, w)$. Let P_1, P_2 be two non-empty subsets of P such that $P_1 \cap P_2 \neq \emptyset$ and let $s_1 = \bigvee_{q \in P_1} q$ and $s_2 = \bigvee_{q \in P_2} q$ correspond to the bitmaps of the right languages of the states P_1 and P_2, respectively. Suppose that $s_1 = s_2$, so P_1 and P_2 cover the same bitmaps. Then, $\mathcal{C}_r \setminus P_2$ would also cover the set $\mathcal{B}_r$, hence $\mathcal{C}_r$ would not be minimal. □

From Lemmas 7 and 8, we obtain the following.

Theorem 3. *The construction of an NFA from a bitmap B of a block language $L \subseteq \Sigma^\ell$ results in a minimal NFA $\mathcal{A}$ such that $\mathcal{L}(\mathcal{A}) = L$.*

Example 8. Let $L \subseteq \{a, b\}^4$ be the language of Example 1 with bitmap B = 1011011100011110. A correspondent NFA is depicted in Fig. 2. One has $\mathcal{B}_1 = \{01, 11, 10\}$ but $\mathcal{C}_1 = \{01, 10\}$ is a minimal cover. Thus, only two states are needed in rank 1 of the NFA. Then, $\mathcal{B}_2 = \{1011, 0111, 0001, 1110\}$, and let $\mathcal{C}_2 = \{1010, 0110, 0001\}$. We have $1011 = 0001 \vee 1010$, $0111 = 0110 \vee 0001$, and $1110 = 1010 \vee 0110$. In rank 3 two states are needed and rank 4 has only the initial state.

The problem of obtaining a minimal NFA from the bitmap is NP-COMPLETE, as proved in the following result.

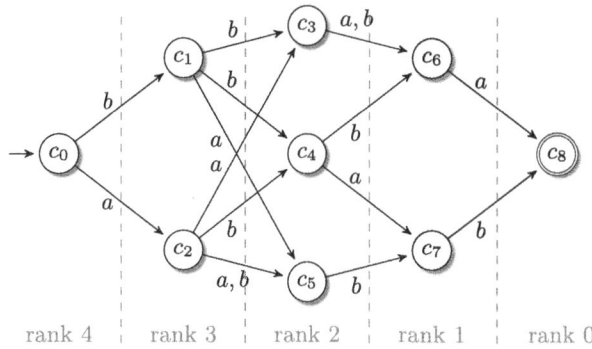

Fig. 2. A minimal NFA accepting the language of Example 1, where $c_8 = 1$, $c_7 = 01$, $c_6 = 10$, $c_5 = 0001$, $c_4 = 0110$, $c_3 = 1010$, $c_2 = 10110111$, $c_1 = 00011110$, and $c_0 = 1011011100011110$.

Theorem 4. *Let* B *be a bitmap of length* k^ℓ. *Given a set of factors* $\mathcal{B}_r$, *each with length* k^r *where* $r \in [\ell]$, *the problem of finding the minimal cover* $\mathcal{C}_r$ *for* $\mathcal{B}_r$ *is* NP-COMPLETE.

Proof. The problem we aim to solve is characterized as:

- *Instance*: Collection of factors $\mathcal{B}_r$ and a positive integer $n \leq |\mathcal{B}_r|$.
- *Question*: Is there a collection of subsets $\mathcal{C}_r$ of size n such that, for each $s \in \mathcal{B}_r$, there is a sub collection of $\mathcal{C}_r$ whose bitwise disjunction is exactly s?

By Lemma 1, each bitmap factor corresponds to a quotient of the language L. Therefore, the bitwise disjunction over factors corresponds to the union over sets. Thus, this problem is the SET-BASIS problem and Stockmeyer proved that it is NP-COMPLETE by reduction to the VERTEX-COVER problem [16].   □

One can use an SMT-solver [10] to find a cover of size n of a set, wherein every factor of size k^r can be represented by a bit vector of the same size, and then use binary search on the solution to determine the minimal one.

5.1   Maximal Size of Minimal NFAs for Block Languages

The width of each rank of the NFA given by the construction described above is bounded by the width of the same rank on the DFA. Also, to cover factors of length k^r, we show that a cover of k^r elements is sufficient. These bounds are formally stated in the following lemma.

Lemma 9. *Let* $\mathcal{A} = \langle Q, \Sigma, \delta, q_0, \{q_f\} \rangle$ *be an NFA for a block language* $L \subseteq \Sigma^\ell$ *over* Σ *of size* k *and* $\ell \geq 0$, *given by the construction of NFAs from bitmaps. Let* $Q = Q_0 \cup \ldots \cup Q_\ell$, *such that* Q_r *is the rank* r, *for* $r \in [\ell]$. *Then, the width of rank* Q_r *is bounded by* $|Q_r| \leq \min(k^{\ell-r}, k^r)$, *for all* $r \in [\ell]$.

Proof. The bound $|Q_r| \leq k^{\ell-r}$, for $r \in [\ell]$, refers to the size of $\mathcal{B}_r$, that is, the number of unique factors of length k^r. We showed in Lemma 2 that $|\mathcal{B}_r| \leq k^{\ell-r}$. On the other hand, we have that $|Q_r| \leq k^r$ since the set $\mathcal{B}_r$ can be covered by the set of unit factors $\{u_i\}_{i \in [k^r-1]}$ of size k^r, where u_i represents the factor filled with zeros, apart from the i-th position which is set to 1. $\qquad\square$

Lemma 9 allows us to determine the exact maximal size of a minimal NFA for a block language, and easily prove the following theorem.

Theorem 5. *The maximal size of a minimal NFA for a block language $L \subseteq \Sigma^\ell$, with $\ell \geq 0$ and $|\Sigma| = k$, is $nsc(L) \leq 2 \cdot \dfrac{k^{\frac{\ell}{2}} - 1}{k-1} + k^{\frac{\ell}{2}}$ if ℓ is even, and $nsc(L) \leq 2 \cdot \dfrac{k^{\lceil \frac{\ell}{2} \rceil} - 1}{k-1}$, otherwise.*

Proof. If ℓ is even, there is an odd number of ranks and the width of the minimal NFA with maximal size is achieved by the rank $\frac{\ell}{2}$. So the maximal number of states is given by $2 \cdot \sum_{r=0}^{\frac{\ell}{2}-1} k^{\ell-r} + k^{\frac{\ell}{2}}$. If ℓ is odd, the width of the minimal NFA with maximal size is reached both in rank $\lceil \frac{\ell}{2} \rceil - 1$ and $\lceil \frac{\ell}{2} \rceil$. So the NFA has at most $2 \cdot \sum_{r=0}^{\lceil \frac{\ell}{2} \rceil - 1} k^{\ell-r}$ states. $\qquad\square$

References

1. Almeida, M., Moreira, N., Reis, R.: Exact generation of minimal acyclic deterministic finite automata. Int. J. Found. Comput. Sci. **19**(4), 751–765 (2008). https://doi.org/10.1142/S0129054108005930
2. Almeida, M., Moreira, N., Reis, R.: Finite automata minimization algorithms. In: Wang, J. (ed.) Handbook of Finite State Based Models and Applications, pp. 145–170. CRC Press (2012)
3. Câmpeanu, C., Ho, W.H.: The maximum state complexity for finite languages. J. Autom. Lang. Comb. **9**(2–3), 189–202 (2004)
4. Dudzinski, K., Konstantinidis, S.: Formal descriptions of code properties: decidability, complexity, implementation. Int. J. Found. Comput. Sci. **23**(1), 67–85 (2012). https://doi.org/10.1142/S0129054112400059
5. Elvey Price, A., Fang, W., Wallner, M.: Compacted binary trees admit a stretched exponential. J. Comb. Theory Ser. A **177**, 105306 (2021). https://doi.org/10.1016/J.JCTA.2020.105306
6. Karhumäki, J., Kari, J.: Finite automata, image manipulation, and automatic real functions. In: Pin, J. (ed.) Handbook of Automata Theory, pp. 1105–1143. European Mathematical Society (2021). https://doi.org/10.4171/AUTOMATA-2/8
7. Karhumäki, J., Okhotin, A.: On the determinization blowup for finite automata recognizing equal-length languages. In: Calude, C.S., Freivalds, R., Kazuo, I. (eds.) Computing with New Resources. LNCS, vol. 8808, pp. 71–82. Springer, Cham (2014). https://doi.org/10.1007/978-3-319-13350-8_6
8. Kjos-Hanssen, B., Liu, L.: The number of languages with maximum state complexity. In: Gopal, T.V., Watada, J. (eds.) TAMC 2019. LNCS, vol. 11436, pp. 394–409. Springer, Cham (2019). https://doi.org/10.1007/978-3-030-14812-6_24

9. Konstantinidis, S., Moreira, N., Reis, R.: Randomized generation of error control codes with automata and transducers. RAIRO **52**, 169–184 (2018)
10. Kroening, D., Strichman, O.: Decision Procedures: An Algorithmic Point of View. Springer, Cham (2016). https://doi.org/10.1007/978-3-662-50497-0
11. Meyer, A.R., Fischer, M.J.: Economy of description by automata, grammars, and formal systems. In: 12th Annual Symposium on Switching and Automata Theory, Los Alamitos, pp. 188–191. IEEE (1971)
12. Priez, J.B.: Enumeration of minimal acyclic automata via generalized parking functions. In: 27th FPSAC. DMTCS, vol. 2471 (2015). https://doi.org/10.46298/dmtcs. 2471
13. Revuz, D.: Minimisation of acyclic deterministic automata in linear time. Theor. Comput. Sci. **92**(1), 181–189 (1992)
14. Salomaa, K., Yu, S.: NFA to DFA transformation for finite languages over arbitrary alphabets. J. Autom. Lang. Comb. **2**(3), 177–186 (1997)
15. Shankar, P., Dasgupta, A., Deshmukh, K., Rajan, B.S.: On viewing block codes as finite automata. Theor. Comput. Sci. **290**(3), 1775–1797 (2003). https://doi.org/ 10.1016/S0304-3975(02)00083-X
16. Stockmeyer, L.: Set basis problem is NP-complete. Technical report. Report No. RC-5431, IBM Research Center (1976)
17. Stockmeyer, L., Meyer, A.R.: Word problems requiring exponential time: preliminary report. In: 5th Annual ACM Symposium on Theory of Computing, pp. 1–9. ACM (1973)

From Relation to Emulation and Interpretation: Computer Algebra Implementation of the Covering Lemma for Finite Transformation Semigroups

Attila Egri-Nagy[1]([✉]) and Chrystopher L. Nehaniv[2]

[1] Akita International University, Akita, Japan
`egri-nagy@aiu.ac.jp`
[2] University of Waterloo, Waterloo, Canada
`cnehaniv@uwaterloo.ca`

Abstract. We give a practical computer algebra implementation of the Covering Lemma for finite transformation semigroups. The lemma states that given a surjective relational morphism $(X, S) \rightarrow (Y, T)$, we can establish emulation by a cascade product (subsemigroup of the wreath product): $(X, S) \hookrightarrow (Y, T) \wr (Z, U)$. The dependent component (Z, U) contains the kernel of the morphism, the information lost in the map.

The implementation complements the existing tools for the holonomy decomposition algorithm. It gives an incremental method to get a coarser decomposition when computing the complete skeleton for holonomy is not feasible. Here, we describe a simplified and generalized algorithm for the lemma and compare it to the holonomy method. Incidentally, the kernel-based method could be the easiest way of understanding the hierarchical decompositions of transformation semigroups and thus the celebrated Krohn-Rhodes theory.

Keywords: transformation semigroup · relational morphism · cascade product

1 Introduction

The Krohn-Rhodes theory [7] is about how to decompose automata and build them from elementary components. As a fundamental theory, its applicability goes beyond algebra and automata theory [13]. For applications, we need computational tools. The `SgpDec` [4] package for the `GAP` system [6] provides an implementation for the holonomy decomposition method [5]. This computer algebra tool enabled the investigation of several applications of the decompositions (see, e.g., [10]). However, the limitations of the holonomy algorithm also became clear. It has exponential resource usage: the number of images of the state set is bounded by 2^n for n states. At this price, it gives a 'highest resolution' decomposition, which is desired in some situations, but not in others. Hence, there is a need for a more flexible and incremental decomposition algorithm.

© The Author(s), under exclusive license to Springer Nature Switzerland AG 2024
S. Z. Fazekas (Ed.): CIAA 2024, LNCS 15015, pp. 138–152, 2024.
https://doi.org/10.1007/978-3-031-71112-1_10

Here, we use an alternative proof of the Krohn-Rhodes Theorem, the Covering Lemma from [9] and simplify it for a computer algebra implementation. Starting with a surjective homomorphism, we build a two-level decomposition. The top level component contains the information preserved by the morphism, and the bottom component the information left out. The latter is the *kernel* of a homomorphism. For groups, it is a subgroup, which is, in general, the inverse image of the identity. For semigroups, we have a more general, possibly partial algebraic structure, a category or a semigroupoid. This kernel, called the 'derived category', is an important tool in advanced semigroup theory [15]. In [9], the kernel maps functorially to sets and characterizes the minimal computation needed to "undo" a relational morphism of transformation semigroups. Here, we take a more elementary approach, omitting the category theoretical concepts.

2   An Algebraic View of Computation – Definitions

We model finite state transition structures (computers) algebraically, as transformation semigroups [3]. We focus on *how* the computation is done, the dynamics, instead of only its result. We do not specify initial and accepting states, thus to avoid confusion, we do not call transformation semigroups "automata". The following definitions will describe what is a computational structure, how to build more complex ones out of simpler components, and how to compare their computational capabilities.

2.1   Transformation and Cascade Semigroups

A *transformation semigroup* (X, S) is a finite nonempty set of *states* X and a set S of total transformations of X, i.e., functions of type $X \to X$, closed under composition. We use the symbol $\cdot$ both for the semigroup action $X \times S \to X$, e.g., $x \cdot s$, and for the semigroup multiplication $S \times S \to S$, e.g., $s \cdot t$. We also extended the operations for sets $A \cdot B = \{a \cdot b \mid a \in A, b \in B\}$. Alternatively, we omit the symbol, so juxtaposition is acting on the right when computing with transformations. For representing states we use positive integers. We denote constant maps as c_j, where j is the unique point in its image set. The action is *faithful* if $(\forall x \in X, xs = xs') \implies s = s'$.

We say that $A \subseteq S$ generates S, denoted by $\langle A \rangle = S$, if S is the smallest semigroup containing A. In other words, S consists of the products of elements of A. In automata terminology, the generators of the semigroup correspond to the input symbols. For efficient computations, we aim to work with a small set of generators only, avoiding the complete enumeration of semigroup elements.

2.2   Cascade Products

The *wreath product* $(X, S) \wr (Y, T)$ of transformation semigroups is the *cascade transformation semigroup* $(X \times Y, W)$ where

$$W = \{(s, d) \mid s \in S, d \in T^X\},$$

whose elements map $X \times Y$ to itself as follows

$$(x, y) \cdot (s, d) = (x \cdot s, y \cdot d(x))$$

for $x \in X, y \in Y$. We call $d : X \to T$ a dependency function. Note that it maps states to transformations. It shows how the action in the bottom component depends on the top level state. The term 'cascade' refers to the flow of control information from the top level component to the bottom level component: the state x determines the action in T. Here, T^X is the semigroup of all functions d from X to T, thus the wreath product has exponential size, and it is unsuitable for computer implementations.

We call a (proper) subsemigroup of the wreath product a *cascade product*. We denote such a product operation by $\wr_\cdot$, indicating that it is a substructure and also hinting the direction of the control flow:

$$(X, S) \wr (Y, T) \supset (X, S) \wr_\cdot (Y, T).$$

The wreath product is the full cascade product, and it is uniquely determined by its components. On the other hand, there are cascade products as many as subsemigroups of the wreath product. Thus, the cascade product notation could be ambiguous, and the context should make it clear which one are we talking about. In practice, we specify a cascade product by the set of transformations, or by a generator set, or as an image of a relation.

2.3   Relational Morphisms

Here we define algebraic structure-preserving maps, but in terms of relations, not functions. In semigroup theory there are several reasons to do this [12,14]. Since we work with transformation semigroups, we need to define two relations: one for states and the other for transformations.

Definition 1 (Relational Morphism). *A relational morphism of transformation semigroups* $(X, S) \xrightarrow{\theta, \varphi} (Y, T)$ *is a pair of relations* $(\theta : X \to Y, \varphi : S \to T)$ *that are fully defined, i.e.,* $\theta(x) \neq \emptyset$ *and* $\varphi(s) \neq \emptyset$, *and satisfy the condition of compatible actions for all* $x \in X$ *and* $s \in S$:

$$y \in \theta(x), t \in \varphi(s) \implies y \cdot t \in \theta(x \cdot s),$$

or more succinctly: $\theta(x) \cdot \varphi(s) \subseteq \theta(x \cdot s)$, *which can be depicted by a subcommutative diagram.*

$$
\begin{array}{ccc}
X \times S & \xrightarrow{\quad \cdot \quad} & X \\
\downarrow{\scriptstyle \theta \times \varphi} & \subseteq & \downarrow{\scriptstyle \theta} \\
Y \times T & \xrightarrow{\quad \cdot \quad} & Y
\end{array}
$$

We will generally be assuming that φ is a relational morphism of semigroups, i.e., $\varphi(s) \cdot \varphi(s') \subseteq \varphi(ss')$. The relations can also be expressed as set-valued functions: $\theta : X \to 2^Y$, $\varphi : S \to 2^T$. In this paper we refer to relational morphisms simply as morphisms, and to avoid misunderstanding we never abbreviate the word 'homomorphism'.

A morphism is *surjective* if $\bigcup_{x \in X} \theta(x) = Y$ and $\bigcup_{s \in S} \varphi(s) = T$. In computational implementations surjectivity is often given, as the image transformation semigroup is defined by the morphism. A surjective morphism can lose information, thus we call the target an *approximation* of the source. A morphism is *injective* if $\theta(x) \cap \theta(x') \neq \varnothing$ implies $x = x'$, and analogously $\varphi(s) \cap \varphi(s') \neq \varnothing$ implies $s = s'$. In other words, the relations map to disjoint sets. Injective morphism is the same as a *division*. Injectivity guarantees that no information is lost during the map, therefore the target of the morphism can compute at least as much as the source. Thus, we call injective morphisms *emulation* or *covering*. We use the standard arrow notations, $\hookrightarrow$ for injective, and $\twoheadrightarrow$ for surjective morphisms.

For relations, being surjective and injective at the same time does not imply isomorphism, but it implies being the inverse of a homomorphism. On the other hand, we have the ability to turn around surjective morphisms to get an injective ones and vice-versa. Relational morphisms preserve subsemigroups both directions.

3   The General Algorithm

First we describe the general mechanism of how parts of the semigroup action in (X, S) are transferred into the newly constructed (Z, U).

3.1   Problem Description

Our goal is to understand the inner workings of a transformation semigroup (X, S). We would like to build a hierarchical decomposition, a cascade product that emulates (X, S). We get simpler components by distributing the computation over the levels. In general, our task is to find the unknowns in

$$(X, S) \hookrightarrow (?, ?) \wr (?, ?) \wr \ldots \wr (?, ?).$$

Requiring only the emulation leaves the solution space huge. The holonomy decomposition uses constraints from the detailed examination of how S acts on the subsets of X and produces a more limited set of solutions [5]. They are essentially the same up to choosing equivalence class representatives and moving the components around when their dependencies allow some wiggle room.

Here, we take a more flexible approach. We define the top level component,

$$(X, S) \hookrightarrow (Y, T) \wr (?, ?),$$

and leave it to the algorithm to define the bottom level component, recovering all the information about (X, S) that is not represented in (Y, T). The decomposition has two levels, but the process can be iterated to have a more fine-grained solution.

3.2   Constructing the Emulation

Given a relational morphism as an input we build an emulation as output:

$$\text{from} \ \ (X, S) \xrightarrow{R(\theta, \varphi)} (Y, T) \ \ \text{to} \ \ (X, S) \xrightarrow{E(\psi, \mu)} (Y, T) \wr (Z, U).$$

We need to construct the transformation semigroup (Z, U), the dependency functions $Y \to U$ for defining the cascade transformations. In practice, we will define only the emulation relational morphism $E(\psi, \mu)$ and only map the generators to define the cascade product.

What are the states in Z? We need to specify one or more coordinate pairs for $x \in X$. The top level coordinates are elements of $\theta(x) \subseteq Y$. Each such top level state y gives a context for the bottom level states. In each context, what a particular $z \in Z$ refers to is given by a *labelling* mechanism, a partial one-to-one mapping $w_y : X \to Z$, that depends on y. More precisely, we only need to map $\theta^{-1}(y) \subseteq X$ to Z. The states in Z can be reused in the different contexts, we need at least $|Z| = \max_y |\theta^{-1}(y)|$. The coordinate pair (y, z) gives the context y and the encoded state z as lift for x. Formally,

$$\psi(x) = \{(y, z) \mid y \in \theta(x), z = x \cdot w_y\}.$$

How to construct transformations in U? Using the labelling above, we simply *transfer* the relevant bits of s acting on X: Formally, to see where a single state z should go respecting the original action of $s \in S$, we do $z \mapsto z \cdot w_y^{-1} \cdot s \cdot w_{yt}$, where $w_y^{-1} : Z \to \theta^{-1}(y)$ is a partial function, $s : X \to X$, and $w_y : \theta^{-1}(y) \to Z$. The net result of this composition is a map $Z \to Z$. If we want to identify the main trick of the method, then this is it: decode the states according to the current context, perform the original action, then encode the states for the new context. We have a 'window' through $\theta^{-1}(y)$ into X, or using another metaphor, $\theta^{-1}(y)$ sheds light on some part of X, so we can see the action of S there.

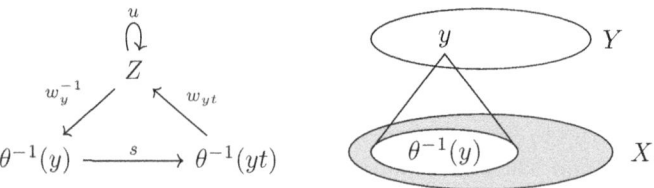

How is it possible at all that differently encoded local actions compose nicely when multiplying cascade transformations? Firstly, in the same context, for a fixed y, they compose naturally. For each y we have a differently behaving clone of the component (a view advocated in [1]), but as long as the top level is fixed, we stay in the same clone transformation semigroup. Secondly, the context switching is built-in in $u = w_y^{-1} s w_{yt}$, since the result of s is encoded for the yt context. In short, when lifting s into the cascade product we have

$$\mu(s) = \{(t, w_y^{-1} s w_{yt}) \mid t \in \varphi(s), y \in \operatorname{Im}\theta\}.$$

Now, if we consider a sequence of actions, it is easy to see how the (partial) labelling function in the middle matches its inverse, thus cancelling out.

$$\mu(ss') = \{(tt', w_y^{-1} s w_{yt} w_{yt}^{-1} s' w_{ytt'} = w_y^{-1} s s' w_{ytt'}) \mid t \in \varphi(s), t' \in \varphi(s'), y \in \operatorname{Im}\theta\}.$$

This forms a valid cascade transformation. On the top level $t \in T$ by the definition of φ. The bottom level action $w_y^{-1} s w_{yt}$ has the form of a dependency function $Y \to U$, since for an (s, t) pair the expression's value depends on y, and $w_y^{-1} s w_{yt}$ as a function has the type $Z \to Z$ by the definition of the labelling. Therefore, we will have a cascade transformation for each image t of s in φ, and each cascade will be defined for every possible state image y by θ. In practice, we simply have $Y = \operatorname{Im}\theta$.

To summarize, *how does this method recover the information lost through R?* For collapsed states, we go back to the preimages of y in θ and see how s acts on those states. By the same token, for collapsed transformations, we go through the preimages of t in φ, and construct a separate cascade transformations for each. Thus, all the different actions in $\varphi^{-1}(t)$ contribute to U.

Here we state the decomposition part of Covering Lemma without the derived category. The main idea is the context-dependent partial transfer of the original semigroup action, thus the proof merely checks that the resulting constructions satisfies the definition of emulation.

Theorem 1 (Covering Lemma). *If $R(\theta, \varphi) : (X, S) \twoheadrightarrow (Y, T)$ is a surjective relational morphism, then there exists an emulation, an injective relational morphism $E(\psi, \mu) : (X, S) \hookrightarrow (Y, T) \wr (Z, U)$.*

Proof. Let's choose a state $x \in X$ and a transformation $s \in S$. The lifted states are coordinate pairs $\psi(x) = \{(y, x w_y) \mid y \in \theta(x)\}$. The lifted transformations are transformation cascades $\mu(s) = \{(t, w_y^{-1} s w_{yt}) \mid t \in \varphi(s)\}$, where $w_y : X \to Z$ is any partial one-to-one function defined on $\theta^{-1}(y)$ for all $y \in \operatorname{Im}\theta$. Then $\psi(x) \cdot \mu(s) = \{(yt, x w_y \cdot w_y^{-1} s w_{yt}) \mid y \in \theta(x), t \in \varphi(s)\}$ according cascade multiplication. Thus, we have the set of coordinate pairs $(yt, x s w_{yt})$ ranging through all the lifts of x and s.

On the other hand, $\psi(x \cdot s) = \{(y', x s w_{y'}) \mid y' \in \theta(xs)\}$. Since $R(\theta, \varphi)$ is a relational morphism, we have $\theta(x) \cdot \varphi(s) \subseteq \theta(x \cdot s)$. Consequently, the set of yt top level states above is a subset of the set of these y' states. Therefore, $\psi(x) \cdot \mu(s) \subseteq \psi(x \cdot s)$, and $E(\psi, \mu)$ is a relational morphism.

For injectivity, pick two states $x_1, x_2 \in X$. Assume that injectivity does not hold for ψ and we have a coordinate pair such that $(y, z) \in \psi(x_1) \cap \psi(x_2)$. Now, z can arise in two ways, $x_1 \mapsto (y, x_1 w_y)$ and $x_2 \mapsto (y, x_2 w_y)$. Since w_y is one-to-one, $x_1 w_y = w_2 w_y \implies x_1 = x_2$. Therefore, E is injective on states.

Let's pick two transformations, s_1, s_2. Assume that $\mu(s_1) \cap \mu(s_2) \neq \varnothing$. Then choose a cascade transformation $(t, w_y^{-1} s w_{yt})$ from the intersection. This again can arise in two ways: $s_1 \mapsto w_y^{-1} s_1 w_{yt}$ and $s_2 \mapsto w_y^{-1} s_2 w_{yt}$. Since these are just different expressions of the same cascade transformation, $\forall y \in \operatorname{Im}\theta$ we have $w_y^{-1} s_1 w_{yt} = w_y^{-1} s_2 w_{yt}$. This implies $s_1 = s_2$ by faithfulness of (X, S), therefore E is injective on transformations.   □

3.3   From Emulation to Interpretation

When working with decompositions, we often want to find the meaning of some coordinatized computations, i.e., doing the computation in the cascade product and see what it corresponds to in the original semigroup. We call this use of a surjective morphism an *interpretation*.

Once we have an emulation, then we get the interpretation for free, since we can always reverse a morphism. Reversing an injective one (the emulation) gives us a surjective one, the interpretation:

$$\text{from } (X,S) \xrightarrow{E(\psi,\mu)} (Y,T) \wr (Z,U) \text{ to } (X,S) \xleftarrow{I(\psi^{-1},\mu^{-1})} (Y,T) \wr (Z,U).$$

If we have E computed completely, then it is possibly to invert the relation. However, it is better to compute inverse relations directly:

$$\psi^{-1}(y, z) = z \cdot w_y^{-1},$$
$$\mu^{-1}(t, d_{s,t}) = \{w_y u w_{yt}^{-1} \mid y \in Y = \operatorname{Im}\theta\}$$

We need only a subset Y' of Y such that $\bigcup_{y \in Y'} \theta^{-1}(y) = X$, since the actions are the same on the same points. The fact that EI should be the identity on (X, S) provides a tool for the verification of a computational implementation.

Corollary 1 (Identities from Composite Relations). *Inverting the emulation $E(\psi, \mu)$ gives the surjective relational morphism $I(\psi^{-1}, \mu^{-1}) : (Y,T) \wr (Z,U) \twoheadrightarrow (X,S)$, and IE is the identity on (X,S), and EI is the identity on the cascade product.*

Proof. It is a standard property of relational morphisms that the inverse of an injective morphism is a surjective one. It is also elementary that a relation combined with its inverse gives the identity relation. However, it is instructional to check this with the specific formulas. For states, $IE(x) = I(E(x)) = \psi^{-1}(\psi(x)) = \psi^{-1}((y, xw_y)) = xw_y w_y^{-1} = x, \forall y \in \theta(x)$, and $EI((y, z)) = E(I((y, z))) = \psi(\psi^{-1}((y, z))) = \psi(zw_y^{-1}) = (y, zw_y^{-1}w_y) = (y, z)$. For transformations, $IE(s) = I(E(s)) = \mu^{-1}(\mu(s)) = \mu^{-1}((t, w_y^{-1} s w_{yt})) = w_y w_y^{-1} s w_{yt} w_{yt}^{-1} = s$, and using the local action $u \in U$ we have $EI((t, u)) = E(I((t, u))) = \mu(\mu^{-1}((t, u))) = \mu(w_y u w_{yt}^{-1}) = (t, w_y^{-1} w_y u w_{yt} w_{yt}^{-1}) = (t, u).$ □

3.4   Practicality Working with the Generators

A degree n transformation semigroup can have up to n^n elements, thus an efficient algorithm should work on a small generator set. Here, we check that if $R(\theta, \varphi)$ is defined for $(X, S = \langle A \rangle)$ only on the generators, we can compute $E(\psi, \mu)$ by lifting those generators only.

The relation on states θ and its inverse θ^{-1} should be fully represented, i.e., all $\theta(x)$ values known and stored explicitly. For computing $\psi(x)$, we need θ^{-1} for the labelling.

For lifting a single transformation s, for computing $\mu(s)$, we need in addition $\varphi(s)$ computed. However, we do not need any other values of φ. Thus, we can lift each generator without fully computing φ. Moreover, the same applies to $I(\psi^{-1}, \mu^{-1})$, thus only the relations on states need to be fully enumerated.

3.5   What is U Exactly?

In general, the above construction does not yield a semigroup for U, but a *semigroupoid* with partially defined multiplication. The transformations $u : Z \mapsto Z$ are technically composable, but if we naively compute what they generate we may get a semigroup bigger than what is needed for the emulation. If the labelling functions do not cancel out, they introduce their arbitrary mappings (unrelated to S) into U, violating efficiency and correctness.

We can identify the 'islands of composability' in U and relate them back to S. The compatible elements are defined by the key idea that *labelling functions should cancel out in the middle*. We have a family of transformation semigroups (Z, U_y) for all $y \in \operatorname{Im} \theta$, where $U_y = \{w_y^{-1} s w_y \mid yt = y, s \in \varphi^{-1}(t)\}$. For instance, let $s = s_1 s_2 \cdots s_k$ be expressed as a product of generators, and $t_1 t_2 \cdots t_k$ be a corresponding product, such that $t_i \in \varphi(s_i)$. If $y \cdot t_1 t_2 \cdots t_k = y$, then in the bottom component we may do context switchings, but they all cancel out:
$$w_y^{-1} s_1 w_{yt_1} \cdot w_{yt_1}^{-1} s_2 w_{yt_1 t_2} \cdots w_{t_1 t_2 \cdots t_{k-1}}^{-1} s_k w_y = w_y^{-1} s_1 s_2 \cdots s_k w_y \in U_y.$$

To show that U_y does not contain any maps not present in S, we define $f_y : U_y \hookrightarrow S$ by $f_y(u) = w_y u w_y^{-1}$. Thus, if $u = w_y^{-1} s w_y$, then $f(u) = w_y w_y^{-1} s w_y w_y^{-1} = s$. It is easy to check that $f_y(u_1) \cdot f_y(u_2) = f_y(u_1 \cdot u_2)$. Therefore, we can build an injective relational morphism $V_y(w_y^{-1}, f_y) : (Z, U_y) \hookrightarrow (X, S)$ to verify that U_y embeds into S. This satisfies the symmetry principle of a good decomposition: the cascade product emulates, while its components are emulated by the original semigroup.

We can characterize the images of f_y as the stabilizers of $\theta^{-1}(y)$ in S, since $yt = y$ implies $\theta^{-1}(y)\varphi^{-1}(t) \subseteq \theta^{-1}(yt) = \theta^{-1}(y)$. This is how local computation is expressed in the bottom level component.

4   The Specifics

Here we describe how to create surjective morphisms, the input for the algorithm, and give labelling functions. These specific details are parameters of the main algorithm.

4.1   The $n(n-1)$ Method

We map each state $X = \{1, 2, \ldots, n\}$ to a set of states by $\theta : X \to 2^X$. The default choice in [9] is $\theta(x) = X \setminus \{x\}$. In other words, each state goes to the set of all states but itself. Note that these sets of lifts are overlapping and θ has the property of being a self-inverse, i.e., $\theta = \theta^{-1}$. For transformations, we use

$$\varphi(s) = \begin{cases} \{s\} & \text{if } X \cdot s = X, s \text{ is a permutation,} \\ \{c_j \mid j \notin \operatorname{Im} s\} & \text{if } X \cdot s \subsetneq X, s \text{ collapses some states.} \end{cases}$$

These 'backwards' relations have a simple explanation. By composing functions we can only reduce the size of the image. To make the morphism work, we need a bigger $\varphi(s)$ set for an s with a smaller image set.

Now we have $R(\theta, \varphi)$ defining a surjective morphism, but we still need a corresponding labelling method. Since θ misses only a single point, the labelling of states only needs to deal with just a single hole in the state set. We need to label $n - 1$ states. For a top level state y we define the map $w_y : \theta^{-1}(y) \to Z$ by

$$w_y(x) = \begin{cases} x & \text{if } x < y, \\ x - 1 & \text{if } x > y. \end{cases}$$

This gives a $n(n-1)$ cascade transformation representation of a degree n transformation semigroup, since $|\theta^{-1}(y)| = n - 1$ for all y. The canonical example for this method is the decomposition of the full transformation semigroup. For arbitrary transformation semigroups, it does not produce an efficient separation of computation. Transformations may get repeated on the lower level.

A simple example can demonstrate this problem. Consider the semigroup generated by the transformation $\left(\begin{smallmatrix} 1 & 2 & 3 & 4 & 5 \\ 2 & 3 & 1 & 5 & 4 \end{smallmatrix}\right)$, which is the permutation $(1, 2, 3)(4, 5)$. Thus, we have a group, and all the elements will make it to the top level. At the same time, the 2- and the 3-cycles will also make it to the bottom level, since the labelling has a cut-off at four states. Therefore, we need to find a better method.

4.2   Generalized Labelling: The Squashing Function

For a general surjective morphism $R(\theta, \varphi)$, the preimages of θ can be of different size, thus simply removing a hole is not enough. There may be more missing points in the preimage set, so need to 'squash' the set to remove all the holes.

Let $\theta^{-1}(y) = \{x_1, x_2, \ldots, x_k\}$. Let σ be the permutation that orders these states, so $x_{\sigma(1)} < x_{\sigma(2)} < \ldots < x_{\sigma(k)}$. Then $w_y = \sigma^{-1}$. The number of states used in Z is the size of the preimage. If we need k states, then we use the first k states in Z. Consequently, $|Z| = \max_y |\theta^{-1}(y)|$. For instance, if $\theta^{-1}(y) = \{4, 5, 2\}$ and $\theta^{-1}(yt) = \{5, 1\}$, then three states would be sufficient in Z for the squashing function to operate.

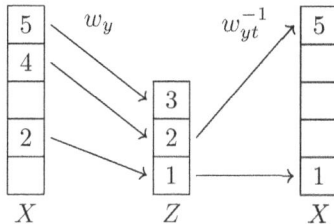

4.3 Getting a Surjective Morphism

Arguably, the Covering Lemma only does half of the decomposition. It needs an input surjective morphism for defining the top level component. Constructing this morphism could already contribute to the understanding of (X, S), and in certain applications it might be defined naturally. The above $n(n-1)$ method gives a working, but not necessarily balanced decomposition (see Sect. 4.1) and we can also project a holonomy decomposition into its top level [5]. Here we describe two more methods: congruences and local monoids.

An equivalence relation $\sim$ on X is a *right congruence* if $x_1 \sim x_2 \implies x_1 s \sim x_2 s$ for all $s \in S$. In other words, the semigroup action is compatible with the equivalence classes. Thus, we have $(X/\sim, S')$ well-defined, where S' is S made faithful, and the morphism to this is injective. In practice, we can find a congruence by a standard method. From the application domain we may have some intuition about which states to identify. We can specify these in disjoint sets of states, and all the remaining states form singleton classes. Then, an algorithm dual to the classical DFA minimization algorithm can compute the minimal (finest) congruence containing the input identified states in equivalence classes, acting as a closure operator. We check the action of the generators for compatibility with the existing classes. If elements from a class go to different classes, then we need to merge those. We iterate this until there is no merging. In general, there are many such congruences for a transformation semigroup, but in special cases we may have only the trivial congruence (X itself is the only class), as for the full transformation semigroup.

The *local transformation monoid* (Xe, eSe) is defined for an idempotent $e = e^2 \in S$. It localizes the semigroup's action to a subset Xe of the state set. It is an excellent tool for finding an interesting subsemigroup by mapping $s \mapsto ese$, but that is not necessarily a homomorphic image: $st \mapsto este$, and ese and ete have product $eseets = esete$, but $set \neq st$ in general. This method works for commutative semigroups and in some special cases.

5 Examples

5.1 Simple Cycle Collapsing

This example provides a minimal but still meaningful demonstration of the action transfer algorithm. We only have three states, thus we can denote the transformation by their image list in condensed format. For instance, $\left(\begin{smallmatrix} 1 & 2 & 3 \\ 1 & 3 & 2 \end{smallmatrix}\right)$ will be written as 132. With this notation we define (X, S) as $X = \{1, 2, 3\}$ and $S = \{e = 123, p = 132, c_1 = 111, c_2 = 222, c_3 = 333\}$. The element p has a non-trivial permutation. The target (Y, T) is defined by $Y = \{1, 2\}$ and $T = \{e' = 12, c_1' = 11, c_2' = 22\}$. Now θ collapses 2 and 3, and fixes 1, i.e., $\theta(2) = \theta(3) = \{2\}$, $\theta(1) = \{1\}$. This determines φ as well, e.g., $\varphi(c_2) = \varphi(c_3) = \{c_2'\}$, $\varphi(p) = \{e'\}$.

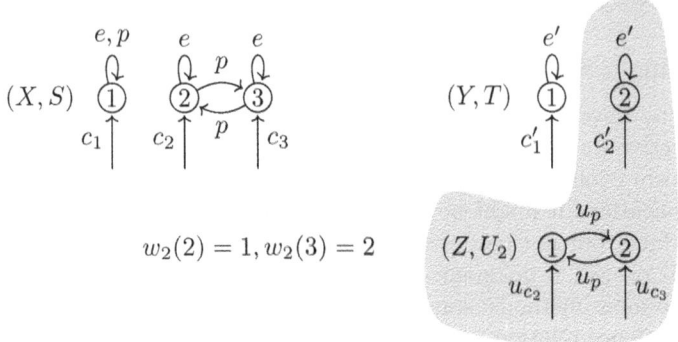

Note that $R(\theta, \varphi)$ is a surjective homomorphism since all the images are singletons. What is the information lost? All the original transformations that had different actions on $\{2, 3\}$ cannot be distinguished any more. Most notably, we should get a permutation on the bottom level. Indeed, when lifting p, we have a non-trivial dependency from $y = 2$ to the permutation $(1, 2)$, as the original cycle $(2, 3)$ is relabelled. Since $|\theta^{-1}(1)| = 1$, there are no fine-grained details attached to this state, U_1 is just the trivial monoid.

5.2 "Bad" Examples

We consider a resulting decomposition "bad" if it does not do a balanced distribution of computation over the components. For the usability of the decomposition we need to follow the 'Divide and conquer!' principle for dealing with complexity. In the extreme cases the decomposition only adds the overhead of the hierarchical form.

For a surjective homomorphism that collapses everything down to the trivial monoid, i.e., $\forall x \in X, \theta(x) = \{1\}$ and $\forall s \in S, \varphi(s) = \{e\}$ (e is the identity in T), we get everything on the second level through a dummy dependency function with a single input value, since $\theta^{-1}(1) = X$. As the other extreme, if we have an isomorphism to start with, we will get a trivial bottom level component. The set of preimages $\theta^{-1}(y)$ serve as the channel for transferring the action in (X, S) to (Z, U). If $|\theta^{-1}(y)| = 1$ for all $y \in Y$, then we completely block the way for any nontrivial action to get in the bottom component. This may seem like a way to craft a counterexample to the Covering Lemma, but here we show that in that case the top level component can emulate (X, S) in itself.

Fact 2 (Blocked action transfer). *If θ is bijective (as a function), i.e., $\theta(x) = \{y\}$ is a singleton $\forall x \in X$, and $\theta(x_1) = \theta(x_2) \implies x_1 = x_2$, so $|X| = |Y|$, then φ is injective.*

Proof. Assume that $t \in \varphi(s_1) \cap \varphi(s_2)$. Since θ^{-1} is also bijective, the action of t on Y uniquely determines the action of X : for all $y \in Y$, for yt we have $\theta^{-1}(y)s_1 = \theta^{-1}(y)s_2$, thus $s_1 = s_2$ by faithfulness of (X, S).   □

6 Software Implementation

The implementation of the algorithms presented here is available in the SgpDec package [2] written for the GAP [6] computer algebra system relying on the foundational Semigroups package [8]. To implement relational morphisms we use the hash-maps provided by the datastructures [11] package. This package has a lightweight reference implementation for hash-maps and comes with a recursive hash code computation for composite data structures, thus transformation cascades in SgpDec can be used as keys. We also use hash-maps when implementing the partial labelling functions and the congruence closure algorithm.

For representing transformations in the local U_y components, we use identities for any undefined maps, instead of implementing them as partial transformations with a sink state. Both methods can introduce unrelated transformations if we multiply incompatible elements. The sink state representation has a second disadvantage: it increases the number of states needed.

The source code closely follows this paper. Function and variable names are as close as possible to the notation in the mathematical text. There are several implementations for θ with their function names starting with the prefix ThetaFor. For the $n(n - 1)$ method there is ThetaForPermutationResets, for the congruences ThetaForCongruence, for local monoids ThetaForLocalMonoid. In case users start with a holonomy decomposition, they can use ThetaForHolonomy. For testing purposes we provide ThetaForConstant, mapping everything to a single state. For all these, we have the matching algorithms for φ, prefixed by PhiFor. Once theta and phi are defined, for computing ψ and μ the user simply calls Psi(theta) and Mu(theta,phi).

Here is a sample session to demonstrate the workflow of investigating of an unknown transformation semigroup (a randomly generated one in this example).

```
S:=Semigroup([Transformation([1,6,11,12,11,10,7,13,7,1,2,1,1]),
              Transformation([2,10,3,3,8,7,2,4,5,6,5,3,4])]);
```

The semigroup has 9221 elements. We apply the congruence based method for finding a surjective homomorphism onto a smaller transformation semigroup. As an arbitrary choice, we try identifying 1 with 2, and 3 with 4.

```
gap> partition := StateSetCongruence(Generators(S), [[1,2],[3,4]]);
[[1,2,6,7,10],[3,4,5,8],[9],[ 11,12,13]]
```

With this state set congruence, the homomorphic image of the semigroup will be acting on 4 points. What is this smaller semigroup? We construct the surjective morphism to see the image.

```
theta := ThetaForCongruence(partition);
phi := PhiForCongruence(partition, Generators(S));
```

Note that we defined φ only for the generators.

```
gap> ImageOfHashMapRelation(phi);
[ Transformation( [ 1, 2, 2, 2 ] ), Transformation( [ 1, 4, 1, 1 ] ) ]
```

The generators produce a semigroup with 5 elements. It is aperiodic, i.e., it does not have a non-trivial subgroup. Therefore, any non-trivial group action should appear on the second level of the decomposition. The generators of the cascade product are given by μ.

```
mu := Mu(theta, phi);  ImageOfHashMapRelation(mu);
[ <trans cascade with 2 levels with (4, 5) pts, 5 dependencies>,
  <trans cascade with 2 levels with (4, 5) pts, 5 dependencies> ]
```

We can work with these cascade transformations instead of the original generators to have a hierarchically structured, more informative computation.

```
Dependency function of depth 1 with 1 dependencies.
[  ] -> Transformation( [ 1, 2, 2, 2 ] )
Dependency function of depth 2 with 4 dependencies.
[ 1 ] -> Transformation( [ 2, 5, 4, 2, 3 ] )
[ 2 ] -> Transformation( [ 1, 1, 4, 2 ] )
[ 3 ] -> Transformation( [ 3, 2, 3 ] )
[ 4 ] -> Transformation( [ 3, 1, 2 ] )

Dependency function of depth 1 with 1 dependencies.
[  ] -> Transformation( [ 1, 4, 1, 1 ] )
Dependency function of depth 2 with 4 dependencies.
[ 1 ] -> Transformation( [ 1, 3, 5, 4, 1 ] )
[ 2 ] -> Transformation( [ 1, 2, 1, 3 ] )
[ 3 ] -> Transformation( [ 4, 2, 3, 4 ] )
[ 4 ] -> Transformation( [ 2, 1, 1 ] )
```

The top level state 3 corresponds to the single original state 9, thus $|\theta^{-1}(3)| = 1$. Still, a non-identity transformation can appear on the second level (only a single map, $1 \mapsto 3$ in the first, and $1 \mapsto 4$ in the second generator), since other top level states can have bigger preimage sets in θ.

To establish and maintain the *correctness* of the software implementation we use a brute force algorithm to verify totally defined relational morphisms. We apply it to a special selection of transformation semigroups. These examples were collected over the years. They include cases that detected earlier bugs and ones that are specially crafted to exhibit special properties. Using the testing framework of GAP, all further changes and developments are verified continuously.

6.1   Comparison: Holonomy Versus Covering Lemma

The holonomy is a monolithic method since it produces, in an all or nothing manner, a most detailed decomposition. The Covering Lemma allows incremental decompositions, and it provides detail on demand only. Therefore, a direct comparison of the time and space complexity of these two algorithms is not possible. Also, using the Covering Lemma method iteratively to produce the same amount detail, as in holonomy, would have very similar resource usage.

Both methods work with generators, but the holonomy decomposition constructs a detailed analysis of the set of images of the semigroup action, requiring to compute possibly as many as 2^n subsets of the state set. This is practically feasible for a few tens of states, but not for thousands of states, required by many applications. The Covering Lemma method makes those possibly by avoiding this exponential computation in order to provide usable decompositions.

The holonomy method uses compression (unlike the "$V \cup T$"-technique used in an alternative proof of the Krohn-Rhodes Theorem [7]). It identifies equivalence classes of subsets of the state set with isomorphic groups action on them. If needed, the Covering Lemma method can also compute these separately for the local transformation semigroups $\{U_y \mid y \in Y\}$. It is even easier than in holonomy, since we can use the state set Y for finding the strongly connected components in the reachability relation under T, instead of working in the set of image sets.

7   Conclusions

Here we extended the computer algebra toolkit for the hierarchical decompositions of finite transformation semigroups by a simplified kernel-based two-level algorithm. In order to meet the requirements of the applications (as opposed to, or in addition to proving the lemma), we needed to make a few innovations: generalized labelling for states (the squashing function, Sect. 4.2); explicit computation of the reversed emulation, the *interpretation* (Sect. 3.3), for typical use cases in applications and for verification purposes; identifying the family of the bottom level transformation semigroups (Sect. 3.5) in lieu of the derived semigroupoid; separation of the general mechanism (Sect. 3) from the specific choices (Sect. 4).

The holonomy decomposition first implemented in `SgpDec` [4] allows one to make the transition from having only hand-calculated small decomposition examples to the machine computed and verified coordinatizations of real-world dynamical systems [10]. Similarly, the now implemented Covering Lemma algorithm (see Appendix 6) makes the transition from the fixed highest resolution decompositions only, monolithic calculation tool to a more flexible one. By iterating decompositions it extends the 'computational horizon' as we can compute and study bigger examples. By allowing an arbitrary surjective morphism as its input, it enables 'precision engineering', constructing tailor-made decompositions for particular discrete dynamical systems. We made effort to present the algorithm in an elementary way accessible to a wider audience. We hope this will allow other researchers and students to apply the decompositions in their projects.

References

1. Bergeron, A., Hamel, S.: From cascade decompositions to bit-vector algorithms. Theor. Comput. Sci. **313**, 3–16 (2004)
2. Egri-Nagy, A., Nehaniv, C.L., Mitchell, J.D.: SGPDEC – software package for Hierarchical Composition and Decomposition of Permutation Groups and Transformation Semigroups, Version 1.0.0 (2024). https://doi.org/10.5281/zenodo.11638444. https://gap-packages.github.io/sgpdec/
3. Egri-Nagy, A.: Finite computational structures and implementations: semigroups and morphic relations. Int. J. Netw. Comput. **7**(2), 318–335 (2017). https://doi.org/10.15803/ijnc.7.2_318
4. Egri-Nagy, A., Mitchell, J.D., Nehaniv, C.L.: SgpDec: cascade (de)compositions of finite transformation semigroups and permutation groups. In: Hong, H., Yap, C. (eds.) ICMS 2014. LNCS, vol. 8592, pp. 75–82. Springer, Heidelberg (2014). https://doi.org/10.1007/978-3-662-44199-2_13
5. Egri-Nagy, A., Nehaniv, C.L.: Computational Holonomy Decomposition of Transformation Semigroups (2015). arXiv:1508.06345 [math.GR]
6. The GAP Group: GAP – Groups, Algorithms, and Programming, Ver. 4.13.0 (2024). https://www.gap-system.org
7. Krohn, K., Rhodes, J.: Algebraic theory of machines. I. Prime decomposition theorem for finite semigroups and machines. Trans. Am. Math. Soc. **116**, 450–464 (1965)
8. Mitchell, J.D., et al.: Semigroups - GAP package, Version 5.3.7 (2024). https://doi.org/10.5281/zenodo.592893
9. Nehaniv, C.L.: From relation to emulation: the Covering Lemma for transformation semigroups. J. Pure Appl. Algebra **107**(1), 75–87 (1996). https://doi.org/10.1016/0022-4049(95)00030-5
10. Nehaniv, C.L., et al.: Symmetry structure in discrete models of biochemical systems: natural subsystems and the weak control hierarchy in a new model of computation driven by interactions. Philos. Trans. R. Soc. A: Math. Phys. Eng. Sci. **373**(2046), 20140223 (2015). https://doi.org/10.1098/rsta.2014.0223
11. Pfeiffer, M., Horn, M., Jefferson, C., Linton, S.: **datastructures**, collection of standard data structures for GAP, Version 0.3.0 (2022). https://gap-packages.github.io/datastructures, GAP package
12. Pin, J.: Mathematical foundations of automata theory. https://www.irif.fr/~jep/PDF/MPRI/MPRI.pdf, version of 18 February 2022
13. Rhodes, J.: Applications of Automata Theory and Algebra via the Mathematical Theory of Complexity to Biology, Physics, Psychology, Philosophy, and Games. World Scientific Press (2009), foreword by Morris W. Hirsch, edited by Chrystopher L. Nehaniv (Original version: UC Berkeley, Mathematics Library, 1971)
14. Rhodes, J., Steinberg, B.: The Q-theory of Finite Semigroups. Springer, Cham (2008)
15. Tilson, B.: Categories as algebra: an essential ingredient in the theory of monoids. J. Pure Appl. Algebra **48**(1–2), 83–198 (1987)

On Pumping Preserving Homomorphisms and the Complexity of the Pumping Problem (Extended Abstract)

Hermann Gruber[1], Markus Holzer[2(✉)], and Christian Rauch[2]

[1] Planerio GmbH, Theresienhöhe 11A, 80339 Munich, Germany
h.gruber@planerio.de
[2] Institut für Informatik, Universität Giessen, Arndtstr. 2, 35392 Giessen, Germany
{holzer,christian.rauch}@informatik.uni-giessen.de

Abstract. This paper complements a recent inapproximability result for the minimal pumping constant w.r.t. a fixed regular pumping lemma for nondeterministic finite automata [H. GRUBER and M. HOLZER and C. RAUCH. The Pumping Lemma for Regular Languages is Hard. *CIAA 2023*, pp. 128–140.], by showing the inapproximability of this problem even for deterministic finite automata, and at the same time proving stronger lower bounds on the attainable approximation ratio, assuming the Exponential Time Hypothesis (ETH). To that end, we describe those homomorphisms that, in a precise sense, preserve the respective pumping arguments used in two different pumping lemmata. We show that, perhaps surprisingly, this concept coincides with the classic notion of star height preserving homomorphisms as studied by McNaughton, and by Hashiguchi and Honda in the 1970s. Also, we gain a complete understanding of the minimal pumping constant for bideterministic finite automata, which may be of independent interest.

1 Introduction

The investigation of combinatorial properties and decision problems dates back to the very early days of automata and formal language theory. One of the most well known combinatorial properties that are satisfied by context-free and regular languages are pumping or interchange lemmata. For regular languages one finds, e.g., the following pumping lemma in Kozen's monograph on automata and computability—the lemma describes a necessary condition for languages to be regular [15, page 70, Theorem 11.1].

Lemma 1. *Let L be a regular language over Σ. Then, there is a constant p (depending on L) such that the following holds: If $w \in L$ and $|w| \geq p$, then there are words $x \in \Sigma^*$, $y \in \Sigma^+$, and $z \in \Sigma^*$ such that $w = xyz$ and $xy^t z \in L$ for $t \geq 0$—it is then said that y can be* pumped *in w.*

© The Author(s), under exclusive license to Springer Nature Switzerland AG 2024
S. Z. Fazekas (Ed.): CIAA 2024, LNCS 15015, pp. 153–165, 2024.
https://doi.org/10.1007/978-3-031-71112-1_11

A lesser-known pumping lemma, attributed to Jaffe [14], characterizes the regular languages, by describing a necessary and sufficient condition for languages to be regular. For other pumping lemmata see, e.g., the annotated bibliography on pumping [18]:

Lemma 2. *A language L is regular if and only if there is a constant p (depending on L) such that the following holds: If $w \in \Sigma^*$ and $|w| = p$, then there are words $x \in \Sigma^*$, $y \in \Sigma^+$, and $z \in \Sigma^*$ such that $w = xyz$ and*[1]

$$wv = xyzv \in L \iff xy^t zv \in L$$

for all $t \geq 0$ and each $v \in \Sigma^*$.

These lemmata are part of the automata and formal language standard toolbox and their combinatorial properties are well understood. But what about their computational properties? Recently this question was answered in [8] by studying the PUMPING-PROBLEM for finite automata. This is, given a finite automaton A and a value p, does the statement of Lemma 1 (or alternatively of Lemma 2) hold for the language $L(A)$ w.r.t. the value p? It turned out that this problem is already intractable for DFAs, i.e., coNP-complete, regardless of whether we check for the pumping property described by Kozen [15] or Jaffe [14]. This is quite remarkable since it is a rare example of a finite automaton problem where already, for a single device, the studied property becomes intractable. The latter pumping property is more complex for NFAs, namely PSPACE-complete, while the former is shown to be coNP-hard and contained in Π_2^P, the second co-level of the polynomial hierarchy, for nondeterministic finite state devices. Moreover, for NFAs also an inapproximability result was shown for both pumping properties unless the Exponential Time Hypothesis (ETH) fails. Whether one can come up with a similar inapproximability result for DFAs was left open, and this is also the starting point of this research.

The exponential time hypothesis—roughly speaking this is an unproven computational hardness assumption which states [13] that the satisfiability of n variable 3-SAT cannot be solved in sub-exponential time $2^{o(n)}$—, as well as related assumptions which are stronger than $P \neq NP$, emerged as a swiss-army knife, which allows for a fine-grained analysis of NP-hard problems. To name a few, it helps to gauge the inherent limits of approximability beyond polynomial time, as well as those of parameterized and exact exponential algorithms [3]. In recent years, numerous algorithmic impossibility results based on the ETH and related hypotheses have been derived [16,21], more recently also for problems related to finite automata and regular expressions, see, e.g., [4,6,9].

The main result of the present paper is an inapproximabilty bound for deterministic finite automata for both pumping properties unless ETH fails. This supersedes the previous inapproximability result for NFAs from [8]. The ingredients for this result are (i) a notion we call pumping preserving homomorphism,

---

[1] Observe that the words $w = xyz$ and $xy^t z$, for all $t \geq 0$, belong to the same Myhill-Nerode equivalence class of the language L. Thus, one can say that the pumping of the word y in w *respects equivalence classes*.

which allows a binary encoding of the problem and (ii) a non-trivial relation between the longest path problem on directed graphs and minimal pumping constants for the two pumping lemmata mentioned above. These two ingredients allow for a reduction from the inapproximability of the longest path problem on directed graphs [2] to the PUMPING-PROBLEM for binary regular languages unless P = NP. Also, assuming ETH, which is a stronger assumption than P $\neq$ NP, it cannot be approximated in polynomial time within an even higher factor. The obtained inapproximability bounds significantly outperform the previously known bounds from [8] for NFAs. Due to space constraints, all proofs can be found in the full version of this paper.

2   Preliminaries

Next we fix some definitions on finite automata—cf. [10]. A *nondeterministic finite automaton* (NFA) is a quintuple $A = (Q, \Sigma, \cdot, q_0, F)$, where Q is the finite set of *states*, Σ is the finite set of *input symbols*, $q_0 \in Q$ is the *initial state*, $F \subseteq Q$ is the set of *accepting states*, and the *transition function* $\cdot$ maps $Q \times \Sigma$ to 2^Q. Here 2^Q refers to the powerset of Q. The *language accepted* by the NFA A is defined as $L(A) = \{ w \in \Sigma^* \mid (q_0 \cdot w) \cap F \neq \emptyset \}$, where the transition function is recursively extended to a mapping $Q \times \Sigma^* \to 2^Q$ in the usual way. An NFA A is said to be *partial deterministic* if $|q \cdot a| \leq 1$ and *deterministic* (DFA) if $|q \cdot a| = 1$ for all $q \in Q$ and $a \in \Sigma$. In these cases, we simply write $q \cdot a = p$ instead of $q \cdot a = \{p\}$. Note that every partial DFA can be made complete by introducing a non-accepting sink state that collects all non-specified transitions. For a DFA, obviously every letter $a \in \Sigma$ induces a mapping from the state set Q to Q by $q \mapsto q \cdot a$, for every $q \in Q$. Finally, a finite automaton is *unary* if the input alphabet Σ is a singleton set, that is, $\Sigma = \{a\}$, for some input symbol a.

The *deterministic state complexity of a finite automaton* A with state set Q is referred to as $\mathrm{sc}(A) := |Q|$ and the *deterministic state complexity of a regular language* L is defined as

$$\mathrm{sc}(L) = \min\{ \mathrm{sc}(A) \mid A \text{ is a DFA accepting } L, \text{ i.e., } L = L(A) \}.$$

A similar definition applies for the *nondeterministic state complexity of a regular language* by changing DFA to NFA in the definition, which we refer to as $\mathrm{nsc}(L)$. It is well known that

$$\mathrm{nsc}(L) \leq \mathrm{sc}(L) \leq 2^{\mathrm{nsc}(L)},$$

for every regular language L.

A finite automaton is *minimal* if its number of states is minimal with respect to the accepted language. It is well known that each minimal DFA is isomorphic to the DFA induced by the Myhill-Nerode equivalence relation. The *Myhill-Nerode* equivalence relation $\sim_L$ for a language $L \subseteq \Sigma^*$ is defined as follows: for $u, v \in \Sigma^*$ let $u \sim_L v$ if and only if $uw \in L \iff vw \in L$, for all $w \in \Sigma^*$. The equivalence class of u is referred to as $[u]_L$ or simply $[u]$ if the language is clear from the context and it is the set of all words that are equivalent to u w.r.t. the relation $\sim_L$, i.e., $[u]_L = \{\, v \mid u \sim_L v \,\}$. Therefore, we refer to the automaton induced by the Myhill-Nerode equivalence relation $\sim_L$ as the minimal DFA for the language L. On the other hand, there may be minimal non-isomorphic NFAs for L.

3   Homomorphisms Preserving the Pumping Property

We investigate the impact of homomorphisms on pumping. To achieve this, we introduce the concept of homomorphisms that maintain pumping (pumping preserving homomorphisms). This concept depends on the way the pumping has to be performed—compare Lemma 1 and Lemma 2– and is closely connected to the idea of star-height preserving homomorphisms [11,17]. The latter have proven to be highly advantageous in exploring issues related to the descriptional complexity of regular expressions—see, e.g., [7]. The property of pumping preservation exhibited by homomorphisms serves a comparable *rôle* in the examination of the computational complexity of pumping problems, as studied in [8].

In fact, our proofs below characterizing the different variants of pumping preserving homomorphisms mirroring similar properties of the star height preserving homomorphisms, culminating in a proof that all these notions coincide. For most of the examples and proofs below, our strategy follows the same outline as in [11]. We try to keep this paper self-contained, but occasionally need some technical lemmata from [11], whose proofs are not essential for understanding the material presented here.

Definition 3. *Let $L \subseteq \Sigma^*$ be a regular language and w be a word in Σ^*. Then we define the following two properties:*

1. *The word w has the pumping property w.r.t. the language L, if w admits a decomposition $w = xyz$ with $|y| \geq 1$ such that $xy^*z \subseteq L$, and*
2. *the word w has the enhanced pumping property w.r.t. the language L, if w admits a decomposition $w = xyz$ with $|y| \geq 1$ such that[2] $xyzv \subseteq L$ if and only if $xy^*zv \subseteq L$, for all words $v \in \Sigma^*$.*

Let $h : \Sigma \to \Gamma^*$ be a homomorphism. Then h is said to preserve the pumping property if and only if the following condition holds: for each regular language L over Σ and each word $w \in \Sigma^*$, the word w has the pumping property w.r.t. the language L if and only if its homomorphic image $h(w)$ has the pumping property w.r.t. $h(L)$. A similar definition applies for a homomorphism to be enhanced pumping preserving by replacing pumping preserving by enhanced pumping preserving everywhere.

The following is immediate from the definition of the pumping properties.

Lemma 4. *Let $L \subseteq \Sigma^*$ be a regular language and w be a word in Σ^*. If the word w does not satisfy the pumping property w.r.t. the language L, then w does*

---

[2] In abuse of notation we write $xyzv \subseteq L$ instead of $\{xyzv\} \subseteq L$.

not *satisfy the enhanced pumping property w.r.t. the same language. Moreover, if the homomorphism h does not preserve the pumping property, then it does not preserve the enhanced pumping property either.*   □

Next we show that one implication in the definition of the (enhanced) pumping preserving property is easily seen to be true.

Lemma 5. *For each ε-free homomorphism $h : \Sigma \to \Gamma^*$ and every regular language L over Σ, and every word $w \in \Sigma^*$ the following holds: if w has the pumping property w.r.t. L, then $h(w)$ has the pumping property w.r.t. $h(L)$. The statement is also valid for the enhanced pumping property.*

What are the homomorphisms such that the reverse implication is satisfied as well? The notion of preserving the (enhanced) pumping property is well defined, although one at first glance might think that its always true for any (ε-free) homomorphism, since regular languages can be pumped and are closed under homomorphisms. This is not the case.[3]

Example 6. Let $\Sigma = \{a, b\}$ let $\Gamma = \{c\}$, and let h denote the unary projection $h(a) = h(b) = c$. Consider the language $L = (aa)^* \cup b(bb)^*$. Then the word $w = b$ cannot be pumped w.r.t. L, since $b^i \notin L$ whenever i is odd. But $h(L) = c^*$, and it is obvious that the word $h(b)$ can be pumped w.r.t. $h(L)$.   □

In fact, all homomorphisms preserving the (enhanced) pumping property are injective, as we shall prove later on. Recall that an injective homomorphism is commonly referred to as *code*, and the homomorphic images of the alphabet letters of its domain are referred to as *codewords*. A code with the property that no codeword is a prefix (suffix, respectively) of another codeword is a *prefix code* (*suffix code*, respectively). A code that is both a prefix code and a suffix code is called a *bifix code*.

Example 7. Let $\Sigma = \{a, b, c\}$, let $\Gamma = \{0, 1\}$. Let h be the prefix code given by $h(a) = 01$, $h(b) = 011$, and $h(c) = 0111$. Consider the language $L_1 = (a \cup bc^*b)^+$. Then the word $w = bb$ does *not* have the pumping property: there are two decompositions $w = xyz$ with $|y| > 0$. But, for these, we have $xy^0z = b$ or $xy^0z = \varepsilon$, which are not in L.

Now let $L_2 = 0(10 \cup 1101)^*1$. Then obviously $0(1101)^i1 \in L_2$, for every value $i \geq 0$, so the word 011011 has the pumping property w.r.t. language L_2. Observe, that for all $i \geq 1$ we have

$$h(a^{i+1}) = (01)^{i+1} = 0 \underbrace{(10) \cdots (10)}_{i \text{ times}} 1$$

as well as

$$h(bc^ib) = 011 \underbrace{(0111) \cdots (0111)}_{i \text{ times}} 011 = 0 \underbrace{(1101) \cdots (1101)}_{i+1 \text{ times}} 1,$$

---

[3] Because of Lemma 4, in the examples to come, it suffices to restrict our attention to the (non-enhanced) pumping property.

for all $i \geq 0$. One can easily observe that these two patterns exhaust the set L_2, so we can conclude that $L_2 = h(L_1)$.   □

The problematic phenomenon illustrated in the above example has been formalized in [11]: Let $h : \Sigma^* \to \Gamma^*$ be a code. Then, we say that h has the *non-crossing property*, if and only if for all $v_1, v_2, w_1, w_2 \in \Gamma^+$ we have that if $(v_1 \cup v_2) \cdot (w_1 \cup w_2) \subseteq h(\Sigma)$, then $v_1 = v_2$ or $w_1 = w_2$.

Lemma 8. *Let h be a code which* does not *have the non-crossing property. Then h is neither pumping preserving nor enhanced pumping preserving.*

If we restrict our attention to bifix codes, the non-crossing property is sufficient:

Lemma 9. *Let h be a bifix code with the non-crossing property, and let L be a regular language. If $w \in L$ does not have the pumping property w.r.t. language L, then $h(w)$ does not have the pumping property w.r.t. $h(L)$. The result is also valid for the enhanced pumping property.*

Now, we are ready to prove that indeed all pumping preserving homomorphisms are codes:

Lemma 10. *Let h be a homomorphism that is* not *injective. Then h is neither pumping preserving, nor enhanced pumping preserving.*

But we observe that in general it is not needed for a pumping preserving homomorphism to be prefix-free.

Lemma 11. *There are (enhanced) pumping preserving codes which are not prefix codes.*

To obtain a necessary and sufficient condition, we use a few more definitions from [11]: Let $h : \Sigma^* \to \Gamma^*$ be a code. Let R_1, R_2 be languages over Γ such that $R_1 \cdot R_2 \subseteq h(\Sigma^*)$. The ordered pair (R_1, R_2) *has the tag* (w.r.t. h) if one of the following holds:

– There exists $r \in \Gamma^*$ and $R_1' \subseteq h(\Sigma^*)$ such that $R_1 = R_1' r$ and $r R_2 \subseteq h(\Sigma^*)$.
– There exists $s \in \Gamma^*$ and $R_2' \subseteq h(\Sigma^*)$ such that $R_2 = s R_2'$ and $R_1 s \subseteq h(\Sigma^*)$.

The shortest word that satisfies the first condition (the second condition, respectively) is called the *suffix tag* (the *prefix tag*, respectively) of (R_1, R_2). A code *has the tag property* if for all $x, x', y, y' \in \Gamma^*$, if $(x \cup x')(y \cup y') \subseteq h(\Sigma^*)$, then the pair $(x \cup x', y \cup y')$ has the tag.

For the proof of the next result, we need a theorem on unique decodability as stated in [11] and attributed to Schützenberger [20].

Theorem 12. *A homomorphism $h : \Sigma^* \to \Gamma^*$ is injective if and only if the following hold:*

– *For all $a \in \Sigma$, $h(a) \notin h(\Sigma \setminus \{a\})^*$.*

– For all $x, y, z \in \Sigma^*$ the following holds: if all of x, xy, yz, and z are in $h(\Sigma^*)$, then $y \in h(\Sigma^*)$.

Now we are equipped for proving the next statement—the proof of the following lemma is similar to that in [11, Theorem 5.1].

Lemma 13. *Let $h : \Sigma^* \to \Gamma^*$ be a code that has the tag property. Then h is pumping and enhanced pumping preserving.*

The converse direction is now easy:

Lemma 14. *If a code h is pumping preserving, then it has the tag property. The implication remains valid if h is enhanced pumping preserving.*

Hence, we have obtained a tight characterization of those homomorphisms that preserve the pumping properties.

Corollary 15. *A homomorphism preserves the (enhanced) pumping property if and only if it has the tag property.*   □

Finally, we can state the main theorem of this section. It was proved in [11] that a homomorphism preserves star height if and only if it has the tag property. Thus, we obtain:

Theorem 16. *A homomorphism preserves the pumping property if and only if it preserves the enhanced pumping property if and only if preserves star height.*

4 Minimal Pumping Constants and Longest Paths in Finite Automata

We consider the pumping constants $\mathrm{mpc}(L)$ and $\mathrm{mpe}(L)$ and their relation to the longest path in the minimal finite automaton accepting the regular language L. Here $\mathrm{mpc}(L)$ ($\mathrm{mpe}(L)$, respectively) refers to the minimal number p satisfying the conditions of Lemma 1 (Lemma 2, respectively), for a regular language L over Σ. Simple facts about these constants can be found in [5, 12]. Concerning the relation between both mpc and mpe we have $\mathrm{mpc}(L) \leq \mathrm{mpe}(L)$, for every regular language L. The relation of $\mathrm{mpe}(L)$ and the state complexities is more subtle, namely for a regular language L over the alphabet Σ we have $\mathrm{mpc}(L) \leq \mathrm{sc}(L)$ and it was shown in [12] that

$$\mathrm{mpe}(L) \leq \mathrm{sc}(L) \leq \sum_{i=0}^{\mathrm{mpe}(L)-1} |\Sigma|^i.$$

The former inequality also holds for nsc, the nondeterministic state complexity, i.e., $\mathrm{mpc}(L) \leq \mathrm{nsc}(L)$, while the latter does not generalize. In fact, mpe and nsc are incomparable [8]. There it was argued that the mpe-measure and the length of the *longest path* of the automaton, that is, a simple directed path of maximum

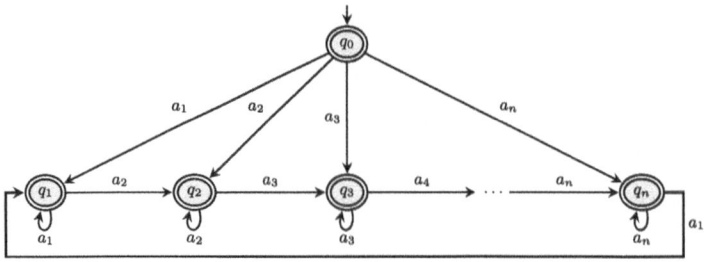

Fig. 1. The deterministic automaton A_n accepting $L_n = L(A_n)$. The non-accepting sink state is not shown. The language L_n satisfying $\mathtt{mpe}(L_n) = 3$ and the longest path in A starting in the initial state q_0 is of length $n + 1$.

length from the initial state of the automaton, are different measures in general, using the witness shown in Fig. 1. Nevertheless, for a restricted class of automata we will show that in fact both pumping measures can be bounded somehow with the length of the longest path of the underlying automaton. Before we state our results for the pumping measures and their relation to the longest path, we introduce two notations: let A be a DFA and q a state of A, not necessarily final. Then ℓ_A ($\ell_A|_q$, respectively) refers to the length, i.e., number of transitions, of the longest simple directed path of maximum length in the automaton A starting in q_0 and ending in any state (in state q, respectively). Here a path is called *simple* if it does not have repeated states/vertices. The following observation is immediate by Jaffe's proof, cf. [14], and was mentioned in [8]:

Lemma 17. *Let A be a DFA and $L := L(A)$. Then $\mathtt{mpe}(L) \le \ell_A + 1$. If L is a unary language, then $\mathtt{mpe}(L) = \mathtt{sc}(L)$.*

Using the witness depicted in Fig. 1, we have a language L where $\mathtt{mpe}(L)$ and $\ell_A + 1$ can be far apart. Nevertheless, for the class of bideterministic finite automata (biDFAs, for short) this is not the case as shown next. Here, a finite automaton A is *bideterministic* if it is both partially deterministic and partially co-deterministic and has a sole accepting state. Moreover, an automaton A is *partially co-deterministic* if the reversed automaton obtained by reversing the transitions of A is partially deterministic. A language L is said to be *bideterministic* if it is accepted by a biDFA A, i.e., $L = L(A)$. Observe that a language L is accepted by a bideterministic finite automaton if and only if the minimal automaton of L is reversible, i.e., deterministic and co-deterministic, and has a unique final state [19]. This leads us to the key property of bideterministic finite automata of which we will make heavy use:

Lemma 18. *Let $A = (Q, \Sigma, \cdot, q_0, F)$ be a biDFA and z a word from Σ^*. Then $p \cdot z = q$, for $p, q \in Q$ implies that $|\{p \mid p \cdot z = q\}| = 1$, i.e., if the z-predecessor state p of q exists, then p is unique.* □

The next lemma uses the above property in the proof and reads as follows:

Lemma 19. *Let L be a bideterministic language accepted by a minimal DFA A and assume $L := L(A)$. Then $\ell_A \leq mpe(L) \leq \ell_A + 1$.*

The bounds in the previous lemma are best possible. This can be seen by the following bideterministic languages: language $L_1 = \{ab\}$ over the binary alphabet $\{a, b\}$ and the unary language $L_2 = \{aa\}$. Both languages are accepted by minimal 4-state DFAs, which are both bideterministic if the non-accepting sink state is removed. Let A_1 (A_2, respectively) be the DFA that accepts L_1 (L_2, respectively)—see Fig. 2. Thus, in both cases we have $\ell_{A_1} = \ell_{A_2} = 3$, but $\texttt{mpe}(L_1) = 3 = \ell_{A_1}$, while $\texttt{mpe}(L_2) = 4 = \ell_{A_2} + 1$. This is seen as follows: first consider the language L_1. The word ab cannot be pumped at all, because any shorter word is not a member of L_1. Hence, $\texttt{mpe}(L_1) \geq 3$. Every word of length at least 3 maps the initial state of the automaton A_1 to the non-accepting sink state. Any word of length at least 3 that starts with the letter b or with the prefix aa visits the non-accepting sink state at least twice and hence can by pumped respecting equivalence classes. In case the word of length at least 3 begins with ab we are left with the prefixes aba or abb. Then, it is easy to see that in the former case the letter b in the middle of the prefix can be pumped, while in the latter case the first letter a is pumpable. The details are left to the reader. Hence, we have $\texttt{mpe}(L) = 3 = \ell_{A_1}$ as required. For the unary language L_2, the argumentation that $\texttt{mpe}(L_2) = 4$ is slightly simpler. Since L_2 is unary, there is only a single word of length 3, but a^3 is not a member of L_2. Shortening this word to any length strictly less than 3, results in a word that belongs to a different Myhill-Nerode equivalence class then the word a^3. Hence a^3 cannot be pumped by respecting equivalence classes. Therefore, $\texttt{mpe}(L_2) \geq 3+1 = 4$ and by Lemma 17 we have $\texttt{mpe}(L) \leq \ell_{A_2} + 1 = 3 + 1 = 4$. Thus, $\texttt{mpe}(L_2) = 4 = \ell_{A_2} + 1$.

Fig. 2. Deterministic finite automata A_1 and A_2 accepting the bideterministic and finite languages $L_1 = \{ab\}$ and $L_2 = \{aa\}$, respectively. In both drawings the non-accepting sink state is not shown. The longest simple path in both automata is of length 3 (including the non-accepting sink state) and the pumping constants of the languages are $\texttt{mpe}(L_1) = 3$ and $\texttt{mpe}(L_2) = 4$.

Next, let us consider the pumping measure $\texttt{mpc}$. Recall that $\texttt{mpc}(L) \leq \texttt{mpe}(L)$, for every regular language L. The gap between path measures can be arbitrarily large even for bideterministic languages. Consider the bideterministic unary language $L_n = (a^n)^*$, for $n \geq 1$, which is accepted by a unary cyclic DFA with n states. Since $\texttt{mpe}$ and $\texttt{sc}$ coincides for unary languages, we have that $\texttt{mpe}(L_n) = n$. On the other hand, $\texttt{mpc}(L_n) = 1$, because $L_n \neq \emptyset$ and each word $w \in L_n$ can be pumped by choosing $x = \lambda$, $y = w$, and $z = \lambda$, since $xy^*z = w^* \subseteq L$. Nevertheless, we can also show a nice relation for the $\texttt{mpc}$-measure with a longest path problem. The result is stated in the following lemma.

Lemma 20. *Let L be a bideterministic language that is accepted by the minimal DFA A and define $L := L(A)$. Then $mpc(L) = \ell_A\big|_{q_f} + 1$, where q_f is the unique final state of A.*

5   The Complexity of Pumping, Revisited

We will consider the following decision problem [8] related to the pumping lemmata stated in the introduction:

> LANGUAGE-PUMPING-PROBLEM or for short PUMPING-PROBLEM:
> INPUT: a finite automaton A and a natural number p, i.e., an encoding $\langle A, 1^p \rangle$.
> OUTPUT: Yes, if and only if the statement from Lemma 1 (Lemma 2, resp.) holds for the language $L(A)$ w.r.t. the value p.

We apply our findings on the relation between the minimal pumping constants from the previous section in order to give a simpler proof for the coNP-completeness of the PUMPING-PROBLEM for DFAs. This will be a corollary to an inapproximability result for DFAs, which significantly improves a previously known inapproximability result for NFAs from [8].

   The LONGEST-DIRECTED-PATH-PROBLEM (LDP-problem) is defined as follows: given a directed graph $G = (V, E)$, find the longest sequence of distinct vertices $v_1, v_2, \ldots, v_k$ such that $(v_i, v_{i+1}) \in E$, for $1 \leq i < k$. Naturally, the LDP-problem gives rise to a NP-complete decision and an approximation problem. The LDP-optimization problem cannot be approximated in polynomial time within a factor of $n^{1-\varepsilon}$ for any constant $\varepsilon > 0$, unless P = NP. Also, assuming the Exponential Time Hypothesis (ETH), which is a stronger assumption than P $\neq$ NP, it cannot be approximated in polynomial time within a factor of $\omega(\frac{n \log \log n}{(\log n)^2})$. Both inapproximability bounds are shown[4] in [2]. By inspecting the proof, one can see that the result in fact applies to digraphs with bounded outdegree [1]. In the next theorem we use the LDP-optimization problem as a starting point for our reduction to the PUMPING-PROBLEM for DFAs.

Theorem 21. *Let A be a DFA with s states over an alphabet of size $O(s)$. Then no deterministic polynomial time algorithm can approximate the minimal pumping constant w.r.t. Lemma 1 (Lemma 2, respectively) within a factor of $s^{1-\varepsilon}$ for any constant $\varepsilon > 0$, unless P = NP. Assuming ETH, the inapproximability factor becomes $\omega(\frac{s \log \log s}{(\log s)^2})$.*

   A direct consequence of the previous proof is that the PUMPING-PROBLEM for both pumping lemmata is intractable; this re-establishes a result from [8]—with an even stronger bound on the approximation ratio—, but it uses a growing-size alphabet. Luckily, we can use a (enhanced) pumping preserving homomorphism

---

[4] The actual inapproximability bound assuming the ETH from [2] is slightly stronger, but involves a function $f(n)$ satisfying some additional tameness constraint. We use the simplified bound for better readability.

to code it down to a binary alphabet. Let $\Sigma = \{a_1, a_2, \ldots, a_r\}$ be an r-letter alphabet and h be the homomorphism given by $a_i \mapsto \text{bin}(i)\text{bin}(i)^R$, for $1 \leq i \leq r$. Here, $\text{bin}(i)$ denotes the $\lceil \log r \rceil$-bit binary encoding of the number i and $\text{bin}(i)^R$ its reversal. As shown in [7], the code h preserves star height, which in turn implies that h also preserves the (enhanced) pumping property by Theorem 16. Now we are ready for the next lemma.

Lemma 22. *Let $r \geq 3$ be an integer, $\Sigma = \{a_1, a_2, \ldots, a_r\}$ be an r-letter alphabet, h be the homomorphism given by $a_i \mapsto \text{bin}(i)\text{bin}(i)^R$, for $1 \leq i \leq r$, and L be a regular language over Σ. Then*

$$mpc(h(L)) = 2b \cdot (mpc(L) - 1) + 1,$$

where $b = \lceil \log r \rceil$.

For the minimal pumping constants w.r.t. Lemma 2 we prove the following lemma in similar vein as Lemma 22.

Lemma 23. *Let $r \geq 3$ be an integer, $\Sigma = \{a_1, a_2, \ldots, a_r\}$ be an r-letter alphabet, h be the homomorphism given by $a_i \mapsto \text{bin}(i)\text{bin}(i)^R$, for $1 \leq i \leq r$, and L be a regular language over Σ. Then*

$$2b \cdot (mpe(L) - 1) < mpe(h(L)) \leq 2b \cdot mpe(L),$$

where $b = \lceil \log r \rceil$.

One may ask whether we can come up with an exact calculation of $\text{mpe}(h(L))$ like for $\text{mpc}(h(L))$. As we will see next this is not possible since the bounds for $\text{mpe}(h(L))$ from the previous lemma are best possible. We begin with a language L that meets the upper bound, i.e., $\text{mpe}(h(L)) = 2b \cdot \text{mpe}(L)$. Define $L = (abc)^*$. Thus, $2b = 4$. Observe that each word of length at least three is pumpable w.r.t. Lemma 2 either (i) by its prefix abc, (ii) by its second letter if the first letter is not an a, (iii) by its first letter if the second letter is not an a, or (iv) by its third letter, otherwise. Thus $\text{mpe}(L) \geq 3$. Moreover, the word ab cannot be pumped w.r.t. Lemma 2, which can easily be seen by concatenating it with $v = c$. Hence, we have $\text{mpe}(L) = 3$. Next, consider $h(L)$. The language L is mapped by h onto $h(L) = (000001101001)^*$. Similarly, as in the above argumentation, one finds that each word in $h(L)$ of length twelve can be pumped w.r.t. Lemma 2 while the word 00000110100 cannot be pumped. Therefore, $\text{mpe}(h(L)) = 12 = 4 \cdot 3 = 2b \cdot \text{mpe}(L)$ as desired. Finally, we consider a language L that meets the lower bound, that is, $\text{mpe}(h(L)) = 2b \cdot (\text{mpe}(L)-1)+1$. Let $L = abc$. Again $2b = 4$. It is easy to see that $\text{mpe}(L) = 4$, since the word abc cannot be pumped w.r.t. Lemma 2 while all words of length four are pumpable. The details are left to the reader. On the other hand, we have $h(L) = 000001101001$, which fulfills $\text{mpe}(h(L)) = 13 = 12+1 = 4 \cdot 3+1 = 4 \cdot (4-1)+1 = 2b \cdot (\text{mpe}(L)-1)+1$. Thus, the bounds proven in Lemma 23 are best possible.

We close this section with the main result on the inapproximability of the PUMPING-PROBLEM for DFAs on binary alphabets.

Theorem 24. *Let A be a DFA with s states over a binary alphabet. Then no deterministic polynomial time algorithm can approximate the minimal pumping constant w.r.t. Lemma 1 within a factor of $s^{1-\varepsilon}$ for any constant $\varepsilon > 0$, unless $\mathsf{P} = \mathsf{NP}$. In case of ETH the inapproximability factor becomes $\frac{s \log \log s}{(\log s)^2}$. Both statements hold true if one considers the approximation of the minimal pumping constant w.r.t. Lemma 2.*

6 Conclusion

In the present paper, we revisited the pumping problem for regular languages. For bideterministic finite automata, the pumping constant is more well-behaved than in the general case. We characterized the pumping constant in terms of the longest path from the start state to an accepting state. Then, we turned our attention to the homomorphisms preserving the pumping property. This is seemingly a more primitive abstraction than that of a homomorphism preserving star height. Interestingly, the two definitions are equivalent. Compared with the results obtained for NFAs from [8], the inapproximability bound we obtained by putting the pieces together is stronger, even though we restricted the input to DFAs, which means that the input is represented less succinctly.

Acknowledgement. The authors would like to thank Andreas Björklund for some fruitful discussion.

References

1. Björklund, A.: Personal communication (2024)
2. Björklund, A., Husfeldt, T., Khanna, S.: Approximating longest directed paths and cycles. In: Díaz, J., Karhumäki, J., Lepistö, A., Sannella, D. (eds.) ICALP 2004. LNCS, vol. 3142, pp. 222–233. Springer, Heidelberg (2004). https://doi.org/10.1007/978-3-540-27836-8_21
3. Cygan, M., et al.: Lower bounds based on the exponential-time hypothesis. In: Cygan, M., et al. (eds.) Parameterized Algorithms, pp. 467–521. Springer, Cham (2015). https://doi.org/10.1007/978-3-319-21275-3_14
4. Czerwinski, W., et al.: Languages given by finite automata over the unary alphabet. In: Bouyer, P., Srinivasan, S. (eds.) Proceedings of the 43rd IARCS Annual Conference on Foundations of Software Technology and Theoretical Computer Science (FSTTCS 2023). LIPIcs, vol. 284, pp. 22:1–22:20. Schloss Dagstuhl–Leibniz-Zentrum für Informatik, Dagstuhl, Germany, IIIT Hyderabad, Telangana, India (2023). https://doi.org/10.4230/LIPIcs.STACS.2008.1354
5. Dassow, J., Jecker, I.: Operational complexity and pumping lemmas. Acta Inform. **59**, 337–355 (2022). https://doi.org/10.1007/s00236-022-00431-3
6. Fernau, H., Krebs, A.: Problems on finite automata and the exponential time hypothesis. Algorithms **10**(1), 24 (2017). https://doi.org/10.3390/a10010024
7. Gruber, H., Holzer, M.: Tight bounds on the descriptional complexity of regular expressions. In: Diekert, V., Nowotka, D. (eds.) DLT 2009. LNCS, vol. 5583, pp. 276–287. Springer, Heidelberg (2009). https://doi.org/10.1007/978-3-642-02737-6_22

8. Gruber, H., Holzer, M., Rauch, C.: The pumping lemma for regular languages is hard. In: Nagy, B. (ed.) CIAA 2023. LNCS, vol. 14151, pp. 128–140. Springer, Cham (2023). https://doi.org/10.1007/978-3-031-40247-0_9
9. Gruber, H., Holzer, M., Wolfsteiner, S.: On minimizing regular expressions without Kleene star. In: Bampis, E., Pagourtzis, A. (eds.) FCT 2021. LNCS, vol. 12867, pp. 245–258. Springer, Cham (2021). https://doi.org/10.1007/978-3-030-86593-1_17
10. Harrison, M.A.: Introduction to Formal Language Theory. Addison-Wesley, London (1978)
11. Hashiguchi, K., Honda, N.: Homomorphisms that preserve star height. Inform. Comput. **30**(3), 247–266 (1976). https://doi.org/10.1016/S0019-9958(76)90511-8
12. Holzer, M., Rauch, C.: On Jaffe's pumping lemma, revisited. In: Bordihn, H., Tran, N., Vaszil, G. (eds.) DCFS 2023. LNCS, vol. 13918, pp. 65–78. Springer, Cham (2023). https://doi.org/10.1007/978-3-031-34326-1_5
13. Impagliazzo, R., Paturi, R.: Complexity of k-SAT. In: Sirakov, B., de Souza, P.N., Viana, M. (eds.) Proceedings of the 14th Annual IEEE Conference on Computational Complexity, Atlanta, Georgia, USA, pp. 237–240. IEEE Computer Society (1999). https://doi.org/10.1109/CCC.1999.766282
14. Jaffe, J.: A necessary and sufficient pumping lemma for regular languages. SIGACT News **10**(2), 48–49 (1978). https://doi.org/10.1145/990524.990528
15. Kozen, D.C.: Automata and Computability. Undergraduate Texts in Computer Science, Springer, Cham (1997). https://doi.org/10.1007/978-1-4612-1844-9
16. Lakshtanov, D., Marx, D., Saurabh, S.: Lower bounds based on the exponential time hypothesis. Bull. Eur. Assoc. Theor. Comput. Sci. **105**, 41–72 (2011)
17. McNaughton, R.: The loop complexity of regular events. Inf. Sci. **1**, 305–328 (1969). https://doi.org/10.1016/S0020-0255(69)80016-2
18. Nijholt, A.: YABBER–yet another bibliography: pumping lemma's. An annotated bibliography of pumping. Bull. EATCS **17**, 34–53 (1982)
19. Pin, J.-E.: On reversible automata. In: Simon, I. (ed.) LATIN 1992. LNCS, vol. 583, pp. 401–416. Springer, Heidelberg (1992). https://doi.org/10.1007/BFb0023844
20. Schützenberger, M.P.: Une théorie algébrique du codage. Séminaire Dubreil. Algèbre et théorie des nombres **9**, 1–24 (1955–1956)
21. Williams, V.V.: On some fine-grained questions in algorithms and complexity. In: Sirakov, B., de Souza, P.N., Viana, M. (eds.) Proceedings of the International Congress of Mathematicians, Rio de Janeiro, Brazil, pp. 3447–3487. World Scientific (2019). https://doi.org/10.1142/9789813272880_0188

Global One-Counter Tree Automata

Luisa Herrmann[1,2] and Richard Mörbitz[1(✉)]

[1] Faculty of Computer Science, TU Dresden, Dresden, Germany
{luisa.herrmann,richard.moerbitz}@tu-dresden.de
[2] Center for Scalable Data Analytics and Artificial Intelligence (ScaDS.AI)
Dresden/Leipzig, TU Dresden, Dresden, Germany

Abstract. We introduce global one-counter tree automata (GOCTA) which deviate from usual counter tree automata by working on only one counter which is passed through the tree in lexicographical order, rather than duplicating the counter at every branching position. We compare the capabilities of GOCTA to those of counter tree automata and obtain that their classes of recognizable tree languages are incomparable. Moreover, we show that the emptiness problem of GOCTA is undecidable while, in stark contrast, their membership problem is in P.

Keywords: one-counter automata · tree automata · global counter

1 Introduction

Similar to the case of finite string automata, there is a long tradition of adding counting mechanisms to finite tree automata in order to increase their expressiveness, resulting in models like *pushdown tree automata* [Gue81], *one-counter tree automata* (OCTA), or *Parikh tree automata* [KR03,Kla04]. OCTA are a special case of pushdown tree automata where the pushdown alphabet consists of only two symbols, one of which being the bottom marker enabling zero-tests during a computation. Thus, OCTA generalize one-counter string automata (OCA) by allowing branching in the input. Although not being explicitly described in the literature, they occur as special case of *regular tree grammars with storage* [Eng86] and are rather folklore. With their classical definition, OCTA read the input top-down where independent subtrees are processed in parallel. At each instance of branching, the counter is duplicated and copies are passed to the successors (cf. the left of Fig. 1). Thus, in contrast to OCA, there is no path among which one instance of the counter traverses the entire input.

On the other hand, Parikh tree automata (PTA) count on their input in a global manner by adding up integer vectors (assigned to symbols) over the whole tree and testing once in the end of the computation if the sum is contained in a semilinear set. While this view of the entire input allows to recognize the language $T_{a=b}$ of all trees where two fixed symbols a and b must have the same number of occurrences, OCTA are not able to do the same. In addition to this global view of the storage, PTA also differ from OCTA by allowing multiple counters, but they lack the capability to test their counters for 0 during a run.

S. Z. Fazekas (Ed.): CIAA 2024, LNCS 15015, pp. 166–179, 2024.
https://doi.org/10.1007/978-3-031-71112-1_12

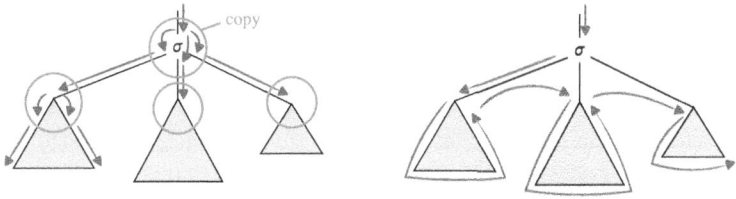

Fig. 1. Storage flow of OCTA (left) vs. GOCTA (right).

[Cas04] introduced a global storage flow for indexed grammars, which are essentially context-free grammars enhanced by a pushdown, introducing global index grammars (GIG). He showed that GIG are suitable for natural language processing (NLP) since they have several properties which are desirable in this area: their membership problem (called parsing in NLP) is in P and their emptiness problem is decidable. To prove the latter statement, he showed that their generated languages are semilinear.

Inspired by GIG and the global counting mode of Parikh tree automata, we introduce global one-counter tree automata (GOCTA). They are essentially OCTA where the counter is passed through the input tree in the lexicographical order of its positions (cf. the right of Fig. 1). Thus, a single instance of the counter reaches the entire input. Indeed, we find that GOCTA can recognize $T_{a=b}$. As the first main contribution of this paper, we explore the expressiveness of GOCTA and we compare the classes of tree languages recognizable by OCTA and by GOCTA, respectively; we find that they are incomparable (Corollary 1). Second, we obtain that the emptiness problem of GOCTA is undecidable (Theorem 1) which casts doubt on the corresponding result for GIG by [Cas04]. Finally, we show that the membership problem of GOCTA is decidable in polynomial time with respect to the size of the input tree (Theorem 2).

2 Preliminaries

In the usual way, we denote by $\mathbb{Z}$ and $\mathbb{N}$ the *set of integers* and *natural numbers*, respectively, and use $[i, j]$ to denote the interval $\{\ell \in \mathbb{Z} \mid i \le \ell \le j\}$ for $i, j \in \mathbb{Z}$. For each $n \in \mathbb{N}$ we set $[n] = [1, n]$.

Given a finite set A, we mean by $|A|$ the number of its elements. For each $k \in \mathbb{N}$ and $a_1, \ldots, a_k \in A$, we call $w = a_1 \cdots a_k$ a *a string (of length k) over A* and, for each $i \in [k]$, we let $w[i]$ denote a_i. The set of all strings of length k over A is denoted by A^k and we let $A^* = \bigcup_{k \in \mathbb{N}} A^k$. We note that $A^0 = \{\varepsilon\}$ where ε is the empty string. We let $\le_{\text{lex}}$ denote the usual lexicographical order on $\mathbb{N}^*$.

Given a finite set Σ and a mapping rk: $\Sigma \to \mathbb{N}$, we call the tuple (Σ, rk) a *ranked alphabet*. Usually we only write Σ and assume rk implicitly. We denote, for each $n \in \mathbb{N}$, $\text{rk}^{-1}(n)$ by $\Sigma^{(n)}$ and by writing $\sigma^{(n)}$ we mean that $\text{rk}(\sigma) = n$.

Let Σ be a ranked alphabet and let H be some set. The set $T_\Sigma(H)$ of *trees (over Σ and indexed by H)* is the smallest set T such that $H \subseteq T$ and

$\sigma(\xi_1, \ldots, \xi_n) \in T$ for each $n \in \mathbb{N}$, $\sigma \in \Sigma^{(n)}$, and $\xi_1, \ldots, \xi_n \in T$. We write T_Σ if $H = \emptyset$. Given $\sigma \in \Sigma^{(n)}$ and $L_1, \ldots, L_n \subseteq T_\Sigma$, we let $\sigma(L_1, \ldots, L_n) = \{\sigma(\xi_1, \ldots, \xi_n) \mid \xi_i \in L_i\}$. Each subset $L \subseteq T_\Sigma$ is called a *tree language*.

Now let $\xi \in T_\Sigma(H)$. As usual, $\mathrm{pos}(\xi) \subseteq \mathbb{N}^*$ is the set of *positions* of ξ, $\xi(\varrho) \in \Sigma$ refers to the *label of ξ at position* ϱ, and we set $|\xi| = |\mathrm{pos}(\xi)|$ standing for the *size* of ξ. For each $\sigma \in \Sigma$ we let $|\xi|_\sigma = |\{\varrho \in \mathrm{pos}(\xi) \mid \xi(\varrho) = \sigma\}|$. If ξ has the form $\gamma_1(\ldots \gamma_k(\#))$ for $\gamma_i \in \Sigma^{(1)}$, $i \in [k]$, we denote it as the string $\gamma_1 \cdots \gamma_k \#$ of length $k+1$. Since each string w can be uniquely considered as such a *monadic* tree, we use notations defined for trees, e.g., $|w|_\sigma$, also for strings. For each $\xi \in T_\Sigma$, we let $\mathrm{ht}(\xi) = \max\{|\varrho| \mid \varrho \in \mathrm{pos}(\xi)\}$.

We fix a countable set $X = \{x_1, x_2, \ldots\}$ of *variables* that is disjoint from each ranked alphabet in this work and let $X_n = \{x_1, \ldots, x_n\}$ for each $n \in \mathbb{N}$. Now let $k \geq 1$ and $\xi \in T_\Sigma(H \cup X_k)$. We say that ξ is a *context* if (1) for each $i \in [k]$ there is exactly one position $\varrho_i \in \mathrm{pos}(\xi)$ with $\xi(\varrho_i) = x_i$ and (2) for each $i_1, i_2 \in [k]$, if $i_1 < i_2$, then $\varrho_i \leq_{\mathrm{lex}} \varrho_j$. We denote the set of all such contexts as $C_\Sigma(H, X_k)$. The *composition* of a tree $\xi \in T_\Sigma(H \cup X_k)$ with trees $\xi_1, \ldots, \xi_k \in T_\Sigma(H)$, denoted by $\xi[\xi_1, \ldots, \xi_k]$, replaces each occurrence of x_i in ξ by ξ_i.

Let $\xi \in T_\Sigma$. We define $\langle\xi\rangle_0 = x_1$ and, for every $i \in [|\xi|]$, we let $\langle\xi\rangle_i$ be the context obtained from $\langle\xi\rangle_{i-1}$ by replacing x_1 (occurring at position ϱ) with $\sigma(x_1, \ldots, x_n)$ if $\xi(\varrho) = \sigma \in \Sigma^{(n)}$ (and, if necessary renaming the other variables to ensure that $\langle\xi\rangle_i$ is a context). Intuitively, $\langle\xi\rangle_i$ contains exactly the i lexicographically first nodes of ξ and $\langle\xi\rangle_{|\xi|} = \xi$. Moreover, we let $\mathrm{w}(i) \in \mathbb{N}$ be the smallest number such that $\xi_i \in C_\Sigma(X_{\mathrm{w}(i)})$.

3   Global One-Counter Tree Automata

Automaton Model. A *global one-counter tree automaton* (GOCTA) is a tuple $\mathcal{A} = (Q, \Sigma, q_0, \Delta)$ where Q is a finite set of states, Σ is a ranked alphabet, $q_0 \in Q$ is the initial state, and Δ is a finite set of transitions of the following two forms:

$$q \xrightarrow{p/\varkappa} q' \qquad\qquad (\varepsilon\text{-transition})$$

$$q \xrightarrow{p/\varkappa} \sigma(q_1, \ldots, q_n) \qquad\qquad (\text{read-transition})$$

where $n \in \mathbb{N}$, $\sigma \in \Sigma^{(n)}$, $q, q', q_1, \ldots, q_n \in Q$, $p \in \{0, >0, \top\}$, and $\varkappa \in \mathbb{Z}$. We denote by Δ_ε and Δ_Σ the sets of all ε-transitions and read-transitions of $\mathcal{A}$, respectively. Given a transition $\tau = q \xrightarrow{p/\varkappa} w \in \Delta$, we refer to $\varkappa$ by $\mathrm{INSTR}(\tau)$ and set $\mathrm{INSTR}(\mathcal{A}) = \bigcup_{\tau \in \Delta} \mathrm{INSTR}(\tau)$. A transition $q \xrightarrow{\top/0} w$ will simply be abbreviated by $q \to w$.

The semantics of a GOCTA $\mathcal{A} = (Q, \Sigma, q_0, \Delta)$ is defined as follows. We denote by $\mathrm{conf}(\mathcal{A})$ the set $T_\Sigma(Q) \times \mathbb{N}$. For each transition $\tau \in \Delta$, we let $\Rightarrow^\tau$ be the binary relation on $\mathrm{conf}(\mathcal{A})$ such that for each $\zeta, \zeta' \in T_\Sigma(Q)$ and $m, m' \in \mathbb{N}$ we have

$$(\zeta, m) \Rightarrow^\tau (\zeta', m')$$

if there are $k \geq 1$, $\hat{\zeta} \in C_\Sigma(X_k)$ and $q, q_1, \ldots, q_{k-1} \in Q$ such that $\zeta = \hat{\zeta}[q, \bar{q}]$ (with $\bar{q} = q_1, \ldots, q_{k-1}$) and either

- $\tau = q \xrightarrow{p/\varkappa} q'$, $p(m)$, $\zeta' = \hat{\zeta}[q', \bar{q}]$, $m + \varkappa \geq 0$, and $m' = m + \varkappa$, or
- $\tau = q \xrightarrow{p/\varkappa} \sigma(p_1, \ldots, p_n)$, $p(m)$, $\zeta' = \hat{\zeta}[\sigma(p_1, \ldots, p_n), \bar{q}]$, $m + \varkappa \geq 0$, and $m' = m + \varkappa$

where $\top(m)$ holds for all $m \in \mathbb{N}$, $0(m)$ iff $m = 0$, and $>0(m)$ iff $m \geq 1$. Thus, transitions are always applied at the lexicographically first position carrying a state. The *computation relation of* $\mathcal{A}$ is the binary relation $\Rightarrow_\mathcal{A} = \bigcup_{\tau \in \Delta} \Rightarrow^\tau$.

A computation is a sequence $t = \zeta_0 \Rightarrow^{\tau_1} \zeta_1 \ldots \Rightarrow^{\tau_n} \zeta_n$ (sometimes abbreviated as $\zeta_0 \Rightarrow^{\tau_1 \cdots \tau_n} \zeta_n$) such that $n \in \mathbb{N}$, $\zeta_0, \ldots, \zeta_n \in \text{conf}(\mathcal{A})$, $\tau_1, \ldots, \tau_n \in \Delta$, and $\zeta_{i-1} \Rightarrow^{\tau_i} \zeta_i$ for each $i \in [n]$. We call t *successful on* $\xi \in T_\Sigma$ if $\zeta_0 = (q_0, 0)$ and $\zeta_n = (\xi, m)$ for some $m \in \mathbb{N}$; the set of all successful computations of $\mathcal{A}$ on ξ is denoted by $\text{comp}_\mathcal{A}(\xi)$. We set $\text{maxcnt}(t) = \max\{m \in \mathbb{N} \mid \zeta_i = (\hat{\zeta}, m), i \in [n]\}$.

Remark 1. Given a tree $\xi \in T_\Sigma$, each computation $t \in \text{comp}_\mathcal{A}(\xi)$ is of the form

$$(q_0, 0) \Rightarrow^{\theta_1 \tau_1} (\langle \xi \rangle_1 [q_1, \bar{q}_1], m_1) \ldots \Rightarrow^{\theta_{n-1} \tau_{n-1}} (\langle \xi \rangle_{n-1} [q_{n-1}, \bar{q}_{n-1}], m_{n-1})$$
$$\Rightarrow^{\theta_n \tau_n} (\langle \xi \rangle_n, m_n)$$

where $n = |\xi|$, $\theta_1, \ldots, \theta_n \in \Delta_\varepsilon^*$, $\tau_1, \ldots, \tau_n \in \Delta_\Sigma$, $m_1, \ldots, m_n \in \mathbb{N}$, and for each $i \in [n-1]$ we have $q_i \in Q$ and $\bar{q}_i = p_1, \ldots, p_{w(i)-1}$ with $p_1, \ldots, p_{w(i)-1} \in Q$. ◁

Remark 2. We note that computations of GOCTA may at first glance look more like derivations of grammars than like runs of automata, since we do not carry the complete input in each computation step for the sake of readability. However, this notational peculiarity can be easily remedied. Furthermore, a slight difference between our model and grammars is that, in our case, at most one terminal is produced per computation step. This is why we speak of "automata" instead of "grammars". ◁

The *language recognized by* $\mathcal{A}$ is the set $\mathcal{L}(\mathcal{A}) = \{\xi \in T_\Sigma \mid \text{comp}_\mathcal{A}(\xi) \neq \emptyset\}$. We say that a tree language $L \subseteq T_\Sigma$ is recognizable by a GOCTA, if there exists a GOCTA $\mathcal{A}$ with $\mathcal{L}(\mathcal{A}) = L$. Moreover, we let the *language 0-recognized by* $\mathcal{A}$ be the set $\mathcal{L}_0(\mathcal{A}) = \{\xi \in T_\Sigma \mid (q_0, 0) \Rightarrow_\mathcal{A}^* (\xi, 0)\}$.

We obtain the well-known *finite-state state tree automata* (FTA) as a special case of GOCTA: a GOCTA is an FTA if each of its ε-transitions is of the form $q \xrightarrow{\top/0} q'$ and each of its read-transitions is of the form $q \xrightarrow{\top/0} \sigma(p_1, \ldots, p_n)$. We drop the counter values when we denote computations of FTA.

3.1 Examples

We consider a few examples for tree languages recognizable by GOCTA in order to show their capabilities.

$(\zeta[p, q'], c) \Rightarrow^{\tau_2} (\zeta[\sigma(p, q), q'], c - 1)$

$\qquad \Rightarrow^{\tau_2} (\zeta[\sigma(\sigma(p, q), q), q'], c - 2)$

$\qquad \ldots$

$\qquad \Rightarrow^{\tau_2} (\zeta[\sigma^c((p, q), q^{c-1}), q'], 0)$

$\qquad \Rightarrow^{\tau_3} (\zeta[\sigma^c((\#, q), q^{c-1}), q'], 0)$

$\qquad \Rightarrow^{\tau_1} (\zeta[\sigma^c((\#, \#), q^{c-1}), q'], k)$

$\qquad \Rightarrow^{\tau_1} (\zeta[\sigma^c((\#, \#), \#q^{c-2}), q'], k \cdot 2)$

$\qquad \ldots$

$\qquad \Rightarrow^{\tau_1} (\zeta[\sigma^c((\#, \#), \#^{c-1}), q'], k \cdot c)$

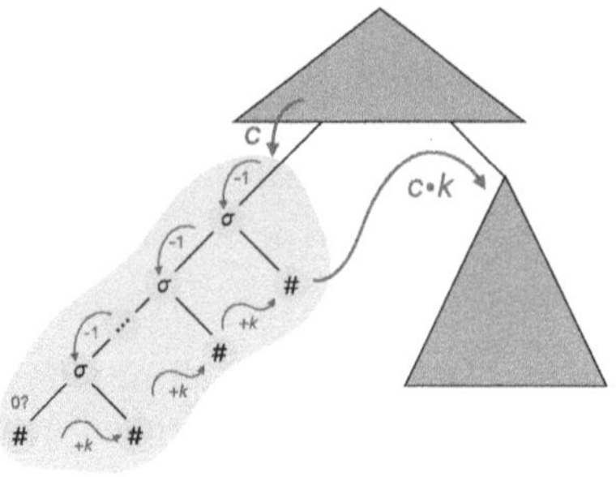

Fig. 2. Left: computation of the GOCTA of Example 2 which multiplies the counter value by k. Right: graphical representation of that computation.

Example 1. Let $\Sigma = \{\sigma^{(3)}, a^{(1)}, b^{(1)}, \#^{(0)}\}$ and consider the tree language

$$L_{a=b} = \{\xi = \sigma(\xi_1, \xi_2, \xi_3) \mid \xi_1, \xi_2, \xi_3 \in T_{\{a,b,\#\}}, |\xi|_a = |\xi|_b\}.$$

This language can be 0-recognized by a GOCTA as follows: Let $\mathcal{A} = (Q, \Sigma, q_0, \Delta)$ where $Q = \{q_0\} \cup \{[u]_v \mid u, v \in \{a, b\}\}$ and Δ consists of the following transitions:

- $q_0 \xrightarrow{\top/0} \sigma([a]_v, [v]_{v'}, [v']_{v''})$ for each $v, v', v'' \in \{a, b\}$,
- $[u]_v \xrightarrow{\top/1} u([u]_v)$, $[u]_v \xrightarrow{>0/-1} u'([u]_v)$, and $[u]_v \xrightarrow{0/1} u'([u']_v)$ for $u, u', v \in \{a, b\}$ with $u \neq u'$, and
- $[u]_u \xrightarrow{\top/0} \#$ for each $u \in \{a, b\}$.

As the counter always stays positive, $\mathcal{A}$ needs to encode in its state whether it currently counts as or bs. It has to guess with its first transition in which counting mode the recognition of ξ_1 (and ξ_2) ends in order to start the recognition of the following subtree in that state. It is not hard to see that with a slightly extended construction we can define a GOCTA $\mathcal{A}'$ with $\mathcal{L}(\mathcal{A}') = \mathcal{L}_0(\mathcal{A})$: it implements a zero-test before reading the rightmost $\#$. ◁

We note that we can easily generalize Example 1 and construct a GOCTA which recognizes the set $T_{a=b}$ of all trees ξ over $\Sigma' \supseteq \{a, b, \#\}$ with $|\xi|_a = |\xi|_b$. However, the restricted form of $L_{a=b}$ will be useful to show in Sect. 4 that this language is not recognizable by a one-counter tree automaton.

Now let us consider two more examples which demonstrate the expressiveness of GOCTA by their ability to multiply the counter value by a constant. In these examples, we will write trees that are combs over binary symbols, i.e., trees of the form $\underbrace{\sigma(\ldots(\sigma(\#, \#), \#), \ldots \#)}_{k \text{ times}}$, $\underbrace{\vphantom{(}}_{k-1 \text{ times}}$ simply as $\sigma^k((\#, \#), \#^{k-1})$.

Example 2. Let Σ be a ranked alphabet, $\sigma^{(2)}, \#^{(0)} \in \Sigma$, and $k \in \mathbb{N}$. We consider a GOCTA $\mathcal{A}$ whose set of transitions contains

$$(\tau_1)\ q \xrightarrow{\top/k} \# \qquad (\tau_2)\ p \xrightarrow{>0/-1} \sigma(p, q) \qquad (\tau_3)\ p \xrightarrow{0/0} \#.$$

Now let $\zeta \in C_\Sigma(X_2)$, $c \in \mathbb{N}$, and q' be a state of $\mathcal{A}$. In Fig. 2, we illustrate how $(\zeta[p, q'], c) \Rightarrow^*_{\mathcal{A}} (\zeta[\sigma^c((\#, \#), \#^{c-1}), q'], k \cdot c)$. Thus, recognizing the subtree $\sigma^c((\#, \#), \#^{c-1})$, the GOCTA $\mathcal{A}$ can multiply the counter value by k.   ◁

Example 3. Now we extend Example 2 as follows: Let $\Sigma = \{\omega^{(2)}, \sigma^{(2)}, \#^{(0)}\}$ and consider the GOCTA $\mathcal{A} = (\{q, p, f, q_0\}, \Sigma, q_0, \Delta)$ where Δ consists of the transitions τ_1, τ_2, τ_3 from Example 2 for $k = 2$ plus the transitions

$$(\tau_0)\ q_0 \xrightarrow{\top/1} \omega(p, f) \qquad (\tau_4)\ f \xrightarrow{\top/0} \omega(p, f) \qquad (\tau_5)\ f \xrightarrow{\top/0} \#.$$

Then $(q_0, 0) \Rightarrow^{\tau_0} (\omega(p, f), 1) \Rightarrow^* (\omega(\zeta_1, f), 2) \Rightarrow^{\tau_4} (\omega(\zeta_1, \omega(p, f)), 2)$ and, by iterating these steps, $(\omega(\zeta_1, \omega(p, f)), 2) \Rightarrow^* (\omega(\zeta_1, \omega(\zeta_2, \ldots, \omega(\zeta_n, \#)\ldots)), 2^n)$ where $\zeta_i = \sigma^{(2^{i-1})}((\#, \#), \#^{(2^{i-1})-1})$. Intuitively, each computation of ζ_i doubles the counter value as outlined in Example 2. Thus, for each tree $\xi \in \mathcal{L}(\mathcal{A})$ we obtain $|\xi|_\sigma = 2^n - 1$ for some $n \in \mathbb{N}$ and, since there is no bound for $|\xi|_\sigma$, this language is not semilinear.   ◁

3.2   Normal Forms and Maximal Counter Values

Now we introduce two normal forms of GOCTA which will simplify the proofs in the remaining work. Moreover, we will also show that, when analysing computations of GOCTA, we can restrict ourselves to those containing counter values polynomial in the size of the computed tree. This insight will later be a key ingredient in our proof that the membership problem of GOCTA is in P.

Let $\mathcal{A}$ be a GOCTA over Σ. We call $\mathcal{A}$ *0-accepting* if $\mathcal{L}_0(\mathcal{A}) = \mathcal{L}(\mathcal{A})$. Furthermore, we say that $\mathcal{A}$ is *normalized* if $\text{INSTR}(\mathcal{A}) \subseteq \{-1, 0, 1\}$ and each of its read-transitions is of the form $q \xrightarrow{\top/0} \sigma(p_1, \ldots, p_n)$.

Lemma 1. *Let $\mathcal{A}$ be a GOCTA. Then there is a 0-accepting GOCTA $\mathcal{A}'$ such that $\mathcal{L}(\mathcal{A}') = \mathcal{L}(\mathcal{A})$.*

Proof. In order to be 0-accepting, the GOCTA $\mathcal{A}'$ we construct needs to empty its counter before reading the rightmost leaf symbol. To do so, the automaton must know whether it is currently operating in the rightmost part of a tree. We encode this information in the states: for each state q of $\mathcal{A}$, we add an auxiliary state $[q]$, which will always be passed to (and only to) the rightmost child starting from the new initial state $[q_0]$.

Formally, let $\mathcal{A} = (Q, \Sigma, q_0, \Delta)$ be a GOCTA. We construct the GOCTA $\mathcal{A}' = (Q', \Sigma, [q_0], \Delta')$ as follows: Let $Q' = Q \cup \{[q] \mid q \in Q\} \cup \{q_\alpha \mid q \in Q, \alpha \in \Sigma\}$ and $\Delta' = \Delta \cup T$ where T consists of the following transitions:

- for each ε-transition $q \xrightarrow{p/x} q' \in \Delta$, the transition $[q] \xrightarrow{p/x} [q']$ is in T,
- for each transition $q \xrightarrow{p/x} \sigma(q_1, \ldots, q_n) \in \Delta$, the transition
  $[q] \xrightarrow{p/x} \sigma(q_1, \ldots, q_{n-1}, [q_n])$ is in T,
- for each transition $q \xrightarrow{p/x} \alpha \in \Delta$, the transition $[q] \xrightarrow{p/x} q_\alpha$ is in T,

– for each $q \in Q$ and $\alpha \in \Sigma$, the transition $q_\alpha \xrightarrow{>0/-1} q_\alpha$ is in T, and

– for each $q \in Q$ and $\alpha \in \Sigma$, the transition $q_\alpha \xrightarrow{0/0} \alpha$ is in T.

Clearly, $\mathcal{A}'$ is 0-accepting and one can show that for each $\xi \in T_\Sigma$, $\xi \in \mathcal{L}(\mathcal{A})$ if and only if $\xi \in \mathcal{L}(\mathcal{A}')$.   □

Observation 1. *Let $\mathcal{A}$ be a GOCTA. Then there is a normalized GOCTA $\mathcal{A}'$ such that $\mathcal{L}(\mathcal{A}') = \mathcal{L}(\mathcal{A})$. Moreover, if $\mathcal{A}$ is 0-accepting, then $\mathcal{A}'$ is 0-accepting.*

Now we show that we can restrict the counter values occurring in computations of GOCTA. We note that a similar proof idea was used in [AJ13] to obtain the pumping lemma for pushdown automata.

Lemma 2. *Let $L \subseteq T_\Sigma$. If L is GOCTA-recognizable, then there is a normalized and 0-accepting GOCTA $\mathcal{A} = (Q, \Sigma, q_0, \Delta)$ with $\mathcal{L}(\mathcal{A}) = L$ and such that for each $\xi \in L$ there is a computation $t \in \text{comp}_\mathcal{A}(\xi)$ with $\text{maxcnt}(t) \leq |\xi| \cdot |Q|^2 + 1$.*

Proof. Let $\mathcal{A} = (Q, \Sigma, q_0, \Delta)$ be a normalized and 0-accepting GOCTA. Let $\xi \in L$ and $t \in \text{comp}_\mathcal{A}(\xi)$ such that $\text{maxcnt}(t) > |\xi| \cdot |Q|^2 + 1$. The idea of the proof is that, for the recognition of ξ, only the counter values up to $|\xi| \cdot |Q|^2 + 1$ contain relevant information, whereas $\mathcal{A}$ can no longer distinguish counter values greater than $|\xi| \cdot |Q|^2 + 1$ from each other: both counting up into and counting down from that range must necessarily contain state cycles within which $\mathcal{A}$ cannot test for 0. We have chosen ξ and t such that $\text{maxcnt}(t)$ is large enough to find two such cycles, one counting up and one counting down, that a) cause the same change (in terms of absolute amount) of the counter and b) occur between the reading of two symbols. We can show that we can cut out both cycles and the resulting computation still recognizes ξ.

We let $m = \text{maxcnt}(t)$. By the assumptions, t is of the form $(q_0, 0) \Rightarrow^\theta (\langle\xi\rangle_i[p, \bar{q}_i], m) \Rightarrow^\omega (\xi, 0)$ for some $i \in \{0, \dots, |\xi| - 1\}$ such that the counter value m is reached after θ for the first time. As $m > |\xi| \cdot |Q|^2 + 1$ and $i < |\xi|$ as well as $\text{INSTR}(\mathcal{A}') \subseteq \{-1, 0, 1\}$, there exist $j \in \{0, \dots, i\}$, $m_j^1 \geq 1, m_j^2 = m_j^1 + (|Q|^2 + 1) \leq m$ and $p_1, p_2 \in Q$ such that θ is of the form $(q_0, 0) \Rightarrow^{\theta_1} (\langle\xi\rangle_j[p_1, \bar{q}_j], m_j^1) \Rightarrow^{\theta_2} (\langle\xi\rangle_j[p_2, \bar{q}_j], m_j^2) \Rightarrow^{\theta_3} (\langle\xi\rangle_i[p, \bar{q}_i], m)$ and we choose the latest such occurrences. Note that $\theta_2 \in \Delta_\varepsilon^*$.

Similarly, there are $l \in \{i, \dots, |\xi| - 1\}$, $m_l^1 \leq m, m_l^2 = m_l^1 - (|Q|^2 + 1) \geq 1$ and $f_1, f_2 \in Q$ such that ω is of the form $(\langle\xi\rangle_i[p, \bar{q}_i], m) \Rightarrow^{\omega_1} (\langle\xi\rangle_l[f_1, \bar{q}_l], m_l^1) \Rightarrow^{\omega_2} (\langle\xi\rangle_l[f_2, \bar{q}_l], m_l^2) \Rightarrow^{\omega_3} (\xi, 0)$ and now we choose the first such occurrences in ω. By choosing θ_2 and ω_2 closest to $(\langle\xi\rangle_i[p, \bar{q}_i], m)$, we obtain:

Property 1. For all counter values m' occurring in configurations of the sub-computation $(\langle\xi\rangle_j[p_2, \bar{q}_j], m_j^2) \Rightarrow^{\theta_3} (\langle\xi\rangle_i[p, \bar{q}_i], m) \Rightarrow^{\omega_1} (\langle\xi\rangle_l[f_1, \bar{q}_l], m_l^1)$ we have $m' > |Q|^2 + 1$.

Now we zoom into the part $t = (\langle\xi\rangle_j[p_1, \bar{q}_j], m_j^1) \Rightarrow^{\theta_2} (\langle\xi\rangle_j[p_2, \bar{q}_j], m_j^2)$: let, for $z \in \{0, \dots, |Q|^2 + 1\}$, $p^z \in Q$ such that $(\langle\xi\rangle_j[p^z, \bar{q}_j], m_j^1 + z)$ is the configuration in t with $m_j^1 + z$ occurring for the last time. Clearly, $p^{|Q|^2+1} = p_2$. Similarly, f^z denotes the state with which $m_l^1 - z$ occurs for the first time in ω_2. We obtain:

Property 2. For each $z \in \{0, \ldots, |Q|^2 + 1\}$ and each counter value m' (except the first one) occurring in the subcomputation $(\langle \xi \rangle_j [p^z, \bar{q}_j], m_j^1 + z) \Rightarrow^*$ $(\langle \xi \rangle_j [p_2, \bar{q}_j], m_j^2)$ of θ_2 we have $m' > m_j^1 + z$. Similarly, for each counter value m'' (except the last one) occurring in the subcomputation $(\langle \xi \rangle_l [f_1, \bar{q}_l], m_l^1) \Rightarrow^*$ $(\langle \xi \rangle_l [p^z, \bar{q}_j], m_j^1 - z)$ of ω_2 we have $m'' > m_l^1 - z$.

Now consider $Z = \{(p^z, f^z) \mid z \in [|Q|^2 + 1]\}$. Clearly, $|Z| \le |Q|^2$. Thus, there are $z_1 < z_2 \in [|Q|^2 + 1]$ with $p^{z_1} = p^{z_2}$ and $f^{z_1} = f^{z_2}$.

But this means that we can find a shorter computation of $\mathcal{A}$ for ξ by cutting out the subcomputations between p^{z_1} and p^{z_2} as well as f^{z_1} and f^{z_2} at the respective position indicated above. We note that, by the above Property 1, all transitions applied in θ_3 and ω_1 can only use as predicate >0 or $\top$. Thus, in the following we mainly need to argue that also after our cut we never reach 0 in the counter during these sequences and, thus, all transitions are still applicable. Let $d = z_2 - z_1$ and $\theta_2 = \theta_{z_1} \theta_{z_1, z_2} \theta_{z_2}$, $\omega_2 = \omega_{z_1} \omega_{z_1, z_2} \omega_{z_2}$ where θ_{z_1, z_2} and ω_{z_1, z_2} are the parts we cut out, respectively. Clearly, $(q_0, 0) \Rightarrow^{\theta_1 \theta_{z_1}} (\langle \xi \rangle_j [p^{z_1}, \bar{q}_j], m_j^1 + z_1)$. By Property 2, $(\langle \xi \rangle_j [p^{z_1}, \bar{q}_j], m_j^1 + z_1) \Rightarrow^{\theta_{z_2}} (\langle \xi \rangle_j [p_2, \bar{q}_j], m_j^2 - d)$ as all transitions are still applicable on counter values decreased by d. With a similar argumentation, by Property 1, $(\langle \xi \rangle_j [p_2, \bar{q}_j], m_j^2 - d) \Rightarrow^{\theta_3 \omega_1} (\langle \xi \rangle_l [f_1, \bar{q}_l], m_l^1 - d)$ as $d \le |Q|^2 + 1$. Finally, by Property 2, $(\langle \xi \rangle_l [f_1, \bar{q}_l], m_l^1 - d) \Rightarrow^{\omega_{z_2}} (\langle \xi \rangle_l [f^{z_2}, \bar{q}_l], m_l^1 - z_2)$ and $(\langle \xi \rangle_l [f^{z_2}, \bar{q}_l], m_l^1 - z_2) \Rightarrow^* (\xi, 0)$.

This procedure can be repeated until no counter value greater than $|\xi| \cdot |Q|^2 + 1$ does occur anymore. □

4   Expressiveness

We investigate the expressiveness of GOCTA in comparison to OCTA; recall that, at each instance of branching, OCTA pass a copy of the counter to every successor. We formalize OCTA as an alternative semantics of GOCTA.

Let $\mathcal{A} = (Q, \Sigma, q_0, \Delta)$ be a GOCTA and denote by ID the set $Q \times \mathbb{N}$. For each transition $\tau \in \Delta$ we let $\Rightarrow^\tau_{\text{copy}}$ be the binary relation on the set $T_\Sigma(\text{ID})$ such that for each $\zeta_1, \zeta_2 \in T_\Sigma(\text{ID})$ we have $\zeta_1 \Rightarrow^\tau \zeta_2$ if there are $\hat{\zeta} \in C_\Sigma(\text{ID}, X_1)$, $\hat{\zeta}_1, \hat{\zeta}_2 \in T_\Sigma(\text{ID})$ such that $\zeta_1 = \hat{\zeta}[\hat{\zeta}_1]$, $\zeta_2 = \hat{\zeta}[\hat{\zeta}_2]$, and either

- $\tau = q \xrightarrow{p/\varkappa} q'$, $\hat{\zeta}_1 = (q, m)$, $p(m)$, $m + \varkappa \ge 0$, and $\hat{\zeta}_2 = (q', m + \varkappa)$, or
- $\tau = q \xrightarrow{p/\varkappa} \sigma(p_1, \ldots, p_n)$, $\hat{\zeta}_1 = (q, m)$, $p(m)$, $m + \varkappa \ge 0$, and $\hat{\zeta}_2 = \sigma((q_1, m + \varkappa), \ldots, (q_n, m + \varkappa))$.

Thus, transitions can be applied at any position carrying a state. We let $\Rightarrow_{\text{copy}} = \bigcup_{\tau \in \Delta} \Rightarrow^\tau_{\text{copy}}$ and $\mathcal{L}_{\text{copy}}(\mathcal{A}) = \{\xi \in T_\Sigma \mid (q_0, 0) \Rightarrow^*_{\text{copy}} \xi\}$. A tree language L can be *recognized by an OCTA* if there exists a GOCTA $\mathcal{A}$ with $\mathcal{L}_{\text{copy}}(\mathcal{A}) = L$.

Example 4. Let $\Sigma = \{\sigma^{(2)}, a^{(1)}, b^{(1)}, c^{(1)}, \#^{(0)}\}$ and consider the tree language $L_{a\sigma bc} = \{a^n(\sigma(b^n \#, c^n \#)) \mid n \in \mathbb{N}\}$. This language can be recognized by the OCTA $\mathcal{A} = (\{q_0, q_1, q_2\}, \Sigma, q_0, \Delta)$ where Δ contains the transitions

$q_0 \xrightarrow{\top/1} a(q_0)$, $q_0 \xrightarrow{\top/0} \sigma(q_1, q_2)$, $q_1 \xrightarrow{>0/-1} b(q_1)$, $q_2 \xrightarrow{>0/-1} c(q_2)$, $q_1 \xrightarrow{0/0} \#$, and $q_2 \xrightarrow{0/0} \#$.

$\triangleleft$

Lemma 3. *The tree language $L_{a\sigma bc}$ is not recognizable by a GOCTA.*

Proof. The proof is based on the following intuition. In order to check whether a tree ξ is in $L_{a\sigma bc}$, a GOCTA $\mathcal{A}$ must traverse ξ in lexicographical order. Thus, it effectively checks whether $h(\xi)$ is in $L = \{a^n \sigma b^n \# c^n \# \mid n \in \mathbb{N}\}$ where h is the tree homomorphism induced by $h(\sigma) = \sigma x_1 x_2$, $h(a) = ax_1$, $h(b) = bx_1$, $h(c) = cx_1$, and $h(\#) = \#$. We show that there exists a one-counter string automaton $\mathcal{A}'$ such that $\mathcal{A}'$ recognizes L if and only if $\mathcal{A}$ recognizes $L_{a\sigma bc}$. Since L is not context-free, this entails that no GOCTA can recognize $L_{a\sigma bc}$.

Formally, assume that there exists a normalized and 0-accepting GOCTA $\mathcal{A} = (Q, \Sigma, q_0, \Delta)$ such that $L(\mathcal{A}) = L_{a\sigma bc}$. We observe that, for every $\xi \in L(\mathcal{A})$ and $t \in \text{comp}_{\mathcal{A}}(\xi)$, the computation t has the form

$$t = (q_0, 0) \Rightarrow^{\theta_a} (q_1 \to \sigma(q_2, q_3)) \; \theta_b \; (q_4 \to \#) \; \theta_c \; (q_5 \to \#) \; (\xi, 0) \tag{1}$$

where $\theta_a, \theta_b, \theta_c \in \Delta^*$ only contain a, b, and c, respectively, in their read transitions. Without loss of generality, we can assume that the sets of states occurring in θ_a, θ_b, and θ_c are pairwise disjoint (thus, e.g., q_1 can only occur in θ_a).

We only sketch the definition of the one-counter string automaton $\mathcal{A}'$; its set of transitions Δ' is obtained from Δ as follows. For each transition $q_1 \to \sigma(q_2, q_3)$, we add $q_1 \to \sigma\langle q_2, q_3\rangle$ to Δ' where $\langle q_2, q_3\rangle$ becomes a state of $\mathcal{A}'$; then, for every transition $q_4 \to \#$ and for every transition of the form $q \to b(q')$ or $q \xrightarrow{p/x} q'$ that can occur in θ_b, we add $\langle q_4, q_3\rangle \to \#q_3$, $\langle q, q_3\rangle \to b\langle q', q_3\rangle$, or $\langle q, q_3\rangle \xrightarrow{p/x} \langle q', q_3\rangle$ to Δ', resp. Thus, intuitively, we encode the single instance of branching in $\mathcal{A}$ into the state behaviour of $\mathcal{A}'$. All other transitions of $\mathcal{A}$ are added to Δ' unchanged.

One can show that, for every $\xi \in L(\mathcal{A})$ and $t \in \text{comp}_{\mathcal{A}}(\xi)$, there exists a matching $t' \in \text{comp}_{\mathcal{A}'}(h(\xi))$. Vice versa, for every $w \in L(\mathcal{A}')$ and $t \in \text{comp}_{\mathcal{A}'}(w)$, there exists $\xi \in L(\mathcal{A})$ and a matching $t' \in \text{comp}_{\mathcal{A}}(\xi)$ such that $h(\xi) = w$. Thus, we obtain that $h(L_{a\sigma bc}) = L = L(\mathcal{A}')$. Since L is not context-free, this is a contradiction. Thus, $\mathcal{A}'$ cannot exist and, hence, $\mathcal{A}$ does not recognize $L_{a\sigma bc}$. $\square$

Lemma 4. $L_{a=b}$ *cannot be recognized by OCTA.*

Proof. If there exists an OCTA $\mathcal{A} = (Q, \Sigma, q_0, \Delta)$ with $\mathcal{L}_{\text{copy}}(\mathcal{A}) = L_{a=b}$, we can assume that there are disjoint $Q_1, Q_2 \subseteq Q$ such that $Q_1 \cup Q_2 = Q$ and each computation of $\mathcal{A}$ has the form $(q_0, 0) \Rightarrow^*_{\mathcal{A}} (q, m) \Rightarrow_{\mathcal{A}} \sigma((q_1, m), (q_2, m), (q_3, m)) \Rightarrow^*_{\mathcal{A}} \sigma(\xi_1, \xi_2, \xi_3)$ with $q_1, q_2, q_3 \in Q_2$. Moreover, for each $q \in Q_2$, no state in Q_1 is reachable. For every $n \in \mathbb{Z}$, we let $L_n = \{w\# \mid w \in \{a, b\}^*, |w|_a = |w|_b + n\}$, and, for every $c \in \mathbb{N}$ and $q \in Q$, we let $L_c^q = \{\xi \in T_\Sigma \mid (q, c) \Rightarrow^*_{\mathcal{A}} \xi\}$. We obtain:

Property 3. For every $(q, c) \in \text{ID}$ with $q \in Q_2$ that occurs in a successful computation, there exists $n \in \mathbb{Z}$ such that $L_c^q \subseteq L_n$.

If Property 3 did not hold, then $\mathcal{A}$ could derive $w_1, w_2 \in \{a,b\}^*\#$ such that $|w_1|_a - |w_1|_b \neq |w_2|_a - |w_2|_b$ from (q,c). Since (q,c) occurs in a successful computation, there would exist a $\xi \in \mathcal{L}_{copy}(\mathcal{A})$ with $|\xi|_a \neq |\xi|_b$; hence $\mathcal{L}_{copy}(\mathcal{A}) \neq L_{a=b}$.

We observe that, for every $\xi \in L_{a=b}$, there exist unique $n_1, n_2, n_3 \in \mathbb{Z}$ such that $\xi \in \sigma(L_{n_1}, L_{n_2}, L_{n_3})$. We let $k \in \mathbb{N}$ and consider the set

$$F(k) = \{(n_1, n_2, n_3) \mid \exists \xi \in L_{a=b}: \xi \in \sigma(L_{n_1}, L_{n_2}, L_{n_3}), |n_1|, |n_2|, |n_3| \leq k\}$$
$$\overset{\star}{=} \{(n_1, n_2, n_3) \mid n_1, n_2, n_3 \in \{-k, \ldots, +k\}, n_1 + n_2 + n_3 = 0\}$$

where $\star$ holds by definition of $L_{a=b}$. Note that, if $ht(\xi) \leq k+1$, then $n_1, n_2, n_3 \leq k$, and, for every $(n_1, n_2, n_3) \in F(k)$, there is a $\xi \in L_{a=b} \cap \sigma(L_{n_1}, L_{n_2}, L_{n_3})$ with $ht(\xi) \leq k+1$. Using combinatorial arguments, one can show $|F(k)| = 3k^2 + 3k + 1$.

Let $\xi, \xi' \in L_{a=b}$ and $n_1, n_2, n_3, n_1', n_2', n_3' \in \mathbb{Z}$ such that $\xi \in \sigma(L_{n_1}, L_{n_2}, L_{n_3})$ and $\xi' \in \sigma(L_{n_1'}, L_{n_2'}, L_{n_3'})$. By definition of $\mathcal{A}$, there exist $q_1, q_2, q_3, q_1', q_2', q_3' \in Q_2$ and $c, c' \in \mathbb{N}$ such that $\xi \in \sigma(L_c^{q_1}, L_c^{q_2}, L_c^{q_3})$ and $\xi' \in \sigma(L_{c'}^{q_1'}, L_{c'}^{q_2'}, L_{c'}^{q_3'})$. Now, by Property 3, we obtain the following. If $(n_1, n_2, n_3) \neq (n_1', n_2', n_3')$, then also $(c, q_1, q_2, q_3) \neq (c', q_1', q_2', q_3')$. We define $D(k)$ to be the set

$$\{(c, q_1, q_2, q_3) \mid (q_0, 0) \Rightarrow_{\mathcal{A}}^* (q, c) \Rightarrow_{\mathcal{A}} \sigma((q_1, c), (q_2, c), (q_3, c)) \Rightarrow_{\mathcal{A}}^* \xi, ht(\xi) \leq k+1\}.$$

Thus, $|D(k)|$ must be at least $|F(k)|$, i.e., $|D(k)| \geq 3k^2 + 3k + 1$. Then there exists $(c, q_1, q_2, q_3) \in D(k)$ with $c \geq \frac{3k^2+3k+1}{|Q|^3}$. Using a method like in Lemma 2, we can show that, if $\xi \in \mathcal{L}_{copy}(\mathcal{A})$, then there exist $m \leq ht(\xi) \cdot |Q|^4 + 1$ and a computation of the form $(q_0, 0) \Rightarrow^* (q, m) \Rightarrow \sigma((q_1, m), (q_2, m), (q_3, m)) \Rightarrow^* \xi$. Thus, for each $(c, q_1, q_2, q_3) \in D(k)$, we can assume $c \leq (k+1) \cdot |Q|^4 + 1$. This is a contradiction for sufficiently large k. □

Corollary 1. *The classes of tree languages recognizable by OCTA and GOCTA are incomparable.*

5 Decision Problems

Now we turn to decidability results. In the following, we examine the emptiness problem and the membership problem of GOCTA in this regard.

5.1 Emptiness

First we investigate the question whether the emptiness problem is decidable for GOCTA. Unfortunately, the answer is negative – we can relate GOCTA to a similar string formalism: *indexed counter grammars with a global counter semantics* (ICG) [DMP92, Sec. 5] (called *R-mode derivation*). This model can be understood as a context-free grammar with a global counter controlling its derivations and was shown to be Turing-complete [DMP92, Thm. 5.1]. We note that, in contrast, with GOCTA we cannot perform the same calculations without reading symbols. However, we can show that, given an ICG G, we can construct a GOCTA $\mathcal{A}$ recognizing all computation trees of G. Thus, we obtain that the emptiness problem of ICG can be reduced to the emptiness problem of GOCTA.

Theorem 1. *Given a GOCTA $\mathcal{A}$, it is undecidable whether $\mathcal{L}(\mathcal{A}) = \emptyset$.*

An intuition on this result is given by Example 2: In [DMP92, Thm. 5.1] it was shown that with help of multiplication, a counter can be used to simulate a pushdown storage by representing each of its configurations as a number with base m (where m is the size of the pushdown alphabet). Since a context-free grammar using a global pushdown is Turing-complete, this entails the result.

Relation to GIG. The above undecidability result seems to be in contrast with [Cas04], where it was shown that the languages generated by Global Index Grammars (GIG) are semilinear, from which it was deduced that the emptiness problem of GIG is decidable. In essence, a GIG is a context-free grammar enhanced with a global pushdown storage where, additionally, each pushing rule must start its right-hand side with a terminal.

We briefly recall the definition of GIG [Cas04, Definition 6]. A GIG is a tuple $G = (N, T, I, S, \bot, P)$ where N, T, and I are pairwise disjoint alphabets (*nonterminals, terminals,* and *stack indices,* respectively), $S \in N$ (*start symbol*), $\bot \notin N \cup T \cup I$ (*stack start symbol*), and P is a finite set (*productions*), each of which having one of the following forms:

$$(i.1)\ A \underset{\varepsilon}{\to} \alpha \qquad (i.2)\ A \underset{[y]}{\longrightarrow} \alpha \qquad (ii.)\ A \underset{x}{\to} a\alpha \qquad (iii.)\ A \underset{\bar{x}}{\to} \alpha$$

where $x \in I$, $y \in I \cup \{\bot\}$, $A \in N$, $\alpha \in (N \cup T)^*$, and $a \in T$. Intuitively, $(i.1)$ is a context-free production, $(i.2)$ checks if the topmost pushdown symbol is y (without changing the pushdown), $(ii.)$ pushes x, and $(iii.)$ pops x.

By assuming GIG with only one pushdown symbol (plus the bottom marker), we can relate this formalism to GOCTA. For instance, we can define a GIG $\mathcal{A}'$ with $I = \{\gamma, \bot\}$ which generates strings encoding the trees of the GOCTA $\mathcal{A}$ from Example 3. Therefore, we let $\mathcal{A}'$ contain the transitions

- $q_0 \underset{\gamma}{\to} \omega(p, f)$ for $q_0 \xrightarrow{\top/1} \omega(p, f)$,

- $p \underset{\bar{\gamma}}{\to} \sigma(p, q)$ for $p \xrightarrow{>0/-1} \sigma(p, q)$,

- $p \underset{[\bot]}{\longrightarrow} \#$ for $p \xrightarrow{0/0} \#$, and

- $f \underset{\varepsilon}{\to} \omega(p, f)$ and $f \underset{\varepsilon}{\to} \#$ for $f \xrightarrow{\top/0} \omega(p, f)$ and $f \xrightarrow{\top/0} \#$, respectively,

by regarding the symbols of $\mathcal{A}$ (including "(", ")", and ",") as string terminals. Moreover, for the transition $\tau_1 = (q \xrightarrow{\top/2} \#)$, which uses a 2-increment, we add two rules to $\mathcal{A}'$ that each push a γ. Formally, we add the symbol $\beta^{(1)}$ to Σ, use the new state q_1, and add the rules $q \underset{\gamma}{\to} \beta(q_1)$ and $q_1 \underset{\gamma}{\to} \#$ to $\mathcal{A}'$. Finally, we note that in the original definition of GIG derivations are required to end with the pushdown consisting of $\bot$. This corresponds to 0-accepting GOCTA and, clearly, changing $\mathcal{A}'$ accordingly does not violate the constraints of GIG.

Since $L(\mathcal{A})$ is not semilinear, it is not hard to see that also $L(\mathcal{A}')$ is not semilinear. This falsifies the claim that GIG generate only semilinear languages [Cas04, Chap. 4, Thm. 5] and also casts doubt on the emptiness result.

Conjecture 1. *The emptiness problem of GIG [Cas04] is undecidable.*

5.2 Membership

Surprisingly, in contrast to emptiness, the (uniform) membership problem for GOCTA is not only decidable but also turns out to be rather easy: we will show with help of the polynomial bound on the counter values in computations we obtained in Lemma 2, that we can decide membership in polynomial time.

Definition 1. *Let $\mathcal{A} = (Q, \Sigma, q_0, \Delta)$ be a normalized GOCTA and $k \in \mathbb{N}$. The k-bounded behaviour automaton of $\mathcal{A}$ is the FTA $\mathcal{A}' = (Q', \Sigma, q_0', \Delta')$ where $Q' = Q \times [0, k] \times [0, k]$, $q_0' = (q_0, 0, 0)$, and Δ' is defined as follows. For every $n \geq 1$, $q \to \sigma(q_1, \ldots, q_n)$ in Δ, and $i_0, \ldots, i_n \in [0, k]$, we let $(q, i_0, i_n) \to \sigma((q_1, i_0, i_1), \ldots, (q_n, i_{n-1}, i_n))$ in Δ', for every $q \to \alpha$ in Δ and $i \in [0, k]$, we let $(q, i, i) \to \alpha$ in Δ', and, for every $q \xrightarrow{p/x} q'$ in Δ and $i_0, i_1 \in [0, k]$, if $p(i_0)$ holds and $0 \leq i_0 + x \leq k$, then we let $(q, i_0, i_1) \to (q', i_0 + x, i_1)$ in Δ.*

Observation 2. $|Q'| \leq |Q| \cdot (k+1) \cdot (k+1)$ *and* $|\Delta'| \leq |\Delta| \cdot (k+1)^{(\max \mathrm{rk}(\Sigma)+1)}$.

Lemma 5. *Let $\mathcal{A} = (Q, \Sigma, q_0, \Delta)$ be a GOCTA and $\xi \in T_\Sigma$. If $\mathcal{A}'$ is the $(|\xi| \cdot |Q|^2 + 1)$-bounded behaviour automaton of $\mathcal{A}$, then $\xi \in \mathcal{L}(\mathcal{A}) \iff \xi \in \mathcal{L}(\mathcal{A}')$.*

Proof. We assume that $\mathcal{A}$ is normalized and 0-accepting. Clearly, these constructions are polynomial wrt. the size of $\mathcal{A}$. Let $m = |\xi| \cdot |Q|^2 + 1$ and $\mathcal{A}' = (Q', \Sigma, (q_0, 0, 0), \Delta')$. In order to show that $\xi \in \mathcal{L}(\mathcal{A}) \implies \xi \in \mathcal{L}(\mathcal{A}')$, we let $\xi \in L(\mathcal{A})$. Then, by Lemma 2, there exists $t \in \mathrm{comp}_{\mathcal{A}}(\xi)$ such that $\mathrm{maxcnt}(t) \leq m$. We let $t = (\zeta_0, c_0) \Rightarrow^{\tau_1 \cdots \tau_n} (\zeta_n, c_n)$ and we note that $\zeta_0 = q_0$, $\zeta_n = \xi$, and $c_0 = c_n = 0$. The idea of the proof is to construct $t' \in \mathrm{comp}_{\mathcal{A}'}(\xi)$, denoted by $t' = \zeta_0' \Rightarrow^{\tau_1' \cdots \tau_n'} \zeta_n'$ such that, for every $i \in \{0, \ldots, n\}$, the following invariant holds. If $\zeta_i = \langle \xi \rangle_k [q, \bar{q}]$, then $\zeta_i' = \langle \xi \rangle_k [(q, c_i, c'), \bar{p}]$ with $|\bar{q}| = |\bar{p}|$. We note that most of this construction is straightforward and only highlight two interesting cases. Let $i \in [n]$, $\zeta_{i-1} = \langle \xi \rangle_k [q, \bar{q}]$, and $\zeta_{i-1}' = \langle \xi \rangle_k [(q, c_{i-1}, c'), \bar{p}]$.

If $\tau_i = q \to \sigma(p_1, \ldots, p_\ell)$ with $\ell > 1$, then, for each $j \in [2, \ell]$, we encode the counter value of $\mathcal{A}$ when it derives p_j into the j-th state of τ_i'. For this, let $u_j \in [n]$ be the step when p_j is derived in the computation of ξ in $\mathcal{A}$. Then

$$\tau_i' = (q, c_{i-1}, c') \to \sigma((p_1, c_{i-1}, c_{u_2}), (p_2, c_{u_2}, c_{u_3}), \ldots, (p_\ell, c_{u_\ell}, c')) \in \Delta'. \quad (2)$$

Clearly, $\zeta_{i-1}' \Rightarrow^{\tau_i'} \zeta_i' = \langle \xi \rangle_{k+1} [(p_1, c_{i-1}, c_{u_2}), (p_2, c_{u_2}, c_{u_3}), \ldots, (p_\ell, c_{u_\ell}, c'), \bar{p}]$.

If $\tau_i = q \to \alpha$ and $|\bar{p}| \neq 0$, then there exists $i' \in [i-1]$ such that $\tau_{i'}$ is of form (2) and the current state (q, c_{i-1}, c') is derived from some state (p, c_1', c_2') in the right-hand side of $\tau_{i'}$. We choose the largest such i'. By definition of Δ', the derivation of (q, c_{i-1}, c') from (p, c_1', c_2') does not change the third component, hence $c' = c_2'$. Moreover, the state which occurs right of (p, c_1', c_2') in $\tau_{i'}$ has the form (p', c_2', c_3') and this is also the first state in $\bar{p}$. Then, $c_2' = c_i$ (cf. the previous case) and $c_{i-1} = c_i$ because $\mathcal{A}$ is normalized, hence $\tau_i' = (q, c_{i-1}, c') \to \alpha$ is in Δ'. Clearly, $\zeta_{i-1}' \Rightarrow^{\tau_i'} \zeta_i' = \langle \xi \rangle_{k+1}[\bar{p}]$.

In order to show that $\xi \in \mathcal{L}(\mathcal{A}') \implies \xi \in \mathcal{L}(\mathcal{A})$, we let $\xi \in \mathcal{L}(\mathcal{A}')$. Then there exists $t \in \text{comp}_{\mathcal{A}'}(\xi)$; we let $t = \zeta_0 \Rightarrow^{\tau_1 \cdots \tau_n} \zeta_n$ and we note that $\zeta_0 = (q_0, 0, 0)$ and $\zeta_n = \xi$. Now we construct $t' = (\zeta_0', c_0) \Rightarrow^{\tau_1' \cdots \tau_n'} (\zeta_n', c_n)$ in $\text{comp}_{\mathcal{A}}(\xi)$ analogously to the other direction of the proof, but now we have to show that the state behaviour of $\mathcal{A}'$ guarantees a valid counter behaviour of $\mathcal{A}$.   □

Due to the following well-known result, we can construct an ε-transition free FTA $\mathcal{A}''$ such that $\mathcal{L}(\mathcal{A}') = \mathcal{L}(\mathcal{A}'')$ in polynomial time.

Lemma 6 (cf., e.g., [Com+08, Theorem 1.1.5]). *For every FTA $\mathcal{A}$ there exists an ε-transition free FTA $\mathcal{A}'$ with $\mathcal{L}(\mathcal{A}') = \mathcal{L}(\mathcal{A})$ and $\mathcal{A}'$ can be constructed in polynomial time wrt. the size of $\mathcal{A}$.*

Theorem 2. *For every GOCTA $\mathcal{A}$ over Σ and $\xi \in T_\Sigma$ it can be decided in polynomial time whether $\xi \in \mathcal{L}(\mathcal{A})$.*

Proof. Let Q be the set of states of $\mathcal{A}$. We let $m = |\xi| \cdot |Q|^2 + 1$ and we construct the m-bounded behaviour automaton of $\mathcal{A}$ and call it $\mathcal{A}'$. By Observation 2, the size of $\mathcal{A}'$ is polynomial in m and the size of $\mathcal{A}$. Clearly, the construction of $\mathcal{A}'$ has the same time bound. Now, by Lemma 6, there exists an ε-transition free FTA $\mathcal{A}''$ with $\mathcal{L}(\mathcal{A}'') = \mathcal{L}(\mathcal{A}')$. Moreover, $\mathcal{A}''$ can be computed in polynomial time wrt. the size of $\mathcal{A}'$. We can compute whether $\xi \in \mathcal{L}(\mathcal{A}'')$ in polynomial time wrt. the size of $\mathcal{A}''$ and ξ (cf., e.g., [Com+08, Theorem 1.7.3]), and, thus, in polynomial time wrt. m. Finally, it holds that

$$\xi \in \mathcal{L}(\mathcal{A}) \overset{\text{L. 5}}{\Longleftrightarrow} \xi \in \mathcal{L}(\mathcal{A}') \overset{[\text{Com+08, Th. 1.1.5}]}{\Longleftrightarrow} \xi \in \mathcal{L}(\mathcal{A}'').$$

□

6   Conclusion

We have introduced global one-counter tree automata (GOCTA) and shown that they extend the capabilities of one-counter tree automata (OCTA) in certain ways, but are less expressive in other regards. We have shown that the membership problem of GOCTA is in P while, in contrast to expectations raised by literature on global index grammars, their emptiness problem is undecidable.

Future work should investigate closure properties of the class of tree languages recognizable by GOCTA. While closure under Boolean operations will likely follow the string case, i.e., closure under union holds but closure under intersection does not, closure under different restrictions of (inverse) tree homomorphisms is an interesting topic for further investigations.

Acknowledgments. This work was supported by the European Research Council through the ERC Consolidator Grant No. 771779 (DeciGUT).

We thank Johannes Osterholzer and Sebastian Rudolph for their valuable ideas which have pointed us into the right direction to obtain several of the results in this work.

References

[AJ13] Amarilli, A., Jeanmougin, M.: A proof of the pumping lemma for context-free languages through pushdown automata (2013). arXiv:1207.2819

[Cas04] Castano, J.: Global index languages. Ph.D. thesis. Brandeis University (2004). www.cs.brandeis.edu/~jcastano/thesis3.pdf

[Com+08] Comon, H., et al.: Tree automata techniques and applications, p. 262 (2008). https://inria.hal.science/hal-03367725

[DMP92] Duske, J., Middendorf, M., Parchmann, R.: Indexed counter languages. RAIRO - Theor. Inform. Appl. **26**(1), 93–113 (1992). https://doi.org/10.1051/ita/1992260100931

[Eng86] Engelfriet, J.: Context-free grammars with storage. Technical Report 86-11. University of Leiden (1986). arXiv:1408.0683

[Gue81] Guessarian, I.: On pushdown tree automata. In: Astesiano, E., Böhm, C. (eds.) CAAP 1981. LNCS, vol. 112, pp. 211–223. Springer, Heidelberg (1981). https://doi.org/10.1007/3-540-10828-9_64

[Kla04] Klaedtke, F.: Automata-based decision procedures for weak arithmetics. Ph.D. thesis. University of Freiburg (2004). http://freidok.ub.uni-freiburg.de/volltexte/1439/index.html

[KR03] Klaedtke, F., Rueß, H.: Monadic second-order logics with cardinalities. In: Baeten, J.C.M., Lenstra, J.K., Parrow, J., Woeginger, G.J. (eds.) ICALP 2003. LNCS, vol. 2719, pp. 681–696. Springer, Heidelberg (2003). https://doi.org/10.1007/3-540-45061-0_54

Decision Problems for Subregular Classes

Michal Hospodár[1]([✉])[ID], Viktor Olejár[1,2][ID], and Juraj Šebej[2][ID]

[1] Mathematical Institute, Slovak Academy of Sciences,
Grešákova 6, 040 01 Košice, Slovakia
hospodar@saske.sk
[2] Department of Computer Science, P. J. Šafárik University,
Jesenná 5, 040 01 Košice, Slovakia

Abstract. We study the computational complexity of deciding whether a given deterministic or nondeterministic finite automaton (DFA or NFA) recognizes a language in a given subclass of regular languages. We prove NL-completeness of this problem on both automata models for the classes of comma-free codes, solid codes, and singleton languages. For the classes of combinational, finitely generated left ideal, star, comet, group, and co-finite languages, the membership problem is NL-complete on DFAs and PSPACE-complete on NFAs. We also show that the membership problem on NFAs is NL-complete for the classes of prefix-, suffix-, factor-, and subword-free, singletons, and finite languages and it is PSPACE-hard for symmetric definite languages. Next, we show that deciding whether a given unary partial DFA recognizes an ordered language is L-complete and deciding whether a partial DFA can be ordered is NP-complete. Finally, deciding whether a given DFA (NFA) recognizes an ordered or power-separating language is NL-hard (PSPACE-hard, respectively).

1 Introduction

There are a number of ways in which it is possible to define a subclass of regular languages. We might use some restrictions or properties on the automaton model, restrictions on the regular expressions, through language identities, and through a number of other possible means. Motivation for carefully studying of individual language classes on their own can stem from various theoretical or

M. Hospodár—This publication was supported by the Operational Programme Integrated Infrastructure (OPII) for the project 313011BWH2: "InoCHF – Research and development in the field of innovative technologies in the management of patients with CHF", co-financed by the European Regional Development Fund.
V. Olejár—Parts of this work were conducted during the Erasmus+ mobility program with mobility ID 1409624.
J. Šebej–This research is funded by Slovak Research and Development Agency project under contract No. APVV-21-0336 "Analysis of Judicial Decisions using Artificial Intelligence".
Supported by Slovak Grant Agency for Science (VEGA) under contract 2/0096/23 "Automata and Formal Languages: Descriptional and Computational Complexity".

S. Z. Fazekas (Ed.): CIAA 2024, LNCS 15015, pp. 180–194, 2024.
https://doi.org/10.1007/978-3-031-71112-1_13

more application-oriented sources. In this paper we focus on a collection of previously studied subclasses of regular languages and ask the following question: What is the resource cost of determining, whether a given language (represented by a DFA or NFA) belongs to a given subclass? More specifically, in what computational complexity class can we classify this decision problem for different subclasses and language representations?

These kinds of decision problems (often referred to as language membership problems in the literature) were studied to a certain degree in the past, either focusing on a single subclass or as part of a collective work on several subclasses. We review selected past investigations that employ a collective approach. Hunt and Rosenkrantz [10] developed techniques and sufficient conditions for several complexity lower bounds presented in a unified framework. Cho and Huynh [4] focused on star-free languages, piecewise testable languages, and languages of dot-depth one. Convex languages and some of their subclasses such as prefix-, suffix-, factor-, subword-free, ideal, and closed languages have been treated by Brzozowski et al. in [3]. Investigations regarding definite and symmetric definite languages, finite and co-finite languages, and union-free languages among others were investigated by Holzer and Kutrib in [9]. When it comes to decidability questions, a summary regarding codes is presented in [16, Chapter 8, Table 9.1].

In the present paper, we continue this research and examine the problem of deciding whether a given DFA or NFA recognizes a language in a subregular class. We look at selected classes that either were not investigated in the literature, or the results on them can be found without proofs, or the proofs of such results can be simplified. We show that the membership problem for the classes of comma-free codes, solid codes, finite, and singleton languages is NL-complete on both DFAs and NFAs, while for the classes of combinational, finitely generated left ideal, star, comet, co-finite, and group languages, it is NL-complete on DFAs and PSPACE-complete on NFAs. We show that deciding whether a given NFA recognizes a prefix-free (suffix-, factor-, subword-free) or symmetric definite language is NL-complete and PSPACE-hard, respectively. We also show that deciding whether a unary partial DFA recognizes an ordered language is L-complete, deciding whether a given partial DFA can be ordered is NP-complete, and deciding whether a given DFA (NFA) recognizes an ordered language is NL-hard (PSPACE-hard, respectively).

To get inclusions in the corresponding complexity classes, we provide appropriate algorithms. To get hardness, we give reductions from problems that are known to be complete for the corresponding complexity classes. We use a reduction from non-emptiness of intersection, which has been shown to be PSPACE-complete in [14, Lemma 3.2.3]. From a given input string w and Turing machine M, we get DFAs representing computations of M on w as in [14]. Then if the intersection of languages recognized by such given DFAs is not empty, it is a singleton language containing a string of length at least two (a code of a valid accepting computation of M on w). Hence the NFA for union of complements of these languages recognizes either Σ^* or $\Sigma^* \setminus \{w\}$ where $|w| \geq 2$; the former belonging to many of our subclasses and the latter not. For example, Σ^* is a symmetric definite and a star language, while $\Sigma^* \setminus \{w\}$ with $|w| \geq 2$ is neither symmetric definite (since it contains ε but is not equal to Σ^*) nor a star language (since $w = aw'$ where $a, w' \in \Sigma^* \setminus \{w\}$). As a result, we get a reduction from non-emptiness of intersection to NFA membership problem for the classes of symmetric definite and star languages.

2   Preliminaries

We assume that the reader is familiar with standard notions and notations in formal language theory and complexity theory. For details, we refer to [21].

Let Σ be a non-empty *alphabet* of symbols. Then Σ^* denotes the set of all strings over Σ, including the empty string ε. A *language* over Σ is any subset of Σ^*. The length of a string w is denoted by $|w|$.

A *nondeterministic finite automaton* (NFA) is a quintuple $M = (Q, \Sigma, \cdot, s, F)$ where Q is a finite non-empty set of *states*, Σ is a finite non-empty set of input symbols (*i.e., input alphabet*), $s \in Q$ is the *initial state*, $F \subseteq Q$ is the set of *final (accepting) states*, and $\cdot : Q \times \Sigma \to 2^Q$ is the *transition function* which can be naturally extended to the domain $2^Q \times \Sigma^*$. The language recognized by the NFA M is $L(M) = \{w \in \Sigma^* \mid s \cdot w \cap F \neq \emptyset\}$. We call an NFA a *(complete) deterministic finite automaton* (DFA) if $|q \cdot \sigma| = 1$ for each $q \in Q$ and each $\sigma \in \Sigma$ and if $|q \cdot \sigma| \leq 1$ we call it a *partial DFA*; in such a case, $\cdot$ is a mapping from $Q \times \Sigma$ to Q.

We assume familiarity with the standard definitions of the complexity classes L, P, PSPACE of languages recognizable by deterministic Turing machines in logarithmic space, polynomial time, polynomial space, respectively, and their nondeterministic counterparts NL, NP, NPSPACE. For a complexity class C, we let $coC = \{L^c \mid L \in C\}$. We summarize some known inclusions [17,24]:

$$L \subseteq NL = coNL \subseteq P \subseteq NP \subseteq PSPACE = NPSPACE.$$

Let f be a mapping defined by a Turing machine on input length n that is polynomial time computable. Given languages A and B, we say A is *polynomial time many-one reducible* to B (denoted $\leq_P$) if we have $w \in A$ if and only if $f(w) \in B$. Now let f be a mapping defined by a Turing machine on input length n with read-only input tape, read/write work tape of size $O(\log(n))$, and a write-only output tape that is not allowed to move its head left. Given languages A and B, we say A is *logarithmic space reducible* to B (denoted $\leq_L$) if we have $w \in A$ if and only if $f(w) \in B$. Additionally, we say that A is *one-way logarithmic space reducible* to B (denoted $\leq_{1-L}$) if we have $A \leq_L B$ and the Turing machine for f has the input head reading the input exactly once from left to right. We say a given language L is *hard* for a complexity class C under a given reduction $\leq$ if $K \leq L$ for every $K \in C$. Moreover, we say L is *complete* for C under some reduction if it is hard for C under the same reduction and we have $L \in C$.

In this paper we focus mainly on two versions of decision problems. Given a subregular language class $\mathcal{C}$ and a DFA (NFA) A, does the language $L(A)$ belong to $\mathcal{C}$? We denote this problem as $\mathcal{C}_{\mathrm{DFA}}$ MEMBERSHIP ($\mathcal{C}_{\mathrm{NFA}}$ MEMBERSHIP). To show hardness results, we make use of several known complete problems related to automata (see [8] for a survey). Namely:

- 1GAP [7] (2GAP [23], GAP [12]): Given a directed graph $G = (V, E)$ with vertices having at most one (at most two, arbitrary) outgoing edges and vertices s and t, is there a directed path from s to t in G?
- DFA (NFA) EMPTINESS [12]: Given a DFA (NFA) A, is $L(A) = \emptyset$?
- DFA (NFA) UNIVERSALITY [1]: Given a DFA (NFA) A, is $L(A) = \Sigma^*$?
- INTERSECTION EMPTINESS [14]: Given k DFAs $A_1, A_2, \ldots, A_k$ over the same alphabet Σ, is $L(A_1) \cap L(A_1) \cap \cdots \cap L(A_k) = \emptyset$?

The 1GAP, where G is given as a list of pairs and s and t are encoded in the beginning, is L-complete under $\leq_{1-\mathrm{L}}$ reductions [7, Theorem 2.1]. The NFA UNIVERSALITY and INTERSECTION EMPTINESS problems are PSPACE-complete under $\leq_{\mathrm{P}}$ reductions [14, Lemma 3.2.3], while the other mentioned problems are NL-complete under $\leq_{\mathrm{L}}$ reductions.

A language L over an alphabet Σ is

- a *comma-free code* (CFC) if $\Sigma^+ L \Sigma^+ \cap L^2 = \emptyset$ [13];
- a *solid code* (SOC) if it is factor-free and no proper prefix of a string in L is also a proper suffix of a string in L [5, 13];
- *combinational* (CB): if $L = \Sigma^* H$ for some $H \subseteq \Sigma$;
- *finitely generated left ideal* (FGLID): if $L = \Sigma^* H$ for some finite language H (also called noninitial definite);
- *definite* (DEF): if $L = E \cup \Sigma^* H$ for some finite languages E and H;
- *generalized definite* (GDEF): if $L = E \cup \bigcup_{i=1}^{m} G_i \Sigma^* H_i$ for some integer $m \geq 1$ and finite languages $E, G_1, H_1, G_2, H_2, \ldots, G_m, H_m$ over Σ;
- *symmetric definite* (SYDEF): if $L = G \Sigma^* H$ for some regular languages G, H;
- *star* (STAR): if $L = G^*$ for a regular language G [2] (equivalently, $L = L^*$);
- *comet* (COM): if $L = G^* H$ for some regular languages G, H with $G \notin \{\emptyset, \{\varepsilon\}\}$;
- a *group language* (GRP) if L is recognized by a permutation automaton, i.e., a DFA where each input symbol performs a permutation on the state set [25];
- *singleton* (SINGL): if it consists of one string;
- *finite* (FIN): if it consists of finitely many strings;
- *co-finite* (COFIN): if its complement is finite;
- *ordered* (ORD): if it is recognized by an ordered DFA, i.e., a (possibly non-minimal) DFA where there exists a total order $\preceq$ on the state set such that for every symbol a we have $p \preceq q$ implies $pa \preceq qa$ [19];
- *star-free* (STFR): if L is constructible from finite languages by concatenation, union, and complementation only [18];
- *power-separating* (PSEP): if for every x in Σ^* there exists a positive integer m such that $x^{\geq m} \subseteq L$ or $x^{\geq m} \subseteq L^c$ [20];
- *piecewise testable* (PT) if it is a finite Boolean combination of languages of the form $\Sigma^* a_1 \Sigma^* \cdots \Sigma^* a_n \Sigma^*$, where $a_i \in \Sigma$ [22];
- a language *of dot depth-one* (DD1) if it is a Boolean combination of languages of the form $w_0 \Sigma^* w_1 \Sigma^* \cdots \Sigma^* w_n$, where $w_i \in \Sigma^*$ [22].

3   Subclasses of Convex Languages

Membership problems for the classes of prefix-, suffix-, factor-, subword-free languages, as well as their -closed and -convex variants, with the input specified by DFAs were investigated in [3, Theorem 30]. It was shown that they are NL-complete. As a consequence, since variants of ideal languages are complements of their closed counterparts, we get NL-completeness for the corresponding ideal classes. For inputs specified by NFAs, Theorem 32 in [3] proves PSPACE-completeness for all classes mentioned at the beginning of this paragraph, except for the -free variants. Han et al. [6] provided an algorithm in $O(n^2 + t^2)$ time for the -free variants with inputs represented by NFAs with n states and t transitions. In the following theorem we show that these problems are NL-complete. Subsequently, we consider two subclasses of factor-free languages: comma-free codes (CFC) and solid codes (SOC).

Theorem 1. *Deciding whether a given NFA recognizes a prefix-free (suffix-free, factor-free, subword-free) language is* NL-*complete.*   □

Theorem 2. *Deciding whether a given DFA or NFA recognizes a comma-free code (solid code) is* NL-*complete.*

Proof. We first show the inclusion of $\mathrm{CFC_{DFA}}$ MEMBERSHIP in NL. Let L be a language recognized by an n-state NFA. Then the language $\Sigma^+ L \Sigma^+ \cap L^2$ is recognized by an NFA of $(n+2)(2n)$ states which is a product automaton of an $(n+2)$-state NFA for $\Sigma^+ L \Sigma^+$ and a $2n$-state NFA for L^2. The nondeterministic Algorithm 1 guesses a string of length at most $(n+2)(2n)$ in $\Sigma^+ L \Sigma^+ \cap L^2$. Checking up to the length $(n+2)(2n)$ is sufficient since if a longer string is accepted by this NFA, then the computation on this string contains some loop, visiting a state twice, and this loop can be omitted, resulting in a string satisfying the desired length bound. It requires one counter up to $(n+2)(2n)$, binary encodings of two state pointers, three bits, and a binary encoding of the currently processed symbol. We utilize q_1 and q_2 as the state pointers for simulating an accepting computation of strings in $\Sigma^+ L \Sigma^+$ and L^2, respectively. The bit flag b_1 indicates the nondeterministic decision that a string in Σ^+ has already been processed. Analogously, b_2 indicates the nondeterministic decision that a string in $\Sigma^+ L \Sigma$ has been processed while reading a string in $\Sigma^+ L \Sigma^+$, and b_3 indicates that a string in L has been processed while reading a string in L^2. A similar argument is used whenever we employ counters for accepted string length checking. Hence $\mathrm{CFC_{DFA}}$ MEMBERSHIP is in NL.

Now we show inclusion of $\mathrm{SOC_{DFA}}$ MEMBERSHIP in NL. We first test in NL that $L(A)$ is factor-free. This can be done as in [3, Theorem 30]. An n-state DFA for proper prefixes of $L(A)$ can be obtained from A by making finite those states q of A, from which a non-empty string is accepted. An n-state NFA for proper suffixes of $L(A)$ can be obtained in the same way from A^R, the automaton recognizing the reversal of $L(A)$. It follows that if a string belongs to the intersection of proper prefixes and proper suffixes of $L(A)$, then also a string of length at most n^2 belongs to this intersection.

Algorithm 1: CFC_{DFA} MEMBERSHIP in NL

---

input : A DFA $A = (Q, \Sigma, \cdot, s, F)$
output: Is $L(A) \in \text{CFC}$?

```
1  q₁, q₂ ← s ;                                    // Memory for states
2  b₁, b₂, b₃ ← 0 ;                                // Bit flags
3  for i ← 0 to (n + 2) · (2n) do
4  │   a ← Guess(Σ);
5  │   switch (b₁, b₂) do
6  │   │   case (0, 0) do                          // Reading Σ⁺
7  │   │   └   b₁ ← Guess({0, 1})
8  │   │   case (1, 0) do                          // Reading Σ⁺L
9  │   │   │   if q₁ ∉ F then  q₁ ← q₁ · a ;
10 │   │   │   else
11 │   │   │   │   b₂ ← Guess({0, 1});
12 │   │   │   └   if b₂ = 0 then q₁ ← q₁ · a;
13 │   │   └   case (1, 1) do  q₁ ← q₁ ;           // Reading Σ⁺LΣ⁺
14 │   if b₃ = 0 then                              // Reading L
15 │   │   if q₂ ∉ F then q₂ ← q₂ · a;
16 │   │   else
17 │   │   │   b₃ ← Guess({0, 1});
18 │   │   │   if b₃ = 0 then q₂ ← q₂ · a;
19 │   │   └   else q₂ ← s · a;
20 │   else
21 │   └   q₂ ← q₂ · a;                            // Reading LL
22 └   if (b₁, b₂, b₃) = (1, 1, 1) and q₁ ∈ F and q₂ ∈ F then return False;
23 return True;
```

We initialize two pointers as follows. A prefix pointer pp pointing to the initial state of A and a suffix pointer sp pointing to some guessed reachable state. We keep guessing symbols for a string u of length at most n^2 and track the computation on u by pp and sp (from the initial state and guessed state, respectively). If sp points to a final state, then we check in NL, as in the graph accessibility problem with states as vertices, whether a final state is reachable from the state that pp points to. If this is the case, then we have guessed a string which is a proper prefix and proper suffix of some string in $L(A)$, hence $L(A) \notin \text{SOC}$. Otherwise, we were unable to guess such a string up to length n^2 and thus $L(A) \in \text{SOC}$. Hence, SOC_{DFA} MEMBERSHIP is in NL.

In order to simplify the proof for NL-hardness we first show some inclusions. Let us show that every comma-free code is factor-free. For a contradiction, assume that L is a comma-free code, but not factor-free. Then there exist strings such that $vuw \in L$ and $u \in L$, where $vw \neq \varepsilon$. Since $vuw \in L$, we have $vuwvuw \in L^2$. However, we have that $v \cdot u \cdot wvuw \in \Sigma^+ L \Sigma^+$, if v is non-empty, and $vuwv \cdot u \cdot w \in \Sigma^+ L \Sigma^+$ if w is non-empty. Hence L is not comma-free.

Next let us show that every solid code is comma-free. For a contradiction, assume that L is a solid code, but not comma-free. Then there exists a string w such that $w \in \Sigma^+ L \Sigma^+ \cap L^2$. It follows that $w = uv$ with $u \in L$ and $v \in L$. At the same time we have $w = xyz$ with x and z in Σ^+ and $y \in L$. If $u = xyz'$ for some prefix z' of z or $v = x'yz$ for some suffix x' of x, then L is not factor-free, therefore not a solid code. Otherwise, we have $y = y_1 y_2$, where neither y_1 nor y_2 are empty and $u = xy_1$ and $v = y_2 z$. Then y_1 is a proper prefix of y and a proper suffix of u, which is a contradiction.

Finally, to get NL-hardness we use a logspace reduction from the DFA EMPTI-NESS problem. From an instance A of the DFA EMPTINESS problem we construct a DFA A' by adding a loop on a new symbol # in each state of A. If $L(A) = \emptyset$, then $L(A') = \emptyset$ and thus a comma-free code and a solid code. If some $w \in L(A)$, then $L(A')$ contains the strings w and $w\#$. It follows that the language $L(A')$ is not factor-free, so neither a comma-free code nor a solid code.

The same algorithms as above work if A is an NFA. So the NL procedures guess the strings of length at most n^2 and at most $(n+2)(2n)$ for their respective cases. The only difference is in tracking the computation where instead of deterministic transitions we guess the successive state. Since deciding membership in these two classes is NL-hard for DFAs, it is NL-hard also for NFAs.   □

4   Variants of Definite Languages

We continue with classes related to the class of left ideals, that is, the variants of definite languages. An alternative name for left ideals is ultimate definite languages. This family of classes also contains combinational, finitely generated left ideal (aka noninitial definite), definite, generalized definite, and symmetric definite languages. We start with combinational languages, and then we consider finitely generated left ideals.

Theorem 3. *Deciding whether a given DFA (NFA) recognizes a combinational language is* NL*-complete (*PSPACE*-complete, respectively).*

Proof. To show in nondeterministic logarithmic space whether a given language recognized by an n-state DFA is combinational, we keep guessing symbols for two strings u and v until both are of length at most $n-1$. If we guess a symbol a such that $ua \in L$ and $va \notin L$, then L is not in CB; otherwise it is in CB. This procedure requires binary encodings of two state pointers, a counter up to n, and a binary encoding of a currently processed symbol. Hence it is in NL.

We show NL-hardness by a logspace reduction from the DFA UNIVERSALITY problem. Let $A = (Q, \Sigma, \cdot, s, F)$ be an instance of DFA UNIVERSALITY. We aim to construct a DFA A' such that the language $L(A)$ is universal if and only if the language $L(A')$ is combinational. We construct A' from A by adding a new input symbol #, a new initial non-final state s', and a new (unique) final state f. By the symbol #, each final state of A, as well as the states s' and f, are sent to f. Each non-final state of A has a loop on #. Next, each symbol $a \in \Sigma$ sends both states s' and f to $s \cdot a$. This construction is shown in Fig. 1. If $L(A) = \Sigma^*$, then $L(A') = (\Sigma \cup \{\#\})^*\#$, so $L(A')$ is combinational. If there exists a string w with $w \notin L(A)$, then $w\#\# \notin L(A')$, while $\# \in L(A')$. So there exist two strings ending with $\#$ such that exactly one of them is in $L(A')$. Hence $L(A') \notin CB$.

For NFAs, we prove that CB_{NFA} MEMBERSHIP is in PSPACE. For inclusion in PSPACE, we make use of NPSPACE=PSPACE and provide a nondeterministic procedure which uses polynomial space to test whether a given n-state NFA recognizes a language from CB. We keep guessing symbols for two strings u and v until both are of length at most $2^n - 1$. If we subsequently guess a symbol a with $ua \in L$ and $va \notin L$, then L is not in CB. Otherwise it is in CB. This procedure requires two n-bit masks to keep track of computations for u and v, a counter up to 2^n, and a binary encoding of a currently processed symbol. Hence it is in NPSPACE=PSPACE. For hardness, the same reduction as for DFAs works, but this time from the NFA UNIVERSALITY problem which is PSPACE-hard. Therefore CB_{NFA} MEMBERSHIP is PSPACE-complete.   □

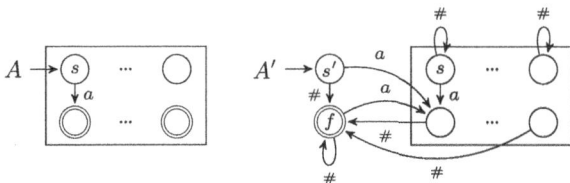

Fig. 1. A reduction from DFA UNIVERSALITY to CB_{DFA} MEMBERSHIP

Theorem 4. *Deciding whether a given DFA (NFA) recognizes a finitely generated left ideal language is* NL-*complete (*PSPACE-*complete, respectively).*   □

We conclude this section with the classes of definite, generalized definite, and symmetric definite languages. The problems DEF_{DFA} MEMBERSHIP and $SYDEF_{DFA}$ MEMBERSHIP have been shown to be NL-complete in [9, Theorems 4 and 7], and $GDEF_{DFA}$ MEMBERSHIP is NL-complete by [11, Theorem 2]. Next, PSPACE-hardness of DEF_{NFA} MEMBERSHIP and $GDEF_{NFA}$ MEMBERSHIP is proved in [10, Theorem 3.4(13)]. Our next result shows that $SYDEF_{NFA}$ MEMBERSHIP is PSPACE-hard.

Proposition 5. *Deciding whether a given NFA recognizes a symmetric definite language is* PSPACE-*hard.*   □

5   Stars, Comets, and Group Languages

In this section, we show that the membership problems for star, comet, and group languages are NL-complete for DFAs and PSPACE-complete for NFAs.

Theorem 6. *Deciding whether a given DFA (NFA) recognizes a star (comet, group) language is* NL-*complete (*PSPACE-*complete, respectively).*

Proof. We only prove NL-hardness for GRP$_{\text{DFA}}$ MEMBERSHIP. We provide a reduction from the DFA UNIVERSALITY problem. Let $A = (Q, \Sigma, \cdot, s, F)$ be an instance of DFA UNIVERSALITY. We construct a DFA A' from A by adding a new symbol $\#$, a new non-final sink state d, and a new final state f which is sent to d on each input symbol including $\#$. For each final state of A, we add a loop on $\#$, and for each non-final state q of A, we add a transition $(q, \#, f)$.

If $L(A)$ is universal, then there is no reachable non-final state in A, so also no reachable non-final state in A'. Hence $L(A')$ is universal, which is a group language. Otherwise, there is a reachable non-final state q of A. Next, state q is sent to f on $\#$, and both f and d are sent to d on $\#$ in A'. Since f and d are reachable and distinguishable in A', the language $L(A')$ is not a group language. Thus $L(A)$ is universal if and only if $L(A')$ is a group language. $\square$

6   Singletons, Finite, and Co-Finite Languages

In this section, we consider the membership problems for the classes of singletons (SINGL), finite languages (FIN), and co-finite languages (COFIN). Recall that a language is a singleton if it contains exactly one string, it is finite if it contains finitely many strings, and it is co-finite if its complement is finite.

We first show that the membership problems on DFAs and NFAs for singletons and finite languages are NL-complete. Then we consider co-finite languages.

Theorem 7. *Deciding whether a given DFA or NFA recognizes a singleton (finite) language is NL-complete.* $\square$

Theorem 8. *Deciding whether a given DFA (NFA) recognizes a co-finite language is NL-complete (PSPACE-complete, respectively).*

Proof. Testing whether a DFA recognizes a co-finite language can be done in NL by inverting finality of all its states and testing the resulting DFA for finiteness. To get NL-hardness, it is enough to invert finality of all states in the DFA A' described in the previous proof. The resulting language is $(\Sigma \cup \{\#\})^*$, so co-finite, if $L(A)$ is empty, while if A accepts a string w, then the resulting language does not contain strings in $w\#\#^*$, so it is not co-finite.

For NFAs, we prove inclusion in PSPACE by guessing a string w of length between 2^n and 2^{n+1} which is rejected by the given NFA. This procedure requires an n-bit mask to keep track of computations for w, a counter up to 2^{n+1}, and a binary encoding of a currently processed symbol. Hence it is in NPSPACE=PSPACE.

To get PSPACE-hardness, we provide a reduction from NFA universality problem. From an NFA N construct an NFA N' by adding one new symbol $\#$, one non-final sink state d, and one final sink state f; the symbol $\#$ sends each non-final state of N to d and it sends each final state of N to f. If $L(N)$ is universal, then $L(N')$ accepts every string over $\Sigma \cup \{\#\}$, so it is co-finite. If a string w is rejected by N, then each string in $w\#\#^*$ is rejected by N', so $L(N')$ is not co-finite. $\square$

7   Ordered, Star-Free, and Power-Separating Languages

In this section, we provide some partial results for the classes of ordered (ORD), star-free (STFR), and power-separating languages (PSEP). We were not able to determine the completeness for the problems of ORD_{DFA} MEMBERSHIP and ORD_{NFA} MEMBERSHIP. However, in the following, we present two slight variations of problems that are related and, we believe, are still interesting.

Proposition 9. *Deciding whether a given unary partial DFA recognizes an ordered language is* L*-complete under* $\leq_{1-L}$ *reductions.*

Proof. We check the number of states and save it in a counter n. Next we initialize two pointers p_1 and p_2. For each reachable state q we let p_1 point to q and we read a n-times, keeping track of the computation by p_2. If q has been visited twice during the reading of a, it means we found a state within a reachable cycle and break the loop. Now we read a n-times again and check with the help of a bit flag whether q and the other states of the cycle have the same finality. If they do then this reachable cycle is equivalent to either a final or non-final sink state and thus this language can be recognized by an ordered DFA. Otherwise we have that p_1 and p_2 point to states of different finality within a reachable cycle in M. Therefore every DFA recognizing $L(M)$ contains a non-trivial cycle and thus we have $L(M) \notin ORD$.

For hardness, we provide a reduction from $\overline{1GAP}$ which is L-complete under $\leq_{1-L}$ reductions since 1GAP is L-complete under $\leq_{1-L}$ reductions and L=coL. Let $G = (s, t, V, E)$ be an instance of $\overline{1GAP}$ based on which we construct a partial DFA A as follows. Each vertex in V is a non-final state in A and in addition, we add a final state f. All states q originally from V except t have transitions on a based on their outdegree in G: if $out(q) = 1$ then we label it a and if $out(q) = 0$ then no transition is added. Finally, we set $t \cdot a = f$ and $f \cdot a = t$. This construction can be done by a single pass of the input from left to right; see Fig. 2.

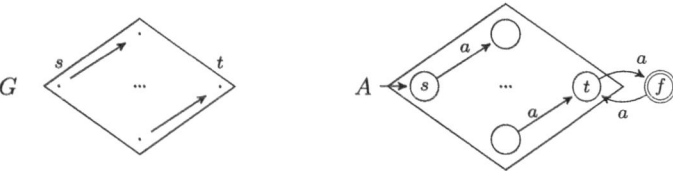

Fig. 2. A $\leq_{1-L}$-reduction from the $\overline{1GAP}$ problem to the decision problem of whether a given unary partial DFA recognizes an ordered language.

If there was no directed path from s to t then $L(A) = \emptyset$ and thus we have $L(A) \in ORD$. Otherwise A contains a reachable cycle with two states of different finality and thus resulting in $L(A) \notin ORD$. □

By working with partial DFAs instead of complete DFAs we are able to obtain NP-completeness for the following problem.

PARTIAL DFA ORDERABILITY
Input: A partial DFA $M = (Q, \Sigma, \cdot, s, F)$.
Question: Is there an order $\leq$ on Q such that $(Q, \leq)$ forms a linearly ordered
set and for each a in Σ and each q_1 and q_2 in Q where $q_1 \cdot a$ and
$q_2 \cdot a$ are defined, we have that $q_1 \leq q_2$ implies $q_1 \cdot a \leq q_2 \cdot a$?

Theorem 10. PARTIAL DFA ORDERABILITY *is* NP-*complete.*

Proof. We nondeterministically guess the order of states and subsequently verify
the required condition. Such verification can be done in $O(|Q|^2 \cdot |\Sigma|)$ time and
thus PARTIAL DFA ORDERABILITY is in NP. To show that this problem is NP-
hard we reduce from the BETWEENNESS problem.

BETWEENNESS [15, Theorem 1]
Input: A finite set A and a finite set C of triples (p, q, r) of distinct
elements from A.
Question: Is there a bijection $f : A \mapsto \{1, 2, \ldots, |A|\}$ such that for
each $(p, q, r) \in C$, we have either $f(p) < f(q) < f(r)$ or $f(r) <$
$f(q) < f(p)$?

Let (A, C) be an instance of BETWEENNESS. To each element of C, we can
assign a unique identifier in $\{1, 2, \ldots, |C|\}$. Let
$\Sigma = \{a_i, b_i \mid$ for each $i \in \{1, 2, \ldots, |C|\}\} \cup \{c_q \mid$ for each $q \in A\}$.
We define the partial DFA $M = (A \cup \{s, t\}, \Sigma, \cdot, s, \{t\})$ with the transition
function defined as follows: let $s \cdot c_q = q$ for each $q \in A$ and let $q \cdot c_q = t$.
Furthermore, for each $(p, q, r) \in C$ let $q \cdot b_i = p$, $q \cdot a_i = r$, $p \cdot a_i = p$, $r \cdot b_i = r$,
where i is the unique identifier assigned to the triple (p, q, r); see Fig. 3.

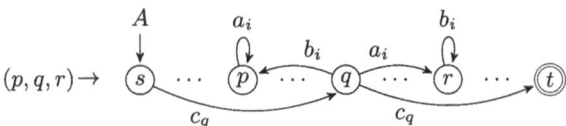

Fig. 3. A gadget for a triple $(p, q, r) \in C$ for a reduction from BETWEENNESS.

Assume that there exists a bijection f for an instance (A, C). Then define
$f' : A \cup \{s, t\} \mapsto \{0, 1, \ldots, |A|, |A| + 1\}$ as the extension of f with $f'(s) = 0$
and $f'(t) = |A| + 1$. Since f is a bijection, we have that $(A \cup \{s, t\}, \preceq)$ is a linearly
ordered set with $\preceq$ defined as $q_1 \preceq q_2$ if and only if $f'(q_1) \leq f'(q_2)$. Furthermore,
by the construction of M, we have that for each a in Σ and each q_1 and q_2
in $A \cup \{s, t\}$ where $q_1 \cdot a$ and $q_2 \cdot a$ are defined, we have $q_1 \preceq q_2$ implies $q_1 \cdot a \preceq q_2 \cdot a$.

Now assume that there is no appropriate bijection for an instance (A, C),
i.e., each bijection $f : A \mapsto \{1, 2, \ldots, |A|\}$ contains at least one triple $(p, q, r) \in$
C where neither $f(p) < f(q) < f(r)$ nor $f(r) < f(q) < f(p)$. Notice that
if $(A \cup \{s, t\}, \preceq)$ is some linearly ordered set where for $(p, q, r) \in C$ we would
have $r \preceq q$ or $q \preceq p$, then by the construction of M we would have $r \cdot b_i \npreceq q \cdot b_i$
or $q \cdot a_i \npreceq p \cdot a_i$, respectively. Thus there is no appropriate order on $A \cup \{s, t\}$,
hence M is not an orderable partial DFA.   □

The STFR$_{\text{DFA}}$ MEMBERSHIP problem was shown to be PSPACE-complete by Cho and Huynh in [4, Theorem 1.17]. The following proposition shows PSPACE-hardness for the respective membership problems on NFAs for the classes of ordered, star-free (cf. [10, Theorem 3.4(19)]), and power-separating languages. Recall that a language L is star-free if L is constructible from finite languages by concatenation, union, and complementation only, it is power-separating if for every x in Σ^* there exists a positive integer m such that $x^{\geq m} \subseteq L$ or $x^{\geq m} \subseteq L^c$; and we have ORD $\subsetneq$ STFR $\subsetneq$ PSEP.

Proposition 11. *Deciding whether a given NFA recognizes an ordered (star-free, power-separating) language is* PSPACE-*hard.*

Proof. To prove this, we provide a reduction from NFA UNIVERSALITY. From an input instance A over Σ, we construct an NFA A' as follows. We add a non-final sink state d (to which we send every undefined transition of A), a final sink state f_1, a final state f_2, and a non-final state q. We also add a new alphabet symbol #. For each final state p of A we add a transition $(p, \#, f_1)$, and for each non-final state p of A we add a transition $(p, \#, q)$. We also add the transitions $(q, \#, f_2)$ and $(f_2, \#, q)$ and add loops on each symbol from Σ for both q and f_2. All states inherited from A are non-final; see Fig. 4.

If $L(A) = \Sigma^*$, then $L(A') = \Sigma^* \#(\Sigma \cup \{\#\})^*$ which is an ordered language, so star-free and power-separating language. Otherwise, there is a string $w \notin L(A)$. Every odd power of $w\#$ is not in $L(A')$, while every even power of $w\#$ is in $L(A')$. It follows that $L(A')$ is not power-separating, and therefore it is neither star-free nor ordered.   □

Notice that the similar reduction from DFA universality can be provided; here the state d would not be added. As a consequence, we have the next result

Proposition 12. *Deciding whether a given DFA recognizes an ordered (star-free, power-separating) language is* NL-*hard.*   □

As shown in [4], STFR$_{\text{DFA}}$ MEMBERSHIP is PSPACE-complete. We think that also for ordered and power-separating languages, the DFA membership problem is not in NL.

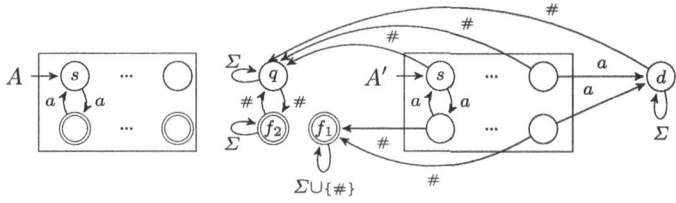

Fig. 4. A reduction from an NFA UNIVERSALITY instance to an instance of deciding whether an input NFA recognizes an ordered (power-separating, star-free) language.

8   Conclusions

We examined the complexity of deciding whether a given DFA or NFA recognizes a language in a given subregular class. Our results, together with some previous results from the literature, are summarized in Table 1. Moreover, we proved that partial DFA orderability problem is NP-complete, and deciding whether a unary partial DFA recognizes an ordered language is L-complete under $\leq_{1-L}$ reductions. Some cells contain question marks, or only hardness, indicating an open problem or that we are not familiar with a corresponding result.

Table 1. The complexity of membership problems for subregular classes.

Class C	C_{DFA} MEMBERSHIP	source	C_{NFA} MEMBERSHIP	source
all free	NL-complete	[3, Thm. 30]	NL-complete	Theorem 1
all convex	NL-complete	[3, Thm. 30]	PSPACE-complete	[3, Thm. 32]
all closed	NL-complete	[3, Thm. 30]	PSPACE-complete	[3, Thm. 32]
all ideal	NL-complete	[3, Thm. 30]	PSPACE-complete	[3, Thm. 32]
comma-free code	NL-complete	Theorem 2	NL-complete	Theorem 2
solid code	NL-complete	Theorem 2	NL-complete	Theorem 2
combinational	NL-complete	Theorem 3	PSPACE-complete	Theorem 3
finitely generated left ideal	NL-complete	Theorem 4	PSPACE-complete	Theorem 4
definite	NL-complete	[9, Thm. 4]	PSPACE-hard	[10, Thm. 3.4]
generalized definite	NL-complete	[11, Thm. 2]	PSPACE-hard	[10, Thm. 3.4]
symmetric definite	NL-complete	[9, Thm. 7]	PSPACE-hard	Proposition 5
star	NL-complete	Theorem 6	PSPACE-complete	Theorem 6
comet	NL-complete	Theorem 6	PSPACE-complete	Theorem 6
group	NL-complete	Theorem 6	PSPACE-complete	Theorem 6
singleton	NL-complete	Theorem 7	NL-complete	Theorem 7
finite	NL-complete	Theorem 7	NL-complete	Theorem 7
co-finite	NL-complete	Theorem 8	PSPACE-complete	Theorem 8
ordered	NL-hard	Proposition 12	PSPACE-hard	Proposition 11
star-free	PSPACE-complete	[4, Thm. 1.17]	PSPACE-hard	Proposition 11
power-eparating	NL-hard	Proposition 12	PSPACE-hard	Proposition 11
piecewise testable	NL-complete	[4, Thm. 2.6]	?	
dot-depth one	NL-complete	[4, Thm. 2.12]	?	

Some subregular classes, like two-sided comets, locally testable languages, or union-free languages, were not considered in this paper, and the membership problems for them could be of interest as well.

Acknowledgments. We express our gratitude to the anonymous referees for their careful reading and useful suggestions for this paper. We also thank Galina Jirásková for fruitful discussions on the topic.

References

1. Aho, A.V., Hopcroft, J.E., Ullman, J.D.: The Design and Analysis of Computer Algorithms. Addison-Wesley (1974)
2. Brzozowski, J.A.: Roots of star events. J. ACM **14**(3), 466–477 (1967). https://doi.org/10.1145/321406.321409
3. Brzozowski, J.A., Shallit, J.O., Xu, Z.: Decision problems for convex languages. Inf. Comput. **209**(3), 353–367 (2011). https://doi.org/10.1016/J.IC.2010.11.009
4. Cho, S., Huynh, D.T.: Finite-automaton aperiodicity is PSPACE-complete. Theor. Comput. Sci. **88**(1), 99–116 (1991). https://doi.org/10.1016/0304-3975(91)90075-D
5. Han, Y.S., Salomaa, K.: Overlap-free languages and solid codes. Int. J. Found. Comput. Sci. **22**(05), 1197–1209 (2011). https://doi.org/10.1142/S0129054111008647
6. Han, Y.S., Wang, Y., Wood, D.: Infix-free regular expressions and languages. Int. J. Found. Comput. Sci. **17**(2), 379–394 (2006). https://doi.org/10.1142/S0129054106003887
7. Hartmanis, J., Immerman, N., Mahaney, S.R.: One-way log-tape reductions. In: FoCS 1978, pp. 65–72. IEEE Computer Society (1978). https://doi.org/10.1109/SFCS.1978.31
8. Holzer, M., Kutrib, M.: Descriptional and computational complexity of finite automata - a survey. Inf. Comput. **209**(3), 456–470 (2011). https://doi.org/10.1016/J.IC.2010.11.013
9. Holzer, M., Kutrib, M.: Structure and complexity of some subregular language families. In: The Role of Theory in Computer Science - Essays Dedicated to Janusz Brzozowski, pp. 59–82. World Scientific (2017). https://doi.org/10.1142/9789813148208_0003
10. Hunt, H.B., Rosenkrantz, D.J.: Computational parallels between the regular and context-free languages. SIAM J. Comput. **7**(1), 99–114 (1978). https://doi.org/10.1137/0207007
11. Iván, S., Nagy-György, J.: On the structure and syntactic complexity of generalized definite languages. CoRR abs/1304.5714 (2013). http://arxiv.org/abs/1304.5714
12. Jones, N.D.: Space-bounded reducibility among combinatorial problems. J. Comput. Syst. Sci. **11**(1), 68–85 (1975). https://doi.org/10.1016/S0022-0000(75)80050-X
13. Jürgensen, H., Staiger, L.: Automata for solid codes. Theor. Comput. Sci. **892**, 25–47 (2021). https://doi.org/10.1016/J.TCS.2021.09.007
14. Kozen, D.: Lower bounds for natural proof systems. In: FoCS 1977, pp. 254–266. IEEE Computer Society (1977). https://doi.org/10.1109/SFCS.1977.16
15. Opatrny, J.: Total ordering problem. SIAM J. Comput. **8**(1), 111–114 (1979). https://doi.org/10.1137/0208008
16. Rozenberg, G., Salomaa, A. (eds.): Handbook of Formal Languages, Volume 1: Word, Language, Grammar. Springer, Cham (1997). https://doi.org/10.1007/978-3-642-59136-5
17. Savitch, W.J.: Relationships between nondeterministic and deterministic tape complexities. J. Comput. Syst. Sci. **4**(2), 177–192 (1970). https://doi.org/10.1016/S0022-0000(70)80006-X
18. Schützenberger, M.P.: On finite monoids having only trivial subgroups. Inf. Control **8**(2), 190–194 (1965). https://doi.org/10.1016/S0019-9958(65)90108-7

19. Shyr, H.J., Thierrin, G.: Ordered automata and associated languages. Tamkang J. Math. **5**, 9–20 (1974)
20. Shyr, H.J., Thierrin, G.: Power-separating regular languages. Math. Syst. Theory **8**(1), 90–95 (1974). https://doi.org/10.1007/BF01761710
21. Sipser, M.: Introduction to the Theory of Computation. PWS Publishing Company (1997)
22. Stern, J.: Complexity of some problems from the theory of automata. Inf. Control **66**(3), 163–176 (1985). https://doi.org/10.1016/S0019-9958(85)80058-9
23. Sudborough, I.H.: On tape-bounded complexity classes and multihead finite automata. J. Comput. Syst. Sci. **10**(1), 62–76 (1975). https://doi.org/10.1016/S0022-0000(75)80014-6
24. Szelepcsényi, R.: The method of forced enumeration for nondeterministic automata. Acta Informatica **26**(3), 279–284 (1988). https://doi.org/10.1007/BF00299636
25. Thierrin, G.: Permutation automata. Math. Syst. Theory **2**(1), 83–90 (1968). https://doi.org/10.1007/BF01691347

State Complexity of the Minimal Star Basis

Jozef Jirásek[1], Galina Jirásková[2(✉)], and Jeffrey Shallit[3]

[1] Institute of Computer Science, Faculty of Science, P. J. Šafárik University,
Jesenná 5, 040 01 Košice, Slovakia
jozef.jirasek@upjs.sk
[2] Mathematical Institute, Slovak Academy of Sciences,
Grešákova 6, 040 01 Košice, Slovakia
jiraskov@saske.sk
[3] School of Computer Science, University of Waterloo,
Waterloo N2L 3G1, Canada
shallit@uwaterloo.ca

Abstract. Let L be a regular language not containing ε. We determine the state complexity of the two operations $L \to LL^+$ and $L \to L \setminus LL^+$. The latter is of interest because $L \setminus LL^+$ is the "minimal star basis", the set of all strings of L that cannot be written as the concatenation of shorter strings of L, a concept first studied by John Brzozowski in 1966.

Dedicated to the memory of our good friend
and colleague, Janusz (John) Brzozowski.

1 Introduction

Given a formal language $L \subseteq \Sigma^*$, we may consider those subsets $S \subseteq L$ whose Kleene closure "covers" L, that is, for which $L \subseteq S^*$. Among all of these S, one stands out: namely, the "smallest" one, defined by

$$M := \bigcap \{ S \subseteq L \ : \ L \subseteq S^* \}.$$

To see that M is indeed the smallest, note that M must contain every element of L that cannot be written as the concatenation of shorter strings in L, and that it suffices to let M be the set of all such strings. It follows that $M = L' \setminus L'(L')^+$, where $L' = L \setminus \{\varepsilon\}$. This was observed by Brzozowski in a classic paper [1].

It now follows that if L is regular, so is its minimal star basis. However, the analogous result does not hold for context-free languages. Let $\texttt{EVENPAL} = \{xx^R : x \in \{0,1\}^+\}$ and define $\texttt{PALSTAR} = \texttt{EVENPAL}^+$. Clearly $\texttt{PALSTAR}$ is context-free. However, we have the following result:

Theorem 1. *Let $P := \texttt{PALSTAR} \setminus \texttt{PALSTAR} \cdot \texttt{PALSTAR}^+$. Then P is not a CFL.*

J. Jirásek—Research supported by VEGA grant 1/0350/22.
G. Jirásková—Research supported by VEGA grant 2/0096/23.

S. Z. Fazekas (Ed.): CIAA 2024, LNCS 15015, pp. 195–207, 2024.
https://doi.org/10.1007/978-3-031-71112-1_14

Proof. In fact, P is the language of so-called "prime palstars", the palstars that cannot be written as the concatenation of shorter palstars [2]. However, it is known that the prime palstars are not context-free [3]. □

In this paper we consider the case where L is regular, and focus on the state complexity of two regular operations: LL^+ and $L \setminus LL^+$. Provided L does not contain ε, the latter is the minimal star basis for L.

We first show that the upper bounds on state complexities of the two operations are $\frac{3}{8}n2^n + \frac{1}{2}2^n - 1$ and $\frac{3}{8}n2^n - \frac{1}{4}2^n + 1$, respectively. Then we prove that both upper bounds are tight, by describing binary witness languages for both operations. We also discuss the unary case, and show that $n^2 - n + 1$ is a tight upper bound on the state complexity of both operations on unary languages.

2   Preliminaries

For a positive integer n, let $\mathbf{n} = \{0, 1, \ldots, n-1\}$. If X is a finite set, then $|X|$ denotes its cardinality, and 2^X its power set.

Let Σ be a finite non-empty alphabet. Then Σ^* denotes the set of all strings over Σ, including the empty string ε. A *language* over Σ is any subset of Σ^*. The *concatenation* of languages K and L is the language $KL = \{uv \mid u \in K, v \in L\}$. The *Kleene closure* of a language L is the language $L^* = \cup_{i \geq 0} L^i$ where $L^0 = \{\varepsilon\}$ and $L^{i+1} = L^i L$ if $i \geq 0$. The *positive closure* of L is the language $L^+ = \cup_{i \geq 1} L^i$.

A *deterministic finite automaton* (DFA) is a quintuple $A = (Q, \Sigma, \cdot, s, F)$ where Q is a finite non-empty *set of states*, Σ is a finite non-empty set of *input symbols*, $\cdot : Q \times \Sigma \to Q$ is the *transition function*, $s \in Q$ is the *initial state*, and $F \subseteq Q$ is the set of *final states*. The transition function $\cdot$ is extended to the domain $Q \times \Sigma^*$ in the natural way. The *language recognized* by the DFA A is the set of strings $L(A) = \{w \in \Sigma^* \mid s \cdot w \in F\}$.

All deterministic finite automata in this paper are assumed to be complete, that is, the transition function is a total function.

We usually write qw instead of $q \cdot w$. We also use $p \xrightarrow{w} q$ to denote that $pw = q$, and we say that the string w sends the state p to the state q. For $S \subseteq Q$ and $w \in \Sigma^*$, we define $Sw = \{qw \mid q \in S\}$ and $wS = \{q \in Q \mid qw \in S\}$.

A state $q \in Q$ is *reachable* in the DFA A if there is a string $w \in \Sigma^*$ such that $q = sw$. Two states p and q are *distinguishable* if there is a string w such that exactly one of the states pw and qw is final. A state $q \in Q$ is a *dead state* if $qw \notin F$ for every string $w \in \Sigma^*$.

A DFA is *minimal* if all its states are reachable and pairwise distinguishable. The *state complexity* of a regular language L, $\mathrm{sc}(L)$, is the number of states in a minimal DFA for L. The *state complexity of a unary regular operation* f is the function given by $n \mapsto \max\{\mathrm{sc}(f(L)) \mid \mathrm{sc}(L) \leq n\}$.

A *nondeterministic finite automaton* (NFA) is a quintuple $N = (Q, \Sigma, \circ, I, F)$ where Q, Σ, and F are the same as for a DFA, $I \subseteq Q$ is the set of initial states, and $\circ : Q \times (\Sigma \cup \{\varepsilon\}) \to 2^Q$ is the transition function. A string w in Σ^* is *accepted* by the NFA N if $w = a_1 a_2 \cdots a_m$ where $a_i \in \Sigma \cup \{\varepsilon\}$ and a sequence of states $q_0, q_1, \ldots, q_m$ exists in Q such that $q_0 \in I$, $q_{i+1} \in q_i \circ a_{i+1}$ for $i =$

$0, 1, \ldots, m - 1$, and $q_m \in F$. The *language recognized* by the NFA N is the set of strings $L(N) = \{w \in \Sigma^* \mid w \text{ is accepted by } N\}$.

Let $N = (Q, \Sigma, \circ, I, F)$ be an NFA. For a set $S \subseteq Q$, let $E(S)$ denote the ε-closure of S; that is, the set of states $\{q \mid q \text{ is reached from a state in } S \text{ through } 0 \text{ or more } \varepsilon\text{-transitions}\}$. The *subset automaton* of the NFA N is the DFA $\mathcal{D}(N) = (2^Q, \Sigma, \cdot, E(I), F')$ where $F' = \{S \in 2^Q \mid S \cap F \neq \emptyset\}$ and $S \cdot a = \cup_{q \in S} E(q \circ a)$ for each $S \in 2^Q$ and each $a \in \Sigma$. The subset automaton $\mathcal{D}(N)$ recognizes the language $L(N)$.

If $\varepsilon \in L$, then $L^+ = L^*$. It follows that $LL^+ = L^*$ and $L \setminus LL^+ = \emptyset$. The state complexity of the star operation is $\frac{3}{4}2^n$ with binary witnesses, and it is $(n-1)^2 + 1$ in the unary case [4]. Therefore, we assume that $\varepsilon \notin L$ in what follows. We also assume that $L \neq \emptyset$ because otherwise $LL^+ = L \setminus LL^+ = \emptyset$. Subsequently, a DFA recognizing L has at least two states.

3   The Case of Alphabet Size at Least Two

In this section we determine the state complexity of LL^+ and $L \setminus LL^+$. To get upper bounds, we construct an NFA N recognizing LL^+, and count the number of reachable and pairwise distinguishable states in the corresponding subset automaton $\mathcal{D}(N)$. Then we make an appropriate change to the set of final states in $\mathcal{D}(N)$ to get a DFA D recognizing $L \setminus LL^+$. We show that some states of $\mathcal{D}(N)$ are equivalent in D, and so derive the second upper bound. Then we provide matching lower bounds, and describe binary witnesses for both operations.

Let us start with the construction of DFAs recognizing the languages resulting from the two considered operations.

Construction 2. *Let L be recognized by a DFA $A = (\mathbf{n}, \Sigma, \cdot, 0, F)$ with $0 \notin F$.*

(1) Construct an NFA B for L^+ from the DFA A by adding an ε-transition from every final state of A to the initial state of A.

(2) Construct an NFA N for LL^+ from the DFA A and NFA B by adding an ε-transition from every final state of A to the initial state of B, by making final states of A non-final, and by making the initial state of B non-initial. Figure 1 illustrates the construction of N.

(3) The subset automaton $\mathcal{D}(N)$ is a DFA for LL^+. Since A is a DFA, every reachable subset of the subset automaton $\mathcal{D}(N)$ consists of exactly one state of the DFA A and several states of the NFA B, so each reachable subset can be represented by a pair (i, S) where $i \in \mathbf{n}$ is a state of A and $S \subseteq \mathbf{n}$ is a set of states of B.

(4) Construct a DFA D from the reachable part of $\mathcal{D}(N)$ by changing the set of final states to $F_D = \{(i, S) \mid i \in F, S \subseteq \mathbf{n}, \text{ and } S \cap F = \emptyset\}$.   □

The next three observations provide some properties of the DFAs $\mathcal{D}(N)$ and D described in Construction 2.

Proposition 3. *Let (i, S) be a reachable state of the subset automaton $\mathcal{D}(N)$. Then*

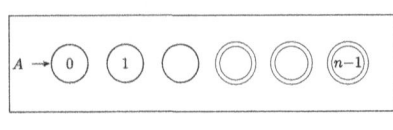
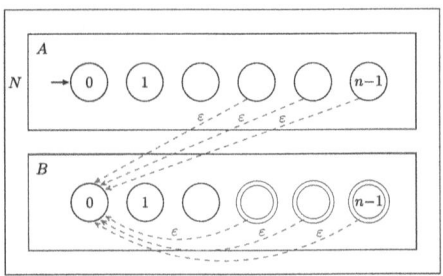

Fig. 1. The construction of an NFA N for the language $L(A) \cdot L(A)^+$.

(a) *if $i \in F$, then $0 \in S$;*
(b) *if $S \cap F \neq \emptyset$, then $0 \in S$;*
(c) *if $\{i, j\} \subseteq S$, then the states (i, S) and (j, S) are equivalent in $\mathcal{D}(N)$.*

Proof.

(a) Let (i, S) be a reachable state of the subset automaton $\mathcal{D}(N)$ such that $i \in F$. Then the NFA N has an ε-transition from the state i of A to the initial state 0 of B. Since (i, S) is reachable in $\mathcal{D}(N)$, we must have $0 \in S$.
(b) Let (i, S) be a reachable state of $\mathcal{D}(N)$ such that $S \cap F \neq \emptyset$. Then there is a state j of B such that $j \in S \cap F$. Since $j \in F$, the NFA N has an ε-transition from the state j of B to the initial state 0 of B. Hence we have $0 \in S$.
(c) Let (i, S) and (j, S) be two reachable states of $\mathcal{D}(N)$ such that $\{i, j\} \subseteq S$. Let us show that each string is accepted by $\mathcal{D}(N)$ from (i, S) if and only if it is accepted by $\mathcal{D}(N)$ from (j, S). Let w be accepted by $\mathcal{D}(N)$ from (i, S). Then w is accepted by N either from a state of B which is in S, or from the state i of A. In the former case, the string w is accepted by $\mathcal{D}(N)$ also from (j, S).
   In the latter case, there must be a factorization $w = uv$ such that, in N, there is a computation on u from the state i of A, to a state f of A with $f \in F$, then the ε-transition sends the state f of A to the initial state of B, and finally, the string v is accepted by N from the initial state of B. It follows from the construction of N, that for each transition in A, the same transition is also in B, and therefore, there is also a computation on u from the state i of B to the state f of B. Since $f \in F$, the state f of B is sent to the initial state of B by the ε-transition, and from the initial state of B, the string v is accepted as above. This means that w is accepted by N also from the state i of B. Since $i \in S$, the string w is accepted by $\mathcal{D}(N)$ from (j, S). The converse is symmetric. □

Proposition 4. *The DFA D recognizes the language $L \setminus LL^+$.*

Proof. Let a state (i, S) be reached from the initial state $(0, \emptyset)$ by a string w in the subset automaton $\mathcal{D}(N)$. Then the string w sends the initial state of A to the state i. Moreover, the string w is in the language LL^+ if and only if $S \cap F \neq \emptyset$.

It follows that $w \in L \setminus LL^+$ if and only if $i \in F$ and $S \cap F = \emptyset$. Hence changing the set of final states to $F_D = \{(i, S) \mid i \in F \text{ and } S \cap F = \emptyset\}$ results in a DFA recognizing $L \setminus LL^+$. $\square$

Proposition 5. *Let (i, S) be a reachable state of the subset automaton $\mathcal{D}(N)$ recognizing the language LL^+. Let $i \in S$. Then (i, S) is a dead state in the DFA D recognizing the language $L \setminus LL^+$.*

Proof. The set of final states of D is $F_D = \{(i, S) \mid i \in F \text{ and } S \cap F = \emptyset\}$. Let (i, S) be a reachable state of $\mathcal{D}(N)$ with $i \in S$. Let us show that every string is rejected by D from (i, S). Let $w \in \Sigma^*$. Since $i \in S$, in the subset automaton $\mathcal{D}(N)$, the string w sends the state (i, S) to a state $(iw, \{iw\} \cup T)$ for some $T \subseteq \mathbf{n}$. The resulting state is non-final in the DFA D since either $iw \notin F$, or $iw \in F$ and then $\{iw\} \cup T$ contains a final state of A. Hence every string is rejected by D from (i, S), and therefore (i, S) is a dead state in D. $\square$

Now we are ready to get upper bounds on the state complexities of LL^+ and $L \setminus LL^+$.

Lemma 6. *Let $n \geq 3$. Let L be a language recognized by an n-state DFA such that $\varepsilon \notin L$. Then LL^+ is recognized by a DFA of at most $\frac{3}{8}n2^n + \frac{1}{2}2^n - 1$ states.*

Proof. Let a language L be recognized by a DFA $A = (\mathbf{n}, \Sigma, \cdot, 0, F)$. Let N be an NFA for LL^+ given by Construction 2. Let us show that the subset automaton $\mathcal{D}(N)$ has at most $\frac{3}{8}n2^n + \frac{1}{2}2^n - 1$ reachable and pairwise distinguishable states. Since L is non-empty and $\varepsilon \notin L$, there is a state f in $F \setminus \{0\}$. First, consider reachable states (i, S) with $i \notin S$. There are three cases to consider.

Case 1: $i = 0$. Since $0 \notin S$, no final state is in S; in particular, we have $f \notin S$. This gives at most 2^{n-2} reachable pairs $(0, S)$ with $0 \notin S$.

Case 2: $i = f$. Since $f \in F$, we must have $0 \in S$, and therefore there are at most 2^{n-2} reachable pairs (f, S) with $f \notin S$.

Case 3: $i \neq 0$ and $i \neq f$. If $0 \in S$, we get at most 2^{n-2} reachable pairs (i, S) with $i \notin S$. If $0 \notin S$, then $f \notin S$, and so there are at most at most 2^{n-3} reachable pairs (i, S) with $i \notin S$. This gives at most $(n-2)(2^{n-2} + 2^{n-3})$ reachable pairs in this case.

In total, there are at most

$$2^{n-2} + 2^{n-2} + (n-2)(2^{n-2} + 2^{n-3}) \tag{1}$$

reachable states (i, S) with $i \notin S$.

Now consider reachable states (i, S) with $i \in S$. By Proposition 3(c), every pair (i, S) with $i \in S$ is equivalent to every pair (j, S) with $j \in S$. All such pairs can be merged and represented by the set S. Since (i, S) is reachable and $i \in S$, the set S is non-empty. Moreover, $f \in S$ implies that $0 \in S$. The number of such sets is at most

$$2^{n-1} + 2^{n-2} - 1 \tag{2}$$

since there are 2^{n-1} sets with $0 \in S$, and at most $2^{n-2} - 1$ non-empty sets with $0 \notin S$ and $f \notin S$.

Finally, adding the number of states in (1) and (2) gives the desired upper bound. □

Lemma 7. *Let $n \geq 3$. Let L be a language recognized by an n-state DFA such that $\varepsilon \notin L$. Then $L \setminus LL^+$ is recognized by a DFA of at most $\frac{3}{8}n2^n - \frac{1}{4}2^n + 1$ states.*

Proof. Let a language L be recognized by a DFA $A = (\mathbf{n}, \Sigma, \cdot, 0, F)$. Let $N, \mathcal{D}(N)$, and D be automata given by Construction 2. By Proposition 4, the DFA D recognizes $L \setminus LL^+$; recall that D is obtained from the reachable part of $\mathcal{D}(N)$ by changing the set of final states to $F_D = \{(i, S) \mid i \in F$ and $S \cap F = \emptyset\}$. By Proposition 5, all states (i, S) with $i \in S$ can be merged into one (dead) state. Moreover, the upper bound on the number of reachable states (i, S) with $i \notin S$ is given in (1). Increasing (1) by one gives the desired upper bound. □

The next two lemmas provide matching lower bounds by describing binary witness languages.

Lemma 8. *Let $n \geq 3$ and L be the language recognized by the n-state DFA A shown in Fig. 2. Then $\mathrm{sc}(LL^+) = \frac{3}{8}n2^n + \frac{1}{2}2^n - 1$.*

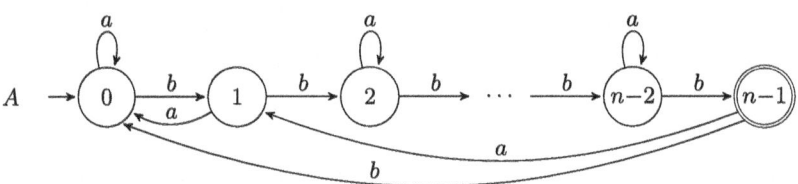

Fig. 2. A binary witness meeting the upper bound $\frac{3}{8}n2^n + \frac{1}{2}2^n - 1$ for LL^+.

Proof. Construct an NFA N for LL^+ as described in Construction 2. Consider the following sets of pairs:

$$R_1 = \{ (i, S) \mid 0 \leq i \leq n - 2, S \subseteq \mathbf{n}, \text{ and } n - 1 \notin S \},$$
$$R_2 = \{ (i, S) \mid 1 \leq i \leq n - 2, S \subseteq \mathbf{n}, \text{ and } \{0, n - 1\} \subseteq S \},$$
$$R_3 = \{ (0, S) \mid S \subseteq \mathbf{n} \text{ and } \{0, 1, n - 1\} \subseteq S \},$$
$$R_4 = \{ (n - 1, S) \mid S \subseteq \mathbf{n} \text{ and } 0 \in S \},$$

and let $\mathcal{R} = \mathcal{R}_1 \cup \mathcal{R}_2 \cup \mathcal{R}_3 \cup \mathcal{R}_4$. Notice that $|R_1| = (n - 1)2^{n-1}$, $|R_2| = (n - 2)2^{n-2}$, $|R_3| = 2^{n-3}$, and $|R_4| = 2^{n-1}$. The family $\mathcal{R}$ contains all pairs (i, S) satisfying the condition that if $i = n - 1$ or $n - 1 \in S$, then $0 \in S$, except for the pairs $(0, S)$ where $\{0, n-1\} \subseteq S$ and $1 \notin S$. However, each such pair $(0, S)$

is equivalent to the pair $(n-1, S)$ in $\mathcal{D}(N)$ by Proposition 3(c), and the latter pair is in $\mathcal{R}_4$.

We can prove that every pair in $\mathcal{R}$ is reachable in $\mathcal{D}(N)$, and that every two distinct pairs (i, S) and (j, T) in $\mathcal{R}$ are distinguishable if $S \neq T$ or $S = T$ and $\{i, j\} \not\subseteq S$. □

Lemma 9. *Let $n \geq 3$ and L be the language recognized by the n-state DFA A shown in Fig. 3. Then $\mathrm{sc}(L \setminus LL^+) = \frac{3}{8}n2^n - \frac{1}{4}2^n + 1$.*

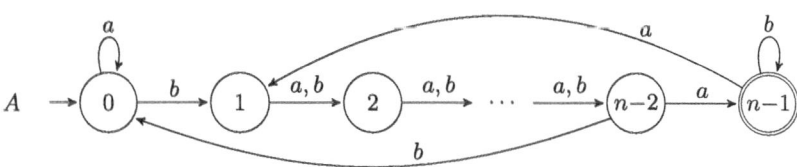

Fig. 3. A binary witness meeting the upper bound $\frac{3}{8}n2^n - \frac{1}{4}2^n + 1$ for $L \setminus LL^+$.

Proof. Since both a and b perform a permutation on $\mathbf{n}$ in the DFA A, we have $|wS| = |S|$ and $(wS)w = S$ for each $w \in \{a, b\}^*$ and $S \subseteq \mathbf{n}$. Construct an NFA N for LL^+ as described in Construction 2, as shown in Fig. 4, and consider the subset automaton $\mathcal{D}(N)$. Each reachable state of $\mathcal{D}(N)$ can be represented by a pair (i, S) where $i \in \mathbf{n}$ is a state in A, and $S \subseteq \mathbf{n}$ is a set of states in B. Moreover, if $i = n - 1$ or $n - 1 \in S$, then $0 \in S$. To get a DFA D for $L \setminus LL^+$ we only need to change the set of final states of $\mathcal{D}(N)$ to $F_D = \{(n-1, S) \mid S \subseteq \mathbf{n} \text{ and } n-1 \notin S\}$. Each state (i, S) with $i \in S$ is dead in D by Proposition 5. Consider the following sets:

$$\mathcal{R}_1 = \{(n-1, S) \mid S \subseteq \mathbf{n}, \, n-1 \notin S, \text{ and } 0 \in S\},$$
$$\mathcal{R}_2 = \{(i, S) \mid 1 \leq i \leq n-2, \, S \subseteq \mathbf{n}, \, i \notin S, \text{ and } 0 \in S\},$$
$$\mathcal{R}_3 = \{(i, S) \mid 1 \leq i \leq n-2, \, S \subseteq \mathbf{n}, \, i \notin S, \, 0 \notin S, \text{ and } n-1 \notin S\},$$
$$\mathcal{R}_4 = \{(0, S) \mid S \subseteq \mathbf{n}, \, 0 \notin S, \text{ and } n-1 \notin S\},$$
$$\mathcal{R}_5 = \{(0, \{0, n-1\})\},$$

and let $\mathcal{R} = \mathcal{R}_1 \cup \mathcal{R}_2 \cup \mathcal{R}_3 \cup \mathcal{R}_4 \cup \mathcal{R}_5$. Notice that the states (i, S) with $i \in S$ are not in $\mathcal{R}$, except for $(0, \{0, n-1\})$. However, such states are equivalent to the dead state $(0, \{0, n-1\})$ in the DFA D. Hence, in order to prove the lemma, we only need to show that each pair in $\mathcal{R}$ is reachable in $\mathcal{D}(N)$, and that every two distinct pairs in $\mathcal{R}$ are distinguishable in D.

First, let us show by induction on $|S|$ that each pair (i, S) in $\mathcal{R}_1 \cup \mathcal{R}_2 \cup \mathcal{R}_3 \cup \mathcal{R}_4$ is reachable in the subset automaton $\mathcal{D}(N)$.

The basis, $|S| = 0$, holds true since $(0, \emptyset)$ is the initial state of $\mathcal{D}(N)$, and it sent by b^i to $(i, \emptyset)$ if $i \leq n-2$. Let $1 \leq k \leq n$ and assume that every pair (i', S')

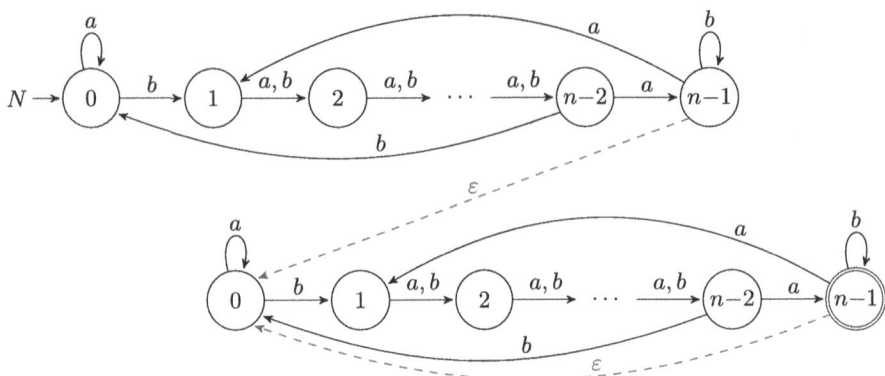

Fig. 4. An NFA for LL^+ where L is recognized by the DFA A from Fig. 3.

in $\mathcal{R}$ with $|S'| = k - 1$ is reachable. Let $|S| = k$ and (i, S) be a pair in $\mathcal{R}$. There are four cases to consider.

Case 1: $(i, S) \in \mathcal{R}_1$. Then $i = n - 1$, $n - 1 \notin S$, and $0 \in S$.

Case 1.a: $1 \in S$. Set $S' = b(S \setminus \{0\})$. Then $|S'| = k - 1$. Moreover, since $n - 1 \notin S$ and $1 \in S$, we have $n - 1 \notin S'$ and $0 \in S'$. It follows that $(n - 1, S')$ is a pair in $\mathcal{R}_1$ which is reachable by the induction hypothesis, and

$$(n - 1, S') = (n - 1, b(S \setminus \{0\})) \xrightarrow{b} (n - 1, \{0\} \cup (S \setminus \{0\})) = (n - 1, S),$$

hence the pair $(n - 1, S)$ is reachable.

Case 1.b: $1 \notin S$. Set $S' = a(S \setminus \{0\})$. Then $|S'| = k - 1$ and $0 \notin S'$. Since $n - 1 \notin S$ and $1 \notin S$, we have $n - 2 \notin S'$ and $n - 1 \notin S'$. Hence $(n - 2, S')$ is a pair in $\mathcal{R}_3$ which is reachable by induction and

$$(n - 2, S') = (n - 2, a(S \setminus \{0\})) \xrightarrow{a} (n - 1, \{0\} \cup (S \setminus \{0\})) = (n - 1, S).$$

Case 2: $(i, S) \in \mathcal{R}_2$. Then $1 \le i \le n - 2$, $i \notin S$, and $0 \in S$. Set $S' = \{0\} \cup a^i(S \setminus \{0\})$. Then $|S'| = k$. Moreover, we have $0 \in S'$, and $n - 1 \notin S'$ since $i \notin S$. Thus $(n - 1, S')$ is a pair in $\mathcal{R}_1$ which has been shown to be reachable in Case 1. Since $0a = 0$, we have

$$(n - 1, S') = (n - 1, \{0\} \cup a^i(S \setminus \{0\})) \xrightarrow{a^i} (i, \{0\} \cup S \setminus \{0\}) = (i, S).$$

Case 3: $(i, S) \in \mathcal{R}_4$. Then $i = 0$, $0 \notin S$, and $n - 1 \notin S$. Set $s = \min S$. Then we have $1 \le s \le n - 2$. Set $S' = \{0\} \cup ba^{s-1}(S \setminus \{s\})$. We have $S \setminus \{s\} \subseteq \{s + 1, s + 2, \ldots, n - 2\}$. It follows that $a^{s-1}(S \setminus \{s\}) \subseteq \{2, 3, \ldots, n - 1 - s\}$ and $ba^{s-1}(S \setminus \{s\}) \subseteq \{1, 2, \ldots, n - 2 - s\}$. It follows that $|S'| = k$, $n - 2 \notin S'$, and $0 \in S'$, so $(n - 2, S')$ is a pair in $\mathcal{R}_2$ which is considered in Case 2, and

$$(n - 2, S') = (n - 2, \{0\} \cup ba^{s-1}(S \setminus \{s\})) \xrightarrow{b} (0, \{1\} \cup a^{s-1}(S \setminus \{s\})) \xrightarrow{a^{s-1}} (0, S).$$

Case 4: $(i, S) \in \mathcal{R}_3$. Then $1 \leq i \leq n - 2$, $i \notin S$, $0 \notin S$, and $n - 1 \notin S$. Then $|b^i S| = k$. Since $i \notin S$ and $n - 1 \notin S$, we have $0 \notin b^i S$ and $n - 1 \notin b^i S$. Hence $(0, b^i S)$ is a pair in $\mathcal{R}_4$ which is considered in Case 3. Next, we have

$$(0, b^i S) \xrightarrow{b^i} (i, S)$$

since after reading each b, the first component is different from $n - 1$, and the second component does not contain the state $n - 1$.

Hence each pair in $\mathcal{R}_1 \cup \mathcal{R}_2 \cup \mathcal{R}_3 \cup \mathcal{R}_4$ is reachable in $\mathcal{D}(N)$. Finally, the unique pair $(0, \{0, n-1\})$ in $\mathcal{R}_5$ is reached by a from the pair $(0, \{n-2\})$ which is in $\mathcal{R}_4$. This proves the reachability of each pair in $\mathcal{R}$ in the subset automaton $\mathcal{D}(N)$.

Recall that we get a DFA D for $L \setminus LL^+$ from the reachable part of $\mathcal{D}(N)$ by changing the set of final states to

$$F_D = \{(n - 1, S) \mid n - 1 \notin S\}.$$

We already know that in D, all pairs (i, S) with $i \in S$ are equivalent to the reachable dead state $(0, \{0, n - 1\})$. Let (i, S) be a pair in $\mathcal{R}$ with $i \notin S$. Let us show that (i, S) is not dead in D. First, let $i \geq 1$. Since $0a = 0$, the string a^{n-1-i} sends the pair (i, S) to the pair $(n - 1, \{0\} \cup Sa^{n-1-i})$ which is in F_D since the state $n - 1$ is not in Sa^{n-1-i}. Hence, the string a^{n-1-i} is accepted from (i, S) by D, so (i, S) is not dead. Now, let $i = 0$. Since $0 \notin S$, we must have $n - 1 \notin S$, and so $n - 1 \notin Sb$. Therefore the string b sends the pair $(0, S)$ to the pair $(1, Sb)$ with $1 \notin Sb$. The latter state is not dead as shown in the first case, and therefore the pair $(0, S)$ is not dead in D.

Now, let (i, S) and (j, T) be two distinct reachable pairs in $\mathcal{R}$ with $i \notin S$ and $j \notin T$. We may assume that $i \leq j$. There are several cases to consider.

Case 1: $i < j$. Consider the string a^{n-1-j}. Since $0a = 0$, we have

$$(i, S) \xrightarrow{a^{n-1-j}} (ia^{n-1-j}, S' \cup Sa^{n-1-j}) \text{ for some } S' \subseteq \{0\},$$

$$(j, T) \xrightarrow{a^{n-1-j}} (n - 1, \{0\} \cup Ta^{n-1-j}).$$

We have $ia^{n-1-j} \neq n-1$ since $i < j$, and $n-1 \notin Ta^{n-1-j}$ since $j \notin T$. It follows that in D, the string a^{n-1-j} is rejected from (i, S) and accepted from (j, T).

Case 2: $i = j$. Then we have pairs (i, S) and (i, T) with $i \notin S$, $i \notin T$, and $S \neq T$. Let $s \in S \setminus T$. Since $s \in S$ and $i \notin S$, we have $i \neq s$.

Case 2.a: $s = n - 1$. Since $i \neq n - 1$ and $n - 1 \in S \setminus T$, for every $k \geq 1$ we have $ib^k \neq n - 1$ and $n - 1 \in Sb^k \setminus Tb^k$. It follows that

$$(i, S) \xrightarrow{b^{n-2-i}} (n - 2, \{0, 1, \ldots, n - 3 - i\} \cup Sb^{n-2-i}) \xrightarrow{b}$$
$$(0, \{0, 1, \ldots, n - 2 - i\} \cup Sb^{n-1-i}),$$

$$(i, T) \xrightarrow{b^{n-2-i}} (n - 2, Tb^{n-2-i}) \xrightarrow{b} (0, Tb^{n-1-i}).$$

Hence b^{n-1-i} sends (i, S) to the dead state $(0, \{0, 1, \ldots, n - 2 - i\} \cup Sb^{n-1-i})$, and it sends (i, T) to the state $(0, Tb^{n-1-i})$ which is not dead since $i \notin T$ implies that $0 \notin Tb^{n-1-i}$.

Case 2.b: $s = n - 2$. Since $i \neq n - 2$ and $n - 2 \in S \setminus T$, we have $ia \neq n - 1$ and $n - 1 \in Sa \setminus Ta$. It follows that

$$(i, S) \xrightarrow{a} (ia, \{0\} \cup Sa),$$
$$(i, T) \xrightarrow{a} (ia, Ta).$$

Since $i \notin S$ and $i \notin T$, we have $ia \notin Sa$ and $ia \notin Ta$. If $i \neq 0$, then $ia \neq 0$, so $ia \notin \{0\} \cup Sa$, and then the resulting pairs are distinguishable as shown in Case 2.a. If $i = 0$, then $ia = 0$. This means that the state $(ia, \{0\} \cup Sa)$ is dead, while the state (ia, Ta) is not dead.

Case 2.c: $1 \leq s \leq n-3$ and $i \neq 0$. Consider the string $w = a^{n-2-s}$. Since $0a = 0$, the string w sends (i, S) and (i, T) to $(iw, S' \cup Sw)$ and $(iw, T' \cup Tw)$, respectively, for some $S', T' \subseteq \{0\}$. Since $s \in S \setminus T$ and $s \neq 0$, $n - 2 \in (S' \cup Sw) \setminus (T' \cup Tw)$. Since $i \notin S$, $i \notin T$, and $iw \neq 0$, we have $iw \notin S' \cup Sw$ and $iw \notin T' \cup Tw$. It follows that the resulting pairs are distinguishable as shown in Case 2.b.

Case 2.d: $1 \leq s \leq n - 3$ and $i = 0$. Then b sends (i, S) and (i, T) to $(1, S' \cup Sb)$ and $(1, T' \cup Tb)$, respectively, for some $S', T' \subseteq \{0\}$. Since the state s is in $S \setminus T$ and $2 \leq sb \leq n - 2$, we have $sb \in (S' \cup Sb) \setminus (T' \cup Tb)$. Since $0 \notin S$ and $0 \notin T$, we have $1 \notin S' \cup Sb$ and $1 \notin T' \cup Tb$. It follows that the resulting pairs are distinguishable by Case 2.b if $sb = n - 2$, and by Case 2.c otherwise.

Case 2.e: $s = 0$. Since $0 \in S \setminus T$, we must have $n - 1 \notin T$, so $n - 1 \notin Tb$. If $n - 1 \in S$, then we have Case 2.a. If $n - 1 \notin S$, then $n - 1 \notin Sb$, and we have

$$(i, S) \xrightarrow{b} (ib, S' \cup Sb),$$
$$(i, T) \xrightarrow{b} (ib, T' \cup Tb),$$

where $S' = T' = \emptyset$ if $ib \neq n-1$, and $S' = T' = \{0\}$ if $ib = n-1$. In both cases, we have $ib \notin S' \cup Sb$ and $ib \notin T' \cup Tb$. Since $0 \in S \setminus T$, we have $1 \in (S' \cup Sb) \setminus (T' \cup Tb)$. It follows that the resulting pairs are distinguishable by Case 2.b if $n = 3$, by Case 2.c if $n \geq 4$ and $ib \neq 0$, and by Case 2.d if $n \geq 4$ and $ib = 0$.   □

The next theorem summarizes the results of this section.

Theorem 10 (State complexity of LL^+ and $L \setminus LL^+$). *Let $n \geq 3$. Let L be a language over an alphabet Σ recognized by a DFA with n states such that $\varepsilon \notin L$. Then*

(1) $\mathrm{sc}(LL^+) \leq \frac{3}{8}n2^n + \frac{1}{2}2^n - 1$,
(2) $\mathrm{sc}(L \setminus LL^+) \leq \frac{3}{8}n2^n - \frac{1}{4}2^n + 1$,

and both upper bounds are tight if $|\Sigma| \geq 2$.   □

4   The Unary Case

The behavior of the state complexity of the operations LL^+ and $L \setminus LL^+$ is different for unary alphabets. In this section we give matching upper and lower bounds for both operations, namely $n^2 - n + 1$. We start with upper bounds, and then show that the unary language $a^{n-1}(a^n)^*$ is a witness for both operations.

Lemma 11. *Let $n \geq 2$. Let L be a unary language recognized by an n-state DFA such that $\varepsilon \notin L$. Then $\mathrm{sc}(LL^+) \leq n^2 - n + 1$.*

Proof. Let a unary language L be recognized by a DFA $A = (Q, \{a\}, \cdot, 0, F)$ where $Q = \mathbf{n}$, $0 \notin F$, and the transitions are given by $i \cdot a = i + 1$ if $i \leq n - 2$, and $(n - 1) \cdot a = \ell$ for some ℓ with $0 \leq \ell \leq n - 1$.

Set $t = \min F$, so $1 \leq t \leq n - 1$. Construct the NFA N for LL^+ as described in Construction 2, and consider the corresponding subset automaton $\mathcal{D}(N)$. For each $j \geq 0$, let (q_j, S_j) be the state of the subset automaton $\mathcal{D}(N)$ that is reached from its initial state $(0, \emptyset)$ by a^{jt}.

First, let us show that $S_j \supseteq S_{j-1}$ for $j \geq 1$. The initial state $(0, \emptyset)$ of the subset automaton $\mathcal{D}(N)$ is sent to $(t, \{0\})$ by a^t. So $S_0 = \emptyset$ and $S_1 = \{0\}$. Let $j \geq 2$. Then

$$(0, \emptyset) \xrightarrow{a^t} (t, \{0\}) \xrightarrow{a^{(j-1)t}} (q_j, S_j), \tag{3}$$

hence, all states of B that are reached by N from the initial state of B by $a^{(j-1)t}$ are in S_j. Now let $p \in S_{j-1}$. Then the state p of B is reached by N from the initial state of A by $a^{(j-1)t}$. But then p is reached by N also from the initial state of B by $a^{(j-1)t}$, and therefore $p \in S_j$ by (3). Hence $\{0\} = S_1 \subseteq S_2 \subseteq \cdots \subseteq S_n$.

Now there are two cases to consider.

Case 1: $S_n = Q$. Then $Q \cdot a = Q$ if $\ell = 0$, and $Q \cdot a = Q \setminus \{0\}$ otherwise. Moreover, the set $Q \setminus \{0\}$ contains a final state of A. This means that in the subset automaton $\mathcal{D}(N)$, we have $(q_n, Q) \xrightarrow{a} (q_n \cdot a, Q)$, and this last state is equivalent to (q_n, Q) by Proposition 3. Hence (q_n, Q) is a final sink state of $\mathcal{D}(N)$.

Case 2: $S_i = S_{i-1}$ for some $i \leq n$. Since $0 \in S_i$ and $(q_{i-1}, S_i) \xrightarrow{a^t} (q_i, S_i)$, we have $0 \cdot a^{kt} \in S_i$ for all $k \geq 0$. Next, we have $q_{i-1} = 0 \cdot a^{(i-1)t}$ and $q_i = 0 \cdot a^{it}$, so both q_{i-1} and q_i are in S_i. By Proposition 3(c), the states (q_{i-1}, S_i) and (q_i, S_i) are equivalent in $\mathcal{D}(N)$.

In both cases, the subset automaton $\mathcal{D}(N)$ has at most $1 + tn \leq n^2 - n + 1$ reachable states. $\square$

Lemma 12. *Let $n \geq 2$. Let L be a unary language recognized by an n-state DFA such that $\varepsilon \notin L$. Then $\mathrm{sc}(L \setminus LL^+) \leq n^2 - n + 1$.*

Proof. Let L be recognized by a unary DFA $A = (\mathbf{n}, \{a\}, \cdot, 0, F)$ with $0 \notin F$. Construct the NFA N for LL^+, and the DFA D for $L \setminus LL^+$ as described in Construction 2. It follows from the construction, that the number of reachable states of $\mathcal{D}(N)$ provides an upper bound on the number of states of D. As shown in the previous proof, the DFA $\mathcal{D}(N)$ has at most $n^2 - n + 1$ reachable states. $\square$

Lemma 13. *Let $n \geq 2$ and $L = a^{n-1}(a^n)^*$ be the unary language recognized by the n-state DFA A shown in Fig. 5. Then $\mathrm{sc}(LL^+) = \mathrm{sc}(L \setminus LL^+) = n^2 - n + 1$.*

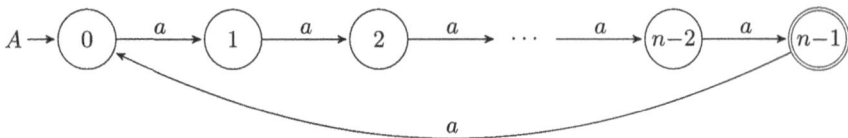

Fig. 5. A unary witness for LL^+ and $L \setminus LL^+$ meeting the upper bound $n^2 - n + 1$.

Proof. Let $L = a^{n-1}(a^n)^*$ be the language recognized by the n-state DFA A shown in Fig. 5. Construct the NFA N as described in Construction 2, and consider the corresponding subset automaton $\mathcal{D}(N)$. Let $t = n-1$ be the unique final state of A. Let $(q, S)_j$ be the state of $\mathcal{D}(N)$ reached from the initial state $(0, \emptyset)$ by the string a^j. By induction on j, we can show that for each j with $1 \leq j \leq n-1$, we have

$$
\begin{aligned}
(q, S)_{jt} &= (t - j + 1, \{0\} \cup \{t - j + 2, t - j + 3, \dots, t\}), \\
(q, S)_{jt-1} &= (t - j, \{t - j + 1, t - j + 2, \dots, t - 1\}), \\
(q, S)_{nt} &= (0, \{0, 1, \dots, t\}), \\
(q, S)_{nt-1} &= (t, \{0, 1, 2, \dots, t - 1\}).
\end{aligned}
$$

It follows that in the subset automaton $\mathcal{D}(N)$, the state $(q, S)_{nt}$ is a final sink state; recall that $t = n - 1$ is a final state of A. On the other hand, the state $(q, S)_{nt-1}$ is non-final since $t \notin \{0, 1, 2, \dots, t - 1\}$. It follows that the DFA $\mathcal{D}(N)$, which has $1 + nt = n^2 - n + 1$ reachable states, is minimal.

Next, the language $L \setminus LL^+$ is recognized by the DFA D obtained from $\mathcal{D}(N)$ by changing the set of final states to $\{(q, S) \mid q = t \text{ and } t \notin S\}$. This means that the state $(q, S)_{nt-1}$ is final in D, while the state $(q, S)_{nt}$ is dead by Proposition 5, since $0 \in \{0, 1, \dots, t\}$. Thus D has $n^2 - n + 1$ reachable and pairwise distinguishable states. This concludes our proof. $\qquad\square$

As a corollary of the three lemmas above, we get the next result.

Theorem 14 (State complexity of LL^+ and $L \setminus LL^+$: the unary case).
Let $n \geq 2$. Let L be a unary language recognized by a DFA with n states such that $\varepsilon \notin L$. Then the languages LL^+ and $L \setminus LL^+$ are recognized by DFAs with at most $n^2 - n + 1$ states, and this upper bound is tight for both operations. $\quad\square$

5   Conclusion

We examined the state complexity of operations $L \rightarrow LL^+$ and $L \rightarrow L \setminus LL^+$, the latter being the "minimal star basis", that is, the set of all strings of L

that cannot be written as the concatenation of shorter strings of L in the case of $\varepsilon \notin L$. We obtained tight upper bounds on the state complexity of these two operations, and to describe witnesses, we used a binary alphabet. We also determined the state complexity of the two operations in the unary case.

Our results are summarized in Table 1, where all the cases with $\varepsilon \in L$, in which case $LL^+ = L^*$ and $L \setminus LL^+ = \emptyset$, are given by known results on the state complexity of the star operation [4].

Table 1. State complexity of LL^+ and $L \setminus LL^+$; the results for $\varepsilon \in L$ are from [4].

$|\Sigma|$	ε?	LL^+	$L \setminus LL^+$
$|\Sigma| \geq 2$	$\varepsilon \notin L$	$\frac{3}{8}n2^n + \frac{1}{2}2^n - 1$	$\frac{3}{8}n2^n - \frac{1}{4}2^n + 1$
	$\varepsilon \in L$	$\frac{3}{4}2^n$	1
$|\Sigma| = 1$	$\varepsilon \notin L$	$n^2 - n + 1$	$n^2 - n + 1$
	$\varepsilon \in L$	$n^2 - 2n + 2$	1

References

1. Brzozowski, J.: Roots of star events. In: 7th Annual Symposium on Switching Automata Theory (SWAT), pp. 88–95. IEEE (1966)
2. Knuth, D.E., Morris, J., Pratt, V.: Fast pattern matching in strings. SIAM J. Comput. **6**, 323–350 (1977)
3. Rampersad, N., Shallit, J., Wang, M.-w: Inverse star, borders, and palstars. Inf. Process. Lett. **111**, 420–422 (2011)
4. Yu, S., Zhuang, Q., Salomaa, K.: The state complexities of some basic operations on regular languages. Theor. Comput. Sci. **125**, 315–328 (1994)

On Properties of Languages Accepted by Deterministic Pushdown Automata with Translucent Input Letters

Martin Kutrib[1], Andreas Malcher[1], Carlo Mereghetti[2](✉),
Beatrice Palano[2], Priscilla Raucci[1,2], and Matthias Wendlandt[1]

[1] Institut für Informatik, Universität Giessen, Arndtstr. 2, 35392 Giessen, Germany
{kutrib,andreas.malcher,matthias.wendlandt}@informatik.uni-giessen.de
[2] Dipartimento di Informatica "G. Degli Antoni", Università degli Studi di Milano,
via Celoria 18, 20133 Milan, Italy
{carlo.mereghetti,beatrice.palano,priscilla.raucci}@unimi.it

Abstract. We study deterministic pushdown automata operating with *translucent* input letters. These devices can be obtained by equipping classical deterministic pushdown automata with a translucency function which, depending on the current state, establishes the set of invisible input symbols: such symbols are skipped in the current move and dealt with in subsequent sweeps, while the first visible symbol from the current input head position is processed. Translucent deterministic pushdown automata can be returning, meaning that a new input sweep starts from the leftmost input symbol immediately after processing a visible symbol, or not. We show some incomparability results between the acceptance capability of returning and non-returning translucent deterministic pushdown automata and that of non-returning translucent deterministic and nondeterministic finite state automata. Then, we prove the non-closure of families of languages accepted by returning and non-returning translucent deterministic pushdown automata under concatenation, Kleene star, length-preserving and inverse homomorphism, reversal, and intersection with regular languages. In particular, arguments used to prove non-closure under this last language operation, enable us to answer a question on non-returning translucent deterministic finite state automata left open in the literature.

Keywords: Translucent input letters · Deterministic pushdown automata · Returning and non-returning computations · Computational capacity · Closure properties

1 Introduction

The main way of processing input on language recognition devices is by reading input strings from left to right, one symbol at a time, and finally providing an accept/reject outcome upon sweeping the whole input. Starting from this basic

S. Z. Fazekas (Ed.): CIAA 2024, LNCS 15015, pp. 208–220, 2024.
https://doi.org/10.1007/978-3-031-71112-1_15

input acquisition mode, more "relaxed" modes have been deeply investigated. So, for instance, two-way motion, the possibility of pausing input reading (stationary or λ-moves) or accepting in the middle of the string, multiple input sweeps, rotating, sweeping and restarting modes, have been considered for a wide variety of machines (the reader is referred to, e.g., [2,4,5,14] for an overview of these and other models). In all these cases, meaningful differences in the accepting and/or descriptional power have been emphasized, stating that the way of acquiring input is truly a computational resource that can be used to tune computational and descriptional power.

Particularly interesting is the *discontinuous* form of parsing, where "jumping" to any position inside the input string is allowed at any move. This paradigm has been suggested for Turing machines in [16], and recently considered for finite state devices. The notion of a *jumping finite automaton* is introduced in [7], where it is proved that the jumping feature increases the accepting power. In fact, even non-context-free languages, such as the language $L = \{ w \in \{a,b,c\}^* \mid |w|_a = |w|_b = |w|_c \}$, can be accepted by deterministic jumping finite automata. The restricted variant of "right one-way jumping" automata has been considered in [1,3]. A way of achieving the jumping feature is by using *translucent letters*. The concept of translucent letters has been introduced by Nagy and Otto in [11] for deterministic and nondeterministic finite automata. The key feature for translucent devices is provided by a translucency function establishing in which states which letters of the input are translucent. At each move, the device skips (by looking through) the translucent portion of the input, from the current input head position up to the first non-translucent letter (thus realizing a jump). After processing the non-translucent symbol, in the *returning mode* the input head returns to the left end of the input while, in the *non-returning mode*, the input head continues to process the input according to its updated current state and the corresponding translucent symbols. In both modes, the input head returns to the left end when the right end of the input is reached. Deterministic and nondeterministic finite automata with translucent letters are deeply investigated in the literature (see, e.g., [9,12,13,15]). However, many questions are still open. Interestingly, the jumping and translucency paradigms have been put together in [8], to form the model of a deterministic or nondeterministic jumping finite automaton.

In this paper, we study *deterministic pushdown automata with translucent letters* in both returning and non-returning mode (DPDAwtl and nrDPDAwtl, respectively). Very roughly speaking, these devices can be obtained by equipping classical deterministic pushdown automata (DPDA) with a translucency function. We point out that returning pushdown automata with translucent letters have been studied by Nagy and Otto in [10,13] in terms of certain cooperating distributed systems of restarting automata with additional pushdown store. This hints a possible application of translucency, as a tool to connect several language definition formalisms, besides to modeling discontinuous (jump) processing paradigms as above discussed. We emphasize that, due to research purposes, the definition of returning pushdown automata with translucent letters provided by Nagy and Otto does not encompass λ-transitions and features acceptance by empty pushdown. Instead, to be closer to the classical definition of DPDA, we here define DPDAwtl and nrDPDAwtl possibly performing λ-transitions, and accepting by accepting states.

In [6], we have begun investigating DPDAwtl and nrDPDAwtl by first formally defining these models. Next, we have assessed their acceptance capabilities, by showing that the families of languages accepted by such devices, namely $\mathscr{L}$(DPDAwtl) and $\mathscr{L}$(nrDPDAwtl), properly lay between deterministic context-free and deterministic context-sensitive languages, with nrDPDAwtl being strictly more powerful than DPDAwtl. To have a better grasp of DPDAwtl and nrDPDAwtl acceptance power, we have also established (incomparability) relations among $\mathscr{L}$(DPDAwtl) and $\mathscr{L}$(nrDPDAwtl) and other well known language classes between deterministic context-free and deterministic context-sensitive languages, such as context-free, growing context-sensitive and Church-Rosser languages. We have than tacked boolean closure properties of $\mathscr{L}$(DPDAwtl) and $\mathscr{L}$(nrDPDAwtl), showing that both these families are closed under complementation but not under union and intersection.

Here, we are going to further deepen the study of both accepting capabilities and closure properties of DPDAwtl and nrDPDAwtl. Concerning the first topic, we show that $\mathscr{L}$(DPDAwtl) and the corresponding language family for non-returning translucent deterministic finite state automata are incomparable. The same incomparability result is obtained, for $\mathscr{L}$(nrDPDAwtl) and the corresponding language family for non-returning translucent nondeterministic finite state automata. These results, together with those in [6], provide a fine-grained analysis of the family of languages sitting between regular, deterministic context-free, and deterministic context-sensitive languages (see Fig. 1, page 10). Therefore, from this viewpoint, translucency might be regarded as a new and promising tool to be applied in the fine analysis of the inner structure of language classes.

Next, we turn to study the closure of $\mathscr{L}$(DPDAwtl) and $\mathscr{L}$(nrDPDAwtl) under very well known language operations, namely: concatenation, Kleene star, length-preserving and inverse homomorphism, and reversal. We obtain that both language classes are not closed under such operations. This, together with results in [6,9,15], yields an almost complete account on the possibility of implementing main language operations on finite state and pushdown devices in presence of translucency (see Table 1, page 12). We feel it worth remarking that, among closure properties, here we also investigate the closure of $\mathscr{L}$(nrDPDAwtl) under intersection with regular languages, obtaining a negative answer. The arguments we use to get this negative result enable us to prove that even the class of languages accepted by non-returning deterministic finite state automata is not closed under intersection with regular languages. This solves an open problem posed by Mráz and Otto in [9].

The paper is organized as follows. Section 2 contains basic definitions and models formal statements. In Sect. 3, we show some preliminary results on the acceptance capabilities of translucent deterministic pushdown automata, useful in our following investigations. In Sect. 4, we compare the accepting power of

DPDAwtl, nrDPDAwtl, and that of corresponding translucent deterministic and nondeterministic finite state automata. Finally, in Sect. 5, we analyze closure properties for the language classes $\mathscr{L}$(DPDAwtl) and $\mathscr{L}$(nrDPDAwtl).

2   Preliminaries

We denote by Σ^* the set of all words on the finite alphabet Σ, including the empty word λ, and let $\Sigma^+ = \Sigma^* \setminus \{\lambda\}$. For any word $w \in \Sigma^*$, we let $|w|$ denote its length, w^R its reversal, and $|w|_a$ the number of occurences of the symbol $a \in \Sigma$ in w. We use $\subseteq$ for inclusions, and $\subset$ for proper inclusion. Given a set S, we denote by 2^S its power set, and by $|S|$ its cardinality. We write S_x to denote the set $S \cup \{x\}$, for a given element $x \notin S$. A language on Σ is any subset $L \subseteq \Sigma^*$. The complement of L is the language $\overline{L} = \Sigma^* \setminus L$, its reversal is $L^R = \{\, w^R \mid w \in L \,\}$. Two language families $\mathscr{L}_1$ and $\mathscr{L}_2$ are said to be incomparable whenever $\mathscr{L}_1$ is not a subset of $\mathscr{L}_2$ and vice versa.

Pushdown automata with translucent letters are extensions of classical pushdown automata that do not have to read their inputs from left to right. Instead, depending on the current state of such devices, some of the input letters may be translucent (invisible). Accordingly, a pushdown automaton with translucent letters either performs a λ-transition without reading an input symbol or reads and processes (by deleting, if not the endmarker) the first visible input letter from left. Here we are particularly interested in deterministic computations.

A *deterministic pushdown automaton with translucent letters* (DPDAwtl, [6]) formally writes as $M = \langle Q, \Sigma, \Gamma, q_0, \lhd, \bot, \tau, \delta \rangle$, where Q is the finite set of states, Σ is the finite set of input symbols, with $\Sigma \cap Q = \emptyset$, Γ is the finite set of pushdown symbols, $q_0 \in Q$ is the initial state, $\lhd \notin \Sigma$ is the endmarker, $\bot \notin \Gamma$ is the bottom-of-pushdown symbol, $\tau : Q \to 2^\Sigma$ is the translucency mapping, and $\delta : Q \times (\Sigma \cup \{\lambda, \lhd\}) \times \Gamma_\bot \to (Q \times \Gamma_\bot^*) \cup \{\texttt{accept}\}$ is the partial transition function. There must never be a choice between using an input symbol and using λ input. So, we require the following property to hold for any $q \in Q$ and $z \in \Gamma_\bot$: if $\delta(q, \lambda, z)$ is defined, then $\delta(q, a, z)$ is undefined for all a in $\Sigma_\lhd$. The translucency mapping τ bears the following meaning: for any state $q \in Q$, the letters from the set $\tau(q)$ are *translucent (invisible) for* q, that is, whenever in q, the automaton M does not see these letters (or equivalently, M can see through such letters).

To simplify matters, we require that at any time during computations the bottom-of-pushdown symbol appears exactly once at the bottom of the pushdown store, meaning that it cannot be deleted nor replicated. Formally, this amounts to require that if $\delta(q, a, z) = (p, \beta)$, then either $z \neq \bot$ and $\beta \in \Gamma^*$, or $z = \bot$ and $\beta = \beta' \bot$ with $\beta' \in \Gamma^*$.

A configuration of M is a pair $(qv \lhd, \gamma)$ or $\texttt{accept}$, where $q \in Q$ is the current state, $v \in \Sigma^*$ is the part of the input left to be processed (recall that, as above hinted, at each step which is neither a λ-transition nor a move on the endmarker, M deletes/processes the first visible input symbol; from this viewpoint, v is the still undeleted/unprocessed part of the input string), and $\gamma \in \Gamma^* \bot$ denotes the current pushdown content, the leftmost symbol being the top of the pushdown store. The initial configuration of M on input $w \in \Sigma^*$ has the form $(q_0 w \triangleleft, \perp)$. Along its computation, M runs through a sequence of configurations. Being in some configuration $(qv\triangleleft, z\gamma)$, a step of M takes place as follows. First, M checks whether $\delta(q, \lambda, z)$ is defined: if this is the case with $\delta(q, \lambda, z) = (p, \beta)$, then the successor configuration is $(pv\triangleleft, \beta\gamma)$. Otherwise, $\delta(q, \lambda, z)$ is undefined, and M determines the next input symbol to be processed by taking the first letter from left that is visible in state q. Precisely, if $v = xay$ with $x \in \tau(q)^*$ and $a \notin \tau(q)$, then M takes a. Now, M halts and rejects whenever $\delta(q, a, z)$ is undefined, otherwise it computes $\delta(q, a, z) = (p, \beta)$ and the successor configuration is $(pxy\triangleleft, \beta\gamma)$. In the case $v \in \tau(q)^*$, then M sees the endmarker $\triangleleft$, and it halts and accepts if and only if $\delta(q, \triangleleft, z) = \mathtt{accept}$, otherwise it rejects. One step from a configuration to its successor configuration is denoted by $\vdash$, which is specified as follows. Let $p, q \in Q$, $a \in \Sigma$, $v, x, y \in \Sigma^*$, $z \in \Gamma_\perp$ and $\beta\gamma, z\gamma \in \Gamma^*\perp$. We set:

1. $(qv\triangleleft, z\gamma) \vdash (pv\triangleleft, \beta\gamma)$, if $(p, \beta) = \delta(q, \lambda, z)$,
2. $(qxay\triangleleft, z\gamma) \vdash (pxy\triangleleft, \beta\gamma)$, if $x \in \tau(q)^*$, $a \notin \tau(q)$, and $(p, \beta) = \delta(q, a, z)$,
3. $(qv\triangleleft, z\gamma) \vdash (pv\triangleleft, \beta\gamma)$, if $v \in \tau(q)^*$ and $(p, \beta) = \delta(q, \triangleleft, z)$,
4. $(qv\triangleleft, z\gamma) \vdash \mathtt{accept}$, if $v \in \tau(q)^*$ and $\mathtt{accept} = \delta(q, \triangleleft, z)$.

We let $\vdash^*$ (resp., $\vdash^+$) denote the reflexive and transitive (resp., transitive) closure of $\vdash$. The language accepted by the DPDAwtl M is the set $L(M)$ of those words in Σ^* for which the computation, beginning in the initial configuration, eventually halts accepting, namely:

$$L(M) = \{\, w \in \Sigma^* \mid (q_0 w \triangleleft, \perp) \vdash^+ \mathtt{accept}\,\}.$$

In general, we denote by $\mathscr{L}(X)$ the family of all languages accepted by some device X. To clarify the behavior of DPDAwtl, we propose the following example.

Example 1. Let the language $L_{dc} \subseteq \{a, b, c, d\}^*$ be defined as $L_{dc} = \{\, uvdc^n \mid u \in \{a, b\}^*, v \in \{a, c\}^*, |uv|_a = |v|_c, n = |u|_b\,\}$. Such a language is presented in [13], where its acceptance is shown, by certain cooperating distributed systems of restarting automata equipped with additional pushdown store. As a matter of fact, such systems can be seen as forerunners of the DPDAwtl model we are proposing in the present paper.

We are now going to design a DPDAwtl A for L_{dc}. We will use this result later, in order to show that the family $\mathscr{L}(\mathsf{DPDAwtl})$ is not closed under inverse homomorphism. Intuitively, the behavior of A on an input string $uvdc^n \in L_{dc}$ consists of the following phases:

- PHASE 1: b symbols are processed, and for each b a symbol B is pushed; thus, at the end of this phase, the input string assumes the form xdc^n, for $x \in \{a, c\}^*$ and $|x| = |u|_a + |v|$, while the pushdown content is $B^{|u|_b}\perp$.
- PHASE 2: a symbols and c symbols sitting before the symbol d are processed, by alternating reading a c and then reading an a, until the prefix x is completely processed; at the end of this phase, the input string has the form dc^n, while the pushdown content remains untouched as $B^{|u|_b}\perp$.

- PHASE 3: once the d symbol has been processed, every c left in the input is matched against a B symbol in the pushdown, in order to check that the number of c is equal to the number of b previously occurring in u.

By suitably tuning translucency and fixing undefined moves, the input strings not belonging to L_{dc} can be rejected by A. Let us now formally implement the above three phases on the DPDAwtl $A = \langle Q, \{a, b, c, d\}, \{B\}, q_0, \triangleleft, \perp, \tau, \delta\rangle$. We let $Q = \{q_0, q_1, q_2, q_3\}$, the translucency $\tau(q_0) = \{a\} = \tau(q_2)$, $\tau(q_1) = \{c\}$, $\tau(q_3) = \emptyset$, and the transition function δ, for $z \in \{B, \perp\}$, as being:

PHASE 1 and PHASE 2	PHASE 3
$\delta(q_0, b, z) = (q_0, Bz)$	$\delta(q_0, d, z) = (q_3, z)$
$\delta(q_0, c, z) = (q_1, z)$	$\delta(q_2, d, z) = (q_3, z)$
$\delta(q_1, a, z) = (q_2, z)$	$\delta(q_3, c, B) = (q_3, \lambda)$
$\delta(q_2, c, z) = (q_1, z)$	$\delta(q_3, \triangleleft, \perp) = \text{accept}$

So, we can conclude that $L_{dc} \in \mathscr{L}(\text{DPDAwtl})$. ∎

Deterministic pushdown automata with translucent letters have been defined in [6] also for the *non-returning* mode. In this mode, after a visible letter is processed, we have that the input head does not return to the left end of the input, but it continues reading from the position of the visible letter just processed. Whenever the endmarker is reached and the transition on the endmarker yields a new state, the computation is continued in this state, with the input head placed at the left end of the remaining input. Such *deterministic pushdown automata with translucent letters working in the non-returning mode* (nrDPDAwtl) generalize non-returning finite state automata with translucent letters [9]. In their formal definition as an 8-tuple, the components are defined as previously stated for DPDAwtl, the difference showing up in the dynamics. Thus, given the nrDPDAwtl $M = \langle Q, \Sigma, \Gamma, q_0, \triangleleft, \perp, \tau, \delta\rangle$, a configuration of M is now a pair $(xqw\triangleleft, \gamma)$ or accept, where $q \in Q$ is the current state, $xw \in \Sigma^*$ is the remaining part of the input with w being to the right and x to the left of the input head, and $\gamma \in \Gamma^* \perp$ is the current pushdown content. The successor configuration yielded by $\vdash$ is now specified as follows. Let $p, q \in Q$, $a \in \Sigma$, $x, u, v, w \in \Sigma^*$, $z \in \Gamma_\perp$, and $\beta\gamma, z\gamma \in \Gamma^* \perp$. Then:

1. $(xqw\triangleleft, z\gamma) \vdash (xpw\triangleleft, \beta\gamma)$, if $(p, \beta) = \delta(q, \lambda, z)$,
2. $(xquav\triangleleft, z\gamma) \vdash (xupv\triangleleft, \beta\gamma)$, if $u \in \tau(q)^*$, $a \notin \tau(q)$, and $(p, \beta) = \delta(q, a, z)$,
3. $(xqw\triangleleft, z\gamma) \vdash (pxw\triangleleft, \beta\gamma)$, if $w \in \tau(q)^*$ and $(p, \beta) = \delta(q, \triangleleft, z)$,
4. $(xqw\triangleleft, z\gamma) \vdash \text{accept}$, if $w \in \tau(q)^*$ and $\text{accept} = \delta(q, \triangleleft, z)$.

The accepted language $L(M)$ can be easily defined. Sometimes, we will be saying that an nrDPDAwtl performs *sweeps*, where a sweep is a sequence of transitions that starts with the input head at the left end of the (remaining) input and ends after the next (if any) return move on the endmarker (move of type (3) above). Let us give an intuition on how a nrDPDAwtl works by the following example.

Example 2. Consider the non-context-free language $L = \{\, x\#y\#z \mid x, y, z \in \{a, c\}^*,\ |x|_a = |y|_a = |z|_c \geq 1 \,\}$. We will use this language in Proposition 16, to show the non-closure under inverse homomorphism of the family $\mathscr{L}(\text{nrDPDAwtl})$. A detailed construction of an nrDPDAwtl M for L is presented in [6]. In what follows, we quickly describe the dynamics of M.

In an initial sweep, M reads the first block x of the input string, while storing in the pushdown as many symbols as the number of a's. Any other symbol in x is simply read and erased. Then, M processes the second and the third block of the input string, deleting all c's in the second block y and all a's in the third block z. Finally, the right endmarker is read and a new sweep starts.

In an accepting computation, the remaining input should have the form $a^n c^n \triangleleft$, and the content of the pushdown store should be $a^n \bot$. At this point, M can check in n sweeps whether the number of a's and c's coincides with the number of a's in the pushdown store: for each a in the first block, M deletes one c in the second block, and pops one symbol of the pushdown. Finally, M accepts when it reads the endmarker with empty pushdown and no symbol left on the tape.   ∎

3   Basic Results on **DPDA** with Translucent Letters

Let us prepare some results on the form and computational power of DPDAwtl and nrDPDAwtl, that will turn out to be useful in the following sections. First, we show a technical lemma stating that, without loss of generality, DPDAwtl and nrDPDAwtl can always be assumed to either pop on λ-transitions only and completely consume accepted inputs. Such assumptions are particularly useful to simplify our proofs.

Lemma 3. *For a given DPDAwtl (resp., nrDPDAwtl), an equivalent DPDAwtl (resp., nrDPDAwtl) can effectively be constructed, in which both the following properties hold:*

(i) pop-moves are performed by λ-transitions only,
(ii) acceptance only takes place after the whole *input has been processed.*

Next, we test DPDAwtl and nrDPDAwtl recognition capabilities on particular language families, that will be used later as witness languages.

Lemma 4. *Let Σ be an alphabet, $c \in \Sigma$, and $\Sigma^* \supseteq L \in \mathscr{L}(DPDAwtl)$ be a language. Both languages $L \cap \{c\}(\Sigma \setminus \{c\})^*$ and $L \cap \{c\}(\Sigma \setminus \{c\})^+$ belong to $\mathscr{L}(DPDAwtl)$.*

Lemma 5. *Let M be an nrDPDAwtl accepting the language $\{c\}L$, where $L \subseteq \Sigma^*$ and $c \notin \Sigma$. Then, an nrDPDAwtl M' accepting the language L can effectively be constructed.*

For the next result, we consider the two languages $L_1 = \{\, b^n \# b^n \mid n \geq 0 \,\}$ and $L_2 = \{\, b^n \# b^{2n} \mid n \geq 0 \,\}$. It has been shown in [6] that the union language $L_1 \cup L_2$ does not belong to $\mathscr{L}(\text{nrDPDAwtl})$.

Lemma 6. *The language* $\{c\}(L_1 \cup L_2)$ *does not belong to* $\mathscr{L}(nrDPDAwtl)$.

Proof. Assume the language $\{c\}(L_1 \cup L_2)$ is accepted by some nrDPDAwtl M. Since $L_1 \cup L_2$ is defined over the alphabet $\{b, \#\}$, by applying Lemma 5 to M we get that $L_1 \cup L_2$ is accepted by some nrDPDAwtl M'. This is a contradiction, since it has been shown in [6] that $L_1 \cup L_2$ is not accepted by any nrDPDAwtl. □

4   Translucent **DPDA** vs. Translucent Finite Automata

Concerning the computational capacities of DPDAwtl, nrDPDAwtl, and that of classical DPDA, the following proper hierarchy is shown in [6], where DCFL denotes the family of deterministic context-free languages and DCSL the family of deterministic context-sensitive languages:

$$\mathsf{DCFL} = \mathscr{L}(\mathsf{DPDA}) \subset \mathscr{L}(\mathsf{DPDAwtl}) \subset \mathscr{L}(\mathsf{nrDPDAwtl}) \subset \mathsf{DCSL}.$$

The computational capacities of deterministic and nondeterministic finite automata with translucent input letters working in the returning and non-returning mode (DFAwtl, NFAwtl, nrDFAwtl, nrNFAwtl) are compared in [9]. In particular, all families of languages accepted by these devices properly include the family of regular languages. Moreover, the proper inclusions $\mathscr{L}(nrDFAwtl) \subset \mathscr{L}(nrNFAwtl)$, $\mathscr{L}(NFAwtl) \subset \mathscr{L}(nrNFAwtl)$, $\mathscr{L}(DFAwtl) \subset \mathscr{L}(NFAwtl)$, and $\mathscr{L}(DFAwtl) \subset \mathscr{L}(nrDFAwtl)$ are also shown. So, the resources *nondeterminism* and *non-returning mode* have an impact on the computational capacities. However, this impact is incomparable if the resources are added to DFAwtl. This raises the natural question on how the resource *pushdown store* compares to *nondeterminism* and *non-returning mode*. In the following, we are going to explore these relationships, enabling us to draw a complete picture depicted in Fig. 1, page 10.

Some of the comparisons among acceptance capabilities can be derived almost immediately from structural reasons and known results. For instance, comparing the acceptance with or without the resource *pushdown store*, yields the incomparability of $\mathscr{L}(\mathsf{DPDA})$ with all of the language families $\mathscr{L}(\mathsf{DFAwtl})$, $\mathscr{L}(\mathsf{NFAwtl})$, $\mathscr{L}(\mathsf{nrDFAwtl})$, and $\mathscr{L}(\mathsf{nrNFAwtl})$. The deterministic context-free language $\{ w\varphi(w^R) \mid w \in \{a, b\}^*, \varphi(a) = c, \varphi(b) = d \}$ is not accepted by any nrNFAwtl [9]. Also, it is not hard to see that the language $\{ b^n \# b^n \mid n \geq 0 \}$ is not accepted by any nrNFAwtl. On the other hand, the family $\mathscr{L}(\mathsf{DFAwtl})$ includes non-context-free languages [11]. Due to the inclusions above recalled, the incomparabilities follow. The same witness languages show that the trivial inclusions $\mathscr{L}(\mathsf{DFAwtl}) \subset \mathscr{L}(\mathsf{DPDAwtl})$ and $\mathscr{L}(\mathsf{nrDFAwtl}) \subset \mathscr{L}(\mathsf{nrDPDAwtl})$ are proper. So, the comparisons between $\mathscr{L}(\mathsf{nrDFAwtl})$ and $\mathscr{L}(\mathsf{DPDAwtl})$, that is, between the resources *non-returning mode* and *pushdown store*, as well as between $\mathscr{L}(\mathsf{nrNFAwtl})$ and $\mathscr{L}(\mathsf{nrDPDAwtl})$, that is, between the resources *non-returning mode and nondeterminism* and *non-returning mode and pushdown store*, are left to be considered in the following Proposition 7 and Proposition 9.

Proposition 7. *The families* $\mathcal{L}(DPDAwtl)$ *and* $\mathcal{L}(nrDFAwtl)$ *are incomparable.*

Proof. As shown in [9], an nrDFAwtl M exists, accepting the context-sensitive language $L = \{\, a^n b^n c^n \mid n \geq 0\,\}$. Very roughly speaking, on any given input string $x \in \{\, a, b, c\,\}^*$, M deletes in succession first a symbol a, then a symbol b, then a symbol c. By iterating this process, we have that M completely deletes x whenever $|x|_a = |x|_b = |x|_c$. In addition, the non-returning mode is easily seen to also ensure that $x \in a^* b^* c^*$.

On the other hand, it is also easy to see that L does not contain any letter-equivalent context-free sub-language. This fact yields that L cannot be accepted by any DPDAwtl, due to Proposition 13 proved in [6] and displayed below. Therefore, the incomparability follows from the incomparability between $\mathcal{L}(DPDA)$ and $\mathcal{L}(nrDFAwtl)$.   □

In preparation for the last incomparability result, we define the language

$$L_{tr} = \{\, w \in \#^+(a\#^+)^*(b\#^+)^*(c\#^+)^* \mid |w|_a = |w|_b = |w|_c \,\}.$$

The words of L_{tr} are essentially words of the form $a^n b^n c^n$ where, at the beginning and after each symbol from $\{a, b, c\}$, arbitrarily many but at least one symbol $\#$ occur. For the language L_{tr}, the following holds:

Theorem 8. *The language* L_{tr} *is not accepted by any nrDPDAwtl.*

By this result, we can obtain

Proposition 9. *The families* $\mathcal{L}(nrDPDAwtl)$ *and* $\mathcal{L}(nrNFAwtl)$ *are incomparable.*

Proof. As above quoted, the deterministic context-free languages $\{\, w\varphi(w^R) \mid w \in \{a, b\}^*,\ \varphi(a) = c,\ \varphi(b) = d \,\}$ and $\{\, b^n \# b^n \mid n \geq 0 \,\}$ are not accepted by any nrNFAwtl. For the converse, we consider the complement of the language L_{tr}. Such a complement consists of the union of the three languages $\{\, w \in \{a, b, c, \#\}^* \mid w \notin \#^+(a\#^+)^*(b\#^+)^*(c\#^+)^* \,\}$, $\{\, w \in \{a, b, c, \#\}^* \mid |w|_a \neq |w|_b \,\}$, and $\{\, w \in \{a, b, c, \#\}^* \mid |w|_a \neq |w|_c \,\}$. All these three languages are accepted by some NFAwtl. So, the complement of L_{tr} is accepted by some NFAwtl as well, since $\mathcal{L}(NFAwtl)$ is closed under union [9]. However, the complement of L_{tr} is not accepted by any nrDPDAwtl, since L_{tr} does not belong to $\mathcal{L}(nrDPDAwtl)$ by Theorem 8 and $\mathcal{L}(nrDPDAwtl)$ is closed under complementation [6].   □

We conclude this section by displaying in Fig. 1, next page, the complete situation of the relationships among language families defined by finite state and pushdown translucent devices, showing their positioning with respect to the well known classes REG and DCSL.

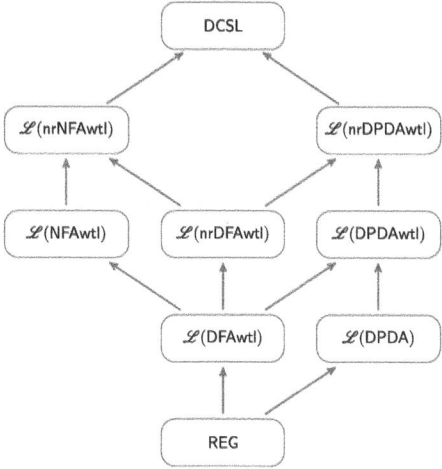

Fig. 1. Relationships between language families. An arrow between families indicates a strict inclusion. Any pair of families not connected by a path is incomparable.

5   Closure Properties

In this section, we investigate some closure properties for the language families $\mathscr{L}(\mathsf{DPDAwtl})$ and $\mathscr{L}(\mathsf{nrDPDAwtl})$, also solving an open problem on nrDFAwtl posed by Mráz and Otto in [9]. We begin by

Proposition 10. *The language families* $\mathscr{L}(\mathsf{DPDAwtl})$ *and* $\mathscr{L}(\mathsf{nrDPDAwtl})$ *are not closed under intersection with regular languages.*

Proof. We consider the following language, with $\sqcup\!\sqcup$ denoting the shuffle operator (see, e.g., [4]):

$$L_{tri} = \{\, w \in \{a, b, c\}^* \mid |w|_a = |w|_b = |w|_c \,\} \sqcup\!\sqcup \{\, \#^n \mid n \geq 0 \,\}.$$

Since $L = \{\, w \in \{a, b, c\}^* \mid |w|_a = |w|_b = |w|_c \,\}$ is accepted by a deterministic finite automaton with translucent letters [12], we have that L can be accepted by a DPDAwtl, and thus by an nrDPDAwtl as well. From these acceptors, modified acceptors of the same type for L_{tri} can be easily constructed. Basically, these new acceptors simulate the old ones, ignoring the symbols #.

Now, we define the regular language $R = \#^+(a\#^+)^*(b\#^+)^*(c\#^+)^*$ and consider the language $L_{tri} \cap R = L_{tr}$. By Theorem 8, the language L_{tr} is not accepted by any nrDPDAwtl, and so it cannot be accepted by any DPDAwtl.   □

We emphasize that, as above suggested, the language L_{tri} can be accepted by some DFAwtl, and so it is accepted by some nrDFAwtl as well. Therefore, the proof of Proposition 10 shows that the family $\mathscr{L}(\mathsf{nrDFAwtl})$ is not closed under intersection with regular languages as well. This solves a problem left open by Mráz and Otto in [9]:

Corollary 11. *The language family $\mathscr{L}(\text{nrDFAwtl})$ is not closed under intersection with regular languages.*

Let us now turn to non-closure under concatenation, Kleene star, and length-preserving homomorphism:

Proposition 12. *The language families $\mathscr{L}(\text{DPDAwtl})$ and $\mathscr{L}(\text{nrDPDAwtl})$ are neither closed under concatenation nor under Kleene star. They are not closed under length-preserving homomorphism, either.*

Proof. Consider the two languages $\{c\}L_1 \cup L_2$ (see Sect. 3 for L_1, L_2) and $\{c\}^*$, which are respectively a deterministic context-free and a regular language. Therefore, they can both be accepted by DPDAwtl. Assume $\mathscr{L}(\text{DPDAwtl})$ is closed under concatenation. Then, the language $(\{c\}^* \cdot (\{c\}L_1 \cup L_2)) \cap \{c\}\{b, \#\}^* = \{c\}(L_1 \cup L_2)$ belongs to $\mathscr{L}(\text{DPDAwtl})$ due to Lemma 4. However, Lemma 6 shows that $\{c\}(L_1 \cup L_2)$ is not accepted by any nrDPDAwtl, whence a contradiction.

For non-closure under Kleene star, consider the language $\{c\}^* \cup \{c\}L_1 \cup L_2$, which is accepted by a DPDAwtl. Assume $\mathscr{L}(\text{DPDAwtl})$ is closed under Kleene star. Then, the language $(\{c\}^* \cup \{c\}L_1 \cup L_2)^* \cap \{c\}\{b, \#\}^+ = \{c\}(L_1 \cup L_2)$ belongs to $\mathscr{L}(\text{DPDAwtl})$ due to Lemma 4, which again contradicts Lemma 6.

For non-closure under length-preserving homomorphism, consider the language $\{c\}L_1 \cup \{d\}L_2$, accepted by a DPDAwtl. By applying to this language the length-preserving homomorphism h such that $h(b) = b$, $h(\#) = \#$, $h(c) = c$, and $h(d) = c$, we obtain as homomorphic image the language $\{c\}(L_1 \cup L_2)$. Thus, assuming $\mathscr{L}(\text{DPDAwtl})$ is closed under length-preserving homomorphism would again contradict Lemma 6.

According to [6], every DPDAwtl can be turned into an equivalent nrDPDAwtl. Hence, we obtain the non-closure results also for $\mathscr{L}(\text{nrDPDAwtl})$. $\square$

For non-closure of $\mathscr{L}(\text{DPDAwtl})$ under inverse homomorphism, we need to recall the following property from [11]:

Proposition 13. *Let M be a DPDAwtl. Then a DPDA M' can effectively be constructed such that $L(M') \subseteq L(M)$ and $L(M')$ is letter-equivalent to $L(M)$.*

With this result in our hands, we are able to show

Proposition 14. *The language family $\mathscr{L}(\text{DPDAwtl})$ is not closed under inverse homomorphism.*

Proof. Consider the language $L_{dc} = \{\, uvdc^n \mid u \in \{a, b\}^*, v \in \{a, c\}^*, |uv|_a = |v|_c, n = |u|_b \,\}$ from Example 1, where it is shown that $L_{dc} \in \mathscr{L}(\text{DPDAwtl})$. On the other hand, from [13], the inverse homomorphic image of L_{dc} by the homomorphism h such that $h(a) = ab$, $h(c) = c$, and $h(d) = d$ is the language $h^{-1}(L_{dc}) = \{\, a^n c^n dc^n \mid n \geq 0 \,\}$. It is not hard to see that this language does not contain any letter-equivalent context-free sub-language. Therefore, by Proposition 13, we can conclude that $h^{-1}(L_{dc}) \notin \mathscr{L}(\text{DPDAwtl})$. $\square$

For showing non-closure of $\mathscr{L}(\text{nrDPDAwtl})$ under inverse homomorphism, we introduce the language $L_{tb} = \{\, b^n \# b^n \# b^n \mid n \geq 1 \,\}$. For this language, we have

Lemma 15. *The language L_{tb} is not accepted by any* nrDPDAwtl.

By this impossibility result, L_{tb} qualifies as a witness language to show

Proposition 16. *The language family $\mathscr{L}(\textit{nrDPDAwtl})$ is not closed under non-erasing inverse homomorphisms.*

Proof. By Example 2, the language $L = \{\, x \# y \# z \mid x, y, z \in \{a, c\}^*, |x|_a = |y|_a = |z|_c > 1 \,\}$ can be accepted by a nrDPDAwtl. We let the non-erasing homomorphism $h\colon \{b, \#\}^* \to \{a, c, \#\}^*$ be defined as $h(b) = ac$ and $h(\#) = \#$. Then, $h^{-1}(L) = L_{tb}$. Since, by Lemma 15, $L_{tb} \notin \mathscr{L}(\text{nrDPDAwtl})$, the non-closure of $\mathscr{L}(\text{nrDPDAwtl})$ under non-erasing inverse homomorphisms follows.   □

Finally, the reversal operation is considered in the following

Proposition 17. *The language families $\mathscr{L}(\textit{DPDAwtl})$ and $\mathscr{L}(\textit{nrDPDAwtl})$ are not closed under reversal.*

In conclusion, for the sake of completeness and reader's ease of mind, we summarize in the following table the complete scenario of closure properties for the language families $\mathscr{L}(\text{DPDAwtl})$ and $\mathscr{L}(\text{nrDPDAwtl})$.

Table 1. A summary of closure properties for the language families $\mathscr{L}(\text{DPDAwtl})$ and $\mathscr{L}(\text{nrDPDAwtl})$. Shaded properties are proved in [6], and can be found in [4] for DCFL (deterministic context-free languages). Bold properties are shown in the present paper.

Language Family	—	∪	∩	∩$_{\text{reg}}$	·	*	$h_{\text{len.pres.}}$	h^{-1}	R
$\mathscr{L}(\text{DPDAwtl})$	✓	✗	✗	✗	✗	✗	✗	✗	✗
$\mathscr{L}(\text{nrDPDAwtl})$	✓	✗	✗	**✗**	✗	✗	✗	**✗**	**✗**
DCFL	✓	✗	✗	✓	✗	✗	✗	✓	✗

Acknowledgements. The authors wish to thank the anonymous referees for their valuable and helpful comments.

References

1. Beier, S., Holzer, M.: Nondeterministic right one-way jumping finite automata. Inform. Comput. **284**, 104687 (2022). https://doi.org/10.1016/J.IC.2021.104687
2. Bensch, S., Bordihn, H., Holzer, M., Kutrib, M.: On input-revolving deterministic and nondeterministic finite automata. Inform. Comput. **207**, 1140–1155 (2009). https://doi.org/10.1016/j.ic.2009.03.002

3. Chigahara, H., Fazekas, S.Z., Yamamura, A.: One-way jumping finite automata. Int. J. Found. Comput. Sci. **27**, 391–405 (2016). https://doi.org/10.1142/S0129054116400165

4. Hopcroft, J.E., Ullman, J.D.: Introduction to Automata Theory, Languages, and Computation. Addison-Wesley, Reading (1979)

5. Jančar, P., Mráz, F., Plátek, M., Vogel, J.: Restarting automata. In: Reichel, H. (ed.) FCT 1995. LNCS, vol. 965, pp. 283–292. Springer, Heidelberg (1995). https://doi.org/10.1007/3-540-60249-6_60

6. Kutrib, M., Malcher, A., Mereghetti, C., Palano, B., Raucci, P., Wendlandt, M.: Deterministic pushdown automata with translucent input letters. In: Day, J., Manea, F. (eds.) DLT 2024. LNCS, vol. 14791, pp. 203–217. Springer, Cham (2024). https://doi.org/10.1007/978-3-031-66159-4_15

7. Meduna, A., Zemek, P.: Jumping finite automata. Int. J. Found. Comput. Sci. **23**, 1555–1578 (2012). https://doi.org/10.1142/S0129054112500244

8. Mitrana, V., Păun, A., Păun, M., Couso, J.R.S.: Jump complexity of finite automata with translucent letters. Theoret. Comput. Sci. **992**, 114450 (2024). https://doi.org/10.1016/j.tcs.2024.114450

9. Mráz, F., Otto, F.: Non-returning deterministic and nondeterministic finite automata with translucent letters. RAIRO Theor. Inform. Appl. **57**, 8 (2023). https://doi.org/10.1051/ita/2023009

10. Nagy, B., Otto, F.: CD-systems of stateless deterministic R(1)-automata governed by an external pushdown store. RAIRO Theor. Inform. Appl. **45**, 413–448 (2011). https://doi.org/10.1051/ITA/2011123

11. Nagy, B., Otto, F.: Finite-state acceptors with translucent letters. In: Bel-Enguix, G., Dahl, V., De La Puente, A. (eds.) International Workshop on AI Methods for Interdisciplinary Research in Language and Biology (BILC 2011), pp. 3–13. SciTePress (2011). https://doi.org/10.5220/0003272500030013

12. Nagy, B., Otto, F.: On CD-systems of stateless deterministic R-automata with window size one. J. Comput. Syst. Sci. **78**, 780–806 (2012). https://doi.org/10.1016/J.JCSS.2011.12.009

13. Nagy, B., Otto, F.: Deterministic pushdown-CD-systems of stateless deterministic R(1)-automata. Acta Inform. **50**, 229–255 (2013). https://doi.org/10.1007/S00236-012-0175-X

14. Otto, F.: Restarting automata and their relations to the Chomsky hierarchy. In: Ésik, Z., Fülöp, Z. (eds.) DLT 2003. LNCS, vol. 2710, pp. 55–74. Springer, Heidelberg (2003). https://doi.org/10.1007/3-540-45007-6_5

15. Otto, F.: A survey on automata with translucent letters. In: Nagy, B. (ed.) CIAA 2023. LNCS, vol. 14151, pp. 21–50. Springer, Cham (2023). https://doi.org/10.1007/978-3-031-40247-0_2

16. Savitch, W.J., Vitányi, P.M.B.: Linear time simulation of multihead turing machines with head-to-head jumps. In: Salomaa, A., Steinby, M. (eds.) ICALP 1977. LNCS, vol. 52, pp. 453–464. Springer, Heidelberg (1977). https://doi.org/10.1007/3-540-08342-1_35

Attributed Tree Transducers for Partial Functions

Sebastian Maneth and Martin Vu[(✉)]

Universität Bremen, Bremen, Germany
{maneth,martin.vu}@uni-bremen.de

Abstract. Attributed tree transducers (atts) have been equipped with regular look-around (i.e., a preprocessing via an attributed relabeling) in order to obtain a more robust class of translations. Here we give further evidence of this robustness: we show that if the class of translations realized by nondeterministic atts with regular look-around is restricted to partial functions, then we obtain exactly the class of translations realized by deterministic atts with regular look-around.

Attributed tree transducers (atts) [9] are a well known formalism for defining tree translations. They are attractive, because they are strictly more expressive than top-down tree transducers and they closely model the behavior of attribute grammars [14,15]. Since attribute grammars are deterministic devices, atts also have typically been studied in their deterministic version.

But atts also have some deficiencies. For instance, they do not generalize the (deterministic) bottom-up tree translations [11]. One possibility to remedy this deficiency is to equip atts with *regular look-around* [2]. The resulting class of translations is more robust in the sense that (i) it does generalized deterministic bottom-up tree translations [4], (ii) it is equivalent to tree-to-dag-to-tree translations that are definable in MSO logic [2], and (iii) it is characterized by natural restrictions of macro tree transducers [13].

In this paper we present another advantage of adding regular look-around: we show that every nondeterministic att (with or without regular look-around) that is *functional* can be realized by a *deterministic* att with regular look-around (and such a transducer can be constructed). In general it is a desirable and convenient property of a nondeterministic translation device that its restriction to functional translations coincides precisely with the translations realized by the corresponding deterministic version of that device. Let us consider some classical examples of translation devices for which this property holds: (1) two-way strings transducers ("2GSM") [6, Theorem 22][18, Theorem 3], (2) top-down tree transducers with look-ahead [5, Theorem 1], and (3) macro tree transducers [7, Corollary 36]. In contrast, these are examples for which the property does *not* hold: (i) one-way string transducers ("GSM"), (ii) top-down and bottom-up tree transducers, and (iii) attributed tree transducers.

To see that functional top-down tree transducers and functional GSMs are strictly more expressive than their deterministic counterparts, consider the following translation: Let monadic input trees of the form $a^n(e)$ be translated to

S. Z. Fazekas (Ed.): CIAA 2024, LNCS 15015, pp. 221–233, 2024.
https://doi.org/10.1007/978-3-031-71112-1_16

$a^n(e)$ while input trees of the form $a^n(f)$ are translated to the single leaf f. A deterministic top-down tree transducer (or a deterministic GSM) cannot realize this translation, because it has to decide whether or not to output a-nodes, before it sees the label of the input leaf (viz. the right end of the string).

Given a relation R, a *uniformizer of R* is a function $f_R \subseteq R$ such that f_R and R have the same domain. Engelfriet [5] showed that for any nondeterministic top-down tree translation R, a uniformizer of R can be constructed as a deterministic top-down tree transducer with look-ahead (dt^R) . Since the latter type of transducer is closed under composition [4], it follows that for any composition of top-down (and bottom-up) tree transductions that is functional, an equivalent dt^R can be constructed. Whenever several rules of the given transducer are applicable, the uniformizer chooses the first one of them (in some order); the regular look-ahead is used to determine which rules are applicable. For an overview over different uniformization results in automata theory see [3].

Now consider a functional attributed tree transducer (att). Its input trees are over the binary labels f and g and the leaf label e and its output trees are over the unary label d and the leaf label e. The att outputs $d^n(e)$ if there exists a first g-node in reverse pre-order and n is the size of the subree rooted at that node (no output is produced, if no such g-node exists). This translation cannot be realized by any deterministic att, even if it uses regular look-ahead. However, there is a deterministic att with *regular look-around* that realizes the translation.

Subsequently, we show that any (possibly circular) nondeterministic att A effectively has a uniformizer that is realized by a deterministic att with look-around. To construct such a deterministic att D with regular look-around, we first construct a nondeterministic (and in general non-functional) top-down relabeling T and a deterministic att D' which operate as follows: Informally, the top-down relabeling T specifies which rules the deterministic att D' should apply. More precisely, given an input tree s, T annotates all nodes of s by rules of A. At every node, the deterministic att D' applies exactly those rules with which the nodes are annotated. In particular, T annotates the nodes of s such that D' simulates a *uniform* translation of A, that is, a translation in which if multiple instance of an attribute α access the same node v, then all these instances of α need to apply the same rule. With the result of [5], a look-around U can be constructed that realizes a uniformizer of the translation of T. With the deterministic att D', this look-around U yields D.

The existence of a uniformizer realized by a deterministic att with regular look-around for any att implies that a composition of n atts (with or without look-around) that is functional can be simulated by a composition of n deterministic atts with look-around. This also implies that for any functional att there is effectively an equivalent deterministic att with look-around.

1   Preliminaries

Denote by $\mathbb{N}$ the set of natural numbers. For $k \in \mathbb{N}$, we denote by $[k]$ the set $\{1, \ldots, k\}$. A set Σ is *ranked* if each symbol of Σ is associated with a *rank*, that is, a non-negative integer. We write σ^k to denote that the symbol σ has rank k. By Σ_k we denote the set of all symbols of Σ which have rank k. For $k' \neq k$, we define that $\Sigma_{k'}$ and Σ_k are disjoint. If the set Σ is finite then we call Σ a *ranked alphabet*.

The set T_Σ of *trees over* Σ is defined as the smallest set of strings such that if $\sigma \in \Sigma_k$, $k \geq 0$, and $t_1, \ldots, t_k \in T_\Sigma$ then $\sigma(t_1, \ldots, t_k)$ is in T_Σ. For $k = 0$, we simply write σ instead of $\sigma()$. The nodes of a tree $t \in T_\Sigma$ are referred to by strings over $\mathbb{N}$. In particular, for $t = \sigma(t_1, \ldots, t_k)$, we define $V(t)$, the set of nodes of t, as $V(t) = \{\epsilon\} \cup \{iu \mid i \in [k] \text{ and } u \in V(t_i)\}$, where ϵ is the *empty string*. For better readability, we add dots between numbers, e.g. for the tree $t = f(a, f(a, b))$ we have $V(t) = \{\epsilon, 1, 2, 2.1, 2.2\}$. For a node $v \in V(t)$, $t[v]$ denotes the label of v, t/v is the subtree of t rooted at v, and $t[v \leftarrow t']$ is obtained from t by replacing t/v by t'. For instance, we have $t[1] = a$, $t/2 = f(a, b)$ and $t[1 \leftarrow b] = f(b, f(a, b))$ for $t = f(a, f(a, b))$. The *size* of a tree t is given by $\text{size}(t) = |V(t)|$. We call the node v a (proper) ancestor of a node $v' \in V(t)$ if v is a (proper) prefix of v'. For a set Λ disjoint with Σ, we define $T_\Sigma[\Lambda]$ as $T_{\Sigma'}$ where $\Sigma'_0 = \Sigma_0 \cup \Lambda$ and $\Sigma'_k = \Sigma_k$ for $k > 0$.

Let $R \subseteq A \times B$ be a relation. We call R a *function* if $(a, b), (a, b') \in R$ implies $b = b'$. We define the *domain of R* by $\text{dom}(R) = \{a \in A \mid \exists b \in B : (a, b) \in R\}$. Analogously, the *range of R* is $\text{range}(R) = \{b \in B \mid \exists a \in A : (a, b) \in R\}$. A function $F \subseteq R$ is called a *uniformizer* of R if $\text{dom}(F) = \text{dom}(R)$. Let $R' \subseteq B \times C$. The *composition* of R and R' is $R \circ R' = \{(a, c) \mid (a, b) \in R, (b, c) \in R'\}$.

2 Attributed Tree Transducers

In the following, we define attributed tree transducers. For an in-depth introduction to attributed tree transducers, we refer to [12].

A *(partial nondeterministic) attributed tree transducer* (or *att* for short) is a tuple $A = (S, I, \Sigma, \Delta, a_0, R)$, where S and I are disjoint finite sets of *synthesized attributes* and *inherited attributes*, respectively, Σ and Δ are ranked alphabets of *input* and *output symbols*, respectively, $a_0 \in S$ is the *initial attribute* and $R = (R_\sigma \mid \sigma \in \Sigma \cup \{\#\})$ is a collection of finite sets of rules. We implicitly assume *atts* to include a unique symbol $\# \notin \Sigma$ of rank 1, the so-called *root marker*, that only occurs at the root of trees.

In the following, we define the rules of an *att*. Let $\sigma \in \Sigma$ be of rank $k \geq 0$. Furthermore, let π be a variable for nodes. Then the set R_σ contains

- arbitrarily many rules of the form $a(\pi) \to \xi$ for every $a \in S$ and
- arbitrarily many rules of the form $b(\pi i) \to \xi'$ for every $b \in I$ and $i \in [k]$,

where $\xi, \xi' \in T_\Delta[\{a'(\pi i) \mid a' \in S, i \in [k]\} \cup \{b'(\pi) \mid b' \in I\}]$. We define the set $R_\#$ analogously with the restriction that $R_\#$ contains *no* rules with synthesized attributes on the left-hand side. Replacing 'arbitrarily many rules' by 'at most one rule' in the definition of the rule sets of R, we obtain the notion of a *(partial) deterministic att* (or *datt* for short). For the *att* A and the attribute $a \in S$, we denote by $\text{RHS}_A(\sigma, a(\pi))$ the set of all right-hand sides of rules in R_σ that are of the form $a(\pi) \to \xi$. For $b \in I$, the sets $\mathrm{RHS}_A(\sigma, b(\pi i))$ with $i \in [k]$ and $\mathrm{RHS}_A(\#, b(\pi 1))$ are defined analogously.

We now define the semantics of A. For a given tree $s \in T_{\Sigma \cup \{\#\}}$, we define $\mathrm{SI}(s) = \{\alpha(v) \mid \alpha \in S \cup I, v \in V(s)\}$. Furthermore, we define that for the node variable π, $\pi 0 = \pi$ and that for a node v, $v.0 = v$. Let $t, t' \in T_\Delta[\mathrm{SI}(s)]$. We write $t \Rightarrow_{A,s} t'$ if t' is obtained from t by substituting a leaf of t labeled by $\gamma(v.i)$ with $i = 0$ if $\gamma \in S$ and $i > 0$ if $\gamma \in I$ by $\xi[\pi \leftarrow v]$, where $\xi \in \mathrm{RHS}_A(s[v], \gamma(\pi i))$ and $[\pi \leftarrow v]$ denotes the substitution that replaces all occurrences of π by the node v. For instance, for $\xi_1 = f(b(\pi))$ and $\xi_2 = f(a(\pi 2))$ where f is a symbol of rank 1, $a \in S$ and $b \in I$, we have $\xi_1[\pi \leftarrow v] = f(b(v))$ and $\xi_2[\pi \leftarrow v] = f(a(v.2))$. Denote by $\Rightarrow_{A,s}^+$ and $\Rightarrow_{A,s}^*$ the transitive closure and the reflexive-transitive closure of $\Rightarrow_{A,s}$, respectively. The *translation realized by* A, denoted by τ_A, is the relation $\{(s,t) \in T_\Sigma \times T_\Delta \mid a_0(1) \Rightarrow_{A,s^\#}^* t\}$, where subsequently $s^\#$ denotes the tree $\#(s)$. If τ_A is a partial function then we say that A is a *functional att*. We define $\mathrm{dom}(A) = \mathrm{dom}(\tau_A)$ and $\mathrm{range}(A) = \mathrm{range}(\tau_A)$.

The reader may wonder why the definition of A and its translations involves the root marker. Informally, the root marker is a technical requirement without which many translations are not possible. To see which role the root marker plays in a translation, consider the following translation which cannot be realized without the root marker.

Example 1. Consider the *att* $A = (S, I, \Sigma, \Delta, a, R)$ where $\Sigma = \{f^2, e^0\}$ and $\Delta = \{d^1, e^0\}$. We define $S = \{a\}$ and $I = \{b\}$. For the symbol f, we define

$$R_f = \{\underbrace{a(\pi) \to d(a(\pi 2))}_{\rho_1}, \; \underbrace{b(\pi 2) \to a(\pi 1)}_{\rho_2}, \; \underbrace{b(\pi 1) \to b(\pi)}_{\rho_3} \}.$$

Finally, the rule set for the symbol e and the root marker are given by

$$R_e = \{\underbrace{a(\pi) \to d(b(\pi))}_{\rho_4}\} \quad \text{and} \quad R_\# = \{\underbrace{b(\pi 1) \to e}_{\rho_5}\},$$

respectively. The tree transformation realized by A is defined as follows: On input $s \in T_\Sigma$, where s is of size n, A outputs the tree $d^n(e)$. We denote by $d^n(e)$ the tree over Δ with exactly n occurrences of d, e.g., $d^4(e)$ denotes the tree $d(d(d(d(e))))$. For instance, for $s = f(e, e)$, the *att* A outputs the tree $d^3(e)$. The corresponding translation is shown in Fig. 1.

Note that by definition *atts* are allowed to be *circular*. We say that an *att* A is *circular* if $s \in T_\Sigma$, $\alpha(v) \in \mathrm{SI}(s^\#)$ and $t \in T_\Delta[\mathrm{SI}(s^\#)]$ exists such that $\alpha(v) \Rightarrow_{A,s^\#}^+ t$ and $\alpha(v)$ occurs in t.

Look-Ahead. To define *attributed tree transducer with look-ahead*, we first define *bottom-up relabelings*. A *bottom-up relabeling* B is a tuple $(P, \Sigma, \Sigma', F, R)$ where P is the set of states, Σ and Σ' are finite ranked alphabets and $F \subseteq P$ is the set of final states. For $\sigma \in \Sigma$ and $p_1, \ldots, p_k \in P$, the set R contains at most one rule of the form $\sigma(p_1(x_1), \ldots, p_k(x_k)) \to p(\sigma'(x_1, \ldots, x_k))$ where $p \in P$ and $\sigma' \in \Sigma'$. The rules of B induce a derivation relation $\Rightarrow_B^*$ defined inductively as follows:

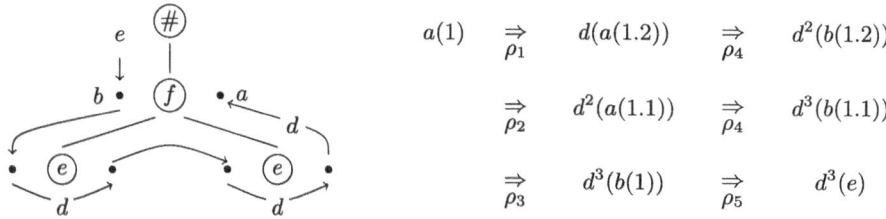

$$
\begin{array}{lllll}
a(1) & \underset{\rho_1}{\Rightarrow} & d(a(1.2)) & \underset{\rho_4}{\Rightarrow} & d^2(b(1.2)) \\[2mm]
& \underset{\rho_2}{\Rightarrow} & d^2(a(1.1)) & \underset{\rho_4}{\Rightarrow} & d^3(b(1.1)) \\[2mm]
& \underset{\rho_3}{\Rightarrow} & d^3(b(1)) & \underset{\rho_5}{\Rightarrow} & d^3(e)
\end{array}
$$

Fig. 1. The translation from $f(e,e)$ to $d^3(e)$ defined by A is pictured on the left. The corresponding transitions are displayed on the right. Each $\Rightarrow$ is annotated with the rule used in the corresponding transition step.

- Let $\sigma \in \Sigma_0$ and $\sigma \to p(\sigma')$ be a rule in R. Then $\sigma \Rightarrow^*_B p(\sigma')$.
- Let $s = \sigma(s_1, \ldots, s_k)$ with $\sigma \in \Sigma_k$, $k > 0$, and $s_1, \ldots, s_k \in T_\Sigma$. For $i \in [k]$, let $s_i \Rightarrow^*_B p_i(s'_i)$. Furthermore, let $\sigma(p_1(x_1), \ldots, p_k(x_k)) \to p(\sigma'(x_1, \ldots, x_k))$ be a rule in R. Then $s \Rightarrow^*_B p(\sigma'(s'_1, \ldots, s'_k))$.

For $s \in T_\Sigma$ and $p \in P$, we write $s \in \mathrm{dom}_B(p)$ if $s \Rightarrow^*_B p(s')$ for some $s' \in T_{\Sigma'}$. The translation realized by B is $\tau_B = \{(s, s') \in T_\Sigma \times T_{\Sigma'} \mid s \Rightarrow^*_B p(s') \text{ with } p \in F\}$.

We define an *attributed tree transducer with look-ahead (or att^R)* as a pair $\hat{A} = (B, A')$ where B is a bottom-up relabeling and A' is an *att*. Recall that translations are relations. The translation realized by $\hat{A}$ is the composition of the translations realized by B and A', that is,

$$
\tau_{\hat{A}} = \{(s, t) \in T_\Sigma \times T_\Delta \mid (s, s') \in \tau_B(s) \text{ and } (s', t) \in \tau_{A'}\}.
$$

Informally, B preprocesses input trees for A' with each node v, that is relabeled by B, providing A' information about the subtree rooted v. We say that $\hat{A}$ is a deterministic att^R ($datt^R$) if A is a *datt*. The domain and the range of $\hat{A}$ are defined in the obvious way.

Look-Around. *Look-around* is in concept similar to look-ahead; it is also a relabeling device that provides additional information to an *att*. However, it is far more expressive than a look-ahead[1]. To define look-around, we first define *top-down relabelings*. A top-down relabeling is a deterministic *att* without inherited attributes $T = (S, \emptyset, \Sigma, \Sigma', a_0, R)$ where for all $\sigma \in \Sigma_k$, $k \geq 0$, the set R_σ only contains rules of the form $q(\pi) \to \sigma'(q_1(\pi 1), \ldots, q_k(\pi k))$ with $\sigma' \in \Sigma'_k$. Note that for a top-down relabeling, we will henceforth write $q(\sigma(x_1, \ldots, x_k)) \to \sigma'(q_1(x_1), \ldots, q_k(x_k))$ instead $q(\pi) \to \sigma'(q_1(\pi 1), \ldots, q_k(\pi k)) \in R_\sigma$. We will also call the attributes of top-down relabeling states. Since a top-down relabeling is an *att*, a *top-down relabeling with look-ahead* is defined in the obvious way.

An *attributed tree transducer with look-around* (or att^U) is a tuple $\check{A} = (U, A')$ where A' is an *att* and U is a top-down relabeling with look-ahead. The translation realized by $\check{A}$ as well as the domain and the range of $\check{A}$ are defined analogously as for att^R. This means that an att^U relabels its input tree in two phases: First the input tree is relabeled in a bottom-up phase. Then the resulting tree

---

[1] The concept of look-around is introduced in [2] where it is called look-ahead.

is relabeled again in a top-down phase before it is processed by A'. We say that $\check{A}$ is a deterministic att^U ($datt^U$) if A is a $datt$. The following results hold.

Proposition 1 *[17, Lemma 27]. Let A be an att. Then an equivalent att $A' = (S, I, \Sigma, \Delta, a_0, R)$ can be constructed such that $R_\#$ contains no distinct rules with the same left-hand side.*

Proposition 2 *[7, Corollary 14]. The domain of an att^U is effectively regular.*

3   Functional Compositions of Atts^U are Determinizable

In the remainder of this paper, we prove the following statement: For any composition of n $atts^U$ that is functional an equivalent composition of n $datts^U$ can be constructed. The overall idea is similar to the one in [5] for top-down tree transducers. More precisely, let C be the composition of the $atts^U$ $A_1, \ldots, A_n$. Subsequently, we show that (a) for $i \in [n]$, a $datt^U$ D_i can be constructed such that D_i realizes a uniformizer of A_i and (b) the composition of $D_1, \ldots, D_n$ is equivalent to C. Obviously our statement implies that for any functional att^U there is an equivalent $datt^U$.

Before we prove our statement, we remark that in the absence of look-around our statement does not hold, i.e., a functional composition of n $atts$ without look-around cannot necessarily be simulated by a composition of n $datts$ without-look-around. In particular, we show that there are functional $atts$ which cannot be simulated by $datts$ without look-around.

3.1   On the Necessity of Look-Around

Subsequently, we show that there are tree translations realizable by functional $atts$ that cannot be realized by any $datt$ even if the $datt$ uses look-ahead.

Example 2. Let $\Sigma = \{f^2, g^2, e^0\}$ and let $\Delta = \{d^1, e^0\}$. Consider the following tree translation. Let $s \in T_\Sigma$ and let v be the first node in reverse pre-order that is labeled by g in s. Note that this means that no proper ancestor of v is labeled by g in s. If such a node v does not exist, i.e., if no input node is labeled by g, then no output is produced. Otherwise, we translate s into $d^m(e)$, where m denotes the *size* of s/v. Recall that $d^m(e)$ denotes the tree over Δ with exactly m occurrences of d, e.g., $d^3(e)$ denotes the tree $d(d(d(e)))$. For instance the tree $f(g(f(e,e),e), f(e, g(e, g(e,e))))$ is translated into $d^5(e)$. Clearly, this tree translation is functional and can be realized by the following functional att.

Let $A = (S, I, \Sigma, \Delta, a, R)$ where $S = \{a, a_g\}$ and $I = \{b, b_g, b'_g\}$. We define $R_\# = \{b'_g(\pi 1) \rightarrow e\}$ and $R_e = \{a(\pi) \rightarrow b(\pi), \; a_g(\pi) \rightarrow d(b_g(\pi))\}$. Let

$$R_f = \{ \begin{array}{llll} a(\pi) & \rightarrow a(\pi 2) & b(\pi 2) \rightarrow a(\pi 1) & b(\pi 1) \rightarrow b(\pi) \\ a_g(\pi) & \rightarrow d(a_g(\pi 2)) & b_g(\pi 2) \rightarrow a_g(\pi 1) & b_g(\pi 1) \rightarrow b_g(\pi) \\ b'_g(\pi 2) & \rightarrow b'_g(\pi) & b'_g(\pi 1) \rightarrow b'_g(\pi) \;\}. \end{array}$$

Furthermore, we define

$$R_g = \{ \ a(\pi) \quad \rightarrow d(a_g(\pi 2)) \quad a_g(\pi) \quad \rightarrow d(a_g(\pi 2)) \quad b_g(\pi 2) \rightarrow a_g(\pi 1)$$
$$b_g(\pi 1) \rightarrow b_g(\pi) \qquad b_g(\pi 1) \rightarrow b'_g(\pi) \ \}.$$

In short, the attributes a and b are used for traversing the input tree in reverse pre-order until a node labeled by g is encountered. Attributes with subscript g produce the output tree if the symbol g is encountered during the traversal. Consider in particular, the rules $b_g(\pi 1) \rightarrow b_g(\pi)$ and $b_g(\pi 1) \rightarrow b'_g(\pi)$ in R_g. With these rules, A guesses whether the node labeled by g that it is currently processing has a proper ancestor labeled by g or not. In particular, applying the rule $b_g(\pi 1) \rightarrow b_g(\pi)$ means that A guesses that such a proper ancestor exists, while applying $b_g(\pi 1) \rightarrow b'_g(\pi)$ means that A guesses the opposite.   □

Though the tree translation in Example 2 can by realized by a functional *att*, no *datt* D can realize it. In the following, we will explain (without a formal proof) why this is the case. The key point is that D is unable to determine for an input tree s and a given node v whether or not a proper ancestor v' of v exists that is also labeled by g *and* in the case that such a v' exists, continue to output symbols afterwards. To specify this statement consider the following. Obviously, D visits every node for which it has to produce a symbol d. W.l.o.g. it can be assumed that D visits all such nodes in reverse pre-order. Assume that D has already produced an output symbol d for every descendant of the node v and that an attribute of D is currently processing v. Now D must determine whether or not any more symbols d need to be produced. In other words, D must determine whether or not an ancestor of v is labeled by g or not. To do so D traverses s from v 'upwards' using rules of the form $\beta(\pi j) \rightarrow \beta'(\pi)$ where β and β' are some inherited attributes. Assume that this upwards traversal yields that a proper ancestor v' of v is labeled by g. To realize the tree translation in Example 2, A must produce a symbol d for every descendant of v'. Now, the question is has D done so for every descendant of v'? If not then D needs to determine for which descendants it still has to produce output. To do so D needs to return to the node v. However D cannot memorize the node v from which it started and thus does not know how to return to v.

Denote by $\mathcal{F}$ the class all functions. Furthermore, denote by ATT^R and $dATT^R$ the classes of tree translations realizable by nondeterministic *atts*^{R} and deterministic *atts*^{R}, respectively. Define ATT and $dATT$ analogously. Examples 2 yields the following.

Proposition 3. $dATT \subsetneq ATT \cap \mathcal{F}$ and $dATT^R \subsetneq ATT^R \cap \mathcal{F}$.

4   Constructing a Uniformizer for an Att

Recall that before we can prove that for a given functional composition of n *atts*^{U} an equivalent composition of n *datts*^{U} can be constructed, we need to prove that for a given *att*^{U} $\hat{A} = (U, A)$, a *datt*^{U} $\hat{D}$ can be constructed such that

$\tau_{\hat{D}}$ is a uniformizer of $\tau_{\hat{A}}$. For the latter, we first show how to construct a $datt^U$ D that realizes a uniformizer of τ_A.

Subsequently, let $A = (S, I, \Sigma, \Delta, a_0, R)$ be fixed. Note that we allow the att A to be circular.

Before we construct the $datt^U$ D, we introduce the following definitions. Let $s \in \mathrm{dom}(A)$. Let $t_1, \ldots, t_n$ be trees in $T_\Delta[\mathrm{SI}(s^\#)]$ and let $\tau = (t_1 \Rightarrow_{A,s^\#} t_2 \Rightarrow_{A,s^\#} \cdots \Rightarrow_{A,s^\#} t_n)$. We call τ a *translation of* A *on input* s if $t_1 = a_0(1)$ and $t_n \in T_\Delta$. If τ is a translation of A on input s, then the *output of* τ is t_n. Furthermore, we say that τ contains a *productive cycle* if $i < j \leq n$, a node u and a proper descendant u' of u as well as $\alpha(\nu) \in \mathrm{SI}(s^\#)$ exist such that $t_i[u] = t_j[u'] = \alpha(\nu)$. On the other hand, if $i < \iota < j$ and a node u exist such that $t_i[u] = t_j[u] = \alpha(\nu)$ but $t_\iota[u] \neq \alpha(\nu)$, then we say that τ contains a *non-productive cycle*. We say that τ is *cycle-free* if τ does not contain a cycle of either type.

Observation 1. Let $s \in \mathrm{dom}(A)$. Then a cycle free translation τ of A on input s exists.

Note that if A is noncircular then any translation of A on input s is cycle-free. Let τ be a cycle-free translation of A on input s. Then clearly, multiple instances of $\alpha'(\nu') \in \mathrm{SI}(s^\#)$ may occur in τ. This means that since A is nondeterministic, at distinct instances of $\alpha'(\nu')$ distinct rules may be applied. In the following, we say that τ is *uniform* if at all such instances of $\alpha'(\nu')$ the same rule is applied. With Observation 1, it is easy to see that the following holds.

Observation 2. Let $s \in \mathrm{dom}(A)$. Then a cycle free translation τ of A on input s that is also uniform exists.

Subsequently, denote by $\breve{\tau}_A$ the set of all pairs $(s, t) \in \tau_A$ for which a uniform translation τ exists whose input is s and whose output is t. By definition of $\breve{\tau}_A$ and due to Observation 2, it should be clear that the following holds.

Lemma 1. $\breve{\tau}_A \subseteq \tau_A$ *and* $\mathrm{dom}(\breve{\tau}_A) = \mathrm{dom}(\tau_A)$.

We remark that $\breve{\tau}_A$ is not necessarily a uniformizer of τ_A because $\breve{\tau}_A$ is not necessarily a function. In the following, we construct a $datt^U$ D that realizes a uniformizer of τ_A on the basis of uniform translations. More precisely, to construct D we proceed as follows; First, we show that given A, a nondeterministic top-down relabeling T and a $datt$ D' can be constructed such that the composition of T and D' simulates all uniform translations of A. In particular, denote by τ_C the tree translation realized by the composition of T and D', that is, the set $\{(s, t) \mid (s, s') \in \tau_T \text{ and } (s', t) \in \tau_{D'}\}$. We can construct T and D' such that $\tau_C = \breve{\tau}_A$. Given T and D', we show how to construct a $datt^U$ D such that D realizes a uniformizer of τ_C. Due to Lemma 1, it follows that D also realizes a uniformizer of τ_A. Before constructing T as well as D' we introduce the following definition. Consider the set R_σ, i.e., the set of all rules of A for the symbol σ. Let $\bar{R} \subseteq R_\sigma$. We call the set $\bar{R}$ *unambiguous* if $\bar{R}$ does not contain two distinct rules with the same left-hand side.

Idea of T and D'. To simulate uniform translations of A, the nondeterministic top-down relabeling T determines for every input node v which rules will be used at the node v. The *datt* D then only has to apply these rules.

More precisely, denote by $\tilde{\Sigma}$ the output alphabet of T and the input alphabet of D'. The alphabet $\tilde{\Sigma}$ consists of symbols of the form $\langle \sigma, \bar{R} \rangle$ where $\sigma \in \Sigma$ and $\bar{R}$ is an unambiguous subset of R_σ. The symbol $\langle \sigma, \bar{R} \rangle$ is of rank k if σ is. By relabeling a node by $\langle \sigma, \bar{R} \rangle$, T signals to D' that only rules in $\bar{R}$ are allowed to be used at that node. Note that since $\bar{R}$ is an unambiguous subset of R_σ, it follows easily that if $(s, \tilde{s}) \in \tau_T$ and τ' is a translation of D' on input $\tilde{s}$, then τ' simulates a translation τ of A on input s that is uniform.

Construction of T. Subsequently, we define the nondeterministic top-down relabeling $T = (\{q\}, \emptyset, \Sigma, \tilde{\Sigma}, q, \bar{R})$. To do so, all we need to do is to define the set $\bar{R}$. Specifically, we define that $q(\sigma(x_1, \ldots, x_k)) \to \langle \sigma, \bar{R} \rangle(q(x_1), \ldots, q(x_k)) \in \bar{R}$ for all $\sigma \in \Sigma_k$, $k \geq 0$ and for all unambiguous $\bar{R} \subseteq R_\sigma$.

Construction of D'. The *datt* D' is constructed in a straight-forward manner. We define $D' = (S, I, \tilde{\Sigma}, \Delta, a_0, R')$. Recall that due to Proposition 1, we can assume that $R_\#$, i.e., the set of rules of A for the root marker, contains no distinct rules with the same left-hand side. In other words, we can assume that $R_\#$ is unambiguous. Therefore, we define $R'_\# = R_\#$. Recall that $\tilde{\Sigma}$ consists of symbols of the form $\langle \sigma, \bar{R} \rangle$ where $\sigma \in \Sigma$ and $\bar{R}$ is an unambiguous subset of R_σ. For the symbol $\langle \sigma, \bar{R} \rangle$ we define $R'_{\langle \sigma, \bar{R} \rangle} = \bar{R}$. Since by definition of $\tilde{\Sigma}$, the set $\bar{R}$ is unambiguous it should be clear that D' is deterministic.

Recall that τ_C denotes the tree translation realized by the composition of T and D' and that $\breve{\tau}_A$ is the set of all pairs $(s, t) \in \tau_A$ for which there is a a uniform translation. We now prove the following statement.

Lemma 2. *The sets $\breve{\tau}_A$ and τ_C are equal.*

Proof. To begin our proof we show that $\tau_C \subseteq \breve{\tau}_A$. Let $(s, \tilde{s}) \in \tau_T$ and let $(\tilde{s}, t) \in \tau_{D'}$. Let $a_0(1) = t_1 \Rightarrow_{D', \tilde{s}\#} t_2 \Rightarrow_{D', \tilde{s}\#} \cdots \Rightarrow_{D', \tilde{s}\#} t_n = t$ be the corresponding translation of D' on input $\tilde{s}$. Let $\langle \sigma, \bar{R} \rangle$ be an arbitrary symbol in $\tilde{\Sigma}$. By construction of D', $R'_{\langle \sigma, \bar{R} \rangle} \subseteq R_\sigma$. Hence, a translation τ of A on input s exists that is of the form $\tau = (a_0(1) = t_1 \Rightarrow_{A, s\#} t_2 \Rightarrow_{A, s\#} \cdots \Rightarrow_{A, s\#} t_n = t)$ Since D' is deterministic, τ is a uniform translation.

To show the converse, that is, to show that $\tau_C \supseteq \breve{\tau}_A$, consider a uniform translation $\tau = (a_0(1) = t_1 \Rightarrow_{A, s\#} t_2 \Rightarrow_{A, s\#} \cdots \Rightarrow_{A, s\#} t_n = t)$ of A on input $s \in T_\Sigma$. Let $v \in V(s)$ and let $s[v] = \sigma \in \Sigma_k$, $k \geq 0$. Let $a \in S$ and $a(\pi) \to \xi \in R_\Sigma$. Recall that for a right-hand side ξ of a rule in R_σ and a node v, $\xi[\pi \leftarrow v]$ denotes the tree obtained by replacing all occurrences of π by v (cf. Sect. 1).

In the following, we say that the rule $a(\pi) \to \xi$ is *used in τ at the input node* v if $i \in [n]$ and a node $u \in V(t_i)$ exist such that $t_i[u] = a(1.v)$ and $t_{i+1} = t_i[u \leftarrow \xi[\pi \leftarrow 1.v]]$. Analogously, we say that the rule $b(\pi j) \to \xi$, where $b \in I$ and $k \in [k]$, is *used in τ at the input node* v if $i \in [n]$ and a node $u \in V(t_i)$ exist such that $t_i[u] = b(1.v.j)$ and $t_{i+1} = t_i[u \leftarrow \xi[\pi \leftarrow 1.v]]$.

Note that the reason why we have $a(1.v)$ and $b(1.v.j)$ in the definitions above is because translations of A always involve the root marker.

Denote by $\tau[v]$ the set of all rules used in τ at the input node v. Note that since τ is a uniform translation, $\tau[v]$ is an unambiguous subset of R_σ. By construction of T, it should be clear that T can transform the tree s into the tree $\tilde{s}$ over $\tilde{\Sigma}$ such that if the node v is labeled by the symbol σ in s then v is labeled by $\langle \sigma, \tau[v] \rangle$ in $\tilde{s}$. By definition of D' it follows that there is a translation of D' on input $\tilde{s}$ that outputs t. Hence, our lemma follows. $\qquad\square$

Before we construct the $datt^U$ $D = (U, D')$ from T and D', note that we can assume that $\mathrm{range}(T) \subseteq \mathrm{dom}(D')$. This is because by Proposition 2 (see also [10]), the domain of the $datt$ D' is effectively regular. Hence a (nondeterministic) top-down automaton M recognizing D' exists. To restrict the range of T to the domain of D', we simply run M in parallel with T.

Furthermore, note by [5], that the following holds for top-down tree transducers, i.e., att without inherited attributes.

Proposition 4. *For any top-down tree transducer M, a deterministic top-down tree transducer M' with regular-look-ahead can be constructed such that M' realizes a uniformizer of τ_M.*

Note that if M is a top-down relabeling, then the top-down tree transducer M' with regular-look-ahead that the procedure in [5] yields is in fact a top-down relabeling with regular-look-ahead, i.e., a look-around.

Construction of the $datt^U$ $D = (U, D')$. To construct the $datt^U$ D, all we need to do is to construct U from T. Note that by definition, T is also a top-down tree transducer. Therefore, due to Proposition 4, let U be constructed as one of the top-down tree transducers with regular-look-ahead that realizes a uniformizer of τ_T. Since T is a top-down relabeling, previous observations yield that U is a look-around. This concludes the construction of the $datt^U$ D.

Lemma 3. *The $datt^U$ $D = (U, D')$ realizes a uniformizer of τ_A.*

Proof. Due to Lemmas 1 and 2, it is obviously sufficient to show that D realizes a uniformizer of τ_C. Note that obviously τ_D is a function since D is deterministic.

By construction of D, it follows obviously that $\tau_D \subseteq \tau_C$. In particular, this follows since by construction U realizes a uniformizer of τ_T and hence $\tau_U \subseteq \tau_T$.

Note that $\tau_D \subseteq \tau_C$ implies $\mathrm{dom}(\tau_C) \subseteq \mathrm{dom}(D)$. To show that $\mathrm{dom}(\tau_C) = \mathrm{dom}(D)$, let $s \in \mathrm{dom}(\tau_C)$. Obviously, this means that $s \in \mathrm{dom}(T)$. Since U realizes a uniformizer of τ_T, it follows that $s \in \mathrm{dom}(U)$. Let $(s, \tilde{s}) \in \tau_U$. Recall our previous assumption that $\mathrm{range}(T) \subseteq \mathrm{dom}(D')$. Obviously, this means that $\mathrm{range}(U) \subseteq \mathrm{dom}(D')$ and hence $\tilde{s} \in \mathrm{dom}(D')$. This in turn yields that $s \in \mathrm{dom}(D)$ and hence $\mathrm{dom}(\tau_C) = \mathrm{dom}(D)$. $\qquad\square$

Due to Lemma 3, the following holds.

Theorem 1. *For any att A a $datt^U$ D can be constructed such that D realizes a uniformizer of τ_A.*

5   From Unformizers for Att to Uniformizers for Att^U

Consider an arbitrary att^U $\hat{A} = (U, A)$. In the following, we show how to construct a $datt^U$ $\hat{D}$ such that $\tau_{\hat{D}}$ is a uniformizer of $\tau_{\hat{A}}$. Consider the underlying att A of $\hat{A}$. Due to Theorem 1, a $datt^U$ $\hat{D}' = (U', D)$ that realizes a uniformizer of τ_A can be constructed. Clearly $\{(s,t) \mid (s,\check{s}) \in \tau_U$ and $(\check{s},t) \in \tau_{\hat{D}'}\} = \{(s,t) \mid (s,\check{s}) \in \tau_U$ and $(\check{s},\hat{s}) \in \tau_{U'}$ and $(\hat{s},t) \in \tau_D\}$ is a uniformizer of $\tau_{\hat{A}}$. Due to Theorem 2.6 of [4] and Theorem 1 of [1] (and their proofs), look-arounds are closed under composition. More precisely, a look-around $\hat{U}$ can be constructed such that $\tau_{\hat{U}} = \{(s,\hat{s}) \mid (s,\check{s}) \in \tau_U$ and $(\check{s},\hat{s}) \in \tau_{U'}\}$. Therefore, the $datt^U$ $(\hat{U}, D)$ realizes a uniformizer of $\tau_{\hat{A}}$. This yields the following.

Theorem 2. *For any att^U $\hat{A}$ a $datt^U$ that realizes a uniformizer of $\tau_{\hat{A}}$ can be constructed.*

6   Final Results

Analogously as in [5] for top-down tree transducers, we obtain the following result using uniformizers.

Theorem 3. *Let C be a composition of n $atts^U$. Then $datts^U$ $D_1, \ldots, D_n$ can be constructed such that C and the composition of $D_1, \ldots, D_n$ are equivalent.*

Proof. Consider the $atts^U$ $A_1, \ldots, A_n$. Let the composition C of $A_1, \ldots, A_n$ be functional, i.e., let $\tau_C = \{(s,t) \mid (s,t_1) \in \tau_{A_1}, (t_{i-1},t_1) \in \tau_{A_i}$ for $1 < i < n$ and $(t_{n-1},t) \in \tau_{A_n}\}$ be a function. For $i < n$, we can assume that $range(A_i) \subseteq dom(A_{i+1})$ holds. In particular, by Proposition 2, $dom(A_{i+1})$ is effectively recognizable, i.e., a (nondeterministic) top-down automaton $T_{A_{i+1}}$ recognizing $dom(A_{i+1})$ can be constructed. By running $T_{A_{i+1}}$ in parallel to A_i, it should be clear that we can restrict the range of A_i to trees in $dom(A_{i+1})$.

By Theorem 2, let $D_1, \ldots, D_n$ be $datts^U$ realizing uniformizers of $\tau_{A_1}, \ldots, \tau_{A_n}$, respectively. Denote by C_d the composition of $D_1, \ldots, D_n$. Obviously $\tau_{C_d} \subseteq \tau_C$. Note that $range(A_i) \subseteq dom(A_{i+1})$ implies $range(D_i) \subseteq dom(D_{i+1})$ for $i < n$. Thus, it follows that C_d and C have the same domain. Since τ_C is a function and $\tau_{C_d} \subseteq \tau_C$, this yields that $\tau_{C_d} = \tau_C$ which in turn yields Theorem 3.   □

Denote by $(ATT^U)^n$ and $(dATT^U)^n$ the classes of tree translations realizable by the composition of n nondeterministic $atts^U$ and n deterministic $atts^U$, respectively. Theorem 3 yields the following.

Theorem 4. $(ATT^U)^n \cap \mathcal{F} = (dATT^U)^n$.

Note that if an att A (with or without look-around) is functional, then any $datt^U$ that realizes a uniformizer of τ_A is in fact equivalent to A. Hence, Theorems 1 and 2 yield the following

Corollary 1. *For any functional att (with or without look-around) an equivalent $datt^U$ can be constructed.*

7   Conclusion

Consider an arbitrary composition C of n attributed tree transducer with look-around that realizes a function. In this paper we have provided a procedure which given C, computes n deterministic attributed tree transducer with look-around $D_1, \ldots, D_n$ such that C and the composition of $D_1, \ldots, D_n$ are equivalent. To do so we have shown that any attributed tree transducer A admits a uniformizer realized a deterministic attributed tree transducer with look-around can be constructed. An obvious question is: Do we always need look-around? One wonders when a uniformizer of A can be implemented by a deterministic attributed tree transducer *without* look-around? This question is addressed in [8] for finite-valued string transducers. Specifically, for such transducers it is decidable whether or not a uniformizer realized by a deterministic string transducer exists.

Given our result another question is: Given a composition of attributed tree transducer C is it decidable whether or not C realizes a function? To the best of our knowledge this is an open problem. Note that whether or not a composition of top-down tree transducers is functional has recently been shown to be decidable [16]. In contrast, even for a single attributed tree transducer it is unknown whether or not functionality is decidable. Note that decidability of the latter would imply that equivalence of deterministic attributed tree transducers is decidable. The latter is a long standing open problem.

References

1. Baker, B.S.: Tree transducers and tree languages. Inf. Control **37**(3), 241–266 (1978)
2. Bloem, R., Engelfriet, J.: A comparison of tree transductions defined by monadic second order logic and by attribute grammars. JCSS **61**(1), 1–50 (2000)
3. Carayol, A., Löding, C.: Uniformization in automata theory. In: Congress of Logic, Methodology and Philosophy of ScienceLogic, Nancy, 19–26 July 2011, pp. 153–178. College Publications, London (2014)
4. Engelfriet, J.: Top-down tree transducers with regular look-ahead. Math. Syst. Theory **10**, 289–303 (1977)
5. Engelfriet, J.: On tree transducers for partial functions. Inf. Process. Lett. **7**(4), 170–172 (1978)
6. Engelfriet, J., Hoogeboom, H.J.: MSO definable string transductions and two-way finite-state transducers. ACM Trans. Comput. Log. **2**, 216–254 (1999)
7. Engelfriet, J., Inaba, K., Maneth, S.: Linear-bounded composition of tree-walking tree transducers: linear size increase and complexity. Acta Inform. **58**(1–2), 95–152 (2021)
8. Filiot, E., Jecker, I., Löding, C., Winter, S.: On equivalence and uniformisation problems for finite transducers. In: ICALP 2016, 11–15 July 2016, Rome, Italy. LIPIcs, vol. 55, pp. 125:1–125:14. Schloss Dagstuhl (2016)
9. Fülöp, Z.: On attributed tree transducers. Acta Cybern. **5**(3), 261–279 (1981)
10. Fülöp, Z., Maneth, S.: Domains of partial attributed tree transducers. Inf. Process. Lett. **73**(5–6), 175–180 (2000)

11. Fülöp, Z., Vágvölgyi, S.: Attributed tree transducers cannot induce all deterministic bottom-up tree transformations. Inf. Comput. **116**(2), 231–240 (1995)
12. Fülöp, Z., Vogler, H.: Syntax-Directed Semantics - Formal Models Based on Tree Transducers. Monographs in Theoretical Computer Science. An EATCS Series. Springer, Heidelberg (1998). https://doi.org/10.1007/978-3-642-72248-6
13. Hashimoto, K., Maneth, S.: Characterizing attributed tree translations in terms of macro tree transducers. Theor. Comput. Sci. **963**, 113943 (2023)
14. Knuth, D.E.: Semantics of context-free languages. Math. Syst. Theory **2**(2), 127–145 (1968). Errata in: [15]
15. Knuth, D.E.: Correction: semantics of context-free languages. Math. Syst. Theory **5**(1), 95–96 (1971)
16. Maneth, S., Seidl, H., Vu, M.: Functionality of compositions of top-down tree transducers is decidable. Inf. Comput. **296**, 105131 (2024)
17. Maneth, S., Vu, M.: Deciding whether an attributed translation can be realized by a top-down transducer. CoRR abs/2306.04326 (2024). https://arxiv.org/abs/2306.04326v3
18. Souza, R.: Uniformisation of two-way transducers. In: Dediu, A.-H., Martín-Vide, C., Truthe, B. (eds.) LATA 2013. LNCS, vol. 7810, pp. 547–558. Springer, Heidelberg (2013). https://doi.org/10.1007/978-3-642-37064-9_48

Translation of Semi-extended Regular Expressions Using Derivatives

Antoine Martin[1]($\boxtimes$)(iD), Etienne Renault[2](iD), and Alexandre Duret-Lutz[1](iD)

[1] LRE, EPITA, Le Kremlin-Bicêtre, France
amartin@lrde.epita.fr
[2] SiPearl, Maisons-Laffitte, France

Abstract. We generalize Antimirov's notion of *linear form* of a regular expression, to the Semi-Extended Regular Expressions typically used in the Property Specification Language or SystemVerilog Assertions. Doing so requires extending the construction to handle more operators, and dealing with expressions over alphabets $\Sigma = 2^{AP}$ of valuations of atomic propositions. Using linear forms to construct automata labeled by Boolean expressions suggests heuristics that we evaluate. Finally, we study a variant of this translation to automata with accepting transitions: this construction is more natural and provides smaller automata.

Keywords: Regular expressions · Automata · PSL · SVA · Derivative

1 Introduction

In this paper we discuss and compare techniques for translating (extended) regular expressions over alphabets $\Sigma = 2^{AP}$ where letters describe valuations of a set of atomic propositions AP. Such alphabets are typically used in formal methods such as *model checking* [3], *runtime verification* [4] or *synthesis* [12]. In our case, we are interested in supporting the regular expression operators of the PSL and SVA standards.

Property Specification Language (PSL) [11] and SystemVerilog Assertions (SVA) [1] are two industrial formal verification languages used in the field of hardware design and verification. These two languages include features for describing linear-time temporal properties or reasoning with clocks, but we restrict ourselves to the (semi-extended) regular expression properties. Both offer a nearly identical set of operators, albeit with different syntaxes.

For instance the expression, "`btn : ((red[=2] && opn[->]) || rst[->])`" is a PSL expression matching any sequence that starts with the `btn` signal on, and in which either the `red` signal is on exactly twice in the interval it takes for the signal `opn` turn on, or in which the `rst` signal turns on. In SVA this expression becomes "`btn ##0 ((red[=2] intersect opn[->1]) or rst[->1])`".

Other logics such as Linear Dynamic Logic (LDL) [14] or Propositional Dynamic Logic (PDL) [13] also use regular expressions over atomic propositions, so our work applies to them too, however, and unlike SVA or PSL these logics are usually defined only with classical regular operators plus a $\varphi?$ operator that is absent from PSL and SVA, and that we do not consider.

© The Author(s), under exclusive license to Springer Nature Switzerland AG 2024
S. Z. Fazekas (Ed.): CIAA 2024, LNCS 15015, pp. 234–248, 2024.
https://doi.org/10.1007/978-3-031-71112-1_17

2 Definitions

In the entire paper, we assume an alphabet $\Sigma = 2^{AP}$ where letters describe valuations of a set of *atomic propositions* AP. For instance, if $AP = \{a, b\}$ then we denote the valuations as $\Sigma = \{\bar{a}\bar{b}, \bar{a}b, a\bar{b}, ab\}$. We use $\Sigma^\star$ to denote the set of finite sequences of valuations. For a sequence $\sigma \in \Sigma^\star$, we denote $|\sigma|$ its length, and we write $\sigma(i)$ for the letter at position $i \in \{0, 1, \ldots, |\sigma| - 1\}$ in σ. Using ";" as concatenation operator, we write $\sigma = \sigma(0) ; \sigma(1) ; \ldots ; \sigma(|\sigma| - 1)$. Finally, for two integers i, j such that $0 \le i \le j < |\sigma|$, we write $\sigma^{i..j}$ the (possibly empty if $i = j$) subsequence $\sigma(i) ; \sigma(i + 1) ; \ldots ; \sigma(j - 1)$. For convenience, we write $\sigma^{i..}$ instead of $\sigma^{i..|\sigma|}$ and $\sigma^{..j}$ instead of $\sigma^{0..j}$.

Definition 1 (Boolean expression). *Any Boolean expression b is built using the following grammar, where $a \in AP$ can be any atomic proposition.*

$$b ::= \bot \mid \top \mid a \mid (b \vee b) \mid (b \wedge b) \mid \neg b$$

For convenience, we omit unnecessary parentheses, and use operators $\rightarrow$ and $\leftrightarrow$ as syntactic sugar with their obvious definitions.

Boolean expression are interpreted over a valuation $v \in \Sigma$ in the obvious way. We write $v \models b$ when the valuation v satisfies b, and $b \equiv b'$ when two Boolean expressions b and b' are satisfied by the same valuations.[1]

We use $\mathbb{B} = \{\bot, \top\}$ to denote the set of Boolean values, and $\mathbb{B}(AP)$ to denote the set of Boolean expressions over AP.

Definition 2 (SERE). *A Semi-Extended Regular Expression (SERE) r is built using the following grammar:*

$$r ::= b \mid \varepsilon \mid (r \, ; r) \mid (r : r) \mid r^\star \mid (r \vee r) \mid (r \wedge r) \mid \mathsf{fm}(r)$$

The symbol "ε" is called the empty *word. Operators "$\vee$" (choice), "$;$" (concatenation) and "*" (Kleene star) are traditional regular operators. SERE extends those with "$\wedge$" (intersection) "$:$" (fusion), and "fm" (SVA's first-match). In practice, we omit parentheses when they are not necessary.*

The set of all SEREs is written SERE.

SEREs are interpreted over a finite sequence $\sigma \in \Sigma^\star$ of valuations defined inductively as follows:

$$
\begin{aligned}
\sigma &\models b & \text{iff} \quad & |\sigma| = 1 \wedge \sigma(0) \models b \\
\sigma &\models \varepsilon & \text{iff} \quad & |\sigma| = 0 \\
\sigma &\models (r_1 \, ; r_2) & \text{iff} \quad & \exists i \ge 0,\ \sigma^{..i} \models r_1 \wedge \sigma^{i..} \models r_2 \\
\sigma &\models (r_1 : r_2) & \text{iff} \quad & \exists i \ge 0,\ \sigma^{..i+1} \models r_1 \wedge \sigma^{i..} \models r_2 \\
\sigma &\models r^\star & \text{iff} \quad & either\,|\sigma| = 0 \ or \ \exists i > 0,\ \sigma^{..i} \models r \wedge \sigma^{i..} \models r^\star \\
\sigma &\models (r_1 \vee r_2) & \text{iff} \quad & \sigma \models r_1 \vee \sigma \models r_2 \\
\sigma &\models (r_1 \wedge r_2) & \text{iff} \quad & \sigma \models r_1 \wedge \sigma \models r_2 \\
\sigma &\models \mathsf{fm}(r) & \text{iff} \quad & (\sigma \models r) \wedge (\forall i < |\sigma|,\ \sigma^{..i} \not\models r)
\end{aligned}
$$

---

[1] Testing $b \equiv b'$ is straightforward if b and b' are represented with BDDs [5].

The language *of a SERE r is the set* $\mathscr{L}(r) = \{\sigma \in \Sigma^\star \mid \sigma \models r\}$ *of all sequences satisfying r (or "matched" by r).*

In the above definition, regular expressions have been extended with three operators: the conjunction "∧" has obvious meaning, the fusion operator ":" ensures that the last letter matching the left operand is also the first letter of matching the right operand (this implies that a fusion can never match the empty word), and finally fm, the first-match operator of SVA, retains only the shortest possible match for a SERE r.

The PSL and SVA specifications defines other SERE operators (such as [=n] or [->n]) that can be seen as syntactic sugar on the above. In our syntax, the expression from the introduction becomes

$$\varphi = btn : \Big(\big(\big((\neg red)^\star; red; (\neg red)^\star; red; (\neg red)^\star\big) \wedge ((\neg opn)^\star; opn)\big) \vee \big((\neg rst)^\star; rst\big)\Big)$$

Definition 3 (Constant Term). *The* constant term *of an expression r, denoted* $\lambda(r)$ *is defined inductively as follows for any Boolean formula b and any SEREs* r_1, r_2.

$$\lambda(b) = \bot \qquad\qquad \lambda(r_1 \vee r_2) = \lambda(r_1) \vee \lambda(r_2)$$
$$\lambda(\varepsilon) = \varepsilon \qquad\qquad \lambda(r_1 \wedge r_2) = \lambda(r_1) \wedge \lambda(r_2)$$
$$\lambda(r_1 : r_2) = \bot \qquad\qquad \lambda(r_1 ; r_2) = \lambda(r_1) ; \lambda(r_2)$$
$$\lambda(r_1^\star) = \varepsilon \qquad\qquad \lambda(\mathsf{fm}(r_1)) = \lambda(r_1)$$

Proposition 1. *With the above notation,* $\lambda(r) = \varepsilon$ *iff* $\varepsilon \models r$.

Definition 4 (Syntactic equivalence). *Given two SEREs* r_1 *and* r_2, *we say that they are syntactically equivalent, denoted* $r_1 \stackrel{\circ}{=} r_2$, *if one can be rewritten into the other using the following so called ACI-rules (associativity, commutativity, and idempotence) and a few others:*

$$(r_1 \odot r_2) \odot r_3 \stackrel{\circ}{=} r_1 \odot (r_2 \odot r_3) \stackrel{\circ}{=} r_1 \odot r_2 \odot r_3 \qquad for\ \odot \in \{;,:,\vee,\wedge\} \qquad \text{(A)}$$
$$r_1 \odot r_2 \stackrel{\circ}{=} r_2 \odot r_1 \qquad for\ \odot \in \{\vee,\wedge\} \qquad \text{(C)}$$
$$r_1 \odot r_1 \stackrel{\circ}{=} r_1 \qquad for\ \odot \in \{\vee,\wedge\} \qquad \text{(I1)}$$

$$r^{\star\star} \stackrel{\circ}{=} r^\star \qquad\qquad \mathsf{fm}(\mathsf{fm}(r)) \stackrel{\circ}{=} \mathsf{fm}(r) \qquad \text{(I2)}$$

$$r \vee \bot \stackrel{\circ}{=} r \qquad\qquad r \wedge \top \stackrel{\circ}{=} r \qquad\qquad r ; \varepsilon \stackrel{\circ}{=} \varepsilon ; r \stackrel{\circ}{=} r \qquad \text{(Z)}$$

$$r \wedge \bot \stackrel{\circ}{=} \bot \qquad\qquad r ; \bot \stackrel{\circ}{=} \bot ; r \stackrel{\circ}{=} \bot \qquad \text{(U1)}$$
$$r \vee \top \stackrel{\circ}{=} \top \qquad\qquad r : \bot \stackrel{\circ}{=} \bot : r \stackrel{\circ}{=} \bot \qquad \text{(U2)}$$
$$r : \varepsilon \stackrel{\circ}{=} \varepsilon : r \stackrel{\circ}{=} \bot \qquad \text{(U3)}$$

$$\mathsf{fm}(r) \stackrel{\circ}{=} \varepsilon\ if\ \varepsilon \models r \qquad \text{(F)}$$

Proposition 2. *Two syntactically equivalent SEREs have the same language. I.e., $(r_1 \stackrel{\circ}{=} r_2) \implies (\mathscr{L}(r_1) = \mathscr{L}(r_2))$.*

From an implementation standpoint, it is straightforward to rewrite any SERE r into a unique representative of its equivalence class $[r]_{\stackrel{\circ}{=}} = \{s \in \mathsf{SERE} \mid r \stackrel{\circ}{=} s\}$. This can be achieved by applying the above rewriting rules during the construction of the syntax tree of r. In particular rule (A) can be implemented by considering these operators as n-ary (rather than binary), and then rules (C) and (I1) can be implemented by sorting operands and removing duplicates. Rule (F) can be implemented at construction as well, by deciding $\varepsilon \models r$ using Proposition 1.

Definition 5 (NFA). *A* nondeterministic finite automaton *is a tuple $\mathcal{A} = \langle Q, \delta, \iota, F \rangle$ where Q is a finite set of states, $\delta \subseteq Q \times \mathbb{B}(AP) \times Q$ is the transition relation, $\iota \in Q$ is the initial state, and $F \subseteq Q$ is the set of final states.*

We write $s \xrightarrow{f} d$ when $(s, f, d) \in \delta$.

A sequence of valuations $\sigma \in \Sigma^n$ of size n is accepted by $\mathcal{A}$ if either $n = 0$ and $\iota \in F$, or $n > 0$ and there exists a sequence of transitions $\rho = s_0 \xrightarrow{f_0} s_1 \xrightarrow{f_1} \cdots \xrightarrow{f_{n-1}} s_n$ such that $s_0 = \iota$, $s_n \in F$, and for each i, $\sigma(i) \models f_i$.

The language *of $\mathcal{A}$, denoted $\mathscr{L}(\mathcal{A})$ is the set of words accepted by $\mathcal{A}$.*

A Deterministic Finite Automaton (DFA) is an NFA where transitions leaving each state have mutually exclusive Boolean expressions. Formally, an automaton is a DFA if for any two different transitions $s \xrightarrow{f} d$ and $s \xrightarrow{f'} d'$ with the same origin, we have $f \wedge f' \equiv \bot$.

Such automata are sometimes called symbolic finite automata [8], however in our case the alphabet is always finite, so they can be handled in a usual way.

3 Building Automata Using Linear Forms

In 1964, Brzozowsky [6] introduced the notion of *derivative* of a regular expression, allowing the construction of an equivalent deterministic finite automaton. This work was extended in 1995 by Antimirov [2], with a notion of *partial derivatives* allowing the construction of a non-deterministic finite automaton. More importantly, Antimirov introduced the concept of *linear form* of a regular expression as a more efficient way to compute the set of *partial derivatives*. An extension of *partial derivatives* was proposed by Caron et al. [7] to handle intersection and complement. Here, we adapt the concept of *linear form*, to SERE with an alphabet over 2^{AP} and their specific operators. In particular, the fact that our alphabet is exponential in the number of atomic propositions makes *linear forms* much more attractive than *partial derivatives*, because using the latter to build an automaton requires iterating over exponentially many letters.

3.1 Linear Forms

Definition 6 (Linear Form). *A* linear form *for a SERE r is a finite set of pairs $\{(p_1, s_1), (p_2, s_2), \ldots\}$ where $p_i \in \mathbb{B}(AP)$, $p_i \not\equiv \bot$, $s_i \in \mathsf{SERE}$ and $s_i \not\equiv \bot$, such that $\bigcup_i \mathscr{L}(p_i ; s_i) = \mathscr{L}(r) \setminus \{\varepsilon\}$.*

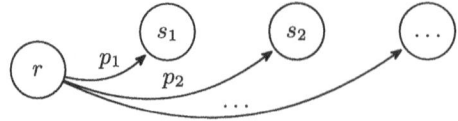

Fig. 1. Automaton view of a linear form $\{(p_1, s_1), (p_2, s_2), \ldots\}$ for a SERE r.

This definition differs from that of Antimirov [2] in that p_i is a satisfiable Boolean expression rather than a letter, and in that we explicitly forbid $s_i \overset{\circ}{=} \bot$. To simplify the upcoming notations we assume that any pair (p_i, s_i) in a linear form we construct is implicitly ignored when $p_i \equiv \bot$; for instance we shall write $\{\ldots, (p_i \wedge \neg p_j, s_i), \ldots\}$ with the implicit assumption that the pair $(p_i \wedge \neg p_j, s_i)$ must be omitted when $p_i \wedge \neg p_j$ is not satisfiable.

As we saw in Proposition 1, an empty sequence may only be matched by a SERE r if $\lambda(r) = \varepsilon$. If a non-empty sequence σ is matched by a SERE r, then Definition 6 implies that a linear form for r will have at least one pair (p_i, s_i) such that $\sigma(0) \models p_i$ and $\sigma^{1\cdots} \models s_i$. As a mental model for a linear form, it is useful to interpret it as the partial automaton shown in Fig. 1: the p_is are Boolean formulas evaluated against $\sigma(0)$, and the p_is tell what SERE should be checked against the suffix $\sigma^{1\cdots}$. The constraints, in Definition 6, that $p_i \not\equiv \bot$, and $s_i \not\equiv \bot$, prevent the creation of paths that will not recognize any word.

A SERE may have multiple linear forms; some will be called *deterministic*.

Definition 7 (Deterministic Linear Form). *A linear form* $\{(p_1, s_1), (p_2, s_2), \ldots\}$ *is* deterministic *if for any* $i \neq j$, $p_i \wedge p_j \equiv \bot$.

Here is a linear form for the formula φ of Sect. 2.

$$L_1 = \{ \big(btn \wedge \neg red \wedge \neg opn, ((\neg red)^\star; red; (\neg red)^\star; red; (\neg red)^\star) \wedge ((\neg opn)^\star; opn)\big),$$
$$\big(btn \wedge red \wedge \neg opn, \quad ((\neg red)^\star; red; (\neg red)^\star) \wedge ((\neg opn)^\star; opn)\big),$$
$$\big(btn \wedge \neg rst, \qquad (\neg rst)^\star; rst\big),$$
$$\big(btn \wedge rst, \qquad \varepsilon\big)\}$$

L_1 is not deterministic because, for instance, $btn \wedge red \wedge \neg opn$ can hold together with $btn \wedge rst$. Here is a deterministic linear form for φ:

$$L_2 = \{ \big(btn \wedge \neg red \wedge \neg opn \wedge \neg rst,$$
$$(((\neg red)^\star; red; (\neg red)^\star; red; (\neg red)^\star) \wedge ((\neg opn)^\star; opn)) \vee ((\neg rst)^\star; rst)\big),$$
$$\big(btn \wedge \neg red \wedge \neg opn \wedge rst,$$
$$(((\neg red)^\star; red; (\neg red)^\star; red; (\neg red)^\star) \wedge ((\neg opn)^\star; opn)) \vee \varepsilon\big),$$
$$\big(btn \wedge red \wedge \neg opn \wedge \neg rst,$$
$$(((\neg red)^\star; red; (\neg red)^\star) \wedge ((\neg opn)^\star; opn)) \vee ((\neg rst)^\star; rst)\big),$$
$$\big(btn \wedge red \wedge \neg opn \wedge rst, \quad (((\neg red)^\star; red; (\neg red)^\star) \wedge ((\neg opn)^\star; opn)) \vee \varepsilon\big)\}$$

Property 1 (determinization of a linear form). Any linear form $\{(p_1, s_1), (p_2, s_2), \ldots\}$ for an expression r can be converted into a deterministic linear form for r.

The idea is that if $p_1 \wedge p_2 \not\equiv \bot$, then $\{(p_1 \wedge \neg p_2, s_1), (\neg p_1 \wedge p_2, s_2), (p_1 \wedge p_2, s_1 \vee s_2), \ldots\}$ is also a linear form for r. This process can be repeated for any $i \neq j$ such that $p_i \wedge p_j \not\equiv \bot$. From now on, we assume the existence of a function det that determinizes a linear form.

3.2   Linearization of SEREs

We now discuss how to convert a SERE into a linear form. For now on, we assume that equations (A)–(F) are always applied, i.e., that we are only working with unique representatives of each equivalence classes of $\stackrel{\circ}{=}$, as discussed in Sect. 2.

To simplify the notations, we extend the concatenation and fusion operators to linear forms: given a linear form $L = \{(p_1, s_1), (p_2, s_2), \ldots, (p_n, s_n)\}$, an operator $\odot \in \{;, :\}$, and a SERE $r \neq \bot$, we write $L \odot r$ instead of $\{(p_1, s_1 \odot r), (p_2, s_2 \odot r), \ldots, (p_n, s_n \odot r)\}$. Additionally, $L \odot \bot = \varnothing$. The notation works similarly for $r \odot L$.

Definition 8 (Linearization of a SERE). *The following* LF *function turns a SERE into a linear form. It mostly extends the "lf" function of Antimirov [2, eq. (45)–(51)] to deal with SERE operators and Boolean formulas.*

$$\mathsf{LF}(\bot) = \varnothing$$
$$\mathsf{LF}(\varepsilon) = \varnothing$$
$$\mathsf{LF}(b) = \{(b, \varepsilon)\}$$
$$\mathsf{LF}(r_1 \vee r_2) = \mathsf{LF}(r_1) \cup \mathsf{LF}(r_2)$$
$$\mathsf{LF}(r^\star) = \mathsf{LF}(r) \,; r^\star$$
$$\mathsf{LF}(r_1 \,; r_2) = (\mathsf{LF}(r_1) \,; r_2) \cup (\lambda(r_1) \,; \mathsf{LF}(r_2))$$
$$\mathsf{LF}(r_1 : r_2) = (\mathsf{LF}(r_1) : r_2) \cup \left\{ (p_i \wedge p_j, s_j) \,\middle|\, \begin{array}{l} (p_i, s_i) \in \mathsf{LF}(r_1), \lambda(s_i) = \varepsilon, \\ (p_j, s_j) \in \mathsf{LF}(r_2) \end{array} \right\}$$
$$\mathsf{LF}(r_1 \wedge r_2) = \{(p_i \wedge p_j, s_i \wedge s_j) \mid (p_i, s_i) \in \mathsf{LF}(r_1), (p_j, s_j) \in \mathsf{LF}(r_2)\}$$
$$\mathsf{LF}(\mathsf{fm}(r)) = \{(p_i, \mathsf{fm}(s_i)) \mid (p_i, s_i) \in \det(\mathsf{LF}(r))\}$$

As noted below Definition 6, we assume that when one of these equations generates a pair (p_i, s_i) with $p_i \equiv \bot$, it is implicitly removed. Since we assume that rules (A)–(F) are always applied, these equations cannot produce a pair (p_i, s_i) such that $s_i \stackrel{\circ}{=} \bot$, but it could nonetheless be the case that $\mathscr{L}(s_i) = \varnothing$ (for instance if $s_i = a \wedge \neg a$).

To understand the definition for $\mathsf{LF}(\mathsf{fm}(r))$, it may be useful to give an intuition of how $\mathsf{fm}(r)$ works. The SERE $\mathsf{fm}(r)$ may match σ if only if σ is the shortest prefix of σ matching r. An easy way to construct an automaton for $\mathsf{fm}(r)$ is therefore to build a DFA for r, and then remove the outgoing edges of all accepting states of that DFA. This is actually what the above definition achieves. The use of $\mathsf{det}(\mathsf{LF}(r))$ is making sure that the linear form is deterministic, and the use of $\mathsf{fm}(s_i)$ serves two purposes: (1) it ensures that upcoming choices will still be deterministic, and (2) more importantly, by applying rule (F), it cuts the successors of s_i when $\varepsilon \models s_i$. For instance, $\mathsf{LF}(\mathsf{fm}(a\,;a^\star)) = \{(a,\varepsilon)\}$ because $\mathsf{fm}(a^\star)$ gets reduced to ε by rule (F).

To use LF in an algorithm for building an automaton that recognizes $\mathscr{L}(r)$, we need two theorems. First, $\mathsf{LF}(r)$ should be a linear form, i.e., it should preserve the language of r, except for the empty word (Theorem 1 below). Then, the number of new expressions that can be created by applying LF recursively has to be finite (Theorem 2 below).

Theorem 1. $\mathsf{LF}(r) = \{(p_1, s_1), (p_2, s_2), \ldots\}$ *is a linear form for* $r \in \mathsf{SERE}$.

Proof (Sketch). The fact hat $\bigcup_i \mathscr{L}(p_i;s_i) = \mathscr{L}(r)\backslash\{\varepsilon\}$ can be shown by induction on the structure of r using Definitions 2 and 8. $\square$

Theorem 2 (Terms). *For* $r \in \mathsf{SERE}$, *let* $\mathrm{Terms}(r)$ *denote the smallest subset of* SERE *such that* $r \in \mathrm{Terms}(r)$ *and for each* $\phi \in \mathrm{Terms}(r)$ *and each* $(p_i, s_i) \in \mathsf{LF}(\phi)$ *we have* $s_i \in \mathrm{Terms}(r)$. *Then the set* $\mathrm{Terms}(r)$ *is finite.*

Proof (Sketch). Our proof, which we omit for brevity, is inspired by a similar theorem by Antimirov [2, Theorem 3.4], however the results differ because of the new operators we support. Specifically, Antimirov did not support operators $\wedge$, :, and fm. (Of these three, : is the least problematic).

We start by addapting Antimirov's notion of *partial derivative* [2, Definition 2.8] to our context. The partial derivative of r with respect to x is defined by:

$$\partial_x r = \{s \mid (p, s) \in \mathsf{LF}(r),\ x \models p\}$$

We extend the notation to support derivation by a nonempty word $w \in \Sigma^+$ with $\partial_w r = \partial_{w^1} .. \partial_{w(0)} r$. Furthermore, we write $\partial_{\Sigma^+} r = \bigcup_{w \in \Sigma^+} \partial_w r$ for the set of all partial derivatives one can obtain using nonempty words of any length. With these conventions we have $\mathrm{Terms}(r) = \partial_{\Sigma^+} r \cup \{r\}$.

Working on the definition of LF and $\partial_x r$, we then establish the following inequalities for two SEREs r_1 and r_2:

$$|\partial_{\Sigma^+}(r_1 \vee r_2)| \leq |\partial_{\Sigma^+} r_1| + |\partial_{\Sigma^+} r_2|$$
$$|\partial_{\Sigma^+} r_1^\star| \leq |\partial_{\Sigma^+} r_1|$$
$$|\partial_{\Sigma^+}(r_1\,;r_2)| \leq |\partial_{\Sigma^+} r_1| + |\partial_{\Sigma^+} r_2|$$
$$|\partial_{\Sigma^+}(r_1 : r_2)| \leq |\partial_{\Sigma^+} r_1| + |\partial_{\Sigma^+} r_2|$$
$$|\partial_{\Sigma^+}(r_1 \wedge r_2)| \leq |\partial_{\Sigma^+} r_1| \times |\partial_{\Sigma^+} r_2|$$
$$|\partial_{\Sigma^+}\mathsf{fm}(r_1)| \leq 2^{|\partial_{\Sigma^+} r_1|}$$

The finiteness of $\partial_{\Sigma^+} r$ and therefore of $\mathrm{Terms}(r)$ follows from the above. $\square$

From the inequalities in the above sketch of proof, one can observe that the added operators do not have the same cost. In particular $\wedge$ incurs a quadratic cost, while fm leads to an exponential blow up.

input : A SERE ϕ
output: An automaton $\mathcal{A}$ such that $\mathscr{L}(\mathcal{A}) = \mathscr{L}(\phi)$

$Q, \delta, F \leftarrow \{\phi\}, \varnothing, \varnothing$;
todo.push(ϕ);
while todo $\neq \varnothing$ **do**
   $f \leftarrow$ todo.pop();
   foreach $(p, s) \in \boxed{\textsf{LF}(f)}$ **do**
      if $s \notin Q$ **then**
        $Q \leftarrow Q \cup \{s\}$;
        todo.push (s);
        if $\lambda(s) = \varepsilon$ **then**
          $F \leftarrow F \cup \{s\}$;
      $\delta \leftarrow \delta \cup \{f \xrightarrow{p} s\}$;
return $\langle Q, \delta, \phi, F \rangle$;

Algorithm 1: Translation of a SERE ϕ to a NFA. Remember that rules (A)–(F) are always applied.

3.3 Automaton Construction

The traditional way to construct a finite automaton from such a linear form is to associate its states to regular expressions. For a state $r \in \text{Terms}(r)$, we interpret the pairs (p_i, s_i) in $\textsf{LF}(r)$ as a transition $r \xrightarrow{p_i} s_i$. Algorithm 1 shows a straightforward implementation of that construction. Final states are those that correspond to expressions that accept the empty word. At the end of this algorithm we naturally have $Q = \text{Terms}(\phi)$ by construction, which ensures termination. The fact that $\mathscr{L}(\langle Q, \delta, \phi, F \rangle) = \mathscr{L}(\phi)$ follows from Theorem 1.

Although $\text{Terms}(\phi)$ is defined as the smallest subset of SERE recursively produced by $\textsf{LF}$, the resulting automaton is not necessarily minimal in terms of number of states. Because we only use syntactic equivalence, the construction can produce two states labeled by SEREs $r_1 \not\equiv r_2$ such that $\mathscr{L}(r_1) = \mathscr{L}(r_2)$.

This algorithm can be altered in several ways in attempt to simplify the resulting automata. In Sect. 4 we present ways to *simplify* the linear forms $\boxed{\textsf{LF}(f)}$ before they get used in Algorithm 1. Then in Sect. 5 we propose larger modifications of Algorithm 1 meant to fuse states with identical linear forms.

4 Linear Form Simplifications

When constructing an automaton from a linear form, it is possible to alter the shape of the automaton constructed by transforming the linear forms it uses into other, equivalent linear forms. In this section we present a few transformations that aim at simplifying linear forms. By "*simplifying*" we mean to reduce the number of pairs in the hope that this results in a smaller automaton. Simplifying a finite automaton can of course be done after its construction using more traditional algorithms like bisimulation-based reductions [18], however it is always good to look for cheap opportunities to keep the intermediate automaton small.

Definition 9 (Unique Suffix and Unique Prefix simplifications). *Let L be a linear form, let* $\mathsf{MergePre}(L, s) = \bigvee_{(p,s) \in L} p$ *be the (Boolean) disjunction of all prefixes sharing a given suffix* s, *and let* $\mathsf{MergeSuf}(L, p) = \bigvee_{(p,s) \in L} s$ *be the (rational) disjunction of all suffixes sharing a given prefix* p.

We define US *(unique suffixes) and* UP *(unique prefixes) as follows:*

$$\mathsf{US}(L) = \{(\mathsf{MergePre}(L, s), s) \mid (p, s) \in L\}$$
$$\mathsf{UP}(L) = \{(p, \mathsf{MergeSuf}(L, p)) \mid (p, s) \in L\}$$

Replacing $\mathsf{LF}(f)$ by $\mathsf{US}(\mathsf{LF}(f))$ in Algorithm 1 is equivalent to merging the edges of the automaton that have the same source and same destination. For instance $\mathsf{US}(\{(a, r), (b, r)\}) = \{(a \vee b, r)\}$.

Replacing $\mathsf{LF}(f)$ by $\mathsf{UP}(\mathsf{LF}(f))$ in Algorithm 1 is merging outgoing edges that share the same label. In Antimorov's setup [2], where prefixes of linear forms are letters, using UP would create a deterministic automaton. However, because in our setup prefixes are Boolean formulas, this is not the case: UP can remove *some* non-determinism, but the result will not necessarily be deterministic. For instance the non-deterministic linear form $\{(a, q_1), (a \wedge b, q_2)\}$ is unchanged by UP. If we wish to construct a deterministic automaton, we can use $\det(\mathsf{LF}(f))$ (see Proposition 1). Our intent with UP is therefore not to produce a deterministic automaton, but to help reduce the size of a non-deterministic result.

We should point out that the equivalent of Theorem 2 still holds when $\mathsf{UP}(\mathsf{LF}(f))$ is used because the terms created by this new variant are disjunctions of terms created by the original construction.

Unfortunately, it is also possible that using UP will introduce new additional states in the automaton. For instance $\mathsf{UP}(\{(a, q_1), (a, q_2), (b, q_1), (\neg b, q_2)\}) = \{(a, q_1 \vee q_2), (b, q_1), (\neg b, q_2)\}$ would be introducing the state $q_1 \vee q_2$ that was not present initially.

Because US only merges edges, it sounds natural to use it together with UP. However, while it is possible to find cases where replacing $\mathsf{LF}(f)$ by $\mathsf{US}(\mathsf{UP}(\mathsf{LF}(f)))$ is better than replacing $\mathsf{LF}(f)$ by $\mathsf{UP}(\mathsf{US}(\mathsf{LF}(f)))$, the opposite also exists.

5   Signature and Transition-Based Variants

We now discuss variants of Algorithm 1, orthogonal to previous simplifications.

Consider the automaton of Fig. 2, where the first two states are labeled by formulas that have the same linear form, and so are the last two states. If two expressions r_1 and r_2 have the same linear form $\mathsf{LF}(r_1) = \mathsf{LF}(r_2)$, it implies that $\mathscr{L}(r_1) \setminus \varepsilon = \mathscr{L}(r_2) \setminus \varepsilon$. Therefore, states that correspond to formulas with the same linear form (i.e., states that have identical sets of outgoing transitions) can be merged if they are both accepting, or both rejecting. Thus, the first two states of Fig. 2 could be merged.

We can obtain such a merge automatically if we modify our translation as in Algorithm 2 to label each state by a pair $(\mathsf{LF}(\varphi), \lambda(\varphi))$ that we call the signature of φ. This gives the automaton of Fig. 3.

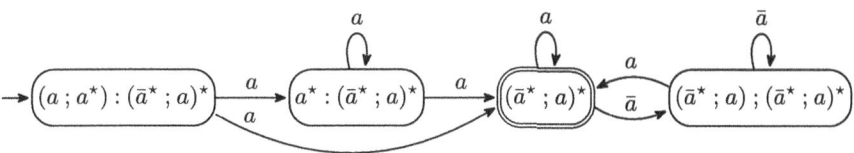

Fig. 2. Automaton for $\varphi = (a \, ; a^*) : (\bar{a}^* \, ; a)^*$. We have $\mathsf{LF}(\varphi) = \mathsf{LF}(a^* : (\bar{a}^* \, ; a)^*) = \{(a, a^* : (\bar{a}^* \, ; a)^*), (a, (\bar{a}^* \, ; a)^*)\}$, and $\mathsf{LF}((\bar{a}^* \, ; a)^*) = \mathsf{LF}((\bar{a}^* \, ; a) \, ; (\bar{a}^* \, ; a)^*) = \{(a, (\bar{a}^* \, ; a)^*), (\bar{a}, (\bar{a}^* \, ; a) \, ; (\bar{a}^* \, ; a)^*)\}$.

Currently, the last two states of Fig. 3 may not be merged because one is accepting while the other is not. We could however merge them by changing our automaton formalism such that the notion of acceptance is carried by the transitions instead of the states. Although finite automata are seldom used with transition-based acceptance [21], ω-automata (i.e., automata over infinite words) with transition-based acceptance have been used for a long time as they often lead to simpler algorithms [15–17, 19, 20, to cite a few]. Let us define a transition-based finite automaton:

Definition 10 (TFA). *A transition-based finite automaton is a tuple* $\mathcal{A} = \langle Q, \delta, \iota, \beta \rangle$ *where* Q *is a finite set of states,* $\delta \subseteq Q \times \mathbb{B}(AP) \times \mathbb{B} \times Q$ *is the transition relation,* $\iota \in Q$ *is the initial state, and* $\beta \in \mathbb{B}$ *is a Boolean indicating whether* ε *should be accepted.*

We write $s \xrightarrow{f,b} d$ *when* $(s, f, b, d) \in \delta$.
A sequence of valuations $\sigma \in \Sigma^n$ *of size* n *is* accepted *by* $\mathcal{A}$ *if either* $n = 0$ *and* $\beta = \top$, *or* $n > 0$ *and there exists a sequence of transitions* $\rho = s_0 \xrightarrow{f_0, b_0} s_1 \xrightarrow{f_1, b_1} \cdots \xrightarrow{f_{n-1}, b_{n-1}} s_n$ *such that* $s_0 = \iota$, $b_{n-1} = \top$, *and for all* i, $\sigma(i) \models f_i$.
The language *of* $\mathcal{A}$, *denoted* $\mathscr{L}(\mathcal{A})$ *is the set of words accepted by* $\mathcal{A}$.

In other words, transitions of a TFA carry an extra Boolean that is used to mark the transition as accepting, and a word is accepted if it is recognized by a run whose last transition is accepting. Graphically, we represent accepting transitions using arrows with double lines. The acceptance of the empty word is indicated by a special Boolean β in the definition, and can be represented graphically by using double lines on the arrow indicating the initial state.

TFA enjoy similar properties as traditional finite automata: they are as expressive as regular expressions, are closed under Boolean operations, etc. [21] However they can be slightly smaller, as we shall see in our evaluation.

Using the above definition, Algorithm 3 generates the automaton of Fig. 4.

In our case, we have additional motivation for using TFAs. The reason we are working on translating SERE to automata is that SERE are part of the PSL and SVA standards. However, the PSL/SVA standards assume a SERE will always match a non-empty word. Therefore, the Boolean β that we added to the definition of a TFA to allow it to recognize ε can simply be ignored. Furthermore, as we translate a PSL formula into an ω-automaton, we build upon

input : A SERE ϕ
output: An NFA $\mathcal{A}$ such that
　　　$\mathscr{L}(\mathcal{A}) = \mathscr{L}(\phi)$

$L, b \leftarrow \mathsf{LF}(\phi), [\lambda(\phi) = \varepsilon]$;
$Q, \delta, F \leftarrow \{(L, b)\}, \varnothing, \varnothing$;
todo.push$((L, b))$;
if b **then**
  $\lfloor$ $F \leftarrow F \cup \{(L, b)\}$;
while todo $\neq \varnothing$ **do**
  $(L, b) \leftarrow$ todo.pop();
  foreach $(p, s) \in L$ **do**
    $L', b' \leftarrow \mathsf{LF}(s), [\lambda(s) = \varepsilon]$;
    if $(L', b') \notin Q$ **then**
      $Q \leftarrow Q \cup \{(L', b')\}$;
      todo.push $((L', b'))$;
      if b' **then**
        $\lfloor$ $F \leftarrow F \cup \{(L', b')\}$;
    $\delta \leftarrow \delta \cup \{(L, b) \xrightarrow{p} (L', b')\}$;
return $\langle Q, \delta, \phi, F \rangle$;

Algorithm 2: Translation that identifies states with identical linear form and identical ε acceptance.

input : A SERE ϕ
output: A TFA $\mathcal{A}$ such that
　　　$\mathscr{L}(\mathcal{A}) = \mathscr{L}(\phi)$

$L, b \leftarrow \mathsf{LF}(\phi), [\lambda(\phi) = \varepsilon]$;
$Q, \delta \leftarrow \{L\}, \varnothing$;
todo.push(L);
while todo $\neq \varnothing$ **do**
  $L \leftarrow$ todo.pop();
  foreach $(p, s) \in L$ **do**
    $L', b' \leftarrow \mathsf{LF}(s), [\lambda(s) = \varepsilon]$;
    if $L' \notin Q$ **then**
      $Q \leftarrow Q \cup \{L'\}$;
      todo.push (L');
    $\delta \leftarrow \delta \cup \{L \xrightarrow{p, b'} L'\}$;
return $\langle Q, \delta, \phi, b \rangle$;

Algorithm 3: Translation to transition-based automata, identifying states with identical linear form regardless of ε acceptance.

translation algorithms that naturally produce transition-based ω-automata [9]. In this context, it seems more natural to have SERE converted into TFA. The fact that TFA are more succinct comes as a bonus.

6　Experimental Evaluation

Our algorithms have been implemented in a development version of Spot [10]. A reproducibility package, archived at https://doi.org/10.5281/zenodo.10799850, contains our implementation, a Jupyter notebook to use it interactively, and scripts to reproduce our experiments.

We are not aware of any existing benchmark of SEREs. Therefore, to evaluate our work, we randomly generated a set of SEREs using Spot's `randltl` tool, with equal probability of occurrence for all SERE operators. We grouped the random SEREs into groups of expressions with equal size (number of nodes in their syntax tree) and equal number of unique atomic propositions, capping each group to 50 expressions. The resulting set has 12500 unique SEREs with sizes ranging from 1 to 35, and between 1 and 15 atomic propositions.

Benchmarks were run on an AMD Ryzen 5 3600 CPU, with 16 GB of RAM, and with core frequency capped at 2.2 GHz to minimize the impact of throttling on timing measurements. For each SERE we evaluated variants of the translation by measuring the number of states of the produced automata, and the time needed to produce them. (We also measured the number of edges, but do not report it here).

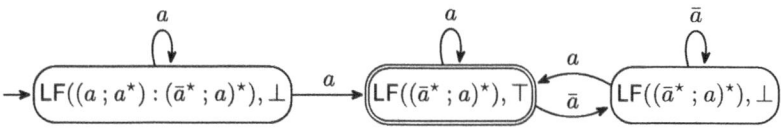

Fig. 3. Automaton obtained by merging states labeled by formulas that have the same linear form and the same acceptance of ε.

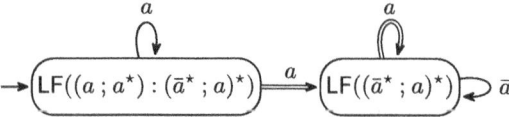

Fig. 4. Transition-based automaton obtained by merging states labeled by formulas that have the same linear form, regardless of the acceptance of ε, since the latter is decided on transitions.

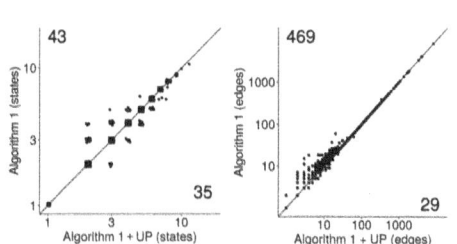
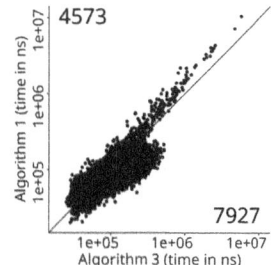

Fig. 5. Effect of UP on Algorithm 1.   **Fig. 6.** Time of Algorithm 1–3.

Scatter plots that show number of states use a jitter of ± 0.4 over their position to distinguish points. The numbers in the top left and bottom right corners of the plots count how many points are strictly above or below the diagonal.

We start by evaluating the impact of the simplification strategies of Sect. 4. Figure 5 presents the impact of UP on the number of states and edges of automata produced by Algorithm 1. As mentioned in Sect. 4, UP has mitigated results: it improves the number of states of the automaton almost as often as it worsens it. However the number of transitions is reduced in general.

Figure 7 shows that Algorithms 2–3 provide a more important reduction of automata sizes compared to Algorithm 1. Impact on translation time, as seen on Fig. 6, is not significant (average speedup is -10%). Small automata have an overhead because of the labeling of states by linear forms rather than formulas, but the savings in size also yields savings in time for larger automata.

Figures 8 and 11 show that applying UP in Algorithm 3 has more effect than it had on Algorithm 1 (compare with Fig. 5 where the impact of UP on the number of states was marginal). Figure 9 shows using UP ∘ US on Algorithm 3 does not yield significant changes in number of states, compared to only using

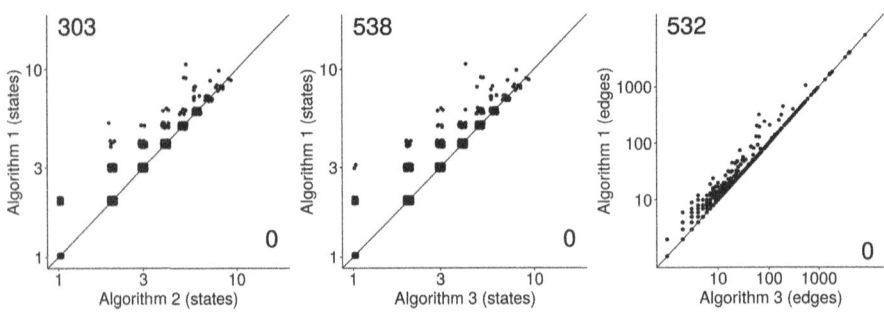

Fig. 7. Comparisons of Algorithms 1, 2, and 3.

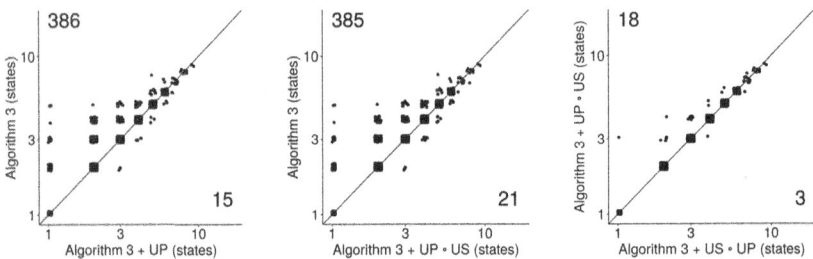

Fig. 8. Impact of UP on Algorithm 3

Fig. 9. Impact of UP ∘ US on Algorithm 3

Fig. 10. Comparison of UP ∘ US and US ∘ UP

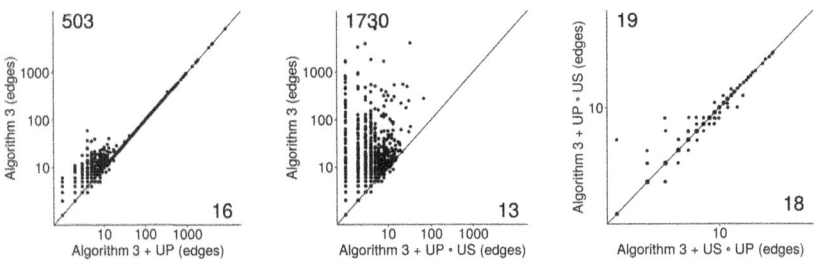

Fig. 11. Impact of UP on Algorithm 3

Fig. 12. Impact of UP∘US on Algorithm 3

Fig. 13. Comparison of UP ∘ US and US ∘ UP

UP. However Fig. 12 shows an impressive reduction in the number of edges. Using US ∘ UP instead of UP would not change the number of states, as this is only merging edges, so we do not compare it to UP. Figures 10 and 13 shows that in practice, the impact of the order of application between UP and US discussed in Sect. 4 is rather limited, producing automata with a different number of states in only 21 cases out of our 12500 formulas.

7   Conclusion

We adapted Antimirov's non-deterministic automata construction based on linear forms, to the semi-extended regular expressions used by the PSL and SVA standards. As these SERE are defined on alphabet of the form 2^{AP}, we introduced some rewritings (UP, US) of these linear forms and evaluated their impact on a large benchmark. We also introduced alternative translation algorithms that use the linear form to simplify the automaton during its construction, or that build a transition-based automaton.

Our evaluation reveals that using transition-based automata, labeling them with linear forms, and simplifying those linear forms with UP are cheap and effective ways of keeping the output small. A compact output matters in applications where the automaton is constructed on-the-fly or only partially, and therefore cannot benefit from subsequent simplifications. (Satisfiability, which cannot be decided syntactically because of the intersection operator, is one such problem).

Finally, we constructed a SERE benchmark dataset, which we hope can be reused in future work to compare different SERE, PSL or SVA translators.

References

1. 1800-2017 - IEEE Standard for SystemVerilog–Unified Hardware Design, Specification, and Verification Language. IEEE (2018). https://doi.org/10.1109/IEEESTD.2018.8299595
2. Antimirov, V.: Partial derivatives of regular expressions and finite automaton constructions. Theoret. Comput. Sci. **155**(2), 291–319 (1996). https://doi.org/10.1016/0304-3975(95)00182-4
3. Baier, C., Katoen, J.-P.: Principles of Model Checking. MIT Press, Cambridge (2008)
4. Bartocci, E., Falcone, Y., Francalanza, A., Reger, G.: Introduction to runtime verification. In: Bartocci, E., Falcone, Y. (eds.) Lectures on Runtime Verification. LNCS, vol. 10457, pp. 1–33. Springer, Cham (2018). https://doi.org/10.1007/978-3-319-75632-5_1
5. Bryant, R.E.: Symbolic Boolean manipulation with ordered binary-decision diagrams. ACM Comput. Surv. **24**(3), 293–318 (1992)
6. Brzozowski, J.A.: Derivatives of regular expressions. J. ACM **11**(4), 481–494 (1964). https://doi.org/10.1145/321239.321249
7. Caron, P., Champarnaud, J.-M., Mignot, L.: Partial derivatives of an extended regular expression. In: Dediu, A.-H., Inenaga, S., Martín-Vide, C. (eds.) LATA 2011. LNCS, vol. 6638, pp. 179–191. Springer, Heidelberg (2011). https://doi.org/10.1007/978-3-642-21254-3_13
8. D'Antoni, L., Veanes, M.: Minimization of symbolic automata. In: Jagannathan, S., Sewell, P. (eds.) Proceedings of the 41st Annual Symposium on Principles of Programming Languages (POPL 2014), pp. 541–554. ACM (2014). https://doi.org/10.1145/2535838.2535849
9. Duret-Lutz, A.: LTL translation improvements in Spot 1.0. Int. J. Crit. Comput.-Based Syst. **5**(1/2), 31–54 (2014). https://doi.org/10.1504/IJCCBS.2014.059594

10. Duret-Lutz, A., et al.: From spot 2.0 to spot 2.10: what's new? In: Shoham, S., Vizel, Y. (eds.) CAV 2022. LNCS, vol. 13372, pp. 174–187. Springer, Cham (2022). https://doi.org/10.1007/978-3-031-13188-2_9

11. Eisner, C., Fisman, D.: A Practical Introduction to PSL. Series on Integrated Circuits and Systems. Springer, Heidelberg (2006). https://doi.org/10.1007/978-0-387-36123-9

12. Finkbeiner, B.: Synthesis of reactive systems. In: Esparza, J., Grumberg, O., Sickert, S. (eds.) Dependable Software Systems Engineering. NATO Science for Peace and Security Series - D: Information and Communication Security, vol. 45, pp. 72–98. IOS Press Ebooks (2016). https://doi.org/10.3233/978-1-61499-627-9-72

13. Fischer, M.J., Ladner, R.E.: Propositional dynamic logic of regular programs. J. Comput. Syst. Sci. **18**(2), 194–211 (1979). https://doi.org/10.1016/0022-0000(79)90046-1. ISSN 0022-0000

14. Giacomo, G.D., Vardi, M.Y.: Linear temporal logic and linear dynamic logic on finite traces. In: Proceedings of the 33rd International Joint Conference on Artificial Intelligence (IJCAI 2013), pp. 854–860. AAAI Press (2013). https://doi.org/10.5555/2540128.2540252. ISBN 9781577356332

15. Giannakopoulou, D., Lerda, F.: From states to transitions: improving translation of LTL formulae to Büchi automata. In: Peled, D.A., Vardi, M.Y. (eds.) FORTE 2002. LNCS, vol. 2529, pp. 308–326. Springer, Heidelberg (2002). https://doi.org/10.1007/3-540-36135-9_20

16. Kurshan, R.P.: Complementing deterministic Büchi automata in polynomial time. J. Comput. Syst. Sci. **35**(1), 59–71 (1987). https://doi.org/10.1016/0022-0000(87)90036-5

17. Le Saëc, B., Litovsky, I.: On the minimization problem for ω-automata. In: Prívara, I., Rovan, B., Ružička, P. (eds.) MFCS 1994. LNCS, vol. 841, pp. 504–514. Springer, Heidelberg (1994). https://doi.org/10.1007/3-540-58338-6_97

18. Lombardy, S., Sakarovitch, J.: Two routes to automata minimization and the ways to reach it efficiently. In: Câmpeanu, C. (ed.) CIAA 2018. LNCS, vol. 10977, pp. 248–260. Springer, Cham (2018). https://doi.org/10.1007/978-3-319-94812-6_21

19. Michel, M.: Algebre de machines et logique temporelle. In: Fontet, M., Mehlhorn, K. (eds.) STACS 1984. LNCS, vol. 166, pp. 287–298. Springer, Heidelberg (1984). https://doi.org/10.1007/3-540-12920-0_26

20. Varghese, T.: Parity and generalized Büchi automata—determinisation and complementation. Ph.D. thesis, University of Liverpool (2014)

21. Xiao, S., Li, J., Zhu, S., Shi, Y., Pu, G., Vardi, M.: On-the-fly synthesis for LTL over finite traces. In: Proceedings of the 35th AAAI Conference on Artificial Intelligence (AAAI 2021), vol. 35, pp. 6530–6537 (2021). https://doi.org/10.1609/aaai.v35i7.16809. Technical Tracks 7

Exact Descriptional Complexity of Determinization of Input-Driven Pushdown Automata

Olga Martynova[✉][iD]

Department of Mathematics and Computer Science, St. Petersburg State University,
7/9 Universitetskaya nab., Saint Petersburg 199034, Russia
st062453@student.spbu.ru

Abstract. The number of states and stack symbols needed to determinize nondeterministic input-driven pushdown automata (NIDPDA) working over a fixed alphabet is determined precisely. It is proved that in the worst case exactly 2^{n^2} states are needed to determinize an n-state NIDPDA, and the proof uses witness automata with a stack alphabet $\Gamma = \{0, 1\}$ working on strings over a 4-symbol input alphabet (Only an asymptotic lower bound was known before in the case of a fixed alphabet). Also, the impact of NIDPDA determinization on the size of stack alphabet is determined precisely for the first time: it is proved that $s(2^{n^2} - 1)$ stack symbols are necessary in the worst case to determinize an n-state NIDPDA working over an input alphabet of size $s + 5$ with s left brackets (The previous lower bound was only asymptotic in the number of states and did not depend on the number of left brackets).

1   Introduction

An input-driven pushdown automaton, also known as a visibly pushdown automaton, is a model of computation equipped with a stack. An automaton has finitely many states. It reads an input string from the left to the right changing its state according to the current state and symbol scanned. The automaton also uses an infinite memory in the form of stack with restricted access. The input alphabet is $\Sigma = \Sigma_0 \cup \Sigma_{+1} \cup \Sigma_{-1}$, with Σ_0 containing neutral symbols, Σ_{+1} left brackets and Σ_{-1} right brackets. When an input-driven pushdown automaton reads a left bracket, it pushes onto the stack a symbol of its stack alphabet; when it sees a right bracket, it looks at the top symbol of the stack and pops it off; and when the automaton processes a neutral symbol, it makes a transition without accessing the stack.

Deterministic input-driven pushdown automata (DIDPDA) were introduced by Mehlhorn [7]. He also proved that every language defined by an n-state DID-PDA can be recognized by an algorithm that uses $O(\frac{(\log n)^2}{\log \log n})$ bits of memory

This work was supported by the Ministry of Science and Higher Education of the Russian Federation, agreement 075-15-2022-287.

S. Z. Fazekas (Ed.): CIAA 2024, LNCS 15015, pp. 249–260, 2024.
https://doi.org/10.1007/978-3-031-71112-1_18

and works in polynomial time. Von Braunmühl and Verbeek [3] considered a nondeterministic version of input-driven pushdown automata (NIDPDA) that at every step may have several possible actions. Such an automaton accepts a string if there is at least one accepting computation on it. A natural question immediately follows: whether deterministic and nondeterministic input-driven pushdown automata are equal in power. Von Braunmühl and Verbeek [3] made the first determinization construction (see the modern variant in the survey by Okhotin and Salomaa [10, Thm. 1]): for an n-state NIDPDA, working over an alphabet $\Sigma = \Sigma_0 \cup \Sigma_{+1} \cup \Sigma_{-1}$, they constructed a DIDPDA with 2^{n^2} states and with $2^{n^2} |\Sigma_{+1}|$ stack symbols recognizing the same language. Also von Braunmühl and Verbeek [3] improved the result by Mehlhorn [7]: they have shown that every language defined by an n-state DIDPDA is recognized by an algorithm working in logarithmic memory. Later Rytter [13] created a simpler algorithm for this task also using $\log n$ bits of memory.

Alur and Madhusudan [1,2] reintroduced the model of input-driven pushdown automata under the name of *visibly pushdown automata* and obtained many important results on these automata. They proved that the class of languages recognized by these automata is closed under intersection, union, concatenation and the Kleene star. They also defined input-driven pushdown automata that work on infinite strings and investigated properties of this variant of the model. Alur and Madhusudan [1,2] established the first lower bound on the number of states needed for NIDPDA determinization: for each n, they constructed an n-state NIDPDA with the fixed input alphabet such that any deterministic automaton recognizing the same language has at least $2^{\Omega(n^2)}$ states; this lower bound is asymptotically tight. Furthermore, Alur and Madhusudan [1,2] studied decidability and complexity of input-driven pushdown automata and showed that universality and inclusion problems for NIDPDA are NEXP-complete. There is some current research on decision problems for NIDPDA. Han, Ko and Salomaa [4] built an algorithm that for a given NIDPDA decides in polynomial time whether its path size is finite, that is, whether a number of leaves in a tree of its computations on every string is bounded by a common constant. They also proved that deciding whether the path size of a given NIDPDA is less than a given number is EXP-complete.

The research on NIDPDA determinization was continued by Okhotin, Piao and Salomaa [9], who bounded from below not only the number of states in the deterministic automaton but also the number of stack symbols it uses. They showed that using $2^{\Omega(n^2)}$ stack symbols can be necessary to determinize an n-state NIDPDA. Furthermore, Okhotin, Piao and Salomaa [9] and later Okhotin and Salomaa [11] studied the state complexity of different operations on DIDPDA. Jirásková and Okhotin [5] continued improving bounds for operations on DIDPDA, and in addition proved that in the worst case one needs 2^{n^2} states to determinize an n-state NIDPDA. This lower bound is precise, however, witness automata used by Jirásková and Okhotin [5] work over an input alphabet of size exponential in n. So the problem of determining the exact state complexity of NIDPDA determinization in the case of a bounded alphabet has remained open.

The state complexity of determinization was investigated for variants of the classical model of input-driven pushdown automata. Nguyen Van Tang and Ogawa [14] introduced event-clock input-driven pushdown automata and proved that these automata can be determinized. Later Ogawa and Okhotin [8] defined a direct determinization construction for these automata and proved a lower bound to this transformation which is asymptotically precise both in the number of states and in the size of a stack alphabet. Rose and Okhotin [12] first considered probabilistic input-driven pushdown automata and determined asymptotically precisely the state complexity of determinization for this model. Kutrib, Malcher and Wendlandt [6] defined a variant of input-driven pushdown automata in which every input string is read twice: first, a deterministic sequential transducer determines the type of each symbol (in this model an alphabet is not initially split into left brackets, right brackets and neutral symbols), and then the string is read by an input-driven pushdown automaton. Kutrib et al. [6] determinized this model, investigated its power and its closure properties.

In this paper I improve the bounds on the complexity of determinization for classical input-driven pushdown automata (NIDPDA) both in the number of states and in the size of the stack alphabet. In Sect. 3, it is proved that in the worst case one needs 2^{n^2} states to determinize an n-state NIDPDA that uses 2 stack symbols and works over a 4-symbol input alphabet. This is the first precise lower bound on the number of states needed for NIDPDA determinization with bounded alphabet.

In Sect. 4, the precise lower bound $|\Sigma_{+1}|(2^{n^2} - 1)$ on the number of stack symbols in a deterministic automaton that recognizes a language defined by an n-state NIDPDA is proved in a special case of only one left bracket ($\Sigma_{+1} = \{<\}$). The witness nondeterministic automata in this proof have stack alphabets growing linearly in n, and the input alphabet is bounded.

Finally, in Sect. 5 I establish the exact lower bound on the complexity of NIDPDA determinization, for any number of left brackets in the input alphabet fewer than 2^{n^2}, both in the number of states and in the number of stack symbols: in the worst case one needs 2^{n^2} states and $|\Sigma_{+1}|(2^{n^2} - 1)$ stack symbols to determinize an n-state NIDPDA. Moreover, these examples of NIDPDA that are hard to determinize work over input alphabets that do not depend on n and use stack alphabets growing linearly in n and logarithmically in the number of left brackets. This lower bound is precise, since the known upper bound of $|\Sigma_{+1}| \cdot 2^{n^2}$ stack symbols is improved to $|\Sigma_{+1}|(2^{n^2} - 1)$ stack symbols in Sect. 2.

2   Input-Driven Pushdown Automata

This paper uses the definition of input-driven pushdown automata given by Alur and Madhusudan [1] but with one difference: Alur and Madhusudan [1] allow computations on ill-nested strings, whereas in this paper input strings must be well-nested, as in the first definitions of these automata. A nondeterministic automaton will be defined first, and a deterministic one is its special case.

Definition 1 (Mehlhorn [7], von Braunmühl and Verbeek [3], Alur and Madhusudan [1]). *A nondeterministic input-driven pushdown automaton (NIDPDA) is a sextuple $A = (\Sigma, Q, \Gamma, Q_0, (\delta_a)_{a \in \Sigma}, F)$, where*

- $\Sigma = \Sigma_0 \cup \Sigma_{+1} \cup \Sigma_{-1}$ *is a finite input alphabet split into three disjoint sets: Σ_0 contains neutral symbols, Σ_{+1} consists of left brackets and Σ_{-1} has right brackets;*
- Q *is a finite set of states of the automaton;*
- Γ *is a finite stack alphabet;*
- $Q_0 \subseteq Q$ *is a set of initial states;*
- $(\delta_a)_{a \in \Sigma}$ *are functions that for each symbol of the alphabet define possible actions of the automaton on this symbol:*
  - *for a neutral symbol $a \in \Sigma_0$, the function $\delta_a \colon Q \to 2^Q$ for each state gives a set of possible next states of the automaton;*
  - *for a left bracket $a \in \Sigma_{+1}$, the function $\delta_a \colon Q \to 2^{Q \times \Gamma}$ for each state specifies the set of pairs (q, s) in which the automaton on a symbol a can make a transition forward in the state q pushing the symbol s onto the stack;*
  - *for a right bracket $a \in \Sigma_{-1}$, the function $\delta_a \colon Q \times \Gamma \to 2^Q$ for a state and for a symbol popped off the stack gives a set of all possible next states;*
- $F \subseteq Q$ *is a set of accepting states.*

Inputs of the automaton A are well-nested strings over an alphabet Σ. A string w is called well-nested if it has as many left brackets as right brackets and if every prefix contains at least as many left brackets as right brackets.

Let $w = a_1 \ldots a_\ell$ be a well-nested string. A computation of the automaton A on the string w is a sequence $(p_0, \alpha_0), \ldots, (p_\ell, \alpha_\ell)$ of pairs of a state and of stack contents, with the neighbouring pairs related to each other as follows. The first pair is initial: $p_0 \in Q_0$, $\alpha_0 = \varepsilon$, and every next pair (p_{i+1}, α_{i+1}) is connected to the previous pair depending on the current symbol a_{i+1} of the string:

- *if $a_{i+1} = c \in \Sigma_0$, then $p_{i+1} \in \delta_c(p_i)$ and $\alpha_{i+1} = \alpha_i$;*
- *if $a_{i+1} = \; < \; \in \Sigma_{+1}$, then $(p_{i+1}, s) \in \delta_<(p_i)$ and $\alpha_{i+1} = \alpha_i s$, for some $s \in \Gamma$;*
- *if $a_{i+1} = \; > \; \in \Sigma_{-1}$, then $\alpha_i = \beta s$, for some $s \in \Gamma$, and $p_{i+1} \in \delta_>(p_i, s)$ and $\alpha_{i+1} = \beta$.*

At the end of the computation the stack is empty: $\alpha_\ell = \varepsilon$, since the string w is well-nested. If the last state of the sequence is accepting ($p_\ell \in F$), then the computation is called accepting. The string w is said to be accepted by the automaton if there is at least one accepting computation on this string. And the automaton A defines the language $L(A)$ consisting of all well-nested strings over the alphabet Σ accepted by the automaton.

An input-driven pushdown automaton is called *deterministic* (DIDPDA) if it has a unique initial state, $|Q_0| = 1$, and if in every situation it has exactly one possible action:

- $|\delta_a(q)| = 1$, for all $q \in Q$ and $a \in \Sigma_0 \cup \Sigma_{+1}$;

– $|\delta_a(q, s)| = 1$, for all $q \in Q$, $s \in \Gamma$ and $a \in \Sigma_{-1}$.

The previous upper bound, that every n-state NIDPDA can be transformed into a deterministic automaton with 2^{n^2} states and $2^{n^2}|\Sigma_{+1}|$ stack symbols, can be slightly improved.

Theorem 1. *Let A be an n-state NIDPDA, working over an alphabet $\Sigma = \Sigma_0 \cup \Sigma_{+1} \cup \Sigma_{-1}$. Then, there is a DIDPDA with at most 2^{n^2} states and $|\Sigma_{+1}|(2^{n^2} - 1)$ stack symbols that recognizes the language $L(A)$.*

The proof is omittted due to space constraints.

3   The Exact Bound on the Number of States

In this section, the exact lower bound of 2^{n^2} states on the state complexity of NIDPDA determinization is proved for the first time in the case of a fixed input alphabet. Moreover, the stack alphabet of nondeterministic automata is also fixed and is of size 2. This is the minimal size of the stack alphabet that allows an automaton to get any information from the stack.

Theorem 2. *For each $n \geqslant 1$, there is an n-state NIDPDA $A_n = (\Sigma, Q, \Gamma, q_0, (\delta_a)_{a \in \Sigma}, F)$, over a 4-symbol input alphabet $\Sigma_{+1} = \{<\}$, $\Sigma_{-1} = \{>\}$, $\Sigma_0 = \{-, \#\}$, and with stack alphabet $\Gamma = \{0, 1\}$, such that any DIDPDA recognizing the language $L(A_n)$ has at least 2^{n^2} states.*

Proof. Let n be fixed. The desired automaton A_n is defined as follows. The set of states is $Q = \{0, \ldots, n-1\}$, with the initial state 0. All states are accepting: $F = Q$. And the symbols of the input alphabet act in the following way.

The automaton A_n has nondeterministic transitions on the symbol $\#$ only. The automaton can move from every state to every state by this symbol:

$$\delta_\#(i) = \{0, \ldots, n-1\}, \qquad \text{for } i \in \{0, \ldots, n-1\}.$$

The automaton works deterministically on all other symbols of Σ. The symbol '$-$' decreases the state of A_n by 1 modulo n:

$$\delta_-(i) = \{(i-1) \bmod n\}, \qquad \text{for } i \in \{0, \ldots, n-1\}.$$

By the left bracket $<$, the automaton does not change the state and pushes the information, whether the current state is 0 or not, onto the stack, that is, for a state i, it pushes $\operatorname{sgn} i$, which is 0 if the state is 0, and 1 otherwise:

$$\delta_<(i) = \{(i, \operatorname{sgn} i)\}, \qquad \text{for } i \in \{0, \ldots, n-1\}.$$

The automaton rejects on the right bracket if the current state and the symbol at the top of the stack both equal 0, and continues in the same state otherwise.

$$\delta_>(i, s) = \{i\}, \qquad \text{for } i \in \{0, \ldots, n-1\}, \ s \in \{0, 1\}, \ (i, s) \neq (0, 0).$$

Now the automaton A_n has been defined. It should be proved that a deterministic automaton must use at least 2^{n^2} states to recognize the language $L(A_n)$. First, strings with specific behaviour of the automaton A_n are constructed.

Claim 1. *For each relation $R \subseteq Q \times Q$, there is a well-nested string $w_R \in \Sigma^*$, such that for all $i, j \in Q$, the automaton A_n can start reading the string w_R in the state i and finish reading it in the state j if and only if $(i, j) \in R$.*

For each state $i \in Q$, there is such a well-nested string $y_i \in \Sigma^*$, that if the automaton A_n enters y_i in some state other than i, then it rejects, and if A_n enters the string y_i in the state i, then it can leave the string in the state i and cannot leave the string y_i in any other state.

There are two strings $x \in \{-, <\}^*$ and $x' \in \{-, >\}^*$, such that x has as many left brackets as there are right brackets in x'. If the automaton A_n begins reading x in the state $i \in Q$, then it leaves this string in the same state i and pushes some string α_i onto the stack. And if A_n enters the string x' in a state $i' \in Q$ having the string α_i on the top of the stack, then, if $i' \neq i$, it rejects, and if $i' = i$, it leaves the string x' in the state i.

The idea of the proof of Theorem 2 is to build strings with the special properties defined in Claim 1, and then to consider computations of the automaton A_n on strings of the form $\#x w_R y_j \# x' y_i$, for $R \subseteq Q \times Q$ and $i, j \in Q$. It will be proved that the automaton A_n accepts a string $\#x w_R y_j \# x' y_i$ if and only if $(i, j) \in R$, and that each deterministic automaton needs a lot of states to do the same.

The conditions on strings w_R in Claim 1 can be reformulated as follows.

For a well-nested string $w \in \Sigma^*$, one can define a *behaviour relation* $R(w) \subseteq Q \times Q$: the pair (i, j) is in $R(w)$ if and only if there is a computation of the nondeterministic automaton A_n on the string w, that begins in the state i and ends by leaving the string in the state j. In these terms, for each relation $R \subseteq Q \times Q$, one wants to construct a well-nested string w_R with the behaviour relation $R(w_R) = R$.

For example, for a full behaviour relation, one can take a string $\#$ as $w_{Q \times Q}$, because $R(\#) = Q \times Q$. To construct strings with all possible relations, it is enough to learn how to eliminate an arbitrary pair of states from the relation of a string. This is done in the following claim.

Claim 2. *There are strings $u_i \in \{-, <\}^*$, for all $i \in Q$, and $v_j \in \{-, >\}^*$, for all $j \in Q$, such that for each well-nested string $w \in \Sigma^*$, and for all states $i, j \in Q$, the equality $R(u_i w v_j) = R(w) \setminus \{(i, j)\}$ holds.*

Furthermore, the strings u_i and v_j satisfy the following conditions.

- *Each string u_i, for $i \in Q$, has exactly one left bracket, and there is exactly one right bracket in each string v_j, for $j \in Q$.*
- *If the automaton A_n begins reading the string u_i, for $i \in Q$, or v_j, for $j \in Q$, in some state, then it can leave the string only in this state.*

Strings u_i and v_j are defined as follows:

$$u_i = (-)^i <(-)^{n-i}, \qquad \text{for } i \in Q;$$
$$v_j = (-)^j >(-)^{n-j}, \qquad \text{for } j \in Q.$$

The idea of the proof of Claim 2 is that the automaton cannot enter a string $u_i w v_j$ in the state i and leave it in the state j, because otherwise it will push 0 onto the stack on the left bracket in the substring u_i, and it will get to the right bracket in the substring v_j in the state 0, pop the symbol 0 off the stack and reject. And since strings u_i and v_j have n symbols '−' each, these strings never change the state of the automaton, and therefore they cannot affect any computations on the string w beginning in a state other than i or ending in a state other than j. Then, $R(u_i w v_j) = R(w) \setminus \{(i,j)\}$.

Now it is time to construct all special strings in Claim 1. Strings w_R for all $R \subseteq Q \times Q$ can be obtained from the string $\#$ with a full behaviour relation, using Claim 2. Let $R \subseteq Q \times Q$ be an arbitrary relation. Let $(i_1, j_1), \ldots, (i_k, j_k)$ be all pairs of states that are not in R. Then, the string w_R can be defined as:

$$w_R = u_{i_k} u_{i_{k-1}} \cdots u_{i_1} \# v_{j_1} \cdots v_{j_{k-1}} v_{j_k}.$$

Next, the well-nested string $y_i \in \Sigma^*$, for $i \in Q$, can be constructed as:

$$y_i = w_R, \text{ where } R = \{(i,i)\}.$$

There is another way to define the string y_i, explicitly and without symbols $\#$:

$$y_i = (<>-)^i - (<>-)^{n-i},$$

In this construction, each pair of brackets $<>$ forbids the current state to be equal to 0 and as a result it is prohibited to enter the string in any state other than i.

The strings $x \in \{-,<\}^*$ and $x' \in \{-,>\}^*$ are constructed as follows. Consider the diagonal relation $R = \{(0,0),(1,1),\ldots,(n-1,n-1)\}$. The string with this relation is $w_R = u_{i_k} u_{i_{k-1}} \cdots u_{i_1} \# v_{j_1} \cdots v_{j_{k-1}} v_{j_k}$, where $(i_1, j_1), \ldots, (i_k, j_k)$ are all pairs of states that are not in R. Then, the strings x and x' are defined as:

$$x = u_{i_k} u_{i_{k-1}} \cdots u_{i_1},$$
$$x' = v_{j_1} \cdots v_{j_{k-1}} v_{j_k}.$$

It can be shown that x and x' thus defined have all desired properties, the proof of this fact is omitted due to space constraints.

Now all special strings have been constructed.

Claim 3. Let $i, j \in Q$ and $R \subseteq Q \times Q$, let the well-nested strings $w_R, y_i, y_j \in \Sigma^*$ and the strings $x \in \{-,<\}^*$, $x' \in \{-,>\}^*$ be constructed as in Claim 1. Then, the automaton A_n accepts the string $\#x w_R y_j \# x' y_i$ if and only if $(i,j) \in R$.

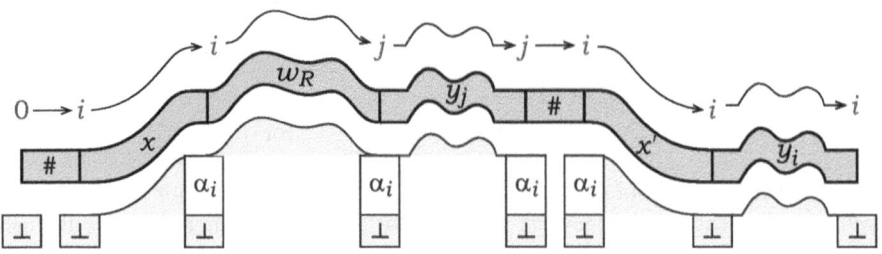

Fig. 1. The accepting computation of the automaton A_n on the string $\#xw_Ry_j\#x'y_i$, where $(i,j) \in R$.

Let (i,j) be in R. Then there is the following accepting computation of the automaton A_n on the string $\#xw_Ry_j\#x'y_i$, illustrated in Fig. 1. The automaton guesses the state i on the first symbol $\#$, then it moves through the substring x pushing the string α_i onto the stack, next it reads w_R changing its state from i to j (this is possible, since $(i,j) \in R$). The stack contents are the same before and after reading w_R. Then the automaton enters the substring y_j in the state j and leaves it in the state j, chooses the state i on the symbol $\#$, pops the string α_i off the stack while reading the substring x', and finally enters y_i in the state i, leaves it in the state i and accepts.

Now let the automaton A_n accept a string $\#xw_Ry_j\#x'y_i$, for some $i,j \in Q$ and $R \in Q \times Q$. It shall be proved that $(i,j) \in R$. Consider an arbitrary accepting computation of the automaton A_n on the string $\#xw_Ry_j\#x'y_i$. Let i' be the state guessed by the automaton on the first symbol $\#$, let j' be the state assumed by A_n after reading the substring w_R, and let the state i'' be chosen on the symbol $\#$ before the substring $x'y_i$. The automaton always gets out of the substring x' in the state in which it enters this substring, and also the automaton rejects if it enters the substring y_i in any state other than i. Therefore, $i'' = i$. Not to reject on the substring x', the automaton should enter it in the same state in which it has entered the substring x, so $i'' = i' = i$. The automaton will reject while reading the substring y_j if it starts reading it in a state other than j. Thus, $j' = j$. However, since the substring x cannot change the state of the automaton, the automaton A_n enters the substring w_R in the state i and leaves it in the state j. This means that $(i,j) \in R$, and Claim 3 has been proved.

It remains to show that every deterministic automaton A recognizing the language $L(A_n)$ has at least 2^{n^2} states. For each relation $R \subseteq Q \times Q$, consider the state q_R in which the deterministic automaton A finishes reading the string $\#xw_R$. There are 2^{n^2} such states. For the sake of a contradiction, let $R_1, R_2 \subseteq Q \times Q$ be two different relations with $q_{R_1} = q_{R_2}$. Then, there is a pair of states $(i,j) \in Q \times Q$ that lies in exactly one of the relations R_1, R_2. Then, the nondeterministic automaton A_n accepts only one of the strings $\#xw_{R_1}y_j\#x'y_i$ and $\#xw_{R_2}y_j\#x'y_i$ by Claim 3, and the automaton A does the same. However, the automaton A on both strings enters the suffix $y_j\#x'y_i$ in the same state $q_{R_1} = q_{R_2}$ and with the same stack contents pushed onto the stack while reading the substring $\#x$. Then, A must give the same answer on both strings, this is a contradiction. Theorem 2 has been proved.   □

4   The Lower Bound on the Number of Stack Symbols: One Left Bracket

In this section, the exact bound on the number of stack symbols needed for NIDPDA determinization is proved in a simpler case of only one left bracket.

Theorem 3. *For each $n \geqslant 1$, there is an n-state NIDPDA $B_n = (\Sigma, Q, \Gamma, q_0, (\delta_a)_{a \in \Sigma}, F)$, with a stack alphabet of size $|\Gamma| = 2n + 2$, working over a 5-symbol input alphabet: $\Sigma_{+1} = \{<\}$, $\Sigma_{-1} = \{>, \gg\}$, $\Sigma_0 = \{-, \#\}$, such that any DIDPDA recognizing a language $L(B_n)$ has at least 2^{n^2} states and at least $2^{n^2} - 1$ stack symbols.*

Proof. (a sketch). The automaton B_n is obtained by extending the automaton A_n from Theorem 2. New stack symbols $\widehat{i}$, for $i \in Q$, and $\overrightarrow{j}$, for $j \in Q$, a double right bracket $\gg$, and also some new transitions on the left bracket $<$ are added. The new transitions allow the automaton B_n to change its state from every state i to every state j, pushing onto the stack either $\widehat{i}$ or $\overrightarrow{j}$. And these new stack symbols can be popped only on the double right bracket $\gg$. If the automaton B_n comes to the double bracket $\gg$ in the state 0, then it can pop a symbol $\widehat{i}$ and move to the state i, and if the automaton comes to $\gg$ in the state 1, then it can pop a symbol $\overrightarrow{j}$ moving to the state j.

The idea is to consider the behaviour of the automaton B_n on special strings of the form $f_{R_1,\ldots,R_m} g_{i,j,k,m}$, where

$$f_{R_1,\ldots,R_m} = {<}w_{R_1}{<}w_{R_2} \ldots {<}w_{R_m}{<},$$

for every sequence of m non-empty relations $R_1, R_2, \ldots, R_m \subseteq Q \times Q$, and

$$g_{i,j,k,m} = (\#{\gg})^{m-k} \#y_0 {\gg} y_j \# y_1 {\gg} y_i (\#{\gg})^{k-1},$$

for all $i, j \in Q$, for each $m \geqslant 1$ and for each $k = 1, \ldots, m$. Here the substrings w_R, for $R \subseteq Q \times Q$, and y_i, for $i \in Q$, are constructed according to Claim 1.

It turns out that the nondeterministic automaton B_n accepts a string $f_{R_1,\ldots,R_m} g_{i,j,k,m}$ if and only if $(i,j) \in R_k$. The condition $(i,j) \in R_k$ is checked as follows. The symbols that the automaton B_n pushes onto the stack on the unmatched left brackets of the substring ${<}w_{R_k}{<}$ of $f_{R_1,\ldots,R_m}$ are popped on the double brackets of the substring $\#y_0 {\gg} y_j \# y_1 {\gg} y_i$ of the string $g_{i,j,k,m}$. The double bracket that matches the first left bracket of the substring ${<}w_{R_k}{<}$ is wrapped in substrings y_1 and y_i; that wrapping requires the automaton to change its state from 1 to i on the double bracket. This can be possible only if the automaton B_n pushes the symbol $\overrightarrow{i}$ onto the stack on the left bracket. Analogously, the automaton B_n should push onto the stack the symbol $\widehat{j}$ on the last bracket of ${<}w_{R_k}{<}$. Then, to push $\overrightarrow{i}$ on the first bracket of the substring ${<}w_{R_k}{<}$ and to push $\widehat{j}$ on the last bracket, the automaton must enter the substring w_{R_k} in the state i and leave it in the state j. And this is possible if and only if $(i, j) \in R_k$.

And every deterministic automaton needs at least $2^{n^2} - 1$ stack symbols to do the same, because it must distinguish all strings of the form $f_{R_1,\ldots,R_m}$ to give correct answers for all suffixes of the form $g_{i,j,k,m}$. And since there are $(2^{n^2} - 1)^m$ strings of the form $f_{R_1,\ldots,R_m}$ for every m, the deterministic automaton should have at least $2^{n^2} - 1$ stack symbols to have at least $(2^{n^2} - 1)^m$ different pairs of a state and of stack contents of length $m + 1$ for each m. $\square$

5   The Lower Bound on the Number of Stack Symbols: Several Left Brackets

The exact lower bound on the number of stack symbols needed to determinize an n-state NIDPDA in the case of several left brackets is proved in the next theorem.

Theorem 4. *For all integers n and s such that $n \geqslant 1$ and $1 \leqslant s \leqslant 2^{n^2}$, there is an n-state NIDPDA $B_{n,s} = (\Sigma, Q, \Gamma, q_0, (\delta_a)_{a \in \Sigma}, F)$, with a stack alphabet of the size $|\Gamma| = 2 + 2n + \lfloor \log_2(2s - 1) \rfloor$ and working over an $(s + 5)$-symbol input alphabet: $\Sigma_{+1} = \{<_0, <_1, \ldots, <_{s-1}\}$, $\Sigma_{-1} = \{>, \gg, \ggg\}$, $\Sigma_0 = \{-, \#\}$, such that every DIDPDA recognizing the language $L(B_{n,s})$ has at least 2^{n^2} states and at least $s(2^{n^2} - 1)$ stack symbols.*

Proof (a sketch). The automaton $B_{n,s}$ is constructed as a more complicated version of the automaton B_n in Theorem 3. It has all transitions of B_n on old symbols of the alphabet: $-$, $\#$, $>$, $\gg$; and on each left bracket $<_\ell$, for $\ell \in \{0, \ldots, s - 1\}$, the automaton $B_{n,s}$ has all transitions that B_n has on the left bracket $<$. New stack symbols are $\widehat{x}$, for $x = 0, \ldots, \lfloor \log_2(s-1) \rfloor$. There are also new transitions on left brackets: on a left bracket $<_\ell$ the automaton can change its state from each state i to each state j, pushing any symbol $\widehat{x}$, such that the coefficient at 2^x in the binary representation of the number ℓ equals 1. Denote that coefficient by $\ell[x]$. There are at most n^2 new stack symbols. The triple right bracket $\ggg$ is used to pop symbols of the form $\widehat{x}$. For each $x = 0, \ldots, \lfloor \log_2(s-1) \rfloor$, there is exactly one transition that pops the symbol $\widehat{x}$:

$$\delta_{\ggg}(x \bmod n, \ \widehat{x}) = \left\{ \left\lfloor \frac{x}{n} \right\rfloor \right\}.$$

The idea is to define more elaborate strings than $f_{R_1,\ldots,R_m}$ from Theorem 3, and then to find a separating string for each pair of such strings. New strings are

$$f_{R_1,\ldots,R_m,\ell_1,\ldots,\ell_{m+1}} = <_{\ell_1} w_{R_1} <_{\ell_2} w_{R_2} \cdots <_{\ell_m} w_{R_m} <_{\ell_{m+1}},$$

for every integer $m \geqslant 1$, for a sequence of m non-empty relations $R_1, R_2, \ldots, R_m \subseteq Q \times Q$ and for a sequence of indices $\ell_1, \ldots, \ell_{m+1} \in \{0, \ldots, s - 1\}$. If two such strings differ in some relation R_k, that is, if (i, j) is in the k-th relation of one string and is not in the k-th relation of the other string, then if one appends the string $g_{i,j,k,m}$ from Theorem 3 to these two strings, the automaton $B_{n,s}$ will accept one string and will reject the other.

But to distinguish all the strings of the form $f_{R_1,\ldots,R_m,\ell_1,\ldots,\ell_{m+1}}$, one also needs to construct such suffixes that will separate every two strings $f_{R_1,\ldots,R_m,\ell_1,\ldots,\ell_{m+1}}$ and $f_{R'_1,\ldots,R'_m,\ell'_1,\ldots,\ell'_{m+1}}$ with different indices of left brackets in some position: $\ell_k \neq \ell'_k$. Let x be a bit number, such that $\ell_k[x] \neq \ell'_k[x]$. Without loss of generality, let $\ell_k[x] = 1$ and $\ell'_k[x] = 0$. Then, the following string can be appended to these two strings:

$$h_{k,x,m} = (\#\gg)^{m-k+1}\#y_{x \bmod n}\gg y_{\lfloor \frac{x}{n} \rfloor}(\#\gg)^{k-1},$$

and the automaton $B_{n,s}$ will give different answers on the strings $f_{R_1,\ldots,R_m,\ell_1,\ldots,\ell_{m+1}}h_{k,x,m}$ and $f_{R'_1,\ldots,R'_m,\ell'_1,\ldots,\ell'_{m+1}}h_{k,x,m}$. On the former string the automaton on the left bracket $<_{\ell_k}$ pushes the symbol that it pops on the triple bracket of the substring $h_{k,x,m}$, and on the latter string it pushes such a symbol on the bracket $<_{\ell'_k}$. Since the triple bracket $\ggg$ is wrapped in substrings $y_{x \bmod n}$ and $y_{\lfloor \frac{x}{n} \rfloor}$, and since each of these substrings allows only one state, the automaton comes to the triple bracket $\ggg$ in the state $(x \bmod n)$ and leaves this bracket in the state $\lfloor \frac{x}{n} \rfloor$. Such a change of a state on the triple bracket $\ggg$ is possible only if the symbol at the top of the stack is $\widehat{x}$. This is possible on the string $f_{R_1,\ldots,R_m,\ell_1,\ldots,\ell_{m+1}}h_{k,x,m}$, since $\ell_k[x] = 1$, and the automaton $B_{n,s}$ accepts this string, and it rejects the other string, since $\ell'_k[x] = 0$.

Then, every deterministic automaton recognizing the same language as $B_{n,s}$ needs at least $s(2^{n^2} - 1)$ stack symbols, because, for each m, it must distinguish all $s^{m+1}(2^{n^2} - 1)^m$ strings of the form $f_{R_1,\ldots,R_m,\ell_1,\ldots,\ell_{m+1}}$.   □

6   Conclusion

There are some open questions on the state complexity of input-driven pushdown automata. The exact lower bound on the number of stack symbols needed for NIDPDA determinization, obtained in this paper, uses witness automata with a stack alphabet linear in n. Is it possible to obtain the same lower bound using a stack alphabet smaller than linear in n?

Jirásková and Okhotin [5] investigated the state complexity of operations on DIDPDA: the state complexity for concatenation of automata with m and n states is between mn^n and $m(n^n + 2^n)$ in the worst case, and for the Kleene star it is between n^n and $n^n + 2^n + 1$. One can try to find the exact state complexity.

One can also try to improve asymptotically precise bounds for determinization of event-clock input-driven pushdown automata obtained by Ogawa and Okhotin [8].

Rose and Okhotin [12] introduced probabilistic input-driven pushdown automata, determinized them and proved the asymptotically precise lower bound on the number of states needed for determinization. It remains an open problem to prove a lower bound on the number of stack symbols needed for determinization of these automata.

Acknowledgement. I am grateful to Alexander Okhotin for his advices on the presentation and for his help in proofreading the text.

References

1. Alur, R., Madhusudan, P.: Visibly pushdown languages. In: ACM Symposium on Theory of Computing (STOC 2004), Chicago, USA, 13–16 June 2004, pp. 202–211 (2004). https://dl.acm.org/doi/10.1145/1007352.1007390
2. Alur, R., Madhusudan, P.: Adding nesting structure to words. J. ACM **56**(3), 1–43 (2009). https://doi.org/10.1145/1516512.1516518
3. von Braunmühl, B., Verbeek, R.: Input driven languages are recognized in log n. Ann. Discrete Math. **24**, 1–20 (1985). https://doi.org/10.1016/S0304-0208(08)73072-X
4. Han, Y.-S., Ko, S.-K., Salomaa, K.: Deciding path size of nondeterministic (and input-driven) pushdown automata. Theor. Comput. Sci. **939**, 170–181 (2023). https://doi.org/10.1016/j.tcs.2022.10.023
5. Jirásková, G., Okhotin, A.: Towards exact state complexity bounds for input-driven pushdown automata. In: Hoshi, M., Seki, S. (eds.) DLT 2018. LNCS, vol. 11088, pp. 441–452. Springer, Cham (2018). https://doi.org/10.1007/978-3-319-98654-8_36
6. Kutrib, M., Malcher, A., Wendlandt, M.: Tinput-driven pushdown, counter, and stack automata. Fundamenta Informaticae **155**(1–2), 59–88 (2017). https://doi.org/10.3233/FI-2017-1576
7. Mehlhorn, K.: Pebbling mountain ranges and its application to DCFL-recognition. In: de Bakker, J., van Leeuwen, J. (eds.) ICALP 1980. LNCS, vol. 85, pp. 422–435. Springer, Heidelberg (1980). https://doi.org/10.1007/3-540-10003-2_89
8. Ogawa, M., Okhotin, A.: On the determinization of event-clock input-driven pushdown automata. In: Kulikov, A.S., Raskhodnikova, S. (eds.) CSR 2022. LNCS, vol. 13296, pp. 256–268. Springer, Cham (2022). https://doi.org/10.1007/978-3-031-09574-0_16
9. Okhotin, A., Piao, X., Salomaa, K.: Descriptional complexity of input-driven pushdown automata. In: Bordihn, H., Kutrib, M., Truthe, B. (eds.) Languages Alive. LNCS, vol. 7300, pp. 186–206. Springer, Heidelberg (2012). https://doi.org/10.1007/978-3-642-31644-9_13
10. Okhotin, A., Salomaa, K.: Complexity of input-driven pushdown automata. ACM SIGACT News **45**(2), 47–67 (2014). https://doi.org/10.1145/2636805.2636821
11. Okhotin, A., Salomaa, K.: State complexity of operations on input-driven pushdown automata. J. Comput. Syst. Sci. **86**, 207–228 (2017). https://doi.org/10.1016/j.jcss.2017.02.001
12. Rose, A., Okhotin, A.: Probabilistic input-driven pushdown automata. In: 48th International Symposium on Mathematical Foundations of Computer Science (MFCS 2023). LIPIcs, vol. 272, pp. 78:1–78:14 (2023). https://doi.org/10.4230/LIPIcs.MFCS.2023.78
13. Rytter, W.: An application of Mehlhorn's algorithm for bracket languages to log n space recognition of input-driven languages. Inf. Process. Lett. **23**, 81–84 (1986). https://doi.org/10.1016/0020-0190(86)90047-5
14. Van Tang, N., Ogawa, M.: Event-clock visibly pushdown automata. In: Nielsen, M., Kučera, A., Miltersen, P.B., Palamidessi, C., Tuma, P., Valencia, F. (eds.) SOFSEM 2009. LNCS, vol. 5404, pp. 558–569. Springer, Heidelberg (2009). https://doi.org/10.1007/978-3-540-95891-8_50

Disproving Termination of Non-erasing Sole Combinatory Calculus with Tree Automata

Keisuke Nakano[1]([⊠])[iD] and Munehiro Iwami[2][iD]

[1] Tohoku University, Sendai, Miyagi, Japan
k.nakano@acm.org
[2] Shimane University, Matsue, Shimane, Japan
munehiro@cis.shimane-u.ac.jp

Abstract. We study the termination of sole combinatory calculus, which consists of only one combinator. Specifically, the termination for non-erasing combinators is disproven by finding a desirable tree automaton with a SAT solver as done for term rewriting systems by Endrullis and Zantema. We improved their technique to apply to non-erasing sole combinatory calculus, in which it suffices to search for tree automata with a final sink state. Our method succeeds in disproving the termination of 8 combinators, whose termination has been an open problem.

Keywords: Combinatory calculus · Non-termination · Tree automata

1 Introduction

Combinatory logic [3,15] has been used in computer science as a theoretical model of computation and also as a basis for the design of functional programming languages [13,18]. It can be viewed as a variant of lambda calculus, in which a limited set of combinators, primitive functions without free variables, is used instead of lambda abstractions.

Combinators in combinatory logic are defined as $Zx_1x_2 \ldots x_n \to e$ where Z is a combinator, $x_1, \ldots, x_n$ are variables, and e is built from the variables with a function application. It is known that a small set of combinators can define a combinatorial calculus that is sufficient to cover all computable functions. Well-known sets of such combinators are $\{\mathbf{S}, \mathbf{K}\}$ and $\{\mathbf{B}, \mathbf{C}, \mathbf{K}, \mathbf{W}\}$ with $\mathbf{S}xyz \to xz(yz)$, $\mathbf{K}xy \to x$, $\mathbf{B}xyz \to x(yz)$, $\mathbf{C}xyz \to xzy$, and $\mathbf{W}xy \to xyy$ [1].

The subject of this paper is *sole combinatory calculus*, which consists of only one combinator. There have been several studies on sole combinatory calculi. Waldmann [19] investigated the $\mathbf{S}$ combinator to provide a procedure that decides whether an $\mathbf{S}$-term, built from $\mathbf{S}$ alone, has a normal form and further showed that the set of normalizing $\mathbf{S}$-terms is recognizable. Probst and Studer [14] proved that the sole combinatory calculus with the $\mathbf{J}$ combinator, defined by $\mathbf{J}xyzw \to xy(xwz)$ is strongly normalizing; that is, no $\mathbf{J}$-term has

© The Author(s), under exclusive license to Springer Nature Switzerland AG 2024
S. Z. Fazekas (Ed.): CIAA 2024, LNCS 15015, pp. 261–275, 2024.
https://doi.org/10.1007/978-3-031-71112-1_19

$$\mathbf{P}xyz \to z(xyz) \quad \mathbf{P}_3xyz \to y(xzy) \quad \mathcal{D}_1xyzw \to xz(yw)(xz) \quad \mathcal{D}_2xyzw \to xw(yz)(xw)$$

$$\mathbf{\Phi}xyzw \to x(yw)(zw) \quad \mathbf{\Phi}_2xyzw_1w_2 \to x(yw_1w_2)(zw_1w_2) \quad \mathbf{S}_1xyzw \to xyw(zw)$$

$$\mathbf{S}_2xyzw \to xzw(yzw) \quad \mathbf{S}_3xyzwv \to xy(zv)(wv) \quad \mathbf{S}_4xyzwv \to z(xwv)(ywv)$$

Fig. 1. Combinators and their reduction rules

an infinite reduction sequence. Ikebuchi and Nakano [6] showed that the sole combinatory calculus with the **B** combinator is strongly normalizing and characterized by equational axiomatization for proving the looping and non-looping properties of repetitive right applications.

This paper concerns the non-termination of sole combinatory calculi, where termination means that no term has an infinite reduction. Let us say that a combinator is *terminating* when the corresponding sole combinatory calculus is terminating. Iwami [8] has investigated the termination of 37 combinators introduced in Smullyan's book [16] and has reported that 10 of them shown in Fig. 1 are of unknown termination.

We disprove the termination of 8 of the 10 combinators. The main idea of disproving the termination is to give a non-empty recognizable set of terms closed under the reduction as Endrullis and Zantema have done [4]. They showed that a SAT solver can find the corresponding tree automaton. We improve their method by showing that it suffices to search for tree automata with a final sink state in our setting and by reducing the number of variables in the SAT problem. Our implementation disproves the termination of 8 combinators.

2   Preliminaries

A *signature* (or *alphabet*) Σ is a non-empty finite set of *function symbols*, each with a fixed natural number called *arity* (or *rank*)[1]. The set of all function symbols of arity n in Σ is written as $\Sigma^{(n)}$. We may write $f^{(n)}$ for $f \in \Sigma^{(n)}$. A function symbol of arity 0 is called a *constant* symbol. A set of *variables* is a countably infinite set disjoint from Σ. For a set $\mathcal{V}$ of variables, a set of *terms* over Σ, denoted by $\mathcal{T}_\Sigma(\mathcal{V})$, is inductively defined as the smallest set S such that $\mathcal{V} \subseteq S$ and $t_1, \ldots, t_n \in S$ implies $f(t_1, \ldots, t_n) \in S$ for every $f \in \Sigma^{(n)}$. The set of variables occurring in $t \in \mathcal{T}_\Sigma(\mathcal{V})$ is denoted by $\mathsf{FV}(t)$. In the rest of the paper, the set $\mathcal{V}$ of variables is fixed and contains $x, x_1, x_2, \ldots$ as its elements. A *substitution* is a finite map from variables to terms. We write $\mathsf{dom}(\alpha)$ for the domain of a substitution α. For a term t and a substitution α, we denote by $t\alpha$ an *instance* of t, a term obtained by replacing every variable x in t with $\alpha(x)$. Substitutions may be represented in the set notation as usual: we write $\{x_1 \mapsto t_1, \ldots, x_n \mapsto t_n\}$ for substitution α when $\mathsf{dom}(\alpha) = \{x_1, \ldots, x_n\}$ and $\alpha(x_i) = t_i$ holds for each i

---

[1] We use the terminology of term rewriting systems, whereas definitions are given alongside that of formal language theory for the readers, e.g., a *tree* and a *rank* in formal language are called a *(ground) term* and an *arity* in term rewriting, respectively.

and, in particular, $\emptyset$ for substitution α when $\mathsf{dom}(\alpha) = \emptyset$. A term containing no variables is called a *ground term*, and the set of ground terms is written as $\mathcal{T}_\Sigma$, i.e., $\mathcal{T}_\Sigma = \mathcal{T}_\Sigma(\emptyset)$. A *context* C is a term over $\Sigma \cup \{\Box\}$ with a constant symbol $\Box$, called a *hole*, such that $\Box$ occurs exactly once in C. For a context C and a term t, we denote by $C[t]$ a term obtained by replacing the hole in C with t. We write $\mathcal{C}_\Sigma(\mathcal{V})$ for the set of contexts; in particular, $\mathcal{V}$ is omitted from the notation if it is empty, i.e., $\mathcal{C}_\Sigma = \mathcal{C}_\Sigma(\emptyset)$. A term $s \in \mathcal{T}_\Sigma$ is a *subterm* of $t \in \mathcal{T}_\Sigma$ if $t = C[s]$ holds with some $C \in \mathcal{C}_\Sigma$. The set of all subterms of t is written as $\mathsf{Sub}(t)$.

A *term rewriting system* (TRS) R over Σ is a set of rewriting rules of the form $l \to r$ with $l, r \in \mathcal{T}_\Sigma(\mathcal{V})$ where $\mathsf{FV}(r) \subseteq \mathsf{FV}(l)$. A TRS R is said to be *non-erasing* if $\mathsf{FV}(r) = \mathsf{FV}(l)$ holds for every rule $l \to r \in R$. A TRS R is said to be *left-linear* if each variable in $\mathsf{FV}(l)$ occurs exactly once in l for every rule $l \to r \in R$. A left-linear TRS R is said to be *orthogonal* if every pair of two (possibly the same) rules in R has no overlapping, which means that the left-hand side of one rule is not unifiable with any non-variable subterm of the left-hand side of the other rule. A *rewrite relation* $\to_R$ over $\mathcal{T}_\Sigma(\mathcal{V})$ induced by R is defined by $\{(C[l\alpha], C[r\alpha]) \mid C \in \mathcal{C}_\Sigma(\mathcal{V}),\ l \to r \in R,\ \alpha : \mathsf{FV}(l) \to \mathcal{T}_\Sigma(\mathcal{V})\}$. We write $\to_R^*$ for the reflexive and transitive closure of $\to_R$. A term t is called a *redex* of R if $t = l\alpha$ holds for some $l \to r \in R$ and substitution α; that is, a redex means a reducible part of a term. A term t is said to be in *normal form* with respect to R if there is no term u such that $t \to_R u$; in other words, no subterm of t is a redex of R. A set of normal forms with respect to R is denoted by $\mathsf{NF}(R)$. A TRS R is *terminating* or *strongly normalizing* if no infinite rewrite sequence $t_0 \to_R t_1 \to_R t_2 \to_R \cdots$ with $t_i \in \mathcal{T}_\Sigma(\mathcal{V})$ and $i \in \mathbb{N}$ exists. A TRS R is *weakly normalizing* if for every term t there exists a term $u \in \mathsf{NF}(R)$ such that $t \to_R^* u$ holds. A rewrite step $s \to_R t$ is *innermost*, denoted by $s \to_R^\mathsf{i} t$, if no proper subterm of the contracted redex is itself a redex [12], that is, the relation $\to_R^\mathsf{i}$ is defined by a subset of $\to_R$ as $\{(C[l\alpha], C[r\alpha]) \mid C \in \mathcal{C}_\Sigma(\mathcal{V}),\ l \to r \in R,\ \alpha : \mathsf{FV}(l) \to \mathcal{T}_\Sigma(\mathcal{V}),\ \mathsf{Sub}(l\alpha) \subseteq \mathsf{NF}(R) \cup \{l\alpha\}\}$. A TRS R is *weakly innermost normalizing* if for every term t there exists a term $u \in \mathsf{NF}(R)$ such that $t \to_R^{\mathsf{i}*} u$ holds. A set S of terms is *closed* under R if for every $t \in S$, $t \to_R u$ implies $u \in S$.

A (non-deterministic bottom-up) *tree automaton* is a quadruple $A = \langle Q, \Sigma, F, \Delta \rangle$ where Q is a finite set of states, $F \subseteq Q$ is a set of final states, and Δ is a set of state transition rules of the form $f(q_1, \ldots, q_n) \rightsquigarrow q$ with $f \in \Sigma^{(n)}$, $q_1, \ldots, q_n, q \in Q$. The set Δ can be viewed as a TRS where $\rightsquigarrow$ is used instead of $\to$. The arrow $\rightsquigarrow_\Delta$ is used for the rewrite relation over $\mathcal{T}_{\Sigma \cup Q}$ induced by the rules in Δ where the subscript Δ may often be omitted if no confusion arises. We write $\rightsquigarrow^*$ for the reflexive and transitive closure of $\rightsquigarrow$. The set $\mathcal{L}(A, q)$ for $q \in Q$ is defined by $\mathcal{L}(A, q) \equiv \{t \in \mathcal{T}_\Sigma \mid t \rightsquigarrow_\Delta^* q\}$. The set of terms *accepted* by A is defined by $\mathcal{L}(A) \equiv \bigcup_{q \in F} \mathcal{L}(A, q)$. A state q is said to be *reachable* if $\mathcal{L}(A, q)$ is not empty. A state q is called a *sink state* when, for every $f \in \Sigma^{(n)}$ ($n > 0$) and $q', q_1, \ldots, q_n \in Q$ with $q_i = q$ for some i, $f(q_1, \ldots, q_n) \rightsquigarrow q \in \Delta$ holds but $f(q_1, \ldots, q_n) \rightsquigarrow q' \in \Delta$ does not hold with $q' \neq q$. When q is a sink state, it is easy to show that $t \in \mathcal{L}(A, q)$ implies $C[t] \in \mathcal{L}(A, q)$ for every context $C \in \mathcal{C}_\Sigma$

and every ground term $t \in \mathcal{T}_\Sigma$. Two tree automata A_1 and A_2 are said to be *equivalent* if $\mathcal{L}(A_1) = \mathcal{L}(A_2)$ holds.

3   Termination of Sole Combinatory Calculus

A combinatory calculus is specified by certain kinds of combinators and reduction rules for each combinator. A reduction rule for a combinator Z has the form $Z \ x_1 \ \ldots \ x_n \to e$ where e is built by combining $x_1, \ldots, x_n$ with function application. One of the most familiar combinatory calculi is given by the **S** and **K** combinators defined by **S** $x_1 \ x_2 \ x_3 \to x_1 \ x_3 \ (x_2 \ x_3)$ and **K** $x_1 \ x_2 \to x_1$. The calculus is well known to be Turing-complete in the sense that these two combinators are sufficient to represent all computable functions. Reductions in combinatory calculus are easily simulated by a TRS using the set of constant symbols for combinators and a binary function symbol @ for function application. For example, reduction rules for the **S** and **K** combinatory calculus can be represented by a TRS consisting of @(@(@(**S**, x_1), x_2), x_3) $\to$ @(@(x_1, x_3), @(x_2, x_3)) and @(@(**K**, x_1), x_2) $\to x_1$, which is obviously non-terminating because of the Turing-completeness of the corresponding reduction system.

   In this paper, we are interested in the question for sole combinatory calculus, which is built only by one combinator. We start with formal definitions of several notions on the sole combinatory calculus in terms of term rewriting. For Z is a combinator, we denote by Σ_Z a signature consisting of a constant Z and binary function symbol @. A Z-*term*, which is built only from Z in combinatory calculus, is represented by a term in $\mathcal{T}_\Sigma$. A *sole combinatory calculus* R_Z induced by a combinator Z is a TRS over $\mathcal{T}_{\Sigma_Z}$ where R_Z is a singleton set of a left-linear rule of the form @($\ldots$@(@(Z, x_1), x_2) $\ldots, x_n$) $\to e$ with a term e built only from @ and variables $x_1, \ldots, x_n$. A combinator Z is said to be *terminating* if R_Z is terminating. It is known that **B** [6] and **J** [14] are terminating while **O** [4,7,9] and **S** [19] are not.

   The following lemma allows us to consider only the case of weakly innermost normalizing instead of terminating since the rule of sole combinatory calculus is orthogonal.

Lemma 1. (*[11, Theorem 11]*). *An orthogonal TRS R is terminating if and only if R is weakly innermost normalizing.*

   The readers might recall a similar result shown by Church [2] that an orthogonal non-erasing TRS R is terminating if and only if R is weakly normalizing. Since we are concerned with non-erasing combinatory calculus, this result may seem more convenient. However, in the context of the present work, we intend to employ the above lemma because the 'innermost' condition plays a crucial role in our method which will be detailed later.

4   Disproving Termination of Combinators by Tree Automata

Endrullis and Zantema have proposed a procedure for disproving weakly and strongly normalizing by finding tree automata that disprove termination of arbitrary forms of left-linear TRSs. This section first explains how to disprove termination using the search for tree automata, and shows that it is sufficient to find a restricted form of tree automata to disprove the termination of non-erasing combinators. Then, we present how to find such tree automata with a SAT solver. Finally, we show the non-termination of 8 combinators with our implementation of the method.

4.1   Disproving Termination with Tree Automata

The idea of disproving termination of a TRS R by Endrullis and Zantema is to find a non-empty set of reducible ground terms (i.e., not in normal form), which is closed under the rules in R. It is easy to see that the existence of such a set implies that R is not weakly normalizing because any reduction from a term in the set is always infinite. Endrullis and Zantema considered the case where the set is recognizable by a tree automaton as defined below, though they did not give a name to it.

Definition 2. A *termination-disproving automaton (TDA)* for a left-linear TRS R is a tree automaton $A = \langle Q, \Sigma, F, \Delta \rangle$ such that (2-1)   there exists a state $q \in F$ that is reachable, (2-2) $\mathcal{L}(A) \cap \mathsf{NF}(R) = \emptyset$, and (2-3) $t \in \mathcal{L}(A, q)$ implies $u \in \mathcal{L}(A, q)$ for all $q \in Q$ and $t, u \in \mathcal{T}_\Sigma$ with $t \to_R u$.

It is easy to see that the set $\mathcal{L}(A)$ for a TDA A for R can disprove weak normalizability of R: the conditions (2-1) and (2-2) allow us to choose a term not in normal form, and the conditions (2-2) and (2-3) force any reduction from the term to be infinite. The condition (2-3) is a bit strong in the sense that it requires every set $\mathcal{L}(A, q)$ with $q \in Q$ to be closed under reductions in R. The last condition might be relaxed to require the closure property only for final states in Q_F. However, we require it for all states in Q to make the SAT solver–based disproof search easier as done by Endrullis and Zantema.

As we will see later, it is sufficient to find a TDA with a sink state as the final state. The restricted form of tree automata is defined as follows.

Definition 3. A *termination-disproving automaton with a final sink (TDA-S)* for an orthogonal TRS R is a tree automaton $A = \langle Q, \Sigma, \{q_F\}, \Delta \rangle$ with $q_F \in Q$ such that (3-1) q_F is sink and reachable, (3-2) $\mathcal{L}(A) \cap \mathsf{NF}(R) = \emptyset$, and (3-3) $t \in \mathcal{L}(A, q)$ implies $u \in \mathcal{L}(A, q) \cup \mathcal{L}(A)$ for all $q \in Q$ and $t, u \in \mathcal{T}_\Sigma$ with $t \to_R^! u$.

We assume here that R is orthogonal, which is stronger than left-linear, because every sole combinatory calculus is orthogonal. A major difference from TDAs is that a TDA-S has a sink state q_F as the unique final state. The sink state is reachable so that $\mathcal{L}(A)$ is non-empty. In addition, it requires the closure property only under the innermost reduction and allows the reduction onto $\mathcal{L}(A)$. The latter relaxation corresponds to a minor improvement done by Endrullis and Zantema where the closure property allows the reduction onto terms accepted by the 'larger' states in the total order over states, which will be used for reachability checking with Lemma 6. The following theorem states that the existence of a TDA-S for an orthogonal TRS R implies that R is not terminating.

Theorem 4. *An orthogonal TRS R is not terminating if a TDA-S for R exists.*

Proof. Let $A = \langle Q, \Sigma, \{q_F\}, \Delta \rangle$ be a TDA-S for an orthogonal TRS R. By Lemma 1, it suffices to show that R is not weakly innermost normalizing. We prove by contradiction that for all $t \in \mathcal{L}(A)$, no reduction sequence from t reaches a normal form, which disproves the weak innermost normalizability of R due to the non-emptiness of $\mathcal{L}(A)$ implied by the (3-1) condition of A. Suppose that there exists a term such that a reduction sequence starting from the term reaches a normal form. Let $t \in \mathcal{L}(A)$ be such a term which has the shortest reduction sequence $t = t_0 \rightarrow^{\mathrm{i}}_R t_1 \rightarrow^{\mathrm{i}}_R \cdots \rightarrow^{\mathrm{i}}_R t_n$ with $t_n \in \mathsf{NF}(R)$. From the (3-2) condition, t is not in normal form, i.e., $n > 0$. From the (3-3) condition with $q = q_F$, $t_1 \in \mathcal{L}(A)$ holds. Since we have a reduction sequence from $t_1 \in \mathcal{L}(A)$ which reaches a normal form, it contradicts the assumption that the reduction sequence from t is the shortest one. □

We will try to find a TDA-S (which has exactly one final state) to disprove the termination of a sole combinatory calculus instead of a TDA. Since the disproof search will be done by fixing the number of states and increasing it iteratively, we have to show that the number of states of a TDA-S is not required to be as large as that of a TDA. Note that we do not need to find a TDA-S equivalent to a TDA if it exists. It is well-known that every non-deterministic tree automaton can be converted into an equivalent one that has exactly one final state (by introducing a fresh final state) but it may have one more state than the original one. The following lemma guarantees that the number of states of a TDA-S is not required to be larger than that of a TDA to disprove the termination of an orthogonal TRS. The proof idea is to construct a TDA-S A from a given TDA A_0 by forcing a final state of A_0 to be the final sink state of A and removing states that accept only terms whose subterm is accepted by the sink state.

Lemma 5. *Let R be an orthogonal TRS. If a TDA $A_0 = \langle Q_0, \Sigma, F_0, \Delta_0 \rangle$ for R exists, then a TDA-S $A = \langle Q, \Sigma, \{q_F\}, \Delta \rangle$ for R also exists with $|Q| \leq |Q_0|$.*

Proof. Let $A_0 = \langle Q_0, \Sigma, F_0, \Delta_0 \rangle$ be a TDA for an orthogonal TRS R, where A_0 satisfies the conditions (2-1), (2-2), and (2-3). Without loss of generality, all states in Q_0 are reachable; otherwise, we could remove unreachable states from Q_0. From (2-1) of A_0, the set F_0 is not empty, hence we can choose a final state $q_F \in F_0$. Let $L_F \subseteq \mathcal{T}_\Sigma$ and $Q_F \subseteq Q_0$ be defined by $L_F = \{C[t] \mid C \in \mathcal{C}_\Sigma, t \in \mathcal{L}(A_0, q_F)\}$ and $Q_F = \{q \in Q_0 \mid \mathcal{L}(A_0, q) \subseteq L_F\}$, respectively. Note that $q_F \in Q_F$ holds in particular. Then we define a tree automaton $A = \langle Q, \Sigma, \{q_F\}, \Delta \rangle$ with $Q = (Q_0 \setminus Q_F) \cup \{q_F\}$ and $\Delta = \Delta_1 \cup \Delta_2$ where

$$\Delta_1 = \{f(q_1, \ldots, q_n) \rightsquigarrow q \in \Delta_0 \mid \{q_1, \ldots, q_n\} \subseteq Q_0 \setminus Q_F, \ q \in Q\} \text{ and}$$

$$\Delta_2 = \{f(q_1, \ldots, q_n) \rightsquigarrow q_F \mid f \in \Sigma^{(n)}, \ q_F \in \{q_1, \ldots, q_n\} \subseteq Q\}$$

so that q_F is to be a sink final state in A. Since $|Q| \leq |Q_0|$ obviously holds, it suffices to show that A is a TDA-S for R.

Before proving that A is a TDA-S for R, we show

(5-1) $t \in L_F$ if and only if $t \in \mathcal{L}(A, q_F)$,

(5-2) $t \in \mathcal{L}(A, q)$ implies $t \in \mathcal{L}(A_0, q)$ for all $q \in Q_0 \setminus Q_F$, and

(5-3) $t \in \mathcal{L}(A_0, q)$ implies $t \in \mathcal{L}(A, q) \cup \mathcal{L}(A, q_F)$ for all $q \in Q_0 \setminus Q_F$,

for all $t \in \mathcal{T}_\Sigma$. These statements can be shown by simultaneous induction on the structure of t. Suppose that $t = f(t_1, \ldots, t_n)$ with $f \in \Sigma^{(n)}$ and $t_1, \ldots, t_n \in \mathcal{T}_\Sigma$. On the (5-1) statement, we examine two cases according to whether $t_i \in L_F$ or not for each i. In the case where $t_i \in L_F$ holds for some $1 \leq i \leq n$, the 'if'-statement of (5-1) is obvious from the definition of L_F, hence we show the 'only if'-statement. Since $t_i \in \mathcal{L}(A, q_F)$ holds from the induction hypothesis, we have $t \in \mathcal{L}(A, q_F)$ using a transition rule in Δ_2. Thus, the 'only if'-statement of (5-1) also holds. In the case where $t_i \notin L_F$ holds for all $1 \leq i \leq n$, we first show the 'if'-statement of (5-1). Assume $t \in \mathcal{L}(A, q_F)$ holds. Then, there exists $q_1, \ldots, q_n \in Q$ such that $f(q_1, \ldots, q_n) \rightsquigarrow q_F \in \Delta$ and $t_i \in \mathcal{L}(A, q_i)$ for every $1 \leq i \leq n$. If $q_i = q_F$ holds for some $1 \leq i \leq n$, then we have $t_i \in L_F$ from the induction hypothesis, hence $t \in L_F$. If $q_i \in Q_0 \setminus Q_F$ holds for all $1 \leq i \leq n$, then the transition rule is in Δ_1 and we have $t_i \in \mathcal{L}(A_0, q_i)$ from the induction hypothesis of (5-2) for all i. Using the same transition rule, we have $t \in \mathcal{L}(A_0, q_F)$, hence $t \in L_F$. Therefore, the 'if'-statement of (5-1) holds. For the 'only if'-statement of (5-1), assume $t \in L_F$. Since $t_i \notin L_F$ holds for all $1 \leq i \leq n$, we have $t \in \mathcal{L}(A_0, q_F)$ owing to the definition of L_F. Then, there exists $q_1, \ldots, q_n \in Q_0$ such that $f(q_1, \ldots, q_n) \rightsquigarrow q_F \in \Delta_0$ and $t_i \in \mathcal{L}(A_0, q_i)$ for every $1 \leq i \leq n$. Note that $t_i \notin L_F$ implies $q_i \in Q_0 \setminus Q_F$ by the definition of Q_F. Thus, the transition rule is in Δ_1 and we have $t_i \in \mathcal{L}(A, q_i) \cup \mathcal{L}(A, q_F)$ from the induction hypothesis of (5-3). When $t_i \in \mathcal{L}(A, q_i)$ holds for all i, we have $t \in \mathcal{L}(A, q_F)$ using the same transition rule. When $t_i \in \mathcal{L}(A, q_F)$ holds for some $1 \leq i \leq n$, we have $t \in \mathcal{L}(A, q_F)$ using the transition rule in Δ_2. Therefore, the 'only if'-statement of (5-1) holds.

On the (5-2) statement, assume $t \in \mathcal{L}(A, q)$ with $q \in Q_0 \setminus Q_F$, that is, $q \neq q_F$. Then, there exist $q_1, \ldots, q_n \in Q$ such that $f(q_1, \ldots, q_n) \rightsquigarrow q \in \Delta$ and $t_i \in \mathcal{L}(A, q_i)$ for every $1 \leq i \leq n$. Note that $q_i \neq q_F$ holds for all $1 \leq i \leq n$ because of the construction of Δ. Since we have $q_i \in Q_0 \setminus Q_F$ for every $1 \leq i \leq n$, the transition rule is in Δ_1 and $t_i \in \mathcal{L}(A_0, q_i)$ holds from the induction hypothesis for every $1 \leq i \leq n$. Using the same transition rule, we have $t \in \mathcal{L}(A_0, q)$. Therefore, (5-2) holds.

On the (5-3) statement, assume $t \in \mathcal{L}(A_0, q)$ with $q \in Q_0 \setminus Q_F$. Then, there exist $q_1, \ldots, q_n \in Q_0$ such that $f(q_1, \ldots, q_n) \rightsquigarrow q \in \Delta_0$ and $t_i \in \mathcal{L}(A_0, q_i)$ for every $1 \leq i \leq n$. When $q_i \in Q_F$ holds for some $1 \leq i \leq n$, we have $t_i \in L_F$ by the definition of Q_F, hence $t_i \in \mathcal{L}(A, q_F)$ holds from the induction hypothesis of (5-1). Using a transition rule in Δ_2, we have $t \in \mathcal{L}(A, q_F)$. When $q_i \in Q_0 \setminus Q_F$ holds for all $1 \leq i \leq n$, the transition rule is in Δ_0 and we have $t_i \in \mathcal{L}(A, q_i) \cup \mathcal{L}(A, q_F)$ from the induction hypothesis for all $1 \leq i \leq n$. If $t_i \in \mathcal{L}(A, q_i)$ holds for all i, then we have $t \in \mathcal{L}(A, q)$ using the same transition rule. If $t_i \in \mathcal{L}(A, q_F)$ holds for some i, then we have $t \in \mathcal{L}(A, q_F)$ using the transition rule in Δ_2. Therefore, (5-3) holds.

Now we are ready to show that A is a TDA-S for R. Concerning the (3-1) condition, the set L_F is not empty since q_F is reachable in A_0 due to the (2-1) condition of A_0. Then we have q_F is reachable also in A owing to (5-1). In addition, q_F is a sink state in A, hence the (3-1) condition holds for A.

The (3-2) condition is shown by contradiction. Suppose that there exists a term $t \in \mathcal{L}(A) \cap \mathsf{NF}(R)$. Then we have $t \in L_F$ from (5-1), hence $t = C[t_0]$ holds for some $C \in \mathcal{C}_\Sigma$ and $t_0 \in \mathcal{L}(A_0, q_F)$. From $t_0 \in \mathcal{L}(A_0, q_F) \subseteq \mathcal{L}(A_0)$ and the (2-2) condition of A_0, we have $t_0 \notin \mathsf{NF}(R)$, and thus $t = C[t_0] \notin \mathsf{NF}(R)$ holds. This contradicts the assumption, hence the (3-2) condition holds for A.

Concerning the (3-3) condition, assume that we have $t \in \mathcal{L}(A, q)$ with $q \in Q$ and $t \to_R^i u$ for some $u \in \mathcal{T}_\Sigma$. In the case of $q = q_F$, we have $t \in L_F$ by (5-1), hence $t = C[t_0]$ holds for some $C \in \mathcal{C}_\Sigma$ and $t_0 \in \mathcal{L}(A_0, q_F)$. Since the t_0 is not in normal form by the (2-2) condition of A_0 and $\to_R^i$ is an innermost relation, we have either $u = C[u_0]$ with $t_0 \to_R^i u_0 \in \mathcal{T}_\Sigma$ or $u = D[t_0]$ with some $D \in \mathcal{C}_\Sigma$. In the former case, we have $u_0 \in \mathcal{L}(A_0, q_F)$ by $t_0 \in \mathcal{L}(A_0, q_F)$ and the (2-3) condition of A_0. Then $u_0 \in \mathcal{L}(A, q_F)$ holds due to $u_0 \in \mathcal{L}(A_0, q_F) \subseteq L_F$ and (5-1). Since q_F is a sink state in A, $u \in \mathcal{L}(A, q_F)$ holds. In the latter case, $u \in \mathcal{L}(A, q_F)$ holds because $t_0 \in \mathcal{L}(A_0, q_F) \subseteq L_F$ implies $t_0 \in \mathcal{L}(A, q_F)$ by (5-1) and q_F is a sink state in A. Therefore, the (3-3) condition holds for A in the case of $q = q_F$. In the case of $q \in Q_0 \setminus Q_F$, we have $t \in \mathcal{L}(A_0, q)$ by (5-2), hence $u \in \mathcal{L}(A_0, q)$ holds by the (2-3) condition of A_0. Since we have $u \in \mathcal{L}(A, q) \cup \mathcal{L}(A, q_F)$ by (5-3), the (3-3) condition holds for A in the case of $q \in Q_0 \setminus Q_F$. □

Since we try to find a TDA-S in the ascending order of the number of states as Endrullis and Zantema have done for a TDA, one of the TDA-Ss for a given TRS with the smallest number of states can be found if it exists. It might be possible to find a TDA-S even with a smaller number of states than a TDA because of the relaxed closure property (3-3). Our experiment results in Sect. 4.3 do not show such a case, though.

4.2 SAT Encoding of Termination-Disproving Tree Automata

Endrullis and Zantema showed that the problem of finding TDAs can be reduced to the boolean satisfiability problem (SAT, for short) by fixing the number of states of tree automata. Although we essentially follow their method, we will present a method which can find a TDA-S more efficiently because of the restriction of its form. The efficiency of the disproof search is improved not only by finding a TDA-S instead of a TDA but also by specializing their method for non-erasing sole combinatory calculus.

We present our SAT encoding method to be self-contained, where the differences from the method by Endrullis and Zantema (*EZ method*, for short) are explicitly explained for each step. We first explain problem settings and definitions used in our method, introduce propositional variables and their meaning, and then show propositional formulas over them that must hold.

Problem Setting and Definitions. Let Z be a non-erasing combinator whose termination is to be disproved. Recall that the reduction rule of Z is represented by a singleton TRS R_Z over $\Sigma_Z = \{Z^{(0)}, @^{(2)}\}$. Let $l_Z \to r_Z$ be the unique rule of R_Z where $l_Z = @(\dots@(@(Z, x_1), x_2)\dots, x_{M_Z})$ with some $M_Z \geq 1$ and r_Z is built from the binary function symbol $@$ and variables $x_1, x_2, \dots, x_{M_Z}$ so that $\mathsf{FV}(l_Z) = \mathsf{FV}(r_Z)$ holds for R_Z to be non-erasing. We write U_Z for the set $\mathsf{Sub}(l_Z) \cup \mathsf{Sub}(r_Z)$. The *left depth* $\mathsf{ld}(t)$ of a Z-term $t \in T_{\Sigma_Z}$ is defined by $\mathsf{ld}(Z) = 0$ and $\mathsf{ld}(@(t_1, t_2)) = 1 + \mathsf{ld}(t_1)$. The left depth of a term is useful in determining whether the term is redex or not. A Z-term t has the left depth M_Z if and only if t is a redex of R_Z.

Let $A = \langle Q, \Sigma, \{q_F\}, \Delta \rangle$ be a TDA-S with a final sink state $q_F \in Q$, which is to be found if it exists. We fix the number of states as $|Q| = N$ in our encoding. We iteratively ask the SAT solver to find a TDA-S increasing N one by one, starting with $N = M_Z + 1$ because no automaton with less than $M_Z + 1$ states can recognize the existence of a redex of R_Z. We use a function $\mathsf{mld}_A : Q \to \mathbb{N}$ defined by $\mathsf{mld}_A(q) = \min_{t \in \mathcal{L}(A,q)} \mathsf{ld}(t)$. The function is total because the TDA-S to be found has only reachable states.

Propositional Variables. Our SAT encoding involves three classes of propositional variables. The first one has the form of either $v_{@(q_1,q_2) \rightsquigarrow q}$ or $v_{Z \rightsquigarrow q}$ with $q_1, q_2, q \in Q$, which identify Δ. These variables are expected to satisfy

$v_{@(q_1,q_2) \rightsquigarrow q}$ is true iff $@(q_1, q_2) \rightsquigarrow q \in \Delta$, for all $q_1, q_2, q \in Q$, and

$v_{Z \rightsquigarrow q}$ is true iff $Z \rightsquigarrow q \in \Delta$, for all $q \in Q$.

This class of variables has been employed in the EZ method.

The second class of propositional variables has the form of either $v_{q,m}$ with $q \in Q \setminus \{q_F\}$ and $m < M_Z$ or $v_{q,\mathsf{rdx}}$ with $q \in Q$. They are expected to satisfy

$v_{q,m}$ is true iff $m = \mathsf{mld}_A(q)$, for all $q \in Q \setminus \{q_F\}$ and $m < M_Z$, and

$v_{q,\mathsf{rdx}}$ is true only if $\mathcal{L}(A, q) \cap \mathsf{NF}(R_Z) = \emptyset$, for all $q \in Q$.

This class of variables is newly introduced in our method in order to check the existence of redexes. Instead, the EZ method employs propositional variables that represent reachability of states of a tree automaton obtained by product construction of two tree automata, A and B, where B accepts all terms in normal form of R_Z.

The third class of propositional variables has the form of $v_{t,\alpha,q}$ with a term $t \in U_Z$, a substitution $\alpha : \mathcal{V}(t) \to Q \setminus \{q_F\}$, and a state $q \in Q$. Note that the number of possible substitutions α is finite because U_Z and Q are finite.

Therefore the number of this class of variables is also finite. These variables are expected to satisfy

$$v_{t,\alpha,q} \text{ is true iff } t\alpha \rightsquigarrow^* q$$

for all $t \in U_Z$, $\alpha : \mathcal{V}(t) \to Q \setminus \{q_F\}$, and $q \in Q$. This class of variables has been employed in the EZ method, with the difference of the domain of substitutions. In their method, each substitution is defined over the set of all states including final states, while our encoding excludes the final state from the domain of substitutions. The difference makes the number of this class of variables much smaller, which will reduce the number of clauses passed to the SAT solver. We will explain later the reason why the final state can be left out in our encoding method.

Besides the three classes of propositional variables, our implementation of the proposed encoding method employs extra variables which is equivalent to conjunction of the other variables. They are introduced in order for the number of clauses to be smaller using a standard technique of Tseitin transformation [17], where sub-formulas are replaced by new propositional variables to avoid exponential brow-up of the number of clauses. The original method by Endrullis and Zantema may also have used the technique thought it is not explicitly mentioned in their article. Their implementation is currently no longer available, so we cannot be sure how they actually do it.

Propositional Formulas. Recall that a TDA-S $A = \langle Q, \Sigma_Z, \{q_F\}, \Delta \rangle$ with $|Q| = N$ is to be found by the SAT solver. The Boolean values of the propositional variables introduced so far exactly identify Δ. It must also be ensured, however, that there is no inconsistency in the valuation of propositional variables, and that A satisfies the TDA-S conditions. The propositional formulas to be satisfied will be shown in sequence condition by condition. It is assumed that they will eventually be combined together by conjunction and passed to the SAT solver to find an appropriate valuation.

Firstly, the final state q_F must be sink according to the (3-1) condition. Hence the following propositional formulas must hold:

$$\bigwedge_{q_F \in \{q_1, q_2\} \subseteq Q} \left(v_{@(q_1,q_2) \rightsquigarrow q_F} \wedge \bigwedge_{q \in Q \setminus \{q_F\}} \neg v_{@(q_1,q_2) \rightsquigarrow q} \right) \tag{1}$$

All of these propositional variables are immediately forced to be assigned to either true or false in the phase of unit propagation of the SAT solver. In addition, the (3-1) condition requires that q_F is reachable. Since we will find a TDA-S with the smallest number of states, we assume that all states in Q are reachable. The reachability of states can be paraphrased as the existence of a total order on Q with appropriate properties as considered in the EZ method. We will employ the following lemma to specify the total order. The proof is similar to that of the EZ method except that the final state is fixed as a sink state in our setting. The details of the proof are found in the full version of this paper [10].

Lemma 6. *Let $A = \langle Q, \Sigma_Z, \{q_F\}, \Delta \rangle$ be a tree automaton with a sink state q_F. Then, all states in Q are reachable if and only if there exists a total order $<$ on Q with maximal element q_F such that for every $q \in Q$ there exists either $Z \rightsquigarrow q \in \Delta$ or $@(q_1, q_2) \rightsquigarrow q \in \Delta$ with $q_1 < q$ and $q_2 < q$.*

Lemma 6 makes it easy to find a TDA-S in which all states are reachable. Without loss of generality, we can fix an ordered sequence of all states in Q as $p_1 < p_2 < \cdots < p_N = q_F$ (recall $|Q| = N$) and force the states to satisfy the property in Lemma 6. The restriction is directly encoded by

$$\bigwedge_{q \in Q} \left(v_{Z \rightsquigarrow q} \vee \bigvee_{q_1, q_2 < q} v_{@(q_1, q_2) \rightsquigarrow q} \right), \tag{2}$$

which has been employed also in the EZ method, although our encoding differs in that the ordered sequence of states should end with the unique final state.

Secondly, A must satisfy the (3-2) condition, which requires that A accepts no terms in normal form. Since $\mathcal{L}(A) = \mathcal{L}(A, q_F)$, a propositional formula

$$v_{q_F, \text{rdx}} \tag{3}$$

(which is just a unit clause) should hold. To ensure that the Boolean value of this variable is valid, all of the propositional variables of the form $v_{q,m}$ or $v_{q,\text{rdx}}$ should be properly assigned as follows. The propositional variables $v_{q,m}$ for $q \in Q \setminus \{q_F\}$ and $m < M_Z$ are expected to satisfy $m = \text{mld}_A(q)$. It is easy to see that $\text{mld}_A(q)$ can be effectively computed for each $q \in Q$ from the transition rules in Δ. The next lemma claims this fact as a logical statement so as to be used for giving appropriate propositional formulas. The proof is found in the full paper [10].

Lemma 7. *Let $A = \langle Q, \Sigma_Z, \{q_F\}, \Delta \rangle$ be a tree automaton with a combinator Z where all states in Q are reachable. Then for every $q \in Q$ and $m \in \mathbb{N}$, (7-1) $\text{mld}_A(q) = 0$ if and only if $Z \rightsquigarrow q \in \Delta$, and (7-2) $\text{mld}_A(q) = m > 0$ if and only if there is neither $Z \rightsquigarrow q \in \Delta$ nor $@(q_1, q_2) \rightsquigarrow q \in \Delta$ with $\text{mld}_A(q_1) < m - 1$ and there exists $@(q_1, q_2) \rightsquigarrow q \in \Delta$ with $\text{mld}_A(q_1) = m - 1$.*

We only need the lemma for non-final states though it holds even for the final state q_F because the propositional variables $v_{q,m}$ are given only for $q \in Q \setminus \{q_F\}$. The statement (7-1) indicates that $v_{q,0}$ for each $q \in Q \setminus \{q_F\}$ has an appropriate Boolean value by the following propositional formula:

$$\bigwedge_{q \in Q \setminus \{q_F\}} (v_{q,0} \Leftrightarrow v_{Z \rightsquigarrow q}). \tag{4}$$

We could use the same propositional variable for $v_{q,0}$ and $v_{Z \rightsquigarrow q}$ in an efficient implementation, though. Additionally, the statement (7-2) indicates that the propositional variable $v_{q,m}$ with $m > 0$ has an appropriate Boolean value by the following propositional formula:

$$\bigwedge_{q \in Q \setminus \{q_F\}} \bigwedge_{m < M_Z} (v_{q,m} \Leftrightarrow (\neg v_{Z \rightsquigarrow q} \wedge P_1(q, m) \wedge P_2(q, m))) \tag{5}$$

where

$$P_1(q, m) = \bigwedge_{0 \le k < m-1} \bigwedge_{q_1, q_2 \in Q \setminus \{q_F\}} \left(v_{@(q_1, q_2) \leadsto q} \Rightarrow \neg v_{q_1, k} \right) \text{ and}$$

$$P_2(q, m) = \bigvee_{q_1, q_2 \in Q \setminus \{q_F\}} \left(v_{@(q_1, q_2) \leadsto q} \wedge v_{q_1, m-1} \right).$$

The propositional variable $v_{q, \mathrm{rdx}}$ for each $q \in Q$ is expected to be true only if $\mathcal{L}(A, q) \cap \mathsf{NF}(R_Z) = \emptyset$. The next lemma is used for giving propositional formulas in order for the variables $v_{q, \mathrm{rdx}}$ to have an appropriate Boolean value.

Lemma 8. *Let $A = \langle Q, \Sigma_Z, \{q_F\}, \Delta \rangle$ be a tree automaton with a sink state q_F, let R_Z be a TRS with a combinator Z, and let $\mathcal{Q} \subseteq Q$ be a set of states such that for all $q \in \mathcal{Q}$, (8-1) $Z \leadsto q \notin \Delta$ holds, and (8-2) $\mathsf{mld}_A(q_1) = M_Z - 1$ holds if there is $@(q_1, q_2) \leadsto q \in \Delta$ with $q_1, q_2 \in Q \setminus \mathcal{Q}$. Then, $\mathcal{L}(A, q) \cap \mathsf{NF}(R_Z) = \emptyset$ holds for every $q \in \mathcal{Q}$.*

The proof is found in the full paper [10]. Let $\mathcal{Q} \subseteq Q$ be the set of states q such that $v_{q, \mathrm{rdx}}$ is true. Lemma 8 guarantees that $v_{q, \mathrm{rdx}}$ is true only if $\mathcal{L}(A, q) \cap \mathsf{NF}(R_Z) = \emptyset$ whenever $\mathcal{Q}$ satisfies the conditions (8-1) and (8-2), that is, the following propositional formulas should be true: for (8-1),

$$\bigwedge_{q \in Q} \left(v_{q, \mathrm{rdx}} \Rightarrow \neg v_{Z \leadsto q} \right); \tag{6}$$

for (8-2),

$$\bigwedge_{q \in Q} \bigwedge_{q_1, q_2 \in Q \setminus \{q_F\}} \left(v_{q, \mathrm{rdx}} \wedge v_{@(q_1, q_2) \leadsto q} \wedge \neg v_{q_1, \mathrm{rdx}} \wedge \neg v_{q_2, \mathrm{rdx}} \Rightarrow v_{q_1, M_Z - 1} \right). \tag{7}$$

Finally, we need to ensure that A is almost closed under R_Z in the sense of the (2-3) condition. As the EZ method does, we use the following lemma which gives a procedure to have the condition. While the EZ method employs the existing results [5, Proposition 12], we give a proof of this lemma in the full paper [10] because the (3-3) condition is different from the closure property.

Lemma 9. *Let $A = \langle Q, \Sigma, \{q_F\}, \Delta \rangle$ be a tree automaton with a sink state q_F, and let R be a left-linear TRS over Σ. Then, $s \in \mathcal{L}(A, q)$ implies $t \in \mathcal{L}(A, q) \cup \mathcal{L}(A)$ for all $q \in Q$ and $s, t \in T_\Sigma$ with $s \to_R^\mathrm{I} t$ if $l\alpha \leadsto_\Delta^* q$ implies $r\alpha \leadsto_\Delta^* q$ or $r\alpha \leadsto_\Delta^* q_F$ for all $q \in Q$, $l \to r \in R$ and $\alpha : \mathsf{FV}(l) \to Q$.*

Recall that the propositional variable $v_{t, \alpha, q}$ is true when a term t is accepted with $q \in Q$ if every variable x in t is substituted with a term accepted with a state $\alpha(x)$. Following the EZ method, the procedure of computing the state of each subterm is simulated by imposing relations among the propositional variables. Unlike their method, we only need to consider the case where the range of α is $Q \setminus \{q_F\}$ since the closure property always holds if $\alpha(x) = q_F$ for some x because q_F is sink and R_Z is non-erasing. This arrangement would substantially reduce the number of propositional variables. For each subterm $t = @(t_1, t_2) \in U_Z$, we should have

$$\bigwedge_{q \in Q} \bigwedge_{\alpha: \mathsf{FV}(t) \to Q \setminus \{q_F\}} \left(v_{t,\alpha,q} \Leftrightarrow \bigvee_{q_1, q_2 \in Q} \left(v_{t_1, \alpha[t_1], q_1} \wedge v_{t_2, \alpha[t_2], q_2} \wedge v_{@(q_1, q_2) \rightsquigarrow q} \right) \right) \quad (8)$$

where $\alpha[t]$ stands for the substition α whose domain is restricted to $\mathsf{FV}(t)$. For each leaf in the subterms U_Z, we should have

$$\bigwedge_{q \in Q} \left((v_{Z, \emptyset, q} \Leftrightarrow v_{Z \rightsquigarrow q}) \wedge \bigwedge_{x \in \mathsf{FV}(l_B)} \left(v_{x, \{x \mapsto q\}, q} \wedge \bigwedge_{q' \in Q \setminus \{q\}} \neg v_{x, \{x \mapsto q'\}, q} \right) \right) \quad (9)$$

where the former part indicates that we could use the same propositional variable for $v_{Z, \emptyset, q}$ and $v_{Z \rightsquigarrow q}$. And then, the (3-3) condition requires the propositional formula

$$\bigwedge_{\alpha: \mathsf{FV}(l_Z) \to Q} (v_{l_Z, \alpha, q} \Rightarrow v_{r_Z, \alpha, q} \vee v_{r_Z, \alpha, q_F}). \quad (10)$$

4.3   Applying to Non-termination of Specific Combinators

We have shown how to construct propositional logic formulas for the existence of TDA-S with fixed number of states for a sole combinatory calculus. We have implemented the construction and input the obtained formulas to a SAT solver to examine termination for combinators shown in Fig. 1, for which termination is unknown. We confirmed that our method can efficiently find a TDA-S compared to the method developed by Endrullis and Zantema for a TDA. Note that their implementation is not currently publicly available, so we have re-implemented it. For fairness, we have also applied the same optimization (e.g., Tseitin transformation) as in the implementation of our construction. Our implementation is written in OCaml and uses the Kissat SAT solver. We also implement the ability to output the smallest term accepted by the found TDA-S, thus allowing the output of a non-normalizable term.

The results of the examination for each combinator are shown in Table 1. Due to page limitation, the details of the TDA-S obtained are found in the full paper [10] with a found counterexample for termination. Here we only show the number of states of the TDA-S, the size of the propositional logic formula used (number of propositional variables and clauses), and the computation time (the sum of encoding and solving time) including the attempts to fewer states. The 'EZ method' column shows the results of (our re-implementation of) the original method by Endrullis and Zantema; the 'Our method' column shows the results of our construction method introduce in the present paper. The EZ method failed to find a TDA for Φ_2 within 24-hour computation. Note that the number of propositional variables and clauses is the same for different combinators. This is because they depend only on the number of states, the number of variables in the left-hand side of the combinator's rule and the number of subterms in the right-hand side. A major improvement in our method is a reduction of the

Table 1. Results of disproving termination of combinators given in Fig. 1.

	EZ method				Our method			
	#Q	#vars	#clauses	time (s)	#Q	#vars	#clauses	time (s)
P	4	15,648	75,237	0.71	4	7,116	33,543	0.22
P$_3$	6	161,544	793,351	11.5	6	96,058	469,233	5.6
$\mathcal{D}_1$	6	936,810	4,594,129	97.0	6	462,528	2,264,943	32.6
$\mathcal{D}_2$	6	936,810	4,594,129	96.6	6	462,528	2,264,943	32.9
Φ	7	1,975,302	9,741,173	260.6	7	1,099,527	5,416,581	108.3
Φ$_2$	–	—	—	—	9	55,695,683	276,052,931	39,268.4
S$_1$	7	1,975,302	9,741,173	262.1	7	1,099,527	5,416,581	111.6
S$_2$	6	745,002	3,655,825	81.6	6	379,278	1,857,693	30.3

number of variables of the form $v_{t,\alpha,q}$ by restricting the range of α, which is the most significant part of the SAT encoding. The number of possible substitutions is reduced from $|Q|^N$ to $(|Q|-1)^N$, where N is the number of variables in the left-hand side of the rule. Our method can indeed generate less propositional logic formulas than the original one, and also succeed in disproving the termination of more combinators. Both method failed to find the disproof of termination for the remaining combinators $\mathbf{S}_3$ and $\mathbf{S}_4$ in Fig. 1. We have confirmed that their termination cannot be disproven by a TDA-S with at most 9 states.

5   Concluding Remark

We have proposed a method to disprove the termination of sole combinatory calculi with tree automata by extending the method proposed by Endrullis and Zantema. Specifically, we have shown that a tree automaton with a final sink state is sufficient to disprove termination of non-erasing combinators. We have succeeded in disproving the termination of 8 combinators which is unknown of their termination found in Smullyan's book. The remaining two combinators $\mathbf{S}_3$ and $\mathbf{S}_4$ still have unknown termination. We may need more improvement to disprove their termination. However, our method in which termination is disproved by a tree automaton with a final sink state may be applicable to other non-erasing term rewriting systems. It would be interesting to investigate how effectual our method is in more general settings.

Acknowledgments. The authors would like to thank the anonymous reviewers for their valuable comments. This work was partially supported by JSPS KAKENHI Grant Numbers, JP21K11744, JP22H00520, and JP22K11904.

References

1. Barendregt, H.P.: The Lambda Calculus – Its Syntax and Semantics, Studies in Logic and the Foundations of Mathematics, vol. 103. North-Holland (1985)
2. Church, A.: The Calculi of Lambda Conversion. (AM-6) (Annals of Mathematics Studies). Princeton University Press, USA (1985)
3. Curry, H.B.: Grundlagen der kombinatorischen Logik. Am. J. Math. **52**(3), 509–536 (1930)
4. Endrullis, J., Zantema, H.: Proving non-termination by finite automata. In: 26th International Conference on Rewriting Techniques and Applications (RTA 2015). LIPIcs, vol. 36, pp. 160–176 (2015). https://doi.org/10.4230/LIPICS.RTA.2015.160
5. Genet, T.: Decidable approximations of sets of descendants and sets of normal forms. In: Nipkow, T. (ed.) RTA 1998. LNCS, vol. 1379, pp. 151–165. Springer, Heidelberg (1998). https://doi.org/10.1007/BFb0052368
6. Ikebuchi, M., Nakano, K.: On properties of B-terms. Log. Meth. Comput. Sci. **16**(2) (2020). https://doi.org/10.23638/LMCS-16(2:8)2020
7. Iwami, M.: On the acyclic and related properties of combinators (in Japanese). IPSJ Trans. Prog. (PRO) **2**(2), 97–104 (2009)
8. Iwami, M.: Non-ω-strong head normalization, non-ground loop and acyclic of several combinators (in Japanese). IPSJ Trans. Prog. (PRO) **16**(3), 14–27 (2023)
9. Klop, J.W.: New fixed point combinators from old. In: Reflections on Type Theory, Lambda-calculus, and the Mind : Essays Dedicated to Henk Barendregt on the Occasion of his 60th Birthday (2007). https://www.cs.ru.nl/barendregt60/essays/klop/
10. Nakano, K., Iwami, M.: Disproving Termination of Non-Erasing Sole Combinatory Calculus with Tree Automata (Full Version). CoRR **abs/2406.14305** (2024). https://arxiv.org/abs/2406.14305
11. O'Donnell, M.J. (ed.): Computing in Systems Described by Equations. LNCS, vol. 58. Springer, Heidelberg (1977). https://doi.org/10.1007/3-540-08531-9
12. Ohlebusch, E.: Advanced Topics in Term Rewriting. Springer, New York, NY (2002). http://www.springer.com/computer/swe/book/978-0-387-95250-5
13. Peyton Jones, S.L.: The Implementation of Functional Programming Languages. Prentice-Hall, Hoboken (1987)
14. Probst, D., Studer, T.: How to normalize the Jay. Theor. Comput. Sci. **254**(1–2), 677–681 (2001). https://doi.org/10.1016/S0304-3975(00)00379-0
15. Schönfinkel, M.: Über die Bausteine der mathematischen Logik. Math. Ann. **92**(3), 305–316 (1924). https://doi.org/10.1007/BF01448013
16. Smullyan, R.: Diagonalization and Self-reference. Clarendon Press, Oxford logic guides (1994)
17. Tseitin, G.S.: Automation of Reasoning 2, chap. On the Complexity of Derivation in Propositional Calculus, pp. 466–483. Springer, Berlin, Heidelberg (1983). https://doi.org/10.1007/978-3-642-81955-1_28
18. Turner, D.A.: A new implementation technique for applicative languages. Softw. Pract. Exp. **9**(1), 31–49 (1979). https://doi.org/10.1002/SPE.4380090105
19. Waldmann, J.: The Combinator S. Information and Computation **159**(1-2), 2–21 (2000). https://doi.org/10.1006/inco.2000.2874

The Equivalence Problem of E-Pattern Languages with Regular Constraints Is Undecidable

Dirk Nowotka and Max Wiedenhöft[(✉)]

Department of Computer Science, Kiel University, Kiel, Germany
{dn,maw}@informatik.uni-kiel.de

Abstract. Patterns are words with terminals and variables. The language of a pattern is the set of words obtained by uniformly substituting all variables with words that contain only terminals. Regular constraints restrict valid substitutions of variables by associating with each variable a regular language representable by, e.g., finite automata. Pattern languages with regular constraints contain only words in which each variable is substituted according to a set of regular constraints. We consider the membership, inclusion, and equivalence problems for erasing and non-erasing pattern languages with regular constraints. Our main result shows that the erasing equivalence problem-one of the most prominent open problems in the realm of patterns-becomes undecidable if regular constraints are allowed in addition to variable equality.

Keywords: Patterns · Pattern Languages · Regular Constraints · Undecidability · Automata · Membership · Inclusion · Equivalence

1 Introduction

A *pattern* is a finite word consisting of symbols from a finite set of letters $\Sigma = \{a_1, ..., a_\sigma\}$, also called terminals, and from an infinite set of variables $X = \{x_1, x_2, ...\}$ with $\Sigma \cap X = \emptyset$. It is a natural and compact device to define formal languages. Words consisting of only terminal symbols are obtained from patterns by a *substitution* h, a terminal preserving morphism which maps all variables from a pattern to words over the terminal alphabet. The *language* of a pattern consists of all words obtainable from that pattern by substitutions.

We differentiate between two kinds of substitutions. Originally, pattern languages introduced by Angluin [1] only consisted of words obtained by *non-erasing substitutions* that required all variables to be mapped to non-empty words. Thus, those languages are also called *NE-pattern languages*. Later, so called *erasing-/extended-* or just *E-pattern languages* have been introduced by Shinohara [24]. In these, substitutions are also allowed to map variables to the empty word. Consider, for example, the pattern $\alpha := x_1\mathsf{ab}x_2x_2$. Then, by mapping x_1 to aaa

M. Wiedenhöft—This work was supported by the DFG project number 437493335.

S. Z. Fazekas (Ed.): CIAA 2024, LNCS 15015, pp. 276–288, 2024.
https://doi.org/10.1007/978-3-031-71112-1_20

and x_2 to ba with a substitution h, we obtain the word $h(\alpha) = $ aaaabbaba. If we consider the E-pattern language of α, we could also map x_2 to the empty word ε with a substitution h' which also maps x_1 to aaa and obtain $h'(\alpha) = $ aaaab.

Due to its practical and simple definition, patterns and their corresponding languages occur in numerous areas regarding computer science and discrete mathematics, including unavoidable patterns [14,17], algorithmic learning theory [1,5,25], word equations [17], theory of extended regular expressions with back references [9], and database theory [7,23].

The main problems regarding patterns and pattern languages are the *membership problem* (and its variations [6,10,11]), the *inclusion problem*, and the *equivalence problem* in both the erasing (E) and non-erasing (NE) cases. The membership problem determines if a word belongs to a pattern's language. This problem is NP-complete for both E- and NE-pattern languages [1,14]. The inclusion problem asks if one pattern's language is included in another's. Jiang et al. [15] showed that it is generally undecidable for E- and NE-pattern languages. Freydenberger and Reidenbach [8], and Bremer and Freydenberger [2] proved its undecidability for all alphabets with size ≥ 2 in both E- and NE-pattern languages. The equivalence problem tests if two patterns generate the same language. It is trivially decidable for NE-pattern languages [1]. Whether its decidable for E-pattern languages is one of the major open problems in the field [15,19–22]. However, for terminal-free patterns, the inclusion and equivalence problems in E-pattern languages have been characterized and shown to be NP-complete [4,15]. The decidability of the inclusion problem for terminal-free NE-pattern languages remains unresolved, though.

Various extensions to patterns and pattern languages have been introduced over time. Some examples are the bounded scope coincidence degree, patterns with bounded treewidth, k-local patterns, and strongly-nested patterns (see [3] and references therein). Koshiba [16] introduced so called *typed patterns* to enhance the expressiveness of pattern languages by restricting substitutions of variables to types, i.e., arbitrary recursive languages. This has recently been extended by Geilke and Zilles [12] who introduced the notion of *relational patterns* and *relational pattern languages*.

We consider a specific class of typed- or relational patterns called *patterns with regular constraints*. Let $\mathcal{L}_{Reg}$ be the set of all regular languages. Then, we say that a mapping $r : X \to \mathcal{L}_{Reg}$ is a *regular constraint* that implicitly defines *languages on variables* $x \in X$ by $L_r(x) = r(x)$. Let $\mathcal{C}_{Reg}$ be the *set of all regular constraints*. A *patterns with regular constraints* $(\alpha, r_\alpha) \in (\Sigma \cup X)^* \times \mathcal{C}_{Reg}$ is a pattern which is associated with a regular constraint. A substitution h is r_α-*valid* if all variables are substituted according to r_α. The language of (α, r_α) is defined analogously to pattern languages with the additional requirement that all substitutions must be r_α-valid.

This paper examines erasing (E) and non-erasing (NE) pattern languages with regular constraints. The membership problem for both is NP-complete, while the inclusion problem is undecidable for the general and terminal-free versions. This immediately follows from known results. The main finding of this paper is that the equivalence problem for erasing pattern languages with regular constraints is indeed undecidable.

2 Preliminaries

Let $\mathbb{N}$ denote the natural numbers $\{1, 2, 3, \dots\}$ and let $\mathbb{N}_0 := \mathbb{N} \cup \{0\}$. For $n, m \in \mathbb{N}$ set $[m, n] := \{k \in \mathbb{N} \mid m \le k \le n\}$. Denote $[n] := [1, n]$ and $[n]_0 := [0, n]$. The powerset of any set A is denoted by $\mathcal{P}(A)$. An *alphabet* Σ is a non-empty finite set whose elements are called *letters*. A *word* is a finite sequence of letters from Σ. Let Σ^* be the set of all finite words over Σ, thus it is a free monoid with concatenation as operation and the empty word ε as natural element. Set $\Sigma^+ := \Sigma^* \setminus \{\varepsilon\}$. We call the number of letters in a word $w \in \Sigma^*$ *length* of w, denoted by $|w|$. Therefore, we have $|\varepsilon| = 0$. If $w = xyz$ for some $x, y, z \in \Sigma^*$, we call x a *prefix* of w, y a *factor* of w, and z a *suffix* of w and denote the sets of all prefixes, factors, and suffixes of w by $\mathrm{Pref}(w)$, $\mathrm{Fact}(w)$, and $\mathrm{Suff}(w)$ respectively. For words $w, u \in \Sigma^*$, let $|w|_u$ denote the number of distinct occurrences of u in w as a factor. Denote $\Sigma^k := \{w \in \Sigma^* \mid |w| = k\}$. For $w \in \Sigma^*$, let $w[i]$ denote w's i^{th} letter for all $i \in [|w|]$. For reasons of compactness, we denote $w[i] \cdots w[j]$ by $w[i \cdots j]$ for all $i, j \in [|w|]$ with $i < j$. Set $\mathrm{alph}(w) := \{\mathsf{a} \in \Sigma \mid \exists i \in [|w|] : w[i] = \mathsf{a}\}$ as w's alphabet.

Let X be a countable set of variables such that $\Sigma \cap X = \emptyset$. A *pattern* is then a non-empty, finite word over $\Sigma \cup X$. The set of all patterns over $\Sigma \cup X$ is denoted by Pat_Σ. For example, $x_1 \mathsf{a} x_2 \mathsf{b} \mathsf{a} x_2 x_3$ is a pattern over $\Sigma = \{\mathsf{a}, \mathsf{b}\}$ with $x_1, x_2, x_3 \in X$. For a pattern $p \in Pat_\Sigma$, let $\mathrm{var}(p) := \{x \in X \mid |p|_x \ge 1\}$ denote the set of variables occurring in p. A *substitution of p* is a morphism $h : (\Sigma \cup X)^* \to \Sigma^*$ such that $h(\mathsf{a}) = \mathsf{a}$ for all $\mathsf{a} \in \Sigma$ and $h(x) \in \Sigma^*$ for all $x \in X$. If we have $h(x) \neq \varepsilon$ for all $x \in \mathrm{var}(p)$, we call h a *non-erasing substitution* for p. Otherwise h is an *erasing substitution* for p. The set of all substitutions w.r.t. Σ is denoted by H_Σ. If Σ is clear from the context, we may write just H. Given a pattern $\alpha \in Pat_\Sigma$, it's erasing pattern language $L_E(\alpha)$ and its non-erasing pattern language $L_{NE}(\alpha)$ are defined respectively by

$$L_E(\alpha) := \{h(\alpha) \mid h \in H, h(x) \in \Sigma^* \text{ for all } x \in \mathrm{var}(\alpha)\}, \text{ and}$$

$$L_{NE}(\alpha) := \{h(\alpha) \mid h \in H, h(x) \in \Sigma^+ \text{ for all } x \in \mathrm{var}(\alpha)\}.$$

Let $\mathcal{L}_{Reg}$ be the set of all regular languages. We call a mapping $r : X \to \mathcal{L}_{Reg}$ a *regular constraint* on X. If not stated otherwise, we always have $r(x) = \Sigma^*$. We denote the *set of all regular constraints* by $\mathcal{C}_{Reg}$. For some $r \in \mathcal{C}_{Reg}$ we define the *language of a variable* $x \in X$ by $L_r(x) = r(x)$. If r is clear by the context, we omit it and just write $L(x)$. A *pattern with regular constraints* is a pair $(p, r_p) \in Pat_\Sigma \times \mathcal{C}_{Reg}$. We denote the *set of all patterns with regular constraints* by $Pat_{\Sigma, \mathcal{C}_{Reg}}$. For some $(p, r_p) \in Pat_{\Sigma, \mathcal{C}_{Reg}}$ and $h \in H$, we say that h is a *r_p-valid substitution* if $h(x) \in L(x)$ for all $x \in \mathrm{var}(p)$. We extend the notion of pattern languages by the following. For any $(p, r_p) \in Pat_{\Sigma, \mathcal{C}_{Reg}}$ we denote by

$$L_E(p, r_p) := \{h(p) \mid h \in H, h(x) \in \Sigma^* \text{ for all } x \in \mathrm{var}(p), h \text{ is } r_p\text{-valid}\}$$

the *erasing pattern language with regular constraints* of (p, r_p) and by

$$L_{NE}(p, r_p) := \{h(p) \mid h \in H, h(x) \in \Sigma^+ \text{ for all } x \in \mathrm{var}(p), h \text{ is } r_p\text{-valid}\}$$

the *non-erasing pattern language with regular constraints* of (p, r_p).

2.1   Nondeterministic 2-Counter Automata

Usually, 2-counter automata are defined over input words utilising an input alphabet and the additional use of two counters. In our setting, we consider a slight variation which assumes that the automaton always runs over an empty input word. A *nondeterministic 2-counter automaton without input* (see e.g. [13]) is a 4-tuple $A = (Q, \delta, q_0, F)$ which consists of a set of states Q, a transition function $\delta : Q \times \{0,1\}^2 \to \mathcal{P}(Q \times \{1,0,+1\}^2)$, an initial state $q_0 \in Q$, and a set of accepting states $F \subseteq Q$. A *configuration* of A is defined as a triple $(q, m_1, m_2) \in Q \times \mathbb{N}_0 \times \mathbb{N}_0$ in which q indicates the current state and m_1 and m_2 indicate the contents of the first and second counter. We define the relation $\vdash_A$ on $Q \times \mathbb{N}_0 \times \mathbb{N}_0$ by δ as follows. For two configurations (p, m_1, m_2) and (q, n_1, n_2) we say that $(p, m_1, m_2) \vdash_A (q, n_1, n_2)$ if and only if there exist $c_1, c_2 \in \{0,1\}$ and $r_1, r_2 \in \{-1, 0, +1\}$ such that

1. if $m_i = 0$ then $c_i = 0$, otherwise if $m_i > 0$, then $c_i = 1$, for $i \in \{1,2\}$,
2. $n_i = m_i + r_i$ for $i \in \{1,2\}$,
3. $(q, r_1, r_2) \in \delta(p, c_1, c_2)$, and
4. we assume if $c_i = 0$ then $r_i \neq -1$ for $i \in \{1,2\}$.

Essentially, the machine checks in every state whether the counters equal 0 and then changes the value of each counter by at most one per transition before entering a new state. A *computation* is a sequence of configurations. An *accepting computation* of A is a sequence $C_1, ..., C_n \in (Q \times \mathbb{N}_0 \times \mathbb{N}_0)^n$ with $C_1 = (q_0, 0, 0)$, $C_i \vdash_A C_{i+1}$ for all $i \in \{1, ..., n-1\}$, and $C_n \in F \times \mathbb{N}_0 \times \mathbb{N}_0$ for some $n \in \mathbb{N}$.

We *encode* configurations of A by assuming $Q = \{q_0, ..., q_e\}$ for some $e \in \mathbb{N}_0$ and defining a function $enc : Q \times \mathbb{N}_0 \times \mathbb{N}_0 \to \{0, \#\}^*$ by

$$enc(q_i, m_1, m_2) := 0^{1+i} \# 0^{5+2m_1} \# 0^{5+2m_2}.$$

Notice that each state q_i is mapped to a word 0^{i+1} and that each number m_i is mapped to an odd number 0^{5+2m_i} where 0^5 denotes 0, 0^7 denotes 1, 0^9 denotes 2 and so on. This is extended to encodings of computations by defining for every $n \geq 1$ and every sequence $C_1, ..., C_n \in Q \times \mathbb{N}_0 \times \mathbb{N}_0$

$$enc(C_1, ..., C_n) := \#\# \ enc(C_1) \ \#\# \ ... \ \#\# \ enc(C_n) \ \#\#.$$

This encoding of configurations and computations is specifically chosen for its utility in proving Theorem 4.

Furthermore, define the set of accepting computations

$$\mathtt{ValC}(A) := \{enc(C_1, ..., C_n) \mid C_1, ..., C_n \text{ is an accepting computation of } A\}$$

and let $\mathtt{InvalC}(A) = \{0, \#\}^* \backslash \mathtt{ValC}(A)$. The emptiness problem for deterministic 2-counter-automata with input is undecidable (cf. e.g. [13,18]), thus it is also undecidable whether a nondeterministic 2-counter automaton without input has an accepting computation [8,15].

2.2 Known Results

The membership problems of both, the erasing and non-erasing pattern languages, have been shown to be NP-complete [1,14]. Hence, we observe the following for patterns with regular constraints.

Corollary 1. *Let $(\alpha, r_\alpha) \in Pat_{\Sigma, \mathcal{C}_{Reg}}$ and $w \in \Sigma^*$. The decision problem of whether $w \in L_X(\alpha, r_\alpha)$ for $X \in \{E, NE\}$ is NP-complete.*

Indeed, we immediately obtain NP-hardness in both cases by the previous results shown in [1,14] for patterns. NP-containment follows by knowing that a valid certificate results in a substitution of α which has at most length $|w|$.

One other notable problem regarding patterns is the inclusion problem. The undecidabilities of the inclusion problems for patterns in the erasing and non-erasing cases have been initially shown by Jiang et al. [15] for unbounded alphabets and have been refined and extended to finite alphabets of sizes greater or equal to 2 in [2,8]. Hence we have the following.

Theorem 2. *[2, 8, 15] Let $\alpha, \beta \in Pat_\Sigma$. In general, for all alphabets Σ with $|\Sigma| \geq 2$, it is undecidable to answer whether*

1. $L_E(\alpha) \subseteq L_E(\beta)$, or
2. $L_{NE}(\alpha) \subseteq L_{NE}(\beta)$.

From that, we immediately obtain the following for patterns with regular constraints.

Corollary 3. *Let $(\alpha, r_\alpha), (\beta, r_\beta) \in Pat_{\Sigma, \mathcal{C}_{Reg}}$. In both, the terminal-free and the non terminal-free cases for α and β we have in general, for all alphabets Σ with $|\Sigma| \geq 2$, that it is undecidable to answer whether*

1. $L_E(\alpha, r_\alpha) \subseteq L_E(\beta, r_\beta)$, or
2. $L_{NE}(\alpha, r_\alpha) \subseteq L_{NE}(\beta, r_\beta)$.

Indeed, the general results follow immediately from Theorem 2. Additionally, in the terminal-free cases, we can reduce the general versions to the terminal free versions by substituting each terminal letter $\mathsf{a} \in \Sigma$ which occurs in a pattern α by a new variable x_a and setting $L(x_\mathsf{a}) = \{\mathsf{a}\}$. This results in effectively the same problem instances without using terminals in the pattern words.

3 Undecidability of E-Pattern Language Equivalence

The main result of this paper considers the equivalence problem for erasing pattern languages with regular constraints. In particular, we show that this problem is undecidable.

Theorem 4. *Let $(\alpha, r_\alpha), (\beta, r_\beta) \in Pat_{\Sigma, \mathcal{C}_{Reg}}$. In general, it is undecidable to decide whether $L_E(\alpha, r_\alpha) = L_E(\beta, r_\beta)$ for all alphabets Σ with $\Sigma \geq 2$.*

The rest of this section is dedicated to show Theorem 4. Roughly based on the idea of the proof of undecidability of the inclusion problem for pattern languages in the case of finite alphabets, given by Freydenerger and Reidenbach [8], we reduce the question whether some non-deterministic 2-counter automaton without input A has some accepting computation to the problem of whether the erasing pattern languages of two patterns with regular constraints are equal. The first is known to be undecidable out of which the undecidability of the second problem follows. In contrast to the proof given in [8], the constructed patterns and predicates (to be explained later) had to be notably adapted to work for the case considered here.

Let $A = (Q, \delta, q_0, F)$ be some non-deterministic 2-counter automaton without input. We construct two patterns with regular constraints $(\alpha, r_\alpha), (\beta, r_\beta) \in Pat_{\Sigma, \mathcal{C}_{Reg}}$ such that $L_E(\alpha, r_\alpha) = L_E(\beta, r_\beta)$ if and only if $\text{ValC}(A) = \emptyset$.

We start with the binary case and assume $\Sigma = \{0, \#\}$. First, we construct (α, r_α). We set the pattern α to

$$\alpha = x_v \, \alpha_1 \, x_v \, \tilde{y}$$

for variables $x_v, \tilde{y}, \alpha_1$. Let $v = 0\#^3 0$. We then define the regular constraint r_α for α by $L_E(x_v) := \{\varepsilon, v\}$, $L_E(\alpha_1) := \{\, 0w0 \in \Sigma^* \mid w \in \Sigma^* \text{ and } |w|_{\#^3} = 0\} \cup \{\varepsilon\}$, and $L_E(\tilde{y}) := \{\, w \in \Sigma^* \mid w \neq vuv \text{ for all } u \in \Sigma^* \text{ with } u \in L_E(\alpha_1) \backslash \{\varepsilon\} \,\}$. Notice that the given regular constraints won't allow $\tilde{y}$ to be substituted to anything we can obtain with $h(x_v \alpha_1 x_v)$ in the case of x_v and α_1 not being substituted by the empty word, but may be substituted to everything else. Next, we construct (β, r_β). We set the pattern β to

$$\beta = \hat{\beta}_1 \, ... \, \hat{\beta}_\mu \, \tilde{z}$$

such that $\tilde{z}$ is a new variable and $\hat{\beta}_1, ..., \hat{\beta}_\mu$ are terminal free patterns defined by $\hat{\beta}_i = x_i \, \gamma_i \, x_i$ for new variables x_i and some later specified terminal free pattern $\gamma_i \in X^*$ for all $i \in [\mu]$. We assume that each variable in $\text{var}(\gamma_i)$ only appears in γ_i and define r_{γ_i} as the set of regular constraints on the variables occurring in γ_i. By the construction that follows we assume that for all $x \in \text{var}(\gamma_i)$ we always have $\varepsilon \in L(x)$. Notice that each x_i occurs 2 times in β for all $i \in [\mu]$ and also notice that for all $x \in \text{var}(\beta)$ we have that $\varepsilon \in L(x)$. We define the regular constraints r_β on β by setting $L(x_i) := \{\varepsilon, v\}$ for all $i \in [\mu]$ and $L(\tilde{z}) := L(\tilde{y})$. Additionally we add all regular constraints defined by r_{γ_i} to r_β for all $i \in [\mu]$. Further, we from now on assume that for all $w \in L_E(\gamma_i, r_{\gamma_i})$ we have either $w = \varepsilon$ or $w = 0u0$ for $u \in \Sigma^*$ with $|u|_{\#^3} = 0$. This assumption holds by the construction that follows.

Using the construction up to this point and the assumptions we made so far, we first show the following property.

Lemma 5. *We have $L_E(\beta, r_\beta) \subseteq L_E(\alpha, r_\alpha)$.*

Proof. Let $w \in L_E(\beta, r_\beta)$. Then, there exists some r_β-valid $h \in H$ such that $h(\beta) = w$. We differentiate between two main cases.

For the first case, assume $h(\beta) = h(\hat{\beta}_1...\hat{\beta}_\mu \tilde{z}) = v0u0v$ for some $u \in \Sigma^*$ with $|u|_{\#^3} = 0$. By that, we know that $h(\beta) \notin L(\tilde{y})$ as $0u0 \in L(\alpha_1) \setminus \{\varepsilon\}$. Let $h' \in H$ be some substitution. Set $h'(x_v) = v$, $h'(\alpha_1) = 0u0$, and $h'(\tilde{y}) = \varepsilon$. We have that h' is r_α-valid. We get $h'(\alpha) = h'(x_v \alpha x_v \tilde{y}) = v0u0v$. So, $h(\beta) = h'(\alpha)$ and by that $h(\beta) \in L_E(\alpha, r_\alpha)$.

In the second case, assume $h(\beta) \neq v0u0v$ for any $u \in \Sigma^*$ with $|u|_{\#^3} = 0$. Then $h(\beta) \in L(\tilde{y})$. Let $h' \in H$ such that $h'(\tilde{y}) = h(\beta)$ and $h'(x_v) = h'(\alpha_1) = \varepsilon$. Then h' is r_α-valid and we get $h'(\alpha) = h(\beta)$. By that, $h(\beta) \in L_E(\alpha, r_\alpha)$ which concludes this lemma. $\qquad\square$

So by now we know that all words in the language of the pattern (β, r_β) are also in the language generated by the pattern (α, r_α). Next, we show the rather immediate result that all words in the language of (α, r_α) that do not follow a specific form are also in the language generated by (β, r_β). This fact is important for the construction that follows.

Lemma 6. *Let $h \in H$ such that $h(\alpha) \in L_E(\alpha, r_\alpha)$. If $h(\alpha) \neq v0u0v$ for all $u \in \Sigma^*$ with $|u|_{\#^3} = 0$, then $h(\alpha) \in L_E(\beta, r_\beta)$.*

Proof. Select $h' \in H$ such that $h'(\tilde{z}) = h(\alpha)$ and $h'(x) = \varepsilon$ for all other $x \in \mathrm{var}(\beta)$ with $x \neq \tilde{z}$. By assumption, we know that $h(\alpha) \in L(\tilde{z})$ and that $\varepsilon \in L(x)$ for all other $x \in \mathrm{var}(\beta)$ with $x \neq \tilde{z}$. Hence, h' is r_β-valid. We get $h'(\beta) = h(\alpha)$, thus we have $h(\alpha) \in L_E(\beta, r_\beta)$. This concludes this lemma. $\qquad\square$

Finally, we show that substitutions of (α, r_α) that follow that specific form can only be obtained from (β, r_α) if and only if there exists some $\hat{\beta}_i$ for which we have $h(\alpha) = h'(\hat{\beta}_i)$.

Lemma 7. *Let $h \in H$ such that $h(\alpha) \in L_E(\alpha, r_\alpha)$. If $h(\alpha) = v0u0v$ for some $u \in \Sigma^*$ with $|u|_{\#^3} = 0$, then $h(\alpha) \in L_E(\beta, r_\beta)$ if and only if there exists some $i \in [\mu]$ and r_β-valid $h' \in H$ with $h'(\hat{\beta}_i) = h(\alpha)$.*

Proof. Let $h \in H$ be given as in the claim. So, we have $h(\alpha) = h(x_v \alpha_1 x_v \tilde{y}) = v0u0v$ for some $u \in \Sigma^*$ with $|u|_{\#^3} = 0$. For the first direction assume $h(\alpha) \in L_E(\beta, r_\beta)$. Then, there exists some r_β-valid substitution $h' \in H$ such that $h'(\beta) = h(\alpha) = v0u0v$. By construction, we know that $h(\alpha) \notin L(\tilde{z})$. Also, we know that for all $i \in [\mu]$ and for all $w \in L_E(\gamma_i, r_{\gamma_i})$ we either have $w = \varepsilon$ or $w = 0u0$ for some $u \in \Sigma^*$ with $|u|_{\#^3} = 0$. Hence, we have $\#^3 \notin \mathrm{Fact}(h'(\gamma_1...\gamma_\mu))$. So, there exists some $i \in [\mu]$ such that $h'(x_i) \neq \varepsilon$ and by that $h'(x_i) = v$. As $|v0u0v|_{\#^3} = 2$ and $|\beta|_{x_i} = 2$, we immediately get that $h'(x_i \gamma_i x_i) = vh'(\gamma_i)v = v0u0v$ and by that $h'(\gamma_i) = w'$. In particular, for all other $x \in \mathrm{var}(\beta)$ with $x \neq x_i$ and $x \notin \mathrm{var}(\gamma_i)$ we have $h'(x) = \varepsilon$. This concludes this direction. The other direction immediately follows by the assumption and by setting all variables $x \in \mathrm{var}(\beta)$ with $x \notin \mathrm{var}(\hat{\beta}_i)$ to $x = \varepsilon$. Then $h'(\hat{\beta}_i) = h(\alpha)$ and by that $h(\alpha) \in L_E(\beta, r_\beta)$. $\qquad\square$

Now, we know that if $h(\alpha)$ has some precise form, that $h(\alpha) \in L_E(\beta, r_\beta)$ if and only if there exists some $\hat{\beta}_i$ which we can use to obtain that specific $h(\alpha)$. By that, we obtain the following for all words which are not in both languages.

Corollary 8. *For some r_α-valid substitution $h \in H$ we have $h(\alpha) \notin L_E(\beta, r_\beta)$ if and only if $h(\alpha) = v\, w'\, v$ for some $w' \in \Sigma^*$ with $w' \in \{\, 0u0 \mid u \in \Sigma^*, |u|_{\#^3} = 0\}$ and for all $i \in [\mu]$ we have $w' \notin L_E(\gamma_i, r_{\gamma_i})$.*

Proof. Immediately follows by Lemma 7 and Lemma 6 and the fact that all other words of $L_E(\alpha, r_\alpha)$ are contained in $L_E(\beta, r_\beta)$. □

We say that a word $w \in \Sigma^*$ is of *good structure* or a *computation* if $w \in L_G$ with $L_G = ((\#\#00^*\#0^5(00)^*\#0^5(00)^*)^+\#\#)$. Otherwise, we say that w is of *bad structure*. Clearly, all encodings of computations of A are words of good structure.

From now on, let $h \in H$ be some r_α-valid substitution and assume $h(\alpha) = v0u0v$ for some $u \in \Sigma^*$ with $|u|_{\#^3} = 0$ and $v = 0\#^30$ as before.

We now have to construct $\hat{\beta}_1$ to $\hat{\beta}_\mu$ such that if u is not an encoding of a valid computation of A, then we have that there exists some $i \in [\mu]$ with $\hat{\beta}_i = x_i\gamma_i x_i$ and $w \in L_E(\gamma_i, r_{\gamma_i})$. Once we have that, we know that for any r_α-valid $h' \in H$ we have $h'(\alpha) \notin L_E(\beta, r_\beta)$ if and only if $h'(\alpha) = v0w_c0v$ for any $w_c \in \mathtt{ValC}(A)$, concluding this reduction.

For all $i \in [\mu]$ we call the pattern with regular constraints (γ_i, r_{γ_i}) a predicate. We construct each predicate independently, hence we omit their specific indexes from now on. Assume each predicate does not share its index with any other predicate and assume the total number of predicates to be $\mu \in \mathbb{N}$. As we will see, the total number of predicates is bound by the number of non-final states $|Q \setminus F|$, the number of invalid transitions not found in δ, and a constant number of predicates considering the basic structure of encodings of computations. Notice, that each constructed predicate ensures that for all r_β-valid $h' \in H$ we have $h'(\gamma) = \varepsilon$ or $h'(\gamma) = 0u'0$ for some $u' \in \Sigma^*$ with $|u'|_{\#^3} = 0$, which satisfies our initial assumption.

(1) First, we construct a predicate which can be used to obtain all substitutions in which $h(u)$ is not of good structure and which does not start with an encoding of the initial configuration $(q_0, 0, 0)$. For that, let $\gamma = y$ for a new and independent variable $y \in X$ and set

$$L(y) := \{\varepsilon\} \cup \{\, 0u'0 \mid u' \in \Sigma^*, |u'|_{\#^3} = 0, u' \in L_{gs}\}$$

for

$$L_{gs} := \overline{L(\ \#\#0\#0^5\#0^5\ (\#\#0^+\#0^5(00)^*\#0^5(00)^*)^*\ \#\#\ )}.$$

Then, if u is not of good structure or does not start with a valid encoding of the initial configuration, we can define a r_β-valid $h' \in H$ such that $h'(\gamma) = 0u0$.

(2) Next, we construct predicates which can be used to obtain all substitutions which end in an encoding of a configuration that is not in a final state. So, for all $q_j \in Q \setminus F$ we define a new and independent predicate $\gamma = y$ for respectively new and independent variables $y \in X$ such that

$$L(y) := \{\varepsilon\} \cup \{\, 0u'0 \mid u' \in \Sigma^*, |u'|_{\#^3} = 0, u' \in \Sigma^* \cdot L(\#\#0^{1+j}\#0^+\#0^+\#\#)\}.$$

Then, if u ends in an encoding of a configuration of A which contains no final state, we can obtain a r_β-valid substitution $h' \in H$ such that $h'(\gamma) = 0u0$.

(3) Now, we have to make sure that in a single step the value of no counter is changed by more than one. For that, we construct four predicates, each corresponding to the value of either the first or second counter being either increased or decreased by more than one (in a single step of an encoding of a computation). First, we construct a new and independent predicate γ which can be used if the first counter is increased by more than one in a single step. Let $\gamma = y_1\, x_1\, y_2\, x_1\, y_3$ for new and independent variables $y_1, y_2, y_3, x_1 \in X$ and set

$$
\begin{aligned}
L(y_1) &:= \{\varepsilon\} \cup \{\, 0u0\#0 \mid u \in \Sigma^*, |u|_{\#^3} = 0\}, \\
L(y_2) &:= \{\varepsilon\} \cup L(0^4\#0^50^*\#\#0^+\#0^4\mathbf{00(00)^+}), \\
L(y_3) &:= \{\varepsilon\} \cup \{\, 0\#0u0 \mid u \in \Sigma^*, |u|_{\#^3} = 0\}, \\
L(x_1) &:= \{00\}^*.
\end{aligned}
$$

Then, if $h(u)$ has a factor $\#0^50^m\#0^50^n\#\#0^{1+j}\#0^50^m\mathbf{00(00)^k}\#$ for $m, n, j, k \in \mathbb{N}$ and $k \geq 2$, which corresponds to a part of an encoding of the first counter being increased by more than one (see bold numbers), we can find a r_β-valid substitution $h' \in H$ for which we have $h'(\gamma) = 0u0$. All other words obtainable from γ are words of bad structure, i.e., they are not in L_G if any of the variables y_1, y_2, or y_3 is substituted by the empty word as $L(y_1)L(x_1)L(x_1) \cap L_G = \emptyset$, $L(x_1)L(y_2)L(x_1) \cap L_G = \emptyset$, $L(x_1)L(x_1)L(y_3) \cap L_G = \emptyset$, $L(y_1)L(x_1)L(y_2)L(x_1) \cap L_G = \emptyset$, $L(y_1)L(x_1)L(x_1)L(y_3) \cap L_G = \emptyset$, and $L(x_1)L(y_2)L(x_1)L(y_3) \cap L_G = \emptyset$. Also, we cannot get $|h'(\gamma)|_{\#^3} > 0$. The cases of the first counter being decreased by more than one, the second counter being increased by more than one, and the second counter being decreased by more than one can all the constructed in an analogue manner, hence we omit their specific constructions here. They only differ in their definition of $L(y_2)$, in particular the placement of either the border $\#\#$ or the position of $00(00)^+$.

By now, only if u corresponds to a word of good structure in which every subsequent pair of encodings of configurations in which either no counter, one counter, or both counters are increased or decreased by at most one, we cannot find a predicate γ and a r_β-valid substitution $h' \in H$ such that $h'(\gamma) = u$. That already contains all encodings of valid computations of A, however we may still get encodings of computations in which two subsequent configurations do not correspond to any valid transition.

(4) So, in a last step, we construct predicates for each invalid pair of consecutive configurations based on the definition δ in A. For all $q_k, q_j \in Q$, $c_1, c_2 \in \{0, 1\}$, and $r_1, r_2 \in \{-1, 0, 1\}$ with $(q_k, r_1, r_2) \notin \delta(q_j, c_1, c_2)$ we define a new and independent predicate γ which can be used to obtain encodings of computations in which such an (invalid) transition is used. We demonstrate the construction using an examplary case by setting $c_1 = 1$, $c_2 = 1$, $r_1 = +1$, and $r_2 = 0$. Let

$$
\gamma = y_1\, x_1\, y_2\, x_2\, y_3\, x_1\, y_4\, x_2\, y_5
$$

for new and independent variables $y_1, ..., y_5, x_1, x_2 \in X$ and set

$L(y_1) := \{\varepsilon\} \cup \{\ u'0\#\#0^{1+j}\#0 \mid u' = \varepsilon \text{ or } u' = 0u'', |u''|_{\#^3} = 0, u', u'' \in \Sigma^*\ \}$,

$L(y_2) := \{\varepsilon\} \cup \{0^6\#0\}$,

$L(y_3) := \{\varepsilon\} \cup \{0^6\#\#0^{1+i}\#0^6\mathbf{00}\}$,

$L(y_4) := \{\varepsilon\} \cup \{0\#0^4\}$,

$L(y_5) := \{\varepsilon\} \cup \{\ 0^3\#\#0u' \mid u' = \varepsilon \text{ or } u' = u''0, |u''|_{\#^3} = 0, u', u'' \in \Sigma^*\}$,

$L(x_1) := \{00\}^*$, and

$L(x_2) := \{00\}^*$.

Then, if $h(u)$ contains a factor

$$\#\#0^{j+1}\#0^50^{2+2m_1}\#0^50^{2+2m_2}\#\#0^{i+1}\#0^50^{2+2m_1}\mathbf{0^2}\#0^50^{2+2m_2}\#\#,$$

which corresponds to $(q_i, r_1, r_2) \notin \delta(q_j, c_1, c_2)$, this predicate can be used to find a r_β-valid substitution $h' \in H$ for which we have $h'(\gamma) = 0u0 = w$. Notice that each counter starts with a value $0^70^{2m_i}$ for $i \in [2]$ and $m_i \in \mathbb{N}_0$ instead of $0^50^{2m_i}$ as we assume both counters not to be zero in this example (by $c_1 = c_2 = 1$). Predicates for all other cases can be constructed analogously by either switching the position of additional $\mathbf{0}'s$ (marked with bold letters in the construction), or removing one or both occurrences of either x_1 or x_2 (and reducing the number of 0's in the corresponding part by 2) if $c_1 = 0$ or $c_2 = 0$ respectively.

Whats left to make sure is that for all r_β-valid $h' \in H$ we have that $h'(\gamma) = 0u0$ for $u \in \Sigma^*$ such that $|u|_{\#^3} = 0$ and $u \notin \mathtt{ValC}(A)$, even if some variables in γ are substituted with the empty word. First, by the way we defined $L(y_i)$ and $L(x_j)$ for $i \in \{1, ..., 5\}$ and $j \in \{1, 2\}$, we cannot obtain words in which $\#^3$ occurs as a factor. Second, notice that substitutions that only map y_2, y_3, or y_4 to nonempty words, directly result in words of bad structure due to their suffixes and prefixes. If we only have $h'(y_1) \neq \varepsilon$ or $h'(y_5) \neq \varepsilon$, then either the suffix or the prefix respectively results in bad structure. The only potentially problematic substitution is if either all variables except y_1, y_2, y_5, and potentially occurrences of x_1 and x_2, or all variables except y_1, y_4, y_5, and potentially occurrences of x_1 and x_2 are substituted by the empty word. Then we get a structure which resembles only one configuration. But then, we notice that either the first or the second counter always has an even number of 0's. This is not a valid encoding of a configuration, i.e., it is a word of bad structure. Hence, we cannot obtain $h'(\gamma) = 0u'0$ with $u' \in \mathtt{ValC}(A)$.

Using all predicates, given some r_α-valid $h \in H$, we can conclude that $h(\alpha) \notin L_E(\beta, r_\beta)$ if and only if $h(\alpha) = v0u0v$ such that $u \in \mathtt{ValvC}(A)$. This decides the problem of whether A has some accepting computation, hence the erasing equivalence problem for pattern languages with regular constraints is undecidable in the binary case. For larger alphabets, we may always restrict the alphabets used in the languages of the variables to the binary case, which allows for an reduction from the binary case to all larger alphabet sizes. This concludes the proof of Theorem 4.

4   Further Discussion

As the constructed patterns in the previous reduction are both terminal-free, we have immediately covered the general and terminal-free case together, as the latter can be easily reduced to the first. We mention the following fact which formalizes the first statement.

Corollary 9. *Let* $(\alpha, r_\alpha), (\beta, r_\beta) \in Pat_{\Sigma, \mathcal{C}_{Reg}}$ *such that* $\alpha, \beta \in X^*$, *i.e.* α *and* β *are terminal-free patterns. In general, it is undecidable to decide whether* $L_E(\alpha, r_\alpha) = L_E(\beta, r_\beta)$ *for all alphabets* Σ *with* $\Sigma \geq 2$.

With that, we obtain undecidability for nearly all problems regarding pattern languages with regular constraints. The only open case is the equivalence problem of non-erasing pattern languages with regular constraints. Using regular constraints, the problem becomes at least as hard as deciding the equivalence of two given regular languages witnessed by the following example.

Example 10. Let $(\alpha, r_\alpha), (\beta, r_\beta) \in Pat_{\Sigma, \mathcal{C}_{Reg}}$ such that $\alpha = x$ and $\beta = y$ for some $x, y \in X$. Then $L_{NE}(\alpha, r_\alpha) = L_{NE}(\beta, r_\beta)$ if and only if $L(x) \setminus \{\varepsilon\} = L(y) \setminus \{\varepsilon\}$.

Despite the most prominent open problem for patterns being undecidable in the case of pattern languages with regular constraints, we see that even this problem, which is trivially decidable for patterns without regular constraints, becomes much harder in this setting. We propose the following open question to which we have no definite conjecture so far. An overview of the current state of patterns with regular constraints can be found in Table 1.

Question 11. Given $(\alpha, r_\alpha), (\beta, r_\beta) \in Pat_{\Sigma, \mathcal{C}_{Reg}}$, is it generally decidable to answer whether $L_{NE}(\alpha, r_\alpha) = L_{NE}(\beta, r_\beta)$?

Table 1. Current state regarding pattern languages with regular constraints

Problem	General	Terminal-Free
E-Membership	NP-complete	NP-complete
E-Inclusion	Undecidable	Undecidable
E-Equivalence	Undecidable	Undecidable
NE-Membership	NP-complete	NP-complete
NE-Inclusion	Undecidable	Undecidable
NE-Equivalence	Open	Open

References

1. Angluin, D.: Finding patterns common to a set of strings. J. Comput. Syst. Sci. **21**(1), 46–62 (1980)
2. Bremer, J., Freydenberger, D.D.: Inclusion problems for patterns with a bounded number of variables. Inf. Comput. **220–221**, 15–43 (2012)
3. Day, J.D., Fleischmann, P., Manea, F., Nowotka, D.: Local patterns. In: Lokam, S., Ramanujam, R. (eds.) FSTTCS 2017. LIPIcs, vol. 93, pp. 24:1–24:14. Schloss Dagstuhl – Leibniz-Zentrum für Informatik, Dagstuhl, Germany (2018)
4. Ehrenfeucht, A., Rozenberg, G.: Finding a homomorphism between two words is NP-complete. Inf. Process. Lett. **9**(2), 86–88 (1979)
5. Fernau, H., Manea, F., Mercaş, R., Schmid, M.L.: Revisiting Shinohara's algorithm for computing descriptive patterns. TCS **733**, 44–54 (2018). special Issue on Learning Theory and Complexity
6. Fleischmann, P., et al.: Matching patterns with variables under Simon's congruence. In: Bournez, O., Formenti, E., Potapov, I. (eds.) RP 2023. LNCS, pp. 155–170. Springer, Cham (2023). https://doi.org/10.1007/978-3-031-45286-4_12
7. Freydenberger, D.D., Peterfreund, L.: The theory of concatenation over finite models. In: Bansal, N., Merelli, E., Worrell, J. (eds.) ICALP 2021, Proceedings. LIPIcs, vol. 198, pp. 130:1–130:17. Schloss Dagstuhl - Leibniz-Zentrum für Informatik (2021)
8. Freydenberger, D.D., Reidenbach, D.: Bad news on decision problems for patterns. Inf. Comput. **208**(1), 83–96 (2010)
9. Freydenberger, D.D., Schmid, M.L.: Deterministic regular expressions with back-references. J. Comput. Syst. Sci. **105**, 1–39 (2019)
10. Gawrychowski, P., Manea, F., Siemer, S.: Matching patterns with variables under hamming distance. In: Bonchi, F., Puglisi, S.J. (eds.) MFCS 2021. Leibniz International Proceedings in Informatics (LIPIcs), vol. 202, pp. 48:1–48:24. Schloss Dagstuhl – Leibniz-Zentrum für Informatik, Dagstuhl, Germany (2021)
11. Gawrychowski, P., Manea, F., Siemer, S.: Matching patterns with variables under edit distance. In: Arroyuelo, D., Poblete, B. (eds.) SPIRE 2022. LNCS, pp. 275–289. Springer, Cham (2022). https://doi.org/10.1007/978-3-031-20643-6_20
12. Geilke, M., Zilles, S.: Learning relational patterns. In: Kivinen, J., Szepesvári, C., Ukkonen, E., Zeugmann, T. (eds.) ALT 2011. LNCS (LNAI), vol. 6925, pp. 84–98. Springer, Heidelberg (2011). https://doi.org/10.1007/978-3-642-24412-4_10
13. Ibarra, O.H.: Reversal-bounded multicounter machines and their decision problems. J. ACM **25**(1), 116–133 (1978)
14. Jiang, T., Kinber, E., Salomaa, A., Salomaa, K., Yu, S.: Pattern languages with and without erasing. Int. J. Comput. Math. **50**(3–4), 147–163 (1994)
15. Jiang, T., Salomaa, A., Salomaa, K., Yu, S.: Decision problems for patterns. J. Comput. Syst. Sci. **50**(1), 53–63 (1995)
16. Koshiba, T.: Typed pattern languages and their learnability. In: Vitányi, P. (ed.) EuroCOLT 1995. LNCS, vol. 904, pp. 367–379. Springer, Heidelberg (1995). https://doi.org/10.1007/3-540-59119-2_192
17. Lothaire, M.: Combinatorics on Words, 2nd edn. Cambridge Mathematical Library, Cambridge University Press, Cambridge (1997)
18. Minsky, M.L.: Recursive unsolvability of Post's problem of "tag" and other topics in theory of Turing machines. Ann. Math. **74**(3), 437–455 (1961)
19. Ohlebusch, E., Ukkonen, E.: On the equivalence problem for e-pattern languages. In: MFCS 1996, pp. 457–468. Springer, Berlin, Heidelberg (1996)

20. Reidenbach, D.: On the equivalence problem for e-pattern languages over small alphabets. In: DLT, pp. 368–380. Springer, Berlin, Heidelberg (2004)
21. Reidenbach, D.: On the learnability of e-pattern languages over small alphabets. In: Learning Theory, pp. 140–154. Springer, Berlin, Heidelberg (2004)
22. Reidenbach, D.: An examination of Ohlebusch and Ukkonen's conjecture on the equivalence problem for E-pattern languages. J. Autom. Lang. Comb. **12**(3), 407-426 (2007)
23. Schmid, M.L., Schweikardt, N.: Document spanners - A brief overview of concepts, results, and recent developments. In: International Conference on Management of Data, PODS 2022, pp. 139–150. ACM (2022)
24. Shinohara, T.: Polynomial Time Inference of Extended Regular Pattern Languages, p. 115-127. Springer, Berlin, Heidelberg (1983)
25. Shinohara, T., Arikawa, S.: Pattern Inference, pp. 259–291. Springer, Berlin, Heidelberg (1995)

Push Complexity: Optimal Bounds and Unary Inputs

Giovanni Pighizzini[✉][ID]

Dipartimento di Informatica, Università degli Studi di Milano, Milan, Italy
`pighizzini@di.unimi.it`

Abstract. The notion of *push complexity* has been recently introduced by Bordhin and Mitrana as a measure of nonregularity for context-free languages. This measure takes into account the number of push operations used to accept inputs of length n. We show that the push complexity of each nonregular context-free language grows at least as a double logarithmic function, with respect to the length of the strings. This lower bound is optimal. Indeed, we prove that there exists a language with push complexity $O(\log \log n)$.

It is known that it cannot be decided whether the number of push operations used by a pushdown automaton is bounded by any constant. We prove that, in the restricted case of pushdown automata with a one-letter input alphabet, this question is decidable. Furthermore, under the same restriction, if the number of push operations used by a pushdown automaton is not bounded by any constant, then it should grow at least as a linear function in the length of the input.

1 Introduction

A classical topic related to complexity aspects of automata and formal languages is the study of computational devices working with very restricted resources. In the pioneering papers by Stearns, Hartmanis, and Lewis [13], and by Hopcroft and Ullman [8], the minimal amount of space necessary for the recognition of nonregular languages has been investigated. This research has been deepened in several papers (e.g., [1,3,4,9]). The computational models considered in these investigations are deterministic, nondeterministic, and alternating Turing machines.

More recently, the same problems have been studied for pushdown and one-counter automata [10,11], measuring, respectively, the amount of pushdown store used to accept, also called the *pushdown height*, and the maximum value reached by the counter. It is clear that if these amounts are bounded by a constant, namely they do not depend on the length of the input, the pushdown and the counter can be incorporated in the finite control. Hence, in this case, only regular languages are recognized. So the question arises about the minimum pushdown

G. Pighizzini—The author is a member of the Gruppo Nazionale Calcolo Scientifico-Istituto Nazionale di Alta Matematica (GNCS-INdAM).

S. Z. Fazekas (Ed.): CIAA 2024, LNCS 15015, pp. 289–301, 2024.
https://doi.org/10.1007/978-3-031-71112-1_21

height (minimum bound on the counter, respectively), as function of the length of the input, which is necessary to recognize nonregular languages.

In [11] it has been proved that if a pushdown automaton does not accept in constant height, then the pushdown height should grow at least as a double logarithmic function. This bound has been proved to be optimal, namely there exists a pushdown automaton accepting a nonregular language using such a height. In the case of counter automata, the optimal bound is logarithmic [10]. We shortly mention that we are referring to the cost of the least expensive accepting computation (weak measure). By considering the costs of all computations and of all accepting computations (respectively, strong and accept measures) all these bounds become linear [10].

In this paper, we make a similar investigation for another complexity measure, recently introduced by Bordihn and Mitrana with the aim of measuring the nonregularity of context-free languages [5]. Even this measure is related to pushdown automata and it is called *push complexity*. It is defined by counting the number of push operations used to accept.

In [5], it has been proved that the push complexity of a language is constant if and only if the language is regular. The authors also presented nonregular languages having push complexity $O(\log n)$ and $O(\sqrt{n})$, and they raise the question of the existence of nonregular languages with push complexity $O(f)$, for some other sublinear functions f.

In this paper we solve this problem, by showing the existence of a nonregular language with push complexity $O(\log \log n)$. This witness language, called REI, has been presented in [2], where it was proved that it is accepted by a real-time pushdown automaton using height $O(\log \log n)$. Observing that, during its computations, such a pushdown automaton switches at most one time from moves that can push symbols to moves that pop off symbols from the pushdown store, we obtain that also the number of push operations on inputs of length n is $O(\log \log n)$. Furthermore, as a consequence of a space lower bound for the recognition of nonregular languages by Turing machines proved in [1], we also conclude that if a pushdown automaton accepts inputs of length n using $o(\log \log n)$ push operations, then the accepted language is regular. This gives a double logarithmic optimal lower bound for the push complexity of non-regular languages. We point out that this is the same bound that has been obtained in [11] for the pushdown height.

So far we mentioned the push complexity measure for languages. However, it is also interesting to study this measure directly referring to machines, regardless of the fact that the accepted languages are regular or not.

In this respect, first of all we have to mention that it cannot be decided if a pushdown automaton accepts in constant push complexity [5]. This result has been obtained by a reduction from the halting problem, suitably encoding Turing machine computations. Such encoding uses an alphabet of at least two symbols.

Here, we prove that in the case of pushdown automata with a one-letter input alphabet, namely accepting *unary languages*, the question is decidable. The result is obtained by adapting and refining the techniques used in [11] to show that it is decidable whether a unary pushdown automaton accepts in constant height.

We remind the reader that all unary context-free languages are regular [6], so unary pushdown automata are no more powerful than finite automata. However, it interesting to consider these devices because for some languages they can be significantly more succinct than finite automata [12].

In [11] it has been proved that if a pushdown automaton accepts a unary language using nonconstant height, then the height should grow, with respect to the input length, at least as a logarithmic function. An example matching this lower bound was also provided. Here we study the same problem for the number of push operations. We obtain a higher lower bound: if a pushdown automaton accepts a unary language with a nonconstant number of push operations, then such a number must grow at least as a linear function.

The paper is organized as follows. After presenting the preliminary notions useful in the paper (Sect. 2), in Sect. 3 we obtain the optimal lower bound for the push complexity of nonregular languages. Section 4 is devoted to the study of the unary case, in particular to prove that it can be decided whether a unary pushdown automaton has constant push complexity. The tools developed to obtain such a result allow us, in Sect. 5, to prove that if a pushdown automaton has nonconstant push complexity, then such a complexity is at least linear. We also provide an example that matches this lower bound.

Due to the lack of space, some proofs and some other material are omitted from this version of the paper.

2   Preliminaries

We assume the reader is familiar with the standard notions of automata and formal language theory. We will use the symbol ε to denote the empty string, and $|x|$ to denote the length of a string x. Given a set S, its cardinality is denoted as $\#S$, and the set of its subsets as 2^S.

A *pushdown automaton* (PDA, for short) is a tuple $\mathcal{M} = \langle Q, \Sigma, \Gamma, \delta, q_I, Z_0, q_F \rangle$ where Q is the finite *set of states*, Σ is the *input alphabet*, Γ is the *pushdown alphabet*, $q_I \in Q$ is the *initial state*, $Z_0 \in \Gamma$ is the *start symbol*, $q_F \in Q$ is the *final state*. Without loss of generality, we make the following assumptions about PDAs:

1. at the start of the computation the pushdown store contains only the start symbol Z_0, being at height 0, the input head scans the first input symbol, and the finite control contains the initial state q_I;
2. the input is accepted if and only if the automaton reaches the final state q_F, the pushdown store contains only Z_0, and all the input symbols have been scanned;

3. when the automaton reads an input symbol, it moves the head to the next symbol, and it does not make any change on the pushdown. Notice that this implies that the contents of the pushdown store can be changed only by ε-moves;
4. every push operation adds exactly one symbol on the pushdown.

The *transition function* δ of a PDA $\mathcal{M}$ in this form can be written as

$$\delta : Q \times (\Sigma \cup \{\varepsilon\}) \times \Gamma \to 2^{Q \times (\{-, \mathtt{pop}\} \cup \{\mathtt{push}(A) \mid A \in \Gamma\})}.$$

In particular, for $q, p \in Q$, $A, B \in \Gamma$, $\sigma \in \Sigma$, $(p, -) \in \delta(q, \sigma, A)$ means that the PDA $\mathcal{M}$, in the state q, with A at the top of the pushdown, by consuming the input σ, can reach the state p without changing the pushdown contents; $(p, \mathtt{pop}) \in \delta(q, \varepsilon, A)$ $((p, \mathtt{push}(B)) \in \delta(q, \varepsilon, A)$, $(p, -) \in \delta(q, \varepsilon, A)$, respectively) means that $\mathcal{M}$, in the state q, with A at the top of the pushdown, without reading any input symbol, can reach the state p by popping off the pushdown the symbol A from the top (by pushing the symbol B onto the top of the pushdown, without changing the pushdown, respectively).

Notice that in any accepting computation the occurrence of the start symbol Z_0 at the bottom of the pushdown is never removed, otherwise the next move would be undefined, so halting in a non-accepting configuration.

Now we present the measures we consider in the paper, namely the *pushdown height* and the *push complexity*. The height of a PDA $\mathcal{M}$ in a given configuration is the number of symbols in the pushdown store, in addition to the occurrence of the start symbol Z_0 at the bottom. Hence, in the initial and in the accepting configurations the height is 0. The *height of a computation $\mathcal{C}$*, denoted as $\mathsf{height}_{\mathcal{M}}(\mathcal{C})$, is the maximum height reached in the configurations occurring in $\mathcal{C}$. By $\mathsf{push}_{\mathcal{M}}(\mathcal{C})$ we denote the total number of push operations occurring in $\mathcal{C}$. We extend these measures to strings by considering, for each $w \in \Sigma^*$, the minimum height and number of push operations on all computations accepting w (0 if w is not accepted) and, for each $n \in \Sigma^*$, the maximum on strings of length n. It is trivial to observe that $0 \leq \mathsf{height}_{\mathcal{M}}(\mathcal{C}) \leq \mathsf{push}_{\mathcal{M}}(\mathcal{C})$.

A PDA $\mathcal{M}$ has *constant push complexity* (*constant height*, resp.), if there exists a constant C such that $\mathsf{push}_{\mathcal{M}}(n) \leq C$ ($\mathsf{height}_{\mathcal{M}}(n) \leq C$, resp.) for each $n \in \mathbb{N}$. In the above notations, we can omit specifying the name of the PDA when it is clear from the context.

Given a function $f : \mathbb{N} \to \mathbb{N}$ and a language $L \subseteq \Sigma^*$, we say that L has push complexity f if there exists a pushdown automaton $\mathcal{M}$ accepting L with $\mathsf{push}_{\mathcal{M}} = f$. In [5] it has been proved that a language L has push complexity $O(1)$ if and only if it is regular.

We point out that PDAs in different forms, as that given in [7] in which any push operation can replace the top of the pushdown by a string of symbols, can be converted in the form used here, by preserving the properties of having constant height or constant push complexity, and by keeping the same growth of them, when they are nonconstant.

We now present some technical notions and results that will be used in the paper. Let $\mathcal{M} = \langle Q, \Sigma, \Gamma, \delta, q_I, Z_0, q_F \rangle$ be a fixed PDA.

A *surface pair* is defined by a state $q \in Q$ and a symbol $A \in \Gamma$, and it is denoted by $[qA]$. The surface pair in a given configuration is defined by the current state and the topmost pushdown symbol, namely the only part of the store which is relevant in order to decide the next move.

A *surface triple* is defined by two states $q, p \in Q$ and a symbol $A \in \Gamma$, and it is denoted by $[qAp]$. Surface triples are used to study parts of computations starting and ending at the same pushdown height and that do not go below that height in between. More precisely, a $[qAp]$-*computation* on a string $x \in \Sigma^*$ is a computation $\mathcal{C}$ which starts from the state q with A on the top of the pushdown at some height h and, after reading x from the input tape, ends in the state p with A on the top of the pushdown at the same height h without reaching pushdown height smaller than h in between. We also say that $\mathcal{C}$ *consumes* the string x. Notice that, at the beginning of $\mathcal{C}$, the input head is on the tape cell containing the leftmost symbol of x, while at the end it is on the cell to the right of the rightmost symbol of x. We point out that, during $\mathcal{C}$, the symbol A at height h is never replaced. Hence, $\mathcal{C}$ does not depend on h and on the symbols stored in the pushdown below A. Notice that the surface pairs at the beginning and at the end of $\mathcal{C}$ are $[qA]$ and $[pA]$, respectively.

We denote by $L_{[qAp]}$ the set of input strings consumed in all possible $[qAp]$-computations. We point out that the set of accepting computations of $\mathcal{M}$ coincides with the set of $[q_I Z_0 q_F]$-computations. Hence, $L_{[q_I Z_0 q_F]}$ is the language accepted by $\mathcal{M}$. Furthermore, for each surface triple $[qAp]$, if we modify $\mathcal{M}$ by using q, A, and p instead of the original initial state, original start pushdown symbol, and the original final state, respectively, we obtain a PDA accepting $L_{[qAp]}$ which, hence, is context free.

A *horizontal loop* on a surface pair $[qA]$ is any $[qAq]$-computation consuming *at least one input symbol*. Notice that such computation starts and ends in the same state q. By considering a computation of 0 moves, we always have $\varepsilon \in L_{[qAq]}$. Hence $[qA]$ *has a horizontal loop* when $L_{[qAq]}$ contains at least one more string, besides ε. A horizontal loop in which the pushdown store is not modified is called *flat loop*. Otherwise, it is called *nonflat loop*. Notice that, by definition, each nonflat loop contains exactly the same number of push and of pop operations.

Lemma 1. *It is decidable if a surface pair $[qA]$ has a flat loop and if it is has a horizontal nonflat loop.*

If a $[qAp]$-computation $\mathcal{C}$ consists of three parts, namely, it begins with a prefix $\mathcal{X}$, followed by a proper $[qAp]$-subcomputation $\mathcal{C}'$ using the *same* triple $[qAp]$, and ends by a suffix $\mathcal{Y}$, such that the middle part $\mathcal{C}'$ starts and ends with pushdown higher than at the beginning of $\mathcal{C}$, then the pair $(\mathcal{X}, \mathcal{Y})$ is called *vertical loop*. Note that, during the execution of $\mathcal{X}$, a nonempty string $A\alpha$ is saved on the top of the pushdown store (the symbol A being at the top) above the symbol A which was on the top at the beginning of $\mathcal{C}$, and this string is popped off during the execution of $\mathcal{Y}$.

3   An Optimal Lower Bound for Push Complexity

In [11] it has been proved that if a PDA M does not accept in constant height, then the height necessary to accept strings of length n should grow at least as $\log \log n$ for infinitely many n's. Furthermore, such a bound is optimal. Here we obtain a similar result for push complexity.

Theorem 1. *Let M be a PDA. If* $\mathsf{push}_M(n) = o(\log \log n)$ *then there is a constant $C > 0$ such that* $\mathsf{push}_M(n) \leq C$, *for each $n > 0$.*

Proof. It is known that if a Turing machine with a two-way read-only input tape accepts in $o(\log \log n)$ space, where the space is measured considering a separate work tape and the least expensive computation, then the machine uses only a constant amount of space [1]. As a special case, we can consider PDAs where the input tape is restricted to be one-way and the work tape is the pushdown store. Since the amount of space on the pushdown store cannot exceed the number of push operations, the result follows.   □

In [2], a language called REI was presented, which is accepted by a PDA M that uses $O(\log \log n)$ height. A close inspection to the proof of such a result shows that, in each accepting computation, M makes at most one turn, namely after executing a pop operation, it does not perform any further push. Hence, the total number of push operations coincides with the height reached by the pushdown store. It can be also observed that the language REI is not regular.

Theorem 2. *There exists a PDA M accepting a nonregular language REI, such that* $\mathsf{push}_M(n) = O(\log \log n)$.

Proof (sketch). The language REI is defined over the alphabet $\{a, b, d, e, 0, 1\}$ and consists of strings which *are not prefixes* of the infinite word

- $\omega = bc_1 ac_2^R bc_2 ac_3^R \cdots bc_k ac_{k+1}^R bc_{k+1} ac_{k+2}^R \cdots$, where
- $c_k = eb_0 db_{k,0} db_0^R \, eb_1 db_{k,1} db_1^R \cdots eb_{\lfloor \log k \rfloor} db_{k,\lfloor \log k \rfloor} db_{\lfloor \log k \rfloor}^R e$ is a counter representation for k, augmented with subcounters, for $k \geq 1$,
- $b_{k,i} \in \{0, 1\}$ is the ith bit in the binary representation of k, and $b_i \in \{0, 1\}^*$ denotes the number i written in binary, for $i \in \{0, 1, \ldots, \lfloor \log k \rfloor\}$.

The nonregularity of REI can be proved using a pumping argument on its complement.

In [2] the authors describe how a PDA M can recognize the language REI. To verify that the input string is not a prefix of the infinite word ω, the machine guesses and checks an error condition, among six possibilities (e.g., the format of the input is not correct, the values of counters are wrong, *etc.*). Some of these conditions can be verified using only the finite control. The other conditions require to compare some input factors. To this aim the pushdown store is used. The authors show that if one of these conditions holds, then there exists a computation verifying it in which height $O(\log \log n)$ is used.

By a close inspection, it can be observed that in this computation only one turn is made, namely after the first pop operation, no more push operations are executed. Hence the number of push operations is equal to the height. This allows to conclude that $\mathsf{push}_{\mathcal{M}}(n) = O(\log \log n)$. □

Since the language REI is not regular, by combining the lower bound in Theorem 1 with upper bound in Theorem 2, we obtain:

Corollary 1. *The language* REI *has push complexity in* $O(\log \log n)$ *but not in* $o(\log \log n)$.

4   Push Complexity in the Unary Case

This section is devoted to prove that it is decidable whether a PDA with an input alphabet of only one letter has constant push complexity. The proof is obtained by adapting the argument used in [11] to show that it is decidable whether a unary PDA accepts in constant height.

We now give a short outline of the argument we use. Any accepting computation on each sufficiently long input should contain some kind of repetitions, i.e., it should contain horizontal or vertical loops. Among them, only the horizontal loops which are flat do not involve push operations. So, to accept in constant push complexity, for each accepted input we have to find an accepting computation that uses a number of vertical and nonflat loops bounded by a constant.

Here, we prove that if a sufficiently long input a^{ℓ} has an accepting computation $\mathcal{C}$ visiting a surface configuration $[rB]$ which has a flat loop, then there is another accepting computation $\mathcal{C}'$ on a^{ℓ} in which the number of vertical and nonflat loops is bounded by a constant. Roughly, the computation $\mathcal{C}'$ is obtained by replacing almost all vertical and nonflat loops in $\mathcal{C}$ with some occurrences of the flat loop on $[rB]$. The replacement is possible since the input is unary, so it is possible to cut some parts of $\mathcal{C}$ consuming certain input factors of total length ℓ', and insert elsewhere some other parts that consume the same length ℓ', in such a way that the resulting computation $\mathcal{C}'$ still accepts the same input a^{ℓ} as $\mathcal{C}$.

This allows us to prove that the push complexity is constant if and only if the set of strings which are accepted only by computations that do not visit any surface configuration having a flat loop is finite. Using the fact that the alphabet is unary and so such a set of strings is a regular language that can be effectively constructed from the given PDA, we will be able to conclude that the problem of deciding if a unary PDA has constant push complexity is decidable.

These results are obtained by using some combinatorial tools, essentially based on pumping arguments. These tools are developed in terms of context-free grammars, which look more manageable than PDA computations. Using a standard construction, we will associate with each PDA a grammar, whose derivation trees describe computations. Furthermore, there is a one-to-one correspondence between push operations and unit productions used in the derivation trees of the resulting grammar. So, the properties of the computations of the PDA will be derived from the properties of the derivation trees.

4.1 Preliminary Tools: Grammars and Trees

In the following, we consider a grammar $G = \langle V, \Sigma, P, S \rangle$ in *binary normal form*, an extension of the Chomsky normal form where, besides productions of the forms $A \to BC$ and $A \to a$, also *unit productions* $A \to B$ and ε-productions $A \to \varepsilon$, are allowed, where $A, B \in V$ and $a \in \Sigma$. As we will see, these grammars will be useful to describe and to study computations of PDAs.

If T is a derivation tree whose root is labeled with a variable $A \in V$ and such that the labels of the leaves, from left to right, form a string $\gamma \in (V \cup \Sigma)^*$, then we write $T : A \overset{\star}{\Rightarrow} \gamma$. Furthermore, we indicate by $\nu(T)$ the set of variables occurring as labels of the nodes in T and by unit(T) the number of times that unit productions are used in T.

A *gap tree* T from a variable $A \in V$ is a tree corresponding to a nonempty derivation of the form $A \overset{+}{\Rightarrow} xAy$, with $x, y \in \Sigma^*$. Gap trees deriving at least one terminal can be used to "pump" derivation trees and they will play an important role to obtain our results. On the one hand, given a derivation tree $T : S \overset{\star}{\Rightarrow} z$ and a variable A occurring in T, i.e., $A \in \nu(T)$, any gap tree $T_A : A \overset{\star}{\Rightarrow} xAy$ can be inserted in T to obtain the derivation tree T' of another string z'. On the other hand, if z is a sufficiently long string, then we can always find a variable $B \in \nu(T)$ which is repeated on a path in T, and a gap tree $T_B : B \overset{+}{\Rightarrow} xBy$, with $xy \neq \varepsilon$. By removing T_B from T, we can find the derivation tree of a shorter string in the same language, as in the proof of classical pumping lemma for context-free languages.

In the case of unary grammars, using these operations, starting from the derivation tree $T : S \overset{\star}{\Rightarrow} a^\ell$ of each sufficiently long string a^ℓ, we can obtain another derivation tree T' of the same string, by pumping a derivation tree $T_0 : S \overset{\star}{\Rightarrow} a^{\ell_0}$, with several occurrences of a gap tree $T_A : A \overset{\star}{\Rightarrow} a^i A a^j$, where A is a variable occurring in T_0 and in T. The tree T_0 is obtained by removing *some* of the gap trees in T. Furthermore, the value ℓ_0 is bounded by a constant which depends on the number of variables of G. The crucial point for our purposes is that if the tree T_A does not involve unit productions, then in the resulting T' the number of times that unit productions are used is the same as in T_0 and, then, it is bounded by a constant. In this way, from the original derivation tree T of a^ℓ, we obtain another tree T' of the same string in which unit(T') is bounded by a constant. This is formalized in the next lemma.

As we will see, gap trees not involving unit productions will correspond, in the constructions presented later, to surface configurations having flat loops. Visiting such configurations in accepting computations will allow to reduce, under a constant value, the number of push operations used to accept a string.

Lemma 2. *For any derivation tree $T : S \overset{\star}{\Rightarrow} a^\ell$ and for any gap tree $T_A : A \overset{\star}{\Rightarrow} a^i A a^j$, with $0 < i + j < 2^{2v-1}$, $A \in \nu(T)$, and unit$(T_A) = 0$, there exists a derivation tree $T' : S \overset{\star}{\Rightarrow} a^\ell$ which is obtained by pumping a tree $T_0 : S \overset{\star}{\Rightarrow} a^{\ell_0}$ such that $A \in \nu(T_0)$, with $k \geq 0$ occurrences of T_A. Furthermore unit$(T') =$ unit$(T_0) = 2^{2^{O(v^2)}}$, i.e., it is bounded by a double exponential function in v^2.*

4.2   Simulating Vertical and Nonflat Loops Using Horizontal Loops

From now on, let us consider a fixed PDA $\mathcal{M} = \langle Q, \Sigma, \Gamma, \delta, q_I, Z_0, q_F \rangle$. We are going to define a context-free grammar $G = \langle V, \Sigma, P, S \rangle$ which generates the same language accepted by $\mathcal{M}$. Unit productions in G will correspond to push operations in $\mathcal{M}$. This will allow us to study the push complexity of $\mathcal{M}$ by considering how many times unit productions are used to derive strings in G.

We give the same construction as in [12], which is a minor variation of that used in classical textbooks (see, e.g., [7]) to present the standard transformation of PDAs into CFGs. The grammar G is defined as follows:

- The set of variables V consists of all triples $[qAp]$, with $q, p \in Q$, $A \in \Gamma$. purpose of the variable $[qAp]$ is to generate the language $L_{[qAp]}$ of strings consumed in all possible $[qAp]$-computations (see Lemma 3 below).
- The set of terminal symbols coincides with the input alphabet Σ of $\mathcal{M}$.
- P contains the following productions:
  1. $[qAp] \rightarrow [qAr][rAp]$, for $q, p, r \in Q$, $A \in \Gamma$;
  2. $[qAp] \rightarrow [q'Bp']$, for $q, q', p, p' \in Q$, $A, B \in \Gamma$ such that $(q', \mathsf{push}(B)) \in \delta(q, \varepsilon, A)$ and $(p, \mathsf{pop}) \in \delta(p', \varepsilon, B)$;
  3. $[qAp] \rightarrow \sigma$, for $q, p \in Q$, $\sigma \in \Sigma \cup \{\varepsilon\}$, $A \in \Gamma$ such that $(p, -) \in \delta(q, \sigma, A)$;
  4. $[qAq] \rightarrow \varepsilon$, for $q \in Q$, $A \in \Gamma$.
- The start symbol S is the triple $[q_I Z_0 q_F]$.

We point out that the grammar G is in binary normal form. Furthermore, the number of variables of G is $v = (\#Q)^2 \cdot \#\Gamma$.

Using a standard induction, it can be proved that each variable $[qAp]$ in the above-defined grammar G generates the language $L_{[qAp]}$, containing each string $x \in \Sigma^*$ such that there exists a $[qAp]$-computation of $\mathcal{M}$ on x, namely a computation $\mathcal{C}$ which starts from the state q with A on the top of the pushdown at some height h and, after reading x from the input tape, ends in the state p with A on the top of the pushdown at the same height h, without reaching pushdown height smaller than h in between. Actually, a stronger result can be proved, relating the number of push operations used in the computation $\mathcal{C}$ on x with the number of unit productions in a derivation tree of x from the variable $[qAp]$:

Lemma 3. *For any* $x \in \Sigma^*$, $q, p \in Q$, $A \in \Gamma$, $\wp \in \mathbb{N}$, *there exists a derivation tree* $T : [qAp] \overset{*}{\rightarrow} x$ *with* $\mathsf{unit}(T) = \wp$ *if and only if there exists a* $[qAp]$-*computation* $\mathcal{C}$ *on* x *with* $\mathsf{push}(\mathcal{C}) = \wp$.

Proof (sketch). The lemma is a variation of Lemma 8 in [11]. The proof can be obtained in a similar way, by induction on the length of the derivation represented by the tree T *(only-if part)* and by induction on the number of moves in the computation $\mathcal{C}$ *(if part)*.   □

As a consequence of Lemma 3 we obtain:

Theorem 3. *For any integer* $\wp \geq 0$ *and for any string* $x \in \Sigma^*$, *there exists an accepting computation path* $\mathcal{C}$ *of* $\mathcal{M}$ *on* x *with* $\mathsf{push}(\mathcal{C}) = \wp$ *if and only if there is a derivation tree* T *of* x *in* G *with* $\mathsf{unit}(T) = \wp$.

The following lemma gives an upper bound on the push complexity of each string:

Lemma 4. *If $z \in \Sigma^*$ is accepted by $\mathcal{M}$, then $\mathsf{push}_{\mathcal{M}}(z) < 2v(|z| + 2^{(v+1)(|z|+2)})$.*

From now, let us consider only unary strings. Using a pigeonhole argument, it is possible to prove the following:

Lemma 5. *If $[rB]$ has a flat loop, then there exists a gap tree $T_{[rBr]} : [rBr] \overset{*}{\Rightarrow} a^m[rBr]$, for some m with $0 < m \leq 2\#Q$, such that $\mathsf{unit}(T_{[rBr]}) = 0$, namely $T_{[rBr]}$ does not involve any unit production.*

Lemma 6. *Let $\mathcal{C}$ be an accepting computation on input a^ℓ that does not contain any flat loop. Then $\mathsf{push}_{\mathcal{M}}(\mathcal{C}) \geq \lfloor \frac{\ell}{2\#Q} \rfloor$.*

The following result will be crucial to obtain decidability in the unary case. Roughly, it states that when an input has an accepting computation that visits a surface configuration having a flat loop, then it has another accepting computation in which the number of push operations is bounded by a value which is at most double exponential in the number of variables of the grammar obtained from the PDA $\mathcal{M}$, while it does not depend on to the input length:

Theorem 4. *Let $\mathcal{C}$ be an accepting computation on input a^ℓ which visits a surface pair $[rB]$ having a flat loop. Then there exists an accepting computation $\mathcal{C}'$ on a^ℓ such that $\mathsf{push}(\mathcal{C}') = 2^{2^{O(v^2)}}$, where $v = (\#Q)^2 \cdot \#\Gamma$.*

Proof. Let G be the above-defined context-free grammar in binary normal form obtained from $\mathcal{M}$. We observe that since $\mathcal{C}$ visits the surface pair $[rB]$, there exists a derivation tree $T : S \overset{*}{\Rightarrow} a^\ell$ with $[rBr] \in \nu(T)$. In fact, one of the triples $[rBs]$ or $[sBr]$, for some $s \in Q$, should appear in the derivation tree corresponding to $\mathcal{C}$. Since G contains the productions $[rBs] \to [rBr][rBs]$, $[sBr] \to [sBr][rBr]$ and $[rBr] \to \varepsilon$, we can suitably modify the tree in order to introduce one occurrence of $[rBr]$, without changing the derived string. According to Lemma 2, we can obtain another tree $T' : S \overset{*}{\Rightarrow} a^\ell$, by pumping a tree $T_0 : S \overset{*}{\Rightarrow} a^{\ell_0}$, such that $[rBr] \in \nu(T_0)$, with $k \geq 0$ occurrences of the gap tree $T_{[rBr]}$ obtained in Lemma 5, and satisfying $\mathsf{unit}(T') = \mathsf{unit}(T_0) = 2^{2^{O(v^2)}}$.

Hence, according to Theorem 3, the PDA $\mathcal{M}$ has an accepting computation on input a^ℓ making $\mathsf{unit}(T')$ push operations. □

4.3   Decidability

Using Theorem 4, we are now able to prove the following result, from which we immediately obtain that it is decidable whether a unary PDA uses a number of push operations bounded by a constant. The argument is similar to the one used to prove a similar result for the height of the pushdown store [11, Thm. 4].

Theorem 5. *Let* $\mathcal{M} = \langle Q, \Sigma, \Gamma, \delta, q_I, Z_0, q_F \rangle$ *be a unary* PDA. *The push complexity of* $\mathcal{M}$ *is bounded by a constant, with respect to the input length, if and only if it is bounded by* $2^{2^{O(v^2)}}$, *where* $v = (\#Q)^2 \cdot \#\Gamma$.

Proof. Given the language L accepted by $\mathcal{M}$, let us consider the following two languages L_f and L_{nf}, whose union is L:

- L_f is the set of strings accepted by the computations of $\mathcal{M}$ which visit at least one surface pair having a flat loop.
- L_{nf} is the set of strings accepted by the computations of $\mathcal{M}$ which visit only surface pairs that do not have flat loops.

If the set $L_{nf} \setminus L_f$ is infinite, then it should contain arbitrarily long strings; as a consequence of Lemma 6, an unbounded number of push operations is required to accept them.

Otherwise, when $L_{nf} \setminus L_f$ is finite, $\mathrm{push}_{\mathcal{M}}(n)$ is bounded by a constant, which is bounded by the number of push operations used to accept strings in L_f and by the number of push operations used to accept strings in $L_{nf} \setminus L_f$. By Theorem 4, the former amount is bounded by $2^{2^{O(v^2)}}$. To estimate the latter one, first we notice that L_{nf} is accepted by a PDA $\mathcal{M}_{nf}$ which can be obtained by just removing from $\mathcal{M}$ all the transitions defined from surface pairs $[rB]$ having flat loops (we remind the reader that this can be decided according to Lemma 1). Hence, L_{nf} is generated by a unary context-free grammar in binary normal form with $v = n^2 m$ variables. According to [12, Theorem 6] (see also [11, Lemma 2]), this grammar can be converted into an equivalent deterministic finite automaton A_{nf} with less than 2^{v^2} states.

From $\mathcal{M}$, we can also build a unary PDA $\mathcal{M}_f$ which accepts L_f by simulating $\mathcal{M}$ and by accepting when the simulated computation is accepting and visits at least one surface pair having a flat loop. To implement $\mathcal{M}_f$, we double the cardinality of the state set, in order to remember if some surface pair having a flat loop has been reached during the computation. That is, for each state q we create a copy q'. Thus, the simulation is straightforward but, *after* visiting a surface pair having a flat loop, $\mathcal{M}$ switches to q' instead of q. Hence, the final state of $\mathcal{M}_f$ is q'_F. From $\mathcal{M}_f$ we can obtain an equivalent grammar in binary normal form with $(2\#Q)^2 \#\Gamma = 4v$ variables. However, in such a grammar, the triples $[q'Ap]$, where q and p are states of $\mathcal{M}$, cannot generate any string (in fact, once a pair having a flat loop is reached, the computation of $\mathcal{M}_f$ can only visit states in the copy of Q). This allows to reduce the number of variables to $3v$. Again, according to [12, Theorem 6], from the grammar derived from $\mathcal{M}_f$ we can obtain an equivalent deterministic finite automaton A_f, with less than 2^{9v^2} states accepting L_f.

Using a standard product construction, from A_f and A_{nf}, we can obtain another deterministic finite automaton A with less than 2^{10v^2} states accepting $L_{nf} \setminus L_f$. Since such a language is finite, the length of each string in it is bounded by the number of states of A, i.e., it is bounded by 2^{10v^2}. By Lemma 4, this implies that $\mathsf{push}_{\mathcal{M}}(w) = 2^{2^{O(v^2)}}$ for each $w \in L_{\mathsf{nf}} \setminus L_{\mathsf{f}}$, namely the same upper bound we obtained for strings in L_{f}.

By summarizing, we can conclude that if $\mathsf{push}_{\mathcal{M}}(n)$ does not depends on n, then it is bounded by $2^{2^{O(v^2)}}$. $\qquad\square$

As a consequence of Theorem 5 we obtain:

Corollary 2. *It is decidable whether a unary PDA accepts using a number of push operations bounded by a constant.*

5   Nonconstant Push Complexity in the Unary Case

In [11] it has been proved that if a unary PDA does not accept in constant height, then the height should grow at least as a logarithmic function in the length of the input. Moreover, such a lower bound is optimal. In the case of the push complexity the optimal lower bound is higher. Indeed, using Lemma 6 and an argument similar to that of the proof of Theorem 5, we can obtain a linear lower bound. We remind the reader that in the case of general alphabets, i.e., of cardinality at least 2, the optimal lower bounds for pushdown height and for push complexity are both double logarithmic with respect to the input length.

Theorem 6. *Let $\mathcal{M}$ be a unary PDA. Then either $\mathsf{push}_{\mathcal{M}}(n)$ is bounded by a constant or there exists $c > 0$ such that $\mathsf{push}_{\mathcal{M}}(n) \geq c \cdot n$ infinitely often.*

It easy to observe that the lower bound in Theorem 6 cannot be improved. Here we give an example in which the logarithmic lower bound for nonconstant height and the linear lower bound for nonconstant push complexity are reached by the same machine and by the same computations.

Theorem 7. *There exists a unary PDA $\mathcal{A}$ that on any input a^{ℓ}, $\ell \geq 0$, has an accepting computation which makes exactly 2ℓ push operations and uses pushdown height exactly $\lfloor \log_2 \ell \rfloor + 1$ (height 0 in the case $\ell = 0$). Furthermore, such computation is the less expensive one, with respect to both measures, namely all the other accepting computations make at least 2ℓ push operations or use at least height $\lfloor \log_2 \ell \rfloor + 1$ (height 0 in the case of the empty word).*

Proof (sketch). The formal definition of the PDA $\mathcal{A}$ can be found in [11, Thm. 8]. Here, we just mention that the computations of $\mathcal{A}$ follow the recursive pattern depicted in Fig. 1, where two recursive calls of the "same" PDA $\mathcal{A}$, with a read of an input symbol in between, are performed. The base of the recursion is an ε-transition from the initial state q_I to the final state q_F.

To read a symbol from the input, the PDA needs to execute two push operations. This gives $\mathsf{push}_{\mathcal{A}}(\ell) \geq 2\ell$, for each $\ell \geq 0$. Furthermore, as proved in [11], pushdown height h is sufficient and necessary to accept each string of length ℓ, with $2^{h-1} \leq \ell < 2^h$.

By induction on $h \geq 0$, it can be proved that, for any ℓ with $2^{h-1} \leq \ell < 2^h$, there exists a computation $\mathcal{C}$ on a^{ℓ} with $\mathsf{push}_{\mathcal{A}}(\mathcal{C}) = 2\ell$ and $\mathsf{height}_{\mathcal{A}}(\mathcal{C}) = h$.

This allows us to conclude that $\mathsf{push}_{\mathcal{A}}(\ell) = \mathsf{push}_{\mathcal{A}}(\mathcal{C}) = 2\ell$ and $\mathsf{height}_{\mathcal{A}}(\ell) = \mathsf{height}_{\mathcal{A}}(\mathcal{C}) = h$, where $h = \lfloor \log_2 \ell \rfloor + 1$ for $\ell > 0$, and $h = 0$ for $\ell = 0$. $\qquad\square$

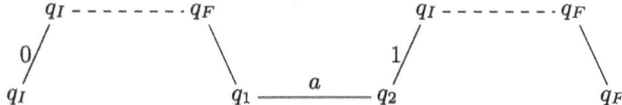

Fig. 1. The evolution of the pushdown store of $\mathcal{A}$ during the recursive subroutine leading from q_I to q_F, when recursive calls are made. The dashed lines should be replaced either by the base case (an ε-move) or by the same pattern.

References

1. Alberts, M.: Space complexity of alternating Turing machines. In: Budach, L. (ed.) FCT 1985. LNCS, vol. 199, pp. 1–7. Springer, Heidelberg (1985). https://doi.org/10.1007/BFb0028785
2. Bednárová, Z., Geffert, V., Reinhardt, K., Yakaryilmaz, A.: New results on the minimum amount of useful space. Int. J. Found. Comput. Sci. **27**(2), 259–282 (2016)
3. Bertoni, A., Mereghetti, C., Pighizzini, G.: An optimal lower bound for nonregular languages. Inf. Process. Lett. **50**(6), 289–292 (1994)
4. Bertoni, A., Mereghetti, C., Pighizzini, G.: Strong optimal lower bounds for Turing machines that accept nonregular languages. In: Wiedermann, J., Hájek, P. (eds.) MFCS 1995. LNCS, vol. 969, pp. 309–318. Springer, Heidelberg (1995). https://doi.org/10.1007/3-540-60246-1_137
5. Bordihn, H., Mitrana, V.: On the degrees of non-regularity and non-context-freeness. J. Comput. Syst. Sci. **108**, 104–117 (2020)
6. Ginsburg, S., Rice, H.G.: Two families of languages related to ALGOL. J. ACM **9**(3), 350–371 (1962)
7. Hopcroft, J.E., Ullman, J.D.: Introduction to Automata Theory, Languages and Computation. Addison-Wesley (1979)
8. Hopcroft, J.E., Ullman, J.D.: Some results on tape-bounded Turing machines. J. ACM **16**(1), 168–177 (1969)
9. Mereghetti, C.: Testing the descriptional power of small Turing machines on non-regular language acceptance. Int. J. Found. Comput. Sci. **19**(4), 827–843 (2008)
10. Pighizzini, G., Prigioniero, L.: Pushdown and one-counter automata: constant and non-constant memory usage. In: Bordihn, H., Tran, N., Vaszil, G. (eds.) DCFS 2023. LNCS, vol. 13918, pp. 146–157. Springer, Cham (2023). https://doi.org/10.1007/978-3-031-34326-1_11
11. Pighizzini, G., Prigioniero, L.: Pushdown automata and constant height: decidability and bounds. Acta Informatica **60**(2), 123–144 (2023)
12. Pighizzini, G., Shallit, J., Wang, M.: Unary context-free grammars and pushdown automata, descriptional complexity and auxiliary space lower bounds. J. Comput. Syst. Sci. **65**(2), 393–414 (2002)
13. Stearns, R.E., Hartmanis, J., Lewis II, P.M.: Hierarchies of memory limited computations. In: 6th Annual Symposium on Switching Circuit Theory and Logical Design, Ann Arbor, Michigan, USA, 6–8 October 1965, pp. 179–190. IEEE Computer Society (1965)

Decision Problems for Reversible and Permutation Automata

Maria Radionova[✉][iD] and Alexander Okhotin[iD]

Department of Mathematics and Computer Science, St. Petersburg State University,
7/9 Universitetskaya nab., Saint Petersburg 199034, Russia
st084881@student.spbu.ru, alexander.okhotin@spbu.ru

Abstract. For different kinds of reversible finite automata, the complexity of decision problems, such as emptiness, universality, equivalence and inclusion, is investigated. For permutation automata, they are all L-complete. For permutation automata with multiple initial states, emptiness is L-complete and the rest are co-NP-complete. For sweeping permutation automata, all are co-NP-complete, whereas length-bounded emptiness is PSPACE-complete. For reversible automata, the results are similar to deterministic automata, but universality is easier: L-complete for one initial state (cf. NL-complete for DFA), co-NP-complete for multiple initial states (cf. PSPACE-complete for multiple-entry DFA) and co-NP-complete in the sweeping case (cf. PSPACE-complete for two-way DFA). The minimality problem and the unary case are also investigated.

1 Introduction

Reversible computation is an active area of theoretical research, which is important, because reversible computers, once constructed, can operate with very low energy consumption. So far, several mathematical models of reversible computations with finite memory have been studied: one-way permutation automata (1PerFA) by Thierrin [22], one-way reversible automata with multiple initial states (MRFA) by Angluin [1] and by Pin [17], and with one initial state (1RFA) by Holzer et al. [9]. The authors [18] have recently studied a new model, sweeping permutation automata (2PerFA). Some recent work on permutation automata was done by Jecker, Mazzocchi and Wolf [12], by Hospodár and Mlynárčik [11], and by Rauch and Holzer [19].

In this paper we investigate the complexity of decision problems, such as emptiness, equivalence, etc., for different models of reversible automata. There are many classical results in the literature. The emptiness problem for one-way deterministic finite automata (1DFA) is a classical NL-complete problem [13], and their equivalence problem is complete in NL too. For one-way non-deterministic finite automata (1NFA) the emptiness problem is NL-complete [13], while the equivalence problem is complete in PSPACE [21].

This work was supported by the Russian Science Foundation, project 23-11-00133.

S. Z. Fazekas (Ed.): CIAA 2024, LNCS 15015, pp. 302–315, 2024.
https://doi.org/10.1007/978-3-031-71112-1_22

Table 1. Complexity of decision problems for automata with unrestricted alphabet.

	emptiness	universality	equivalence	inclusion	minimality
1PerFA	**L** T1, T2	**L** C1	**L** T3	**L** T3	**L** T4
1RFA	NL [8]	**L** T5	NL	NL	NL [9]
1DFA	NL [13]	NL	NL	NL	NL
2PerFA	**co-NP** T6, T7	**co-NP** T8	**co-NP** T8	**co-NP** T8	$\in \Pi_p^2$ T9
sRFA	PSPACE [2]	**co-NP** T10	PSPACE	PSPACE	$\in$ PSPACE
2DFA	PSPACE [14]	PSPACE	PSPACE	PSPACE	$\in$ PSPACE
MPerFA	**L** T13	**co-NP** T14	**co-NP** T14	**co-NP** T14	$\in \Pi_p^2$ T17
MRFA	NL	**co-NP** T15	$\in$ PSPACE	$\in$ PSPACE	$\in$ PSPACE
MDFA	NL	PSPACE [10]	PSPACE	PSPACE	$\in$ PSPACE

The complexity results for one-way deterministic automata with multiple initial states (MDFA) [10] are similar to 1NFA. And for two-way deterministic automata (2DFA) both emptiness and equivalence problems are PSPACE-complete. For one-way reversible automata, the emptiness problem is known to be NL-complete [8]. Also, the intersection emptiness problem was proved to be PSPACE-complete for 1DFA by Kozen [14] and for 1RFA by Birget et al. [2]. The intersection emptiness problem for 1PerFA was studied by Blondin et al. [3]. And the dependence of the complexity of this problem on the number of automata being intersected was investigated by Lange and Rossmanith [15].

One would expect the decision problems for permutation automata to be in a smaller complexity class than for deterministic automata. Indeed, our first result is that the emptiness problem for 1PerFA is L-complete, cf. NL-complete for 1RFA and 1DFA. For sweeping permutation automata, all the decision problems are proved to be co-NP-complete, while for 2DFA they are complete in PSPACE. To solve the problems for 2PerFA in co-NP, we use algorithms for the permutation group by Furst et al. [6]. Among the results of this paper, we note unexpected low complexity of the universality problem for a few types of reversible automata: it is co-NP-complete for sweeping reversible automata (sRFA) and for MRFA, and L-complete for 1RFA.

Whereas the emptiness problem for 2PerFA is in co-NP, along with the intersection emptiness problem for 1PerFA [3], their restricted versions with a bound on the length of the strings, called the bounded emptiness and the bounded intersection emptiness problems, turn out to be harder: they are complete for PSPACE. In the literature, the bounded universality problem was studied by Cho and Huynh [4].

The complexity of decision problems for different kinds of automata is compared in Tables 1 and 2, with the results of this paper emphasized in bold.

Table 2. Complexity of decision problems for automata with unary alphabet.

	emptiness	universality	equivalence	inclusion	minimality
1PerFA	**L** T2	**L** C1	**L** T3	**L** T3	**L** T4
1RFA	L	L	L	L	L
1DFA	L	L	L	L	L
2PerFA	**co-NP** T7	**co-NP** T8	**co-NP** T8	**co-NP** T8	$\in \Pi_p^2$
sRFA	co-NP	co-NP	co-NP	co-NP	$\in \Pi_p^2$
2DFA	co-NP [7]	co-NP	co-NP	co-NP	$\in \Pi_p^2$
MPerFA	**L** T13	**co-NP** T14	**co-NP** T14	**co-NP** T14	$\in \Pi_p^2$ T17
MRFA	L	co-NP	co-NP T16	co-NP T16	$\in \Pi_p^2$ T17
MDFA	L	co-NP	co-NP	co-NP	$\in \Pi_p^2$

2   Definitions

Let us introduce the automata for which we study the decision problems. Permutation automata were first defined by Thierrin [22], and we consider them both with one and with multiple initial states.

Definition 1. *A one-way permutation automaton with multiple initial states (MPerFA) is a quintuple* $\mathcal{A} = (\Sigma, Q, Q_0, \langle \delta_a \rangle_{a \in \Sigma}, F)$, *where*

- Σ *is the input alphabet;*
- Q *is the finite set of states;*
- $Q_0 \subseteq Q$ *is the set of initial states;*
- *for each* $a \in \Sigma$, *the function* $\delta_a \colon Q \to Q$ *is the bijective transition function;*
- $F \subseteq Q$ *is the set of accepting states.*

The computation of MPerFA on a string $w = a_1 a_2 \ldots a_\ell$ *starting in a state* $q_0 \in Q_0$ *is the sequence of states* $q_0, q_1, \ldots, q_\ell$ *where* $q_{i+1} = \delta_{a_{i+1}}(q_i)$ *for all* i. *A string* w *is accepted by MPerFA if there exists a state* $q_0 \in Q_0$ *from which the computation ends with an accepting state* $q_\ell \in F$. *An MPerFA* $\mathcal{A}$ *recognizes the language* $L(\mathcal{A}) = \{\, w \in \Sigma^* \mid w \text{ is accepted by } \mathcal{A} \,\}$.

A one-way reversible automaton with multiple initial states (MRFA) [1,17] is defined similarly to the MPerFA, but it is allowed to have an injective partial transition function δ_a by each symbol $a \in \Sigma$.

A classical one-way permutation automaton (1PerFA) of Thierrin [22] is an MPerFA with one initial state. Similarly, a reversible one-way automaton (1RFA) is an MRFA with one initial state (Holzer et al. [9] call it REV-DFA).

We also consider reversible classes of *sweeping automata*, that is, two-way automata that change the direction of motion only on the end-markers. Sweeping permutation automata (2PerFA) were recently defined by the authors [18]: these are the only two-way variant of sweeping automata (whence the acronym 2PerFA), and they turn out to be equal in power to 1PerFA.

Definition 2 ([18])**.** *A sweeping permutation automaton (2PerFA) is a 9-tuple* $(\Sigma, Q_+, Q_-, q_0, \langle \delta_a^+ \rangle_{a \in \Sigma}, \langle \delta_a^- \rangle_{a \in \Sigma}, \delta_\vdash, \delta_\dashv, F)$, *where*

- Σ is the alphabet;
- $Q_+ \cup Q_-$ is the set of states, where $Q_+ \cap Q_- = \varnothing$;
- $q_0 \in Q_+$ is the initial state;
- for each symbol $a \in \Sigma$, $\delta_a^+ : Q_+ \rightarrow Q_+$ and $\delta_a^- : Q_- \rightarrow Q_-$ are bijective transition functions;
- the transition functions at the end-markers $\delta_\vdash : (Q_- \cup \{q_0\}) \rightarrow Q_+$, $\delta_\dashv : Q_+ \rightarrow Q_-$ are partially defined and injective on their respective domains;
- $F \subseteq Q_+$ is the set of accepting states.

The computation on a string $w = a_1 a_2 \ldots a_\ell$ is the sequence of configurations (q, i), where $q \in Q_+ \cup Q_-$, and $i \in \{0, 1, \ldots, \ell + 1\}$ is the position in w, with positions 0 and $\ell + 1$ representing the end-markers. The initial configuration of a 2PerFA is $(q_0, 0)$, from which the automaton makes a transition to $(\delta_\vdash(q_0), 1)$. In a configuration (q, i) with $q \in Q_+$ and $i \in \{1, \ldots, \ell\}$, it moves right to the next configuration $(\delta_{a_i}^+(q), i + 1)$. Once the automaton is in a configuration $(q, \ell + 1)$, it accepts if $q \in F$, or moves left to $(\delta_\dashv(q), \ell)$ if $\delta_\dashv(q)$ is defined, and rejects otherwise. In a configuration (q, i) with $q \in Q_-$ and $i \in \{1, \ldots, \ell\}$, the automaton moves left to $(\delta_{a_i}^-(q), i - 1)$. Finally, in a configuration $(q, 0)$ with $q \in Q_-$, the automaton turns back to $(\delta_\vdash(q), 1)$ or rejects if this transition is undefined.

A *sweeping reversible automaton* (sRFA) is defined as 2PerFA but it is allowed to have injective partial functions δ_a^+ and δ_a^-. A more general *sweeping (deterministic) automaton* (sDFA) may use arbitrary partial functions δ_a^+ and δ_a^-.

The classical *two-way automata*, which may change the direction of motion over the input string at any time, are also considered in the paper in their deterministic (2DFA) and reversible (2RFA) variants.

3   Problems for One-Way Automata

The first problem we consider in the paper is the emptiness problem. Given an automaton, an algorithm determines whether it accepts an empty language.

Lemma 1. *For every 1PerFA $\mathcal{A} = (\Sigma, Q, q_0, \delta, F)$, and for every two states s, t, there is a path from s to t in the directed transition graph $G = (Q, \{(p, q) \mid \exists a \in \Sigma : \delta_a(p) = q\})$ if and only if s and t are connected by an undirected path in G.*

Proof. The non-trivial part of the proof is that an undirected path implies a directed path. We consider each edge in the undirected path from s to t one by one, constructing a directed path. If an edge is co-directed to the path, then we append it to the directed path as it is. Otherwise, if the edge is counter-directed, then we consider the cycle (by the same symbol) to which this transition belongs, and follow this cycle to get to the next vertex. Formally, if the edge is from q to p, but the corresponding transition is $\delta_a(p) = q$, then, as the transition function δ_a is a permutation on the set of states Q, there exists such an i, that $\delta_a^i(q) = p$. $\square$

Theorem 1. *The emptiness problem for 1PerFA is in L (the deterministic logarithmic space).*

Proof. To solve this problem, it is sufficient to check the accessibility of any accepting state from q_0 in the transition graph G, defined in Lemma 1. The algorithm interprets G as an undirected graph, and solves undirected reachability (USTCON) by the result of Reingold [20]. By Lemma 1 this is equivalent to directed reachability. □

To prove L-hardness, we consider the unary case and use a result on a decision problem about permutations.

Theorem 2. *The emptiness problem for unary 1PerFA is L-hard.*

Proof. The proof is by a reduction of the following decision problem proved to be L-complete by Cook and McKenzie [5]: given a permutation written as a sequence of numbers from 0 to $n-1$, and numbers i and j, all given in binary, determine whether i and j belong to the same cycle of the permutation. The reduction is as follows: the permutation now is the transition function of a unary 1PerFA with n states, the numbers i and j are initial and accepting states, respectively. Then the language of the constructed 1PerFA is non-empty if and only if i and j are on the same cycle of the permutation. □

Next, we consider the universality problem: given an automaton, determine whether it accepts all strings over the input alphabet.

Corollary 1. *The universality problem for 1PerFA is L-complete and remains so in the unary case.*

Proof. The universality problem is reduced to the emptiness problem and vice versa, it is sufficient to switch accepting and rejecting states. □

The next problems are equivalence and inclusion. In the equivalence problem, given two automata, determine whether they accept the same language. And in the inclusion problem, determine whether the first language is a subset of the second one.

Theorem 3. *Both the equivalence and the inclusion problems for two 1PerFA are L-complete and remain so in the unary case.*

Proof (a sketch). The L-hardness follows by a reduction from the emptiness problem. These problems are in L by solving emptiness for direct products of given two automata. □

The last decision problem we consider for 1PerFA is the minimality problem. Given an automaton, determine whether it is the minimal automaton of the same type recognizing the same language.

Theorem 4. *The minimality problem for 1PerFA is L-complete and remains so for the unary alphabet.*

Proof (a sketch). As shown by Hospodar and Mlynarčik [11, Prop. 3], if a language is recognized by a 1PerFA, then the minimal 1DFA for that language is always a permutation automaton. Hence, we prove the complexity of checking whether the given permutation automaton is a minimal 1DFA.

A logarithmic-space algorithm first checks whether all states are accessible. By Lemma 1, it is sufficient to check undirected reachability.

Now assume that all states of the automaton are accessible. Then it is minimal if and only if there is a separating string for each pair of states.

Accordingly, the algorithm implicitly constructs for each pair (q_1, q_2), with $q_1 \neq q_2$, a permutation automaton $\mathcal{B}_{q_1, q_2}$ recognizing the set of separating strings for these states. It is left to test all these automata for non-emptiness, and this can be done in logarithmic space, as shown in Theorem 1.

The hardness is proved for the unary case by a reduction from the following L-complete problem studied by Cook and McKenzie [5]: "Given a permutation, determine whether it is a single cycle".   □

There are also results for 1RFA. The NL-completeness of the emptiness problem was determined by Holzer and Jakobi [8, Proof of Thm.5]. The equivalence and the inclusion problems have the same complexity. The universality problem has unexpected complexity: L-complete both for unrestricted and unary alphabets, and not NL-hard as the other decision problems for 1RFA (assuming L ≠ NL).

Theorem 5. *The universality problem for 1RFA is L-complete and remains so in the unary case.*

Proof. Let us show the solvability of the universality problem for 1RFA. Given a 1RFA $\mathcal{A} = (\Sigma, Q, q_0, \delta, F)$, first, the algorithm checks that there are no accessible states with undefined transitions. If there are any, then the language already does not contain some string. To check this in L, the algorithm tests the undirected accessibility of such states in the transition graph G.

If some state with an undefined transition is accessible in $\mathcal{A}$, then it is accessible by an undirected path. Conversely, if no states with undefined transitions are accessible in $\mathcal{A}$, then we prove that the set S of accessible states is not connected to the rest of the automaton even by undirected paths. Indeed, all transitions in S are defined and lead to S, and hence transitions from $Q \backslash S$ cannot go to S due to reversibility of $\mathcal{A}$. Therefore, $Q \backslash S$ and S are not connected in the undirected graph, and the algorithm reports the absence of states with undefined transitions.

Now assume that all accessible states have no undefined transitions, hence our automaton $\mathcal{A}$ can cast off all unreachable states and become a permutation automaton, for which the universality problem is in L, see Corollary 1.

The L-hardness of the universality problem holds already for 1PerFA.   □

The minimality problem for 1RFA with unrestricted alphabet is NL-complete, see Holzer, Jakobi and Kutrib [9, Thm. 10]. In the unary case, it becomes L-complete.

4   Problems for Two-Way Automata

How is the complexity of decision problems affected if we allow the automata to move over a string in both directions? In this section, sweeping permutation and sweeping reversible automata are investigated.

Theorem 6. *The emptiness problem for 2PerFA is in co-NP.*

Proof. As Furst, Hopcroft and Luks [6] discovered, the following problem is solvable in polynomial time: given a set of generators from the permutation group, and given a permutation, determine whether the permutation belongs to the subgroup. We use it to solve the non-emptiness problem for 2PerFA in NP.

The algorithm guesses non-deterministically a permutation π on the set Q_+ and a permutation σ on Q_-. Assume that there is a string by which, reading it from left to right from an arbitrary state $q \in Q_+$, the automaton moves to $\pi(q) \in Q_+$, and reading from right to left it comes from $p \in Q_-$ to $\sigma(p) \in Q_-$. Then we can trace the computation on this string and determine the states in which the automaton comes to the end-markers, finally learning whether it is accepted. This stage of the algorithm takes linear time in the number of states.

It remains to check whether there is a string $w = a_1 a_2 \ldots a_k$ that implements the permutations π and σ, that is,

$$\pi = \delta^+_{a_k} \circ \delta^+_{a_{k-1}} \circ \ldots \circ \delta^+_{a_1} \qquad \text{and} \qquad \sigma = \delta^-_{a_1} \circ \delta^-_{a_2} \circ \ldots \circ \delta^-_{a_k}.$$

We rewrite the equality for σ:

$$\sigma^{-1} = \left(\delta^-_{a_k}\right)^{-1} \circ \left(\delta^-_{a_{k-1}}\right)^{-1} \circ \ldots \circ \left(\delta^-_{a_1}\right)^{-1}$$

Now, we consider a single permutation ϕ on the set $Q_- \cup Q_+$, which acts on Q_+ as π and acts on Q_- as σ^{-1}. The generators will be permutations $\eta_a : Q_- \cup Q_+ \to Q_- \cup Q_+$ for each $a \in \Sigma$. The function η_a acts on Q_+ as δ^+_a and acts on Q_- as $\left(\delta^-_a\right)^{-1}$. And we check whether the permutation ϕ belongs to the subgroup with generators $\langle \eta_a \rangle_{a \in \Sigma}$.   □

A similar problem, intersection emptiness for multiple 1PerFA, is already known to be co-NP-complete, see Blondin, Krebs and McKenzie [3]. It is tempting to try to reduce this problem to ours by simulating those 1PerFA in a 2PerFA one by one. But this would not work, because a sweeping permutation automaton requires injective transition functions on both end-markers. On the other hand, permutation automata from the intersection non-emptiness problem may have multiple accepting states, and we cannot come from all of them at once to the initial state of the next automaton. So, we need a new proof.

Theorem 7. *The emptiness problem for unary 2PerFA is co-NP-hard.*

Proof. We reduce the 3-SAT problem to the non-emptiness problem. By a given formula Φ in 3-CNF, we construct a 2PerFA working on unary strings. Each string encodes the values of variables, along with some extra information about their substitution into Φ. The automaton accepts the strings that encode satisfying assignments. The value of each variable x_i is encoded by a residual of the string's length modulo the i-th prime p_i. Then the automaton can read the value of each variable by traversing the string and counting its length modulo p_i.

The automaton considers all clauses one by one, verifying each to be true under the substitution of encoded values of variables. While checking a clause, the automaton makes several sweeps across the string and calculates the values of literals until it finds the first true literal. At that moment it either proceeds to the next clause or accepts if the current clause is the last in Φ. The first true literal may be on any of three positions in the clause, and in each case the automaton reads the string to the end and moves to the next clause. That means that for three different possible positive outcomes of reading a string the automaton should proceed with the same action, and do so *reversibly*.

The solution is to encode this outcome (the position of the first true literal in the j-th clause) in the string too, modulo the $(k+j)$-th prime p_{k+j}, where k is the number of variables in the formula. The automaton uses a group of states called a *switch box*, in which it counts the length of the string modulo p_{k+j}. Once the first true literal is found in the j-th clause, the automaton enters the j-th switch box in the state 0, 1 or 2 depending on the literal, and "forgets" this information while traversing the string in the switch box.   □

The following three problems for 2PerFA are co-NP-complete.

Theorem 8. *The universality, equivalence and inclusion problems for 2PerFA are co-NP-complete, and remain so for the unary case.*

Hardness comes from the emptiness problem for 2PerFA by constructing the complement for a given automaton (which is easy, since 2PerFA never loops and never rejects in the middle). Solvability is shown similarly by the membership problem for permutation subgroups [6].

For the minimality problem for 2PerFA, the same methods lead to a solution in the second level of the polynomial hierarchy.

Theorem 9. *The minimality problem for 2PerFA is in Π_2^p.*

Proof. Using universal non-determinism, the algorithm chooses any 2PerFA with fewer states. And by existential non-determinism it solves the non-equivalence problem to ensure that the automata recognize different languages.   □

Another related model is the sweeping reversible automaton (sRFA), which is a 2PerFA with some transitions by symbols of the alphabet possibly omitted. And for this model many decision problems become PSPACE-complete, yet the complexity of the universality problem surprisingly turns out to be the same as for 2PerFA. The solvability in PSPACE holds already for 2DFA. The hardness of the emptiness problem for sRFA follows from the PSPACE-completeness of this problem for intersections of multiple 1RFA, each with a single accepting state, proved by Birget et al. [2]. We construct an sRFA for a given intersection of 1RFA, which simulates them one by one on the way from left to right. And on the way back, the sRFA moves from the accepting state of one automaton to the initial state of the next one. The transition function by end-markers is injective because each automaton from the intersection has one accepting state. The equivalence and inclusion problems for sRFA are also PSPACE-complete.

And what about the universality problem? This time, we cannot reduce emptiness to universality by constructing the complement, as we did for 2PerFA. And there is an unexpected more efficient solution.

Theorem 10. *The universality problem for sRFA is co-NP-complete.*

Proof. To solve the non-universality problem in NP, the algorithm tries to find a string not accepted by the given automaton. The computation on such a string ends with an undefined transition either on a symbol inside the string or on one of the end-markers. First, for each symbol from the alphabet, the algorithm completes the sRFA's partial transition function to a bijection, obtaining a 2PerFA (the absent values are set arbitrarily). To handle strings which are rejected in the middle, the algorithm guesses non-deterministically a symbol a from the input alphabet, on which this happens. A rejected string splits in two substrings: one before a and one after a, and the behavior on each can be described by two permutations (one for the behavior while reading the string from left to right, and the other from right to left). The algorithm guesses these four permutations, tests their membership in the subgroup in polynomial time, and finally traces the computation to an undefined transition.

The hardness in co-NP follows from the case of 2PerFA.   □

Again consider the minimality problem, this time for sRFA. It is solvable in PSPACE by considering all automata of smaller size and by solving the equivalence problem.

Theorem 11. *The minimality problem for sRFA is in PSPACE.*

Next, consider the unary case. It turns out, that all the mentioned problems, except the minimality problem, are co-NP-complete. The hardness comes from the hardness of the same problems for unary 2PerFA (Theorems 7, 8). And these problems are solvable in co-NP already for 2DFA.

In the minimality problem, the algorithm uses universal non-determinism to guess an automaton of smaller size and solves the non-equivalence problem in NP.

Theorem 12. *The minimality problem for unary sRFA is in Π_2^p.*

5   Problems for One-Way Automata with Multiple Initial States

Another model of one-way deterministic finite automata are automata with multiple initial states. In this section, the complexity of decision problems will be shown for such reversible and permutation automata.

For the emptiness problem for MPerFA the algorithm iterates through all initial states and solves the emptiness problem for 1PerFA (Theorem 1) in L. The hardness follows from the same problem for 1PerFA.

Theorem 13. *The emptiness problem for MPerFA is complete in L and remains so in the unary case.*

The universality, equivalence and inclusion problems are complete in co-NP both for unrestricted and unary alphabets. Their solvability is shown similarly to the same problems for 2PerFA. The hardness in co-NP can be easily shown directly using the methods by Stockmeyer and Meyer [21, Th. 6.1], but instead we apply the result about hardness in this class of the intersection emptiness problem for unary 1PerFA proved by Blondin et al. [3].

Theorem 14. *The universality, equivalence and inclusion problems for MPerFA are co-NP-complete, and remain so for the unary case.*

Let us also consider the problems for MRFA. Emptiness is NL-complete as for 1RFA. Universality is harder.

Theorem 15. *The universality problem for MRFA is complete in co-NP both for unrestricted and unary alphabets.*

Proof. Hardness comes from the universality problem for unary MPerFA (Theorem 14).

Let us solve the non-universality in NP. For this, the algorithm tries to find a string, which is rejected from every initial state. If such a string exists, then the automaton's behavior, illustrated in Fig. 1, is like this: first, computations set forth from all the initial states. While all transitions are defined, the number of computations stays the same. From time to time some of the computations perish due to undefined transitions. In the end all surviving computations come to rejecting states of the MRFA. The algorithm guesses non-deterministically the symbols, on which any computations perish, and the behavior of the MRFA between them. To simulate the behavior, for each symbol of the alphabet, the algorithm completes MRFA's transition function to a bijection and uses the polynomial-time algorithm to test membership in a permutation subgroup [6].

First, the algorithm guesses non-deterministically the behavior of the constructed MPerFA until the given MRFA reaches the first undefined transition. This behavior is encoded in a permutation π_1. Next, the algorithm guesses a symbol a_1, on which there is an undefined transition, and applies the composition of the permutation and the transition function by this symbol $\delta_{a_1}(\pi_1(\cdot))$ to the set of initial states, and at least one state must perish. If all states are still alive, then the algorithm rejects. So, there is a new set of states of smaller size, from which the algorithm continues simulating computations in the same way. The last step is to check that each permutation π_i is implemented on some string w_i, that is, whether it lies in the subgroup generated by transition functions of the MPerFA. If all of them do, then the algorithm accepts.   □

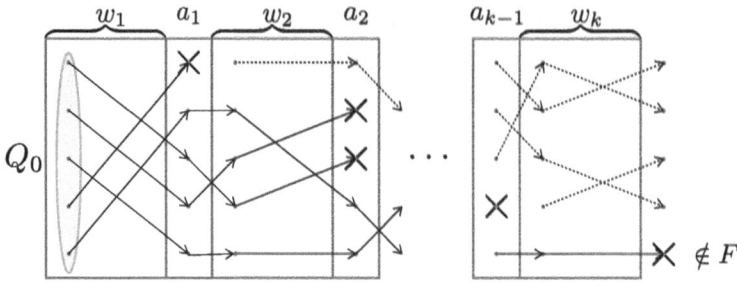

Fig. 1. An MRFA rejecting a string, all computations eventually perish.

The equivalence and inclusion problems are in PSPACE, and co-NP-hard already for unary MPerFA; their exact complexity remains unknown.

In the unary case, the equivalence and inclusion problems are in co-NP, so they become complete in this class.

Theorem 16. *The equivalence and inclusion problems for unary MRFA are complete in co-NP.*

Proof (a sketch). We reduce our problems to the same problems for unary 2DFA, which are in co-NP. Each MRFA can be transformed to a 2DFA, which recognizes the same language.   □

It is known from Holzer et al. [10] that the problem of whether a given MDFA can be reduced to a given size is PSPACE-complete. If the number of initial states is fixed, then, as shown by Malcher [16], the same problem becomes NP-complete.

We consider the *minimality problem* for MDFA, where for a given automaton one should determine whether it is a minimal MDFA for its language. About the minimality problem for MDFA it is only known that it is in PSPACE. For the unary case it can be solved in Π_2^p, by generating a smaller MDFA using universal non-determinism, then by solving the non-equivalence problem in NP— and similarly for MPerFA over unrestricted alphabet.

Theorem 17. *The minimality problem for MRFA with unrestricted alphabet is in PSPACE. For unary MRFA, as well as for unrestricted MPerFA, it is in Π_2^p.*

6   The Bounded Emptiness Problem

The intersection emptiness problem for 1PerFA is in co-NP, and if we allow each 1PerFA to have only one accepting state, then the problem becomes solvable even in NC [3]. And if we add a restriction on the length, the problem becomes PSPACE-complete.

Definition 3. *In the bounded emptiness problem, given an automaton $\mathcal{A}$ (or an intersection thereof) over an input alphabet Σ, and a number ℓ in binary, determine whether $L(\mathcal{A}) \cap \Sigma^{\leqslant \ell} = \varnothing$.*

The complexity is determined in the following theorem.

Theorem 18. *The bounded intersection emptiness problem for 1PerFA, each with a single accepting state, is PSPACE-complete for unrestricted alphabets.*

Proof (a sketch). The complexity of the intersection emptiness problem for 1DFA is well-known to be PSPACE-complete [14]. This result was improved by Birget et al. [2], who showed that the same problem is PSPACE-complete already for 1RFA, each with a single accepting state. Their proof is by a reduction from the membership problem for a fixed reversible Turing machine, and it maps its input string w to polynomially many 1RFA. Together, these 1RFA check their input for being the accepting computation history of the Turing machine on w. Hence, their intersection is a singleton if w is accepted, and is empty otherwise. Each of these 1RFA rejects upon finding any mistake.

In our proof, several 1PerFA are employed to carry out the same checks as each of the automata of Birget et al., but instead of rejecting at an error, they count them modulo different primes. Each automaton accepts if the counter is zero in the end. These primes are chosen so that their product is greater than the maximum possible length of the accepting computation of the Turing machine. The latter length is also taken as the bound ℓ in the bounded emptiness problem.

Thus, the automata read the string in the same way and count errors which occur at the same moments, modulo different primes. If all these automata return their error counters to zeros in the end, then the total number of errors must be at least the product of primes, which is strictly greater than ℓ. Therefore, the string must be at least ℓ symbols long, and hence is not taken into account. And if a short string is accepted, then it must be the valid computation history. $\square$

The similar problem for 2PerFA is PSPACE-complete too.

Corollary 2. *The bounded emptiness problem for 2PerFA is complete in PSPACE.*

Proof. Hardness comes from the hardness of the bounded intersection emptiness problem for 1PerFA, each with a single accepting state. For the given intersection, the reduction constructs a 2PerFA which reads the string in all 1PerFA one by one.

Solvability is shown by the classical method, as for 2DFA. $\square$

References

1. Angluin, D.: Inference of reversible languages. J. ACM **29**(3), 741–765 (1982). https://doi.org/10.1145/322326.322334
2. Birget, J.-C., Margolis, S.W., Meakin, J.C., Weil, P.: PSPACE-complete problems for subgroups of free groups and inverse finite automata. Theor. Comput. Sci. **242**(1–2), 247–281 (2000). https://doi.org/10.1016/S0304-3975(98)00225-4
3. Blondin, M., Krebs, A., McKenzie, P.: The complexity of intersecting finite automata having few final states. Comput. Complex. **25**(4), 775–814 (2016). https://doi.org/10.1007/s00037-014-0089-9
4. Cho, S., Huynh, D.T.: The parallel complexity of finite-state automata problems. Inf. Comput. **97**(1), 1–22 (1992). https://doi.org/10.1016/0890-5401(92)90002-W
5. Cook, S.A., McKenzie, P.: Problems complete for deterministic logarithmic space. J. Algorithms **8**, 385–394 (1987). https://doi.org/10.1016/0196-6774(87)90018-6
6. Furst, M., Hopcroft, J., Luks, E.: Polynomial-time algorithms for permutation groups. In: FOCS, pp. 36–41 (1980). https://doi.org/10.1109/SFCS.1980.34
7. Galil, Z.: Hierarchies of complete problems. Acta Informatica **6**, 77–88 (1976). https://doi.org/10.1007/BF00263744
8. Holzer, M., Jakobi, S.: Minimal and hyper-minimal biautomata. Int. J. Found. Comput. Sci. **27**(2), 161–186 (2016). https://doi.org/10.1142/S0129054116400050
9. Holzer, M., Jakobi, S., Kutrib, M.: Minimal reversible deterministic finite automata. Int. J. Found. Comput. Sci. **29**(2), 251–270 (2018). https://doi.org/10.1142/S0129054118400063
10. Holzer, M., Salomaa, K., Yu, S.: On the state complexity of k-entry deterministic finite automata. J. Automata, Lang. Combinatorics, **6**(4), 453–466 (2001) https://doi.org/10.25596/jalc-2001-453
11. Hospodár, M., Mlynárčik, P.: Operations on permutation automata. In: Jonoska, N., Savchuk, D. (eds.) DLT 2020. LNCS, vol. 12086, pp. 122–136. Springer, Cham (2020). https://doi.org/10.1007/978-3-030-48516-0_10
12. Jecker, I., Mazzocchi, N., Wolf, P.: Decomposing Permutation Automata. In: CONCUR 2021, LIPIcs, vol. 203, pp. 18:1–18:19 (2021). https://doi.org/10.4230/LIPIcs.CONCUR.2021.18
13. Jones, N.D.: Space bounded reducibility among combinatorial problems. J. Comput. Syst. Sci. **11**(1), 68–85 (1975). https://doi.org/10.1016/S0022-0000(75)80050-X
14. Kozen, D.: Lower bounds for natural proof systems. In: FOCS 1977, pp. 254–266 (1977). https://doi.org/10.1109/SFCS.1977.16
15. Lange, K.-J., Rossmanith, P.: The emptiness problem for intersections of regular languages. In: Havel, I.M., Koubek, V. (eds.) MFCS 1992. LNCS, vol. 629, pp. 346–354. Springer, Heidelberg (1992). https://doi.org/10.1007/3-540-55808-X_33
16. Malcher, A.: Minimizing finite automata is computationally hard. Theor. Comput. Sci. **327**(3), 375–390 (2004). https://doi.org/10.1016/j.tcs.2004.03.070
17. Pin, J.E.: On the languages accepted by finite reversible automata. In: Ottmann, T. (ed.) ICALP 1987. LNCS, vol. 267, pp. 237–249. Springer, Heidelberg (1987). https://doi.org/10.1007/3-540-18088-5_19
18. Radionova, M., Okhotin, A.: Sweeping permutation automata. In: Proceedings of the 13th International Workshop on Non-Classical Models of Automata and Applications (NCMA 2023, Famagusta, North Cyprus, 18–19 September 2023), EPTCS, vol. 388, pp. 110–124 (2023). https://doi.org/10.4204/EPTCS.388.11

19. Rauch, C., Holzer, M.: On the accepting state complexity of operations on permutation automata. RAIRO Theor. Inform. Appl. **57**(9) (2023). https://doi.org/10.1051/ita/2023010
20. Reingold, O.: Undirected connectivity in log-space. J. ACM, **55**(4) (2008). https://doi.org/10.1145/1391289.1391291
21. Stockmeyer, L.J., Meyer, A.R.: Word problems requiring exponential time. In: 5th Annual ACM Symposium on Theory of Computing (STOC 1973, Austin, USA, April 30–May 2, 1973), pp. 1–9 (1973). https://doi.org/10.1145/800125.804029
22. Thierrin, G.: Permutation automata. Math. Syst. Theory **2**(1), 83–90 (1968). https://doi.org/10.1007/BF01691347

Benchmarking Regular Expression Matching

Alexander Roodt[1], Brendan Keith Mark Watling[1], Willem Bester[1],
Brink van der Merwe[1,3(✉)], Sicheol Sung[2(✉)], and Yo-Sub Han[2]

[1] Stellenbosch University, Private Bag X1, Matieland,
Stellenbosch 7602, South Africa
bkmwatling@protonmail.com, {whkbester,abvdm}@cs.sun.ac.za
[2] Yonsei University, Seoul, Republic of Korea
{sicheol.sung,emmous}@yonsei.ac.kr
[3] National Institute for Theoretical and Computational Sciences,
Stellenbosch, South Africa

Abstract. In this paper we benchmark the matching time of regular expression matching engines when they use either the Thompson or Glushkov regular expressions to state machine conversion algorithms, with or without using memoisation while matching, and doing matching either by using a lockstep or a Spencer type scheduler. We conduct our empirical investigation by expanding on the virtual machine for regular expressions matching approach, introduced by Russ Cox.

Keywords: Thompson and Glushkov constructions · Spencer regular expression matchers · memoisation · virtual machine

1 Introduction

Although all mainstream programming languages have regular expression matching (regex matching for short) software as part of their standard libraries, and most software developers encounter regular expressions in their software projects on a regular basis, questions remain in terms of how best to support various regular expressions constructs as efficiently as possible in terms of matching time and memory usage. For example, very few developers and researchers have an understanding of the trade-offs made in the latest standard Java regular expression matching library, between memory usage and matching time, in terms of how much memoisation to perform to reduce the prevalence of non-linear matching time without adding an unnecessary high memory overhead, nor is this well-documented or clear how these decisions were made. This points to the fact that researchers are in need of a software framework to investigate these questions, instead of developing custom regular matching software to investigate the various questions individually.

In this work, we investigate regular expression matching time questions experimentally, such as in [13]. We decided not to embed the required code in an industry strength matcher, given the complex nature of such code bases. We went the

S. Z. Fazekas (Ed.): CIAA 2024, LNCS 15015, pp. 316–331, 2024.
https://doi.org/10.1007/978-3-031-71112-1_23

route of modifying and extending the implementation discussed in [5], but we did this in a way to make it convenient for others to use our software framework when expanding on our work, or when investigating related questions. In fact, one of the main contributions of this work is to provide other researchers with a regular expression matching framework (available on GitHub[1]) which is implemented in a way to minimise the effort involved in performing regular expression matching benchmarking related investigations.

We showcase the flexibility of our software framework by expanding on experiments performed in [9] and [13], on how to improve upon worst-case matching time with memoisation schemes, and on how the Glushkov construction can be used to reduce ambiguity (in comparison to the Thompson construction) and thus to improve matching time (in contrast to when a matcher derived from the Thompson NFA construction is used).

Our framework supports both lockstep matching (matching using the on-the-fly subset construction) and matching performed by using an input-directed depth first search (also known as Spencer matching). In our benchmarking experiments, we compare lockstep with Spencer style matching, using both the Thompson and Glushkov constructions for converting regular expressions to state machines, and also make use of two forms of memoisation in the Spencer matcher. These two memoisation schemes are taken from [9], and are the following: (i) IN(E): memoise all states with in-degree at least two, in a state machine constructed from the regular expression E; (ii) CN(E): memoise all closure nodes/states in a state machine constructed from the regular expression E, where these are the nodes to which we have a back-edge/transition when we do a DFS from the initial state of the automaton. For example, in the Thompson state machine for the regular expression a^*, we have a closure node just before the node encountered from which we match "a" (again), and in general, in the Thompson construction, we have a CN node for each Kleene star. Strictly speaking, we should indicate these memoisation schemes by $\text{IN}_T(E)$ or by $\text{IN}_G(E)$, depending on if a Thompson or Glushkov regular expression to state machine construction is used, and similarly for CN(E).

In this paper, we investigate the following question experimentally: What is the (relative) matching time when using Glushkov or Thompson state machines, with a Spencer (making use of IN, CN, or no memoisation) or a lockstep style matcher?

The outline of the paper is as follows. In the next section, we provide the required background. This is followed by a section describing in detail our regex matching framework, after which we present our experimental results, which is followed by our conclusion.

2   Background, Definitions, Related Work

First, we consider the concepts, which include regular expressions over effective Boolean algebras and prioritised transducers, required to follow this paper. We

---

[1] https://github.com/bkmwatling/srvm.

denote by $\mathcal{A} = (\Sigma, \Psi, [\![\_]\!], \bot, \top, \vee, \wedge, \neg)$ an effective Boolean algebra over Σ, where Ψ is a set of predicates that is closed under the Boolean connectives $\vee$, $\wedge$, and $\neg$, with $\bot, \top \in \Psi$ and $[\![\_]\!] : \Psi \to 2^\Sigma$ a denotation function; see D'Antoni and Veanes [7] for a detailed definition. In the sequel, Σ is the standard 32-bit character set of Unicode. We only consider the interval algebra over Σ, which we denote as $\mathfrak{I}$.

Example 1. In $\mathfrak{I}$, the interval [0-9], also denoted by \d, is a predicate such that $[\![\backslash\mathrm{d}]\!] = \{0, 1, 2, \ldots, 9\}$, matching any decimal digit. We also use "." as a predicate equivalent to $\top$, which is to say, $[\![.]\!] = \Sigma$.

A regular expression over $\mathfrak{I}$ is either an element of $\Psi \cup \{\varepsilon\}$, or is constructed inductively from existing subexpressions E and E', using any combination of the following expressions: (1) $\hat{}E$, with $\hat{}$ the zero-width assertion (ZWA) specifying that matching must start at the beginning of the input string, (2) $E\$$, with $\$$ the ZWA specifying that matching must terminate at the end of the input string, (3) union $(E + E')$, (4) concatenation $(E \cdot E')$, (5) greedy Kleene star (E^*), (6) lazy Kleene star $(E^{*?})$, and (7) a capture group $(_iE)_i$, for a nonnegative integer i, where $(_i$ and $)_i$ do not already appear in E. In addition to being used to also indicate the priority of operators, similar to '(' and ')', the subscripted parenthesis '$(_i$' and '$)_i$' are used to indicate that the integer i is associated with the substring of the input string being matched by the subexpression E in $(_iE)_i$, according to PCRE semantics; more on PCRE below.

We assume that any regex, as a whole, is surrounded implicitly or explicitly by $(_0$ and $)_0$, which is to say, the complete match (i.e. when the regex starts and ends with $\hat{}$ and $\$$ respectively) or partial match when using a given regex on an input string w, can be accessed via the zeroth capture. When it is not required to distinguish between any parentheses used to indicate precedence of operators, capturing, or noncapturing subexpressions, it is mostly standard to number open parenthesis from left to right, starting at 1, and to number the closing parenthesis correspondingly.

When matching with the greedy Kleene star E^*, the subexpression E is used as many times as possible (while still matching with the complete regex if possible), whereas with the lazy Kleene star $E^{*?}$, E is used as few times as possible (with preference in terms of greediness or laziness given to earlier appearing Kleene starred subexpressions). On its own (with no $\$$), $E^{*?}$ will always match the empty string at the beginning of an input string. Should a regex E contain no occurrence of $\hat{}$ or $\$$, we can model partial matching with E via full matching, by replacing E with $\hat{}(_s.^{*?})_s(_0E)_0(_f.^*)_f\$$, assuming the indices s and f do not appear elsewhere in $(_0E)_0$. Some parentheses can be dropped (assuming they are not indexed to retrieve capturing information) with the rule that a (lazy or greedy) Kleene star takes precedence over concatenation, which takes precedence over union. Furthermore, outermost parentheses can be dropped, and $E \cdot E'$ can be written as EE'.

We also allow the following syntactic sugar in regular expressions: (1) E^+ denotes EE^*, (2) $E^?$ denotes $E + \varepsilon$, (3) E^n denotes E concatenated with itself

$(n-1)$ times, and (4) $E^{0,n}$, for $n > 0$, is defined inductively as $(EE^{0,n-1})^?$. In particular, $E^{0,1}$ denotes $E^?$, and accordingly, for $0 < m < n$, we obtain the counted repetition $E^{m,n}$ as $E^m E^{0,n-m}$. The idea of handling counted repetition inductively instead of by straightforward expansion—for example, $E^{1,3}$ as $E(E(E)^?)^?$ instead of $EE^? E^?$ is due to Cox [6], who employs it in the RE2 matcher, presumably for this formulation's ability to "skip over" remaining options when matching with another iterated expression E fails.

Example 2. When ˆ or \$ appears in a regex, they do not necessarily appear as the first or last symbol in the regex, respectively: We allow regexes such as $(a\$ + b + ˆc)$. This regex matches the last symbol in the input string if this symbol is an "a"; otherwise, it matches the first "b" in the input string (which might not be the first symbol in the input string); and finally, if none of these two options are satisfied, it will match the first symbol in the input string, if this symbol is a "c".

We consider both full and partial matching of input strings by regular expressions. Also, we do not only associate with a regular expression ˆ$E\$$ the language $\mathcal{L}(E)$ of strings matched by E, or the leftmost "first" substring matched by E when given an input string w, but consider the substrings of w matched by subexpressions in E.

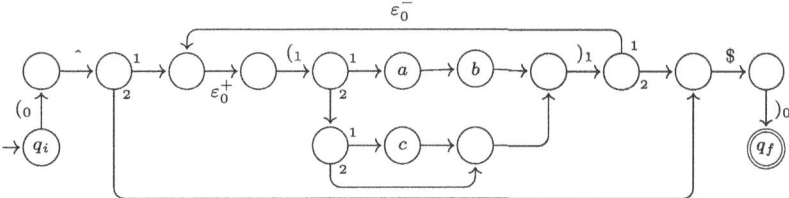

Fig. 1. Thompson's construction for the regex ˆ$(ab + c^?)^*\$$.

We modify the definitions of PCRE semantics by Berglund and Van der Merwe [3], to also allow ˆ and \$, as follows: Similar to capturing parentheses $(_i$ and $)_i$, we regard ˆ and \$ as output symbols, or more accurately, as actions. But in contrast to the indexed parentheses used for capturing, these symbols filter out all output strings over $\Sigma \cup \{(_i,)_i, ˆ, \$\}$, in which ˆ do not appear before any symbols in Σ, and similarly for \$, all output strings in which \$ do not appear after all symbols from Σ. If a regex is of the form ˆ$E\$$ (starts with ˆ and ends with \$), with E not containing the symbols ˆ and \$, we associate a language denoted $\mathcal{L}(E)$, which is a subset of Σ^*, and is obtained by evaluating E as usual, where ε stands for the empty string, and $\psi \in \Psi$ for $[\![\psi]\!]$.

Instead of providing a formal definition of a prioritised transducer—used to model PCRE matching semantics with a state machine, and described in detail in [3]—we use the Thompson-like construction in Fig. 1, for the regular expression $\hat{}(ab + c^?)^*\$$ to explain this formalism. PCRE matching semantics, especially submatching and capturing, and including the semantics of the lazy Kleene star, are defined in Definitions 7 to 9 in [3]. Theorem 2 in this paper also shows that a Thompson-like construction (used by the Java regex matcher) in fact computes PCRE capture semantics.

Example 3. In contrast to the definition in [3], we allow states (as opposed to transitions) in transducers to consume symbols from Σ. There is only one initial state q_i (without incoming transitions) and only one final state q_f (without outgoing transitions), and besides q_i and q_f, each other state will either not be labelled, or be labelled by a predicate over $\Im$ or a symbol from Σ, to be consumed by that state. If a state q consumes a, then this is just shorthand for adding an extra dummy state q', rerouting to q' the transitions originally incoming to q, and adding a transition from q' to q that consumes a. When states have more than one outgoing transition, these transitions are numbered from 1 onwards. This way, we associate a sequence of integers with each path from q_i to q_f and in PCRE semantics, the lexicographic least path (also referred to as the highest priority path) is selected. We number the parentheses in the regex $\hat{}(ab + c^?)^*\$$ with a 1, and the transducer outputs "$(_1$" each time matching enters the subexpression $ab + c^?$, and similarly, "$)_1$" each time matching with $ab + c^?$ ends. The transition leaving q_i is labelled by $\hat{}$, which checks that the the state machine is at the beginning of the input string. Without the $\hat{}$ on the transition from q_i, we need an implicit lazy Kleene star on "." (any input symbol) associated with q_i (which could of course also be added explicitly), which finds the left-most position in the input string from which matching starts. Also, the $\$$ on the transition to q_f verifies that the state machine is at the end of the input, and without it, the state machine is allowed to accept without consuming all input symbols.

The regex $\hat{}(ab + c^?)^*\$$ is also a *problematic regex*, since we have a subexpression that matches the empty string, to which a Kleene star is applied. To remedy the possibility of having arbitrary long ways of matching a given input string, each lexicographically smaller than shorter accepting paths, we have transitions labelled by ε_0^+ and ε_0^- associated with entering and reentering matching with the Kleene-starred subexpression. The purpose is to verify that a character was consumed when taking ε_0^- since the last time that ε_0^+ was taken.

As mentioned in the introduction, we also investigate the IN and CN-memoisation schemes, introduced by Davis et al. [9]. Intuitively, when memoising a particular state q in a state machine, when using a Spencer matcher, the matcher is only allowed to visit q at most once at a given position in the input string. Memoisation was also investigated in detail by Van der Merwe at al. [12], and those results show that either of these schemes will ensure linear matching time, and generally, having at least one state memoised in each loop of a state machine, ensures linear matching time. Note that the number of states memoised with CN is a subset of the number of states memoised with IN. Also, when using the Thompson construction, the number of states memoised using CN is equal to the number of Kleene stars in the associated regex.

3   Overview of Regex Matching Framework

3.1   Symbolic Regular Expressions and Parser

Our framework, BRU (Brendan's Regex Utility), begins processing by taking the regex pattern string and then parses it into a *regex abstract syntax tree (RAST)*, using a recursive descent parser, which recognises all PCRE constructs. However, since we do not support all PCRE constructs (such as backreferences), the parser will fail with appropriate return codes for unsupported constructs. A RAST is a binary AST that encodes all the necessary information of a regex. In the RAST, leaf nodes only representing terminals in the regex, such as characters, zero-width assertions (like ˆ, $, and lookaheads – which we do not discuss in this paper), whilst non-leaf nodes represent non-terminals, for example alternations and concatenations with two children, or capturing groups and iterations (like Kleene star) with a single child. With this basic tree structure, it is easy to make constructions from regexes to our intermediate representation. We support *Symbolic Regular Expressions (SREs)*, with character classes defined over ℑ (see Sect. 2), which is implemented as an array of intervals we can iterate over for membership testing.

3.2   Symbolic Regular Expression Virtual Machine

With the ability to parse SREs, we created a *Symbolic Regular Expression Virtual Machine (SRVM)* that executes compiled SRE programs. Inspired by Cox [5], the virtual machine language is much like assembly, in which we support a basic but expressive set of instructions that cater to matching SREs against input strings, using a *program counter (PC)* and *string pointer (SP)*, combined with standard control flow through jumps. The instructions include `char` and `pred` for matching against single characters and character classes, respectively, `jmp`, `split`, and `tswitch` for unary, binary, and n-ary split execution (for alternations and repetition), respectively, and the `match` instruction which signals a match has been found.

The execution of the SRVM is defined through the standard program control flow of incrementing the PC by one for all non-split instructions, whereas split-style instructions update the PC according to their jump points. For binary and n-ary splits, the SRVM needs to have some way of keeping the necessary multijump information. This is handled through (software) threads, where the SRVM contains a *scheduler* and *thread manager* to control the order of thread execution, as well as storing all the threads to be executed. Different thread managers and schedulers are provided and can be used (including custom-made ones), each using its own defined threads. Therefore, the information stored in the different types of threads varies, but most important is that each thread contains a PC and SP for its current path execution through the SRE program; other fields include an array of capturing information.

The scheduler allows for easy runtime changes of how the SRVM is executed, in other words, how the search tree is explored in matching; the part of the search tree explored by a Spencer matcher is formally defined in Definition 5 by Weideman et al. [15]. We have implemented schedulers for both Spencer-style DFS and Thompson-style lockstep (modified BFS) modes, which enables easy comparisons of these tree traversals in terms of performance and overheads. Note that the scheduler is controlled through a thread manager and *not* by the SRVM directly. For the Spencer scheduler, we maintain a stack of threads from which the SRVM calls (in a FIFO manner), to mimic DFS using recursion.

3.3   State-Machine Intermediate Representation

Initially, we compiled SRE programs directly from RAST, but we found duplication in the code for the Thompson and Glushkov constructions. This led to the *State Machine Intermediate Representation (SMIR)*, which provides the means to construct a state machine (automaton) that can have multiple initial states. States can have a list of *actions* that are to be executed upon reaching that state. An action is either a character comparison, membership test into an interval (predicate function), capture, ZWA (like ^ and \$), memoisation instruction, or ε-loop check. A transition also contains a list of actions instead of simply being an ε or comparing against a character. Although we support much more generality in the SMIR, in this paper, we have a single initial and a single final state, actions on transitions, and we consume characters in states.

One issue going from *Symbolic Nondeterministic Finite-state Automata (SNFAs)* theory to implementation is that an SNFA always picks a correct path through the execution tree, but in practice, nondeterminism must be handled explicitly, taking PCRE semantics [3] into account. As such, a SMIR instance must execute in a predefined order. To output the correct capture information for submatching, we need *prioritised execution*, which is accomplished with implicit priorities coming from the order of the out-transitions defined for a state: how the SMIR instance is created (i.e. the defined order of the transitions) determines the semantics of the submatches and which paths have higher priority.

It is often helpful to create an automaton in steps, using a *transform* to convert one SMIR instance into another. We provide such a transform for memoisation which adds memoisation actions to states that have an in-degree (number of incoming transitions) of two or more (IN-memoisation), or have a back edge coming in that forms a loop in the automaton (CN-memoisation).

3.4   SMIR Thompson and Glushkov Constructions

Having the RAST, SMIR, and SRVM on their own is of no use, and from a RAST, we must construct a SMIR instance, which is then compiled into an SRE program. Custom constructions can easily be made by using the SMIR API, but we provide the two most common constructions, namely Thompson's [14] and Glushkov's [11].

We followed the standard Thompson construction, but since the theory works with SNFAs, which can always pick the correct path and avoid ε-loops, we needed to add explicit actions for these. In particular, we have ε-loop checking actions that we add for all repetition operators. This involves adding an **epsset** action which marks where in the input string the current iteration of the repetition starts. Then when trying to perform another iteration, we have an **epschk** action that checks whether the previous iteration actually moved forward from the marked point in the string and signals failure if it didn't. We also add support for working with captures and ZWAs, which is simple with the SMIR API.

We construct the Glushkov machine from the Thompson machine by using the flattening procedure as described in [3], instead of making use of the standard approach of first, follow and last sets. Note, in [3] the flattening procedure is not referred to as the Glushkov construction. Our approach of obtaining the Glushkov construction was necessitated by having to ensure that the constructed Glushkov automaton will also respect capturing semantics. This observation also indicates why it is not straightforward to incorporate any of the other possible regex to state machine constructions in our experimental investigation.

Each state (in a Glushkov automaton), besides q_i and q_f, consume symbols. To ensure that we label states (besides q_i and q_f) with unique symbols or predicates, we linearise the regex and corresponding Thompson transducer (like in the standard Glushkov construction). We allow more than two outgoing transitions at a given state, and also, we have a sequence of output symbols/actions on transitions. Thus, by doing a DFS on the Thompson construction (between each pair of states), we obtain the Glushkov construction. It happens from time to time that there is more than one sequence of transitions to be added between pairs of states in the Glushkov machine. To explain the flattening procedure, and specifically what happens when we have more than one sequence of output symbols associated with a transition, we briefly discuss how we obtain the Glushkov machine in Fig. 2 from Fig. 1; we leave extending the standard approach of using first, follow and last (to obtain the Glushkov PCRE machine) to our setting as future work.

Example 4. Consider the transition between states q_0 and q_f in Fig. 2. Note in Fig. 1 that it is also possible to go from q_0 to q_f without reading any symbols, producing the output ˆ $. But this path has lower priority than the one on which we output ˆ $(_1)_1$, and thus, gets dropped. The situations in which we drop transitions when going from Thompson to a Glushkov machine might lead to situations where we match with fewer steps when using a Glushkov Spencer matcher (as opposed to using a Thompson Spencer matcher). These are also precisely the situations where there is a difference between weak and strong ambiguity of the regex being considered; see [4].

Example 5. Next, we consider the problematic regex ˆ $(a^*)^*$, a common example that causes ReDoS when using the Thompson construction without memoisation. However, due to the properties of the Glushkov construction, this "double" Kleene star collapses to a single Kleene star, resulting in linear matching time against the failing input string $aa \cdots ab$. Comparing the machines in Fig. 3, it is obvious that the Glushkov construction ensures linear matching time compared to the worst-case exponential matching time for Thompson, which is caused by

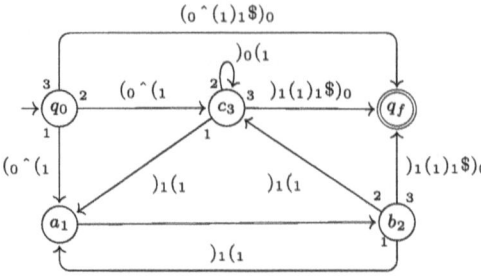

Fig. 2. Glushkov's construction for $\hat{}\,(ab + c^?)^*\,\$$.

the double loop around the single compare-to-"a" state in Thompson. This issue is removed with the Glushkov construction which essentially simplifies the regex to (a^*). However, the capturing semantics of PCRE dictate that the last substring that capture group 1 should match is the empty string since the Thompson machine will try to match with a^* when at the end of the string, and thus, match the empty substring. As such, this operation is replicated in our Glushkov construction by having $)_1(_1)_1\$_0$ instead of just $)_1\$_0$, which ensures the empty string match for capture group 1.

It is important to note that since the inner loop for the Thompson construction in Fig. 3 does not contain a nullable subexpression, the ε-loop check is *not* necessary. However, since the subexpression of the outer loop (the a^*) is nullable, the ε-loop check is necessary. The bad performance of this regex is mitigated through memoisation with the Thompson construction (memoising the inner loop), but is completely unnecessary for Glushkov, since its matching time is linear without memoisation. In fact, the IAR (infinite ambiguity removal) memoisation scheme, described by Van der Merwe et al. [12], is sophisticated enough to figure out that no memoisation is required in this specific case for the Glushkov machine.

3.5   Compilation

Completing the compilation pipeline, with the SMIR constructed, we can at last compile it into an SRE program. This required "serialising" the SNFA into a sequential program, which was done using standard assembly techniques for handling control flow with "while" and "if–else" statements. We compile each state from the SNFA with the out-transitions immediately following the state. A state becomes a list of actions following each other in the program and then a potential `split`/`tswitch` instruction (if there is more than one transition from the state) followed by each transition's actions and jumps to the destination state's code. In Table 1 we list the SRE programs for the regex $(a^*)^*$. See [1] for how to simulate a VM regex program, where the VM instructions are similar to ours.

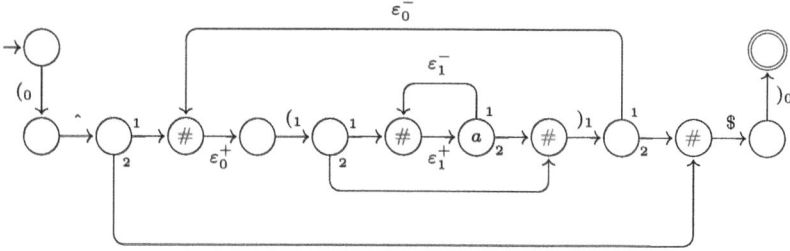

(a) Thompson's construction for $^(a^*)^*\$$.

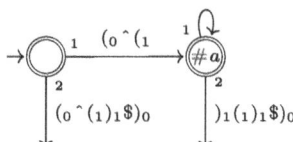

(b) Glushkov's construction for $^(a^*)^*\$$.

Fig. 3. The SMIR instances for the regex $^(a^*)^*\$$. We use the convention in Fig. 3(b) of leaving the unique final state implicit, and when we transition to the implicit final state, we mark the source state as final, and have transitions with no destination state. A blue hash indicates CN-memoisation, and red, IN-memoisation that is not CN.

Table 1. The SRE programs for the regex $(a^*)^*$

(a) Thompson constructed SRE program.

0:	save	0
1:	split	2, 14
2:	epsset	0
3:	save	2
4:	split	5, 10
5:	epsset	8
6:	char	a
7:	split	8, 10
8:	epschk	8
9:	jmp	5
10:	save	3
11:	split	12, 14
12:	epschk	0
13:	jmp	2
14:	save	1
15:	match	

(b) Glushkov constructed SRE program.

0:	split	1, 4
1:	save	0
2:	save	2
3:	jmp	9
4:	save	0
5:	save	2
6:	save	3
7:	save	1
8:	jmp	15
9:	char	a
10:	split	9, 11
11:	save	3
12:	save	2
13:	save	3
14:	save	1
15:	match	

4   Experimental Results

We compare the performance of BRU based on combinations of (1) NFA construction algorithms (Thompson vs Glushkov), (2) memoisation schemes (CN vs IN), and (3) schedulers (Spencer vs lockstep). We use the regex corpus of Davis et al. [8], which consists of 537,806 real-world regexes. We generate at most ten accepting strings for each regex by graph random walking using xeger[2]. We also generate rejecting strings by inserting random characters at random positions, and then we verify that the newly generated strings are indeed rejected. This approach prevents the generation of meaningless results due to early rejection. For each regex, an average is taken over all inputs, when counting matching or predicate steps, as discussed below. We exclude those regexes with (A) extra features such as backreferences that BRU does not support, and (B) star operations of nullable subexpressions (i.e., problematic regexes) which (in some cases) result in different matching priorities and thus capturing semantics between (i) the Thompson construction with memoisation, or (ii) any other construction (which all agree, and also agree with the formulation as in [3]), i.e., Glushkov with or without memoisation or Thompson without memoisation. This complication is caused by the fact that memoisation might incorrectly change capturing semantics (and the order of exploring paths in a state machine) when ε-loops are present. This happens for example on the regex $(a^* + b)^*(b^*)$, on input ab, where the subexpression (b^*) under Thompson with IN-memoisation captures the empty string, but with Glushkov (using no, IN- or CN-memoisation) or Thompson with no or CN-memoisation, it will capture b (which can be verified with BRU). Given these considerations, we used 439,634 regexes in our experiments.

4.1   Comparison of Glushkov and Thompson Under Matching Steps

First, we compare the number of *matching steps* during matching, which is the number of state visits. Matching steps serve as a proxy for the matching time, and ensures that we can ignore the influence of implementation details and runtime environments. We claim that there is a linear relationship between matching time and matching steps, although we do not report results on these experiments. It is important to note that BRU evaluates the validity of a transition only after having reached the target state, similar to other regex implementations such as RE2 [5] and Java's regex matcher [2]. In Fig. 4a, all regexes are on or below the diagonal, indicating a reduction in the number of matching steps required by Glushkov compared to Thompson. Not only does the number of matching steps decrease on average when using the Glushkov construction, but it does so on every pair of pattern and input for the Spencer scheduler with no, IN, and CN-memoisation, and the lockstep scheduler. This finding should be contrasted with the results reported by Sung, Cheon, and Han [13], which observed an increase in matching steps (with Spencer) for some regexes when using the Glushkov (instead of Thompson) construction, which is caused by a lack of proper consideration of matching priorities in their implementation.

---

[2] https://pypi.org/project/xeger/.

Finding 1. The Thompson construction should never be preferred over Glushkov, since compared to when using Thompson, Glushkov keeps the number of matching steps for all regexes the same, or reduces it.

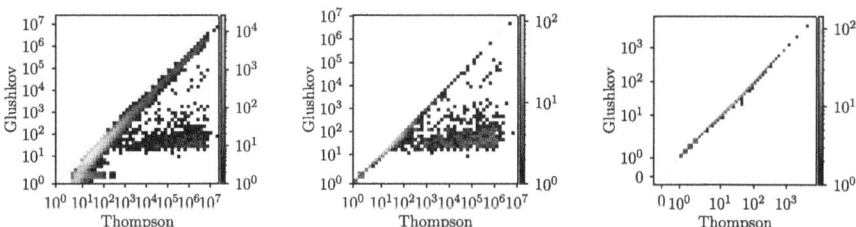

(a) Number of matching steps with no memoisation.   (b) Number of predicate steps with no memoisation.   (c) Number of predicate steps with CN-memoisation.

Fig. 4. A comparison of the number of matching and predicate steps when using the Spencer scheduler. The x- and y-coordinates of each cell denote the number of steps when using the Thompson and Glushkov construction, respectively. Figure 4b and Fig. 4c used only the 1,687 path-eliminated regexes.

4.2 Comparison of Glushkov and Thompson Under Predicate Steps

One could argue that the time cost of each matching step is not the same for all states, as the predicate-labelled states require determining whether an input character satisfies a predicate, whereas ε-states do not require any testing. Thus, we also compute the number of *predicate steps*, which is the number of occurrences of predicate testing (including character comparisons) during matching. Our analysis showed that the Glushkov construction removes at least one path (or transition) with lower priority from the corresponding automaton, for 1,687 of the 439,632 regexes. Thus, the Glushkov automaton for these 1,687 regexes exhibits less ambiguity than their Thompson counterparts [15]. In all other cases, the number of predicate steps will stay the same.

Figure 4b shows a decrease in the number of predicate steps required by the Glushkov construction, compared to Thompson, when the Spencer scheduler does not use memoisation. We found that the Glushkov construction reduces the predicate steps for 1,228 of the 1,687 regexes with reduced ambiguity. However, the situation is different when we use memoisation. The advantage of Glushkov over the Thompson construction diminishes when CN-memoisation is employed, as depicted in Fig. 4c, where the Glushkov construction reduces predicate steps for only 492 regexes. Furthermore, both constructions exhibit exactly the same number of predicate steps for all regexes when using IN-memoisation or the lockstep scheduler.

Finding 2. If the processing time of ε-states is negligible and memoisation is used, the Glushkov construction provides marginal benefits over Thompson. Specifically, the Glushkov construction decreases predicate steps for 1,228 and 492 of 439,634 regexes. An analysis of the relative memory overhead for memoisation is left as future work.

4.3   Comparison of Spencer and Lockstep Schedulers

The experimental results for the lockstep scheduler demonstrate a structural equivalence of the Thompson and Glushkov automata, as noted by Giamarresi et al. [10]. Essentially, the Glushkov automaton is a version of the Thompson automaton that omits ε-transitions. In our setup, in Glushkov automata, all states, apart from the initial and final state, will consume symbols. The outcomes observed with the lockstep scheduler are theoretically predictable, as ε-transitions/states have a minimal influence (i.e., we count them when determining matching steps) on matching efficiency, but they do not cause more paths to be explored. Note that the lockstep scheduler considers each character–state pair only once. Memoisation makes the Spencer scheduler behave similarly to the lockstep scheduler in this regard by preventing the repeated evaluation of the same character–state pair (at those states being memoised).

Finally, we compare the Spencer scheduler when using IN-memoisation with the lockstep scheduler. Figure 5 illustrates that it is better to use the Spencer scheduler with IN-memoisation rather than the lockstep scheduler, both in terms of matching and predicate steps. The results show that the Spencer scheduler with IN-memoisation consistently requires the same or fewer matching steps and predicate steps, compared to the lockstep scheduler. Note that, as pointed out in Sect. 4.2, the number of predicate steps is the same for the Thompson and Glushkov (when using IN-memoisation).

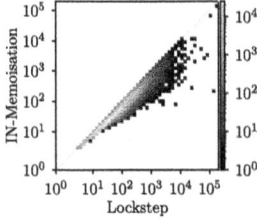 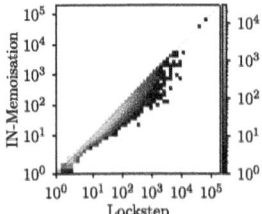 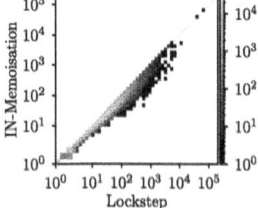

(a) Number of matching steps when using the Thompson construction.

(b) Number of matching steps when using the Glushkov construction.

(c) Number of predicate steps. The results are the same for the Thompson and Glushkov constructions.

Fig. 5. A comparison of the number of matching and predicate steps when using either the Spencer scheduler with IN-memoisation or the lockstep scheduler.

Finding 3. Using Spencer with memoisation instead of lockstep is preferable, when ignoring memory overhead for memoisation, because Spencer supports various extended features, but also, it reduces matching and predicate steps (or in some cases, keeps them the same). An analysis of memory overhead is left as future work.

In Table 2 we give the counts of the `char` matching steps (regarded as predicate steps) when matching with the (linearised) regex $\hat{}(b_1^* + b_2a_3)^*a_4\$$ on input *bbababa*. This example serves two purposes. It provides a concrete example of computing predicate steps, and shows that there are indeed cases where Glushkov performs worse than Thompson. Note, this is an example of a problematic regex, and as mentioned before, we exclude problematic regexes form our results, since under IN-memoisation, when using the Thompson construction, it might lead to incorrect capture semantics.

Table 2. Counts of the incremental `char` matching (and nonmatching) steps (regarded as predicate steps) when matching with the (linearised) regex $\hat{}(b_1^* + b_2a_3)^* a_4\$$ on input *bbababa*. Summing a row gives the total `char` steps at that point, with the sum of the last row the value reported by BRU. The 'Symbols' column indicates which symbols in the regex were under consideration at that point in matching. For Spencer it is always a single symbol, while for Lockstep it is a set of symbols.

	Spencer ($IN_{\{T,G\}}$)							Symbols		Lockstep						
	b	b	a	b	a	b	a			b	b	a	b	a	b	a
Thompson	1	0	0	0	0	0	0	b_1	b_1, b_2, a_4	3	0	0	0	0	0	0
	1	1	0	0	0	0	0	b_1	b_1, b_2, a_3, a_4	3	4	0	0	0	0	0
	1	1	1	0	0	0	0	b_1	b_1, b_2, a_3, a_4	3	4	4	0	0	0	0
	1	1	2	0	0	0	0	b_2	b_1, b_2, a_4	3	4	4	3	0	0	0
	1	1	3	0	0	0	0	a_4	b_1, b_2, a_3, a_4	3	4	4	3	4	0	0
	1	2	3	0	0	0	0	b_2	b_1, b_2, a_4	3	4	4	3	4	3	0
	1	2	4	0	0	0	0	a_3	b_1, b_2, a_3, a_4	3	4	4	3	4	3	4
	1	2	4	1	0	0	0	b_1								
	1	2	4	1	1	0	0	b_1								
	1	2	4	1	2	0	0	b_2								
	1	2	4	1	3	0	0	a_4								
	1	2	4	2	3	0	0	b_2								
	1	2	4	2	4	0	0	a_3								
	1	2	4	2	4	1	0	b_1								
	1	2	4	2	4	1	1	b_1								
	1	2	4	2	4	1	2	b_2								
	1	2	4	2	4	1	3	a_4								
Glushkov	1	0	0	0	0	0	0	b_1	b_1, b_2, a_4	3	0	0	0	0	0	0
	1	1	0	0	0	0	0	b_1	b_1, b_2, a_3, a_4	3	4	0	0	0	0	0
	1	1	1	0	0	0	0	b_1	b_1, b_2, a_3, a_4	3	4	4	0	0	0	0
	1	1	2	0	0	0	0	a_4	b_1, b_2, a_4	3	4	4	3	0	0	0
	1	1	3	0	0	0	0	b_2	b_1, b_2, a_3, a_4	3	4	4	3	4	0	0
	1	2	3	0	0	0	0	a_4	b_1, b_2, a_4	3	4	4	3	4	3	0
	1	3	3	0	0	0	0	b_2	b_1, b_2, a_3, a_4	3	4	4	3	4	3	4
	1	3	4	0	0	0	0	a_3								
	1	3	4	1	0	0	0	b_1								
	1	3	4	1	1	0	0	b_1								
	1	3	4	1	2	0	0	a_4								
	1	3	4	1	3	0	0	b_2								
	1	3	4	2	3	0	0	a_4								
	1	3	4	3	3	0	0	b_2								
	1	3	4	3	4	0	0	a_3								
	1	3	4	3	4	1	0	b_1								
	1	3	4	3	4	1	1	b_1								
	1	3	4	3	4	1	2	a_4								

5   Future Work

Our experimental results focused on comparing the matching steps (or states visited) when using the Thompson or Glushkov constructions but did not take into account the memory overhead of memoisation and the stack for Spencer-style DFS matching with (or without) memoisation, or the queue for lockstep/BFS matching. We aim to leverage our framework to get exact benchmarks for memory overhead and thus offer holistic comparisons of each construction. We also plan to investigate the improvement offered in terms of states visited by caching the computation of subsets when using the lockstep scheduler (which is standard practice in industry strength lockstep matchers).

The capture semantics of real-world PCRE matchers is not standardised, in particular also in cases where one uses the ability of matchers to obtain multiple matches for a given input string by continuing where the previous match ended in the input string. We tested the regex $(a^* + b^* + c^+)^*$ against $cccccccc$ in multimatch mode, and observed that the actual PCRE matcher differs from the Java and C# matchers which also differ from the Rust and JavaScript matchers. For example, the PCRE matcher report the empty string first (at the beginning of the string) then match all the c's on the next match, and finally the empty string on the final match (at the end of the string), while the Java and C# matchers keep matching the empty string (starting the next search one character after the previous match) until they reach the end of the string. Rust reports only a single match, which is all the c's, whilst the JavaScript matcher does not match the first empty string of the PCRE matcher, but has the other two matches from the PCRE matcher. We would like to investigate and classify these inconsistent capture semantics in more detail, and consider which of these should be allowed as options in our matcher, and which could be added with minimal effort to our framework. Also, as pointed out in the previous section, in some cases, memoisation might change capture semantics. Investigating this in detail, is also left as future work.

We would like to implement counters through counter automata (and not through counter expansion) and investigate how to modify the various memoisation schemes in the presence of counters. Also, we want to add additional memoisation schemes such as the IAR (infinite ambiguity removal) memoisation scheme, described in [12], and the memoisation scheme employed in the current version of the Java matcher.

Finally, we would like to reformulate the standard Glushkov construction using first, follow and last sets, so that Glushkov machines with capturing output can be constructed in this way, instead of (as in this paper) first constructing a Thompson state machine, which is then modified to obtain the corresponding Glushkov state machine.

Acknowledgment. Sung and Han were supported by the NRF grant (RS-2023-00208094) and the AI Graduate School Program at Yonsei University (No. RS-2020-II201361) funded by the Korean government (MSIT).

References

1. Barrière, A., Pit-Claudel, C.: Linear matching of javascript regular expressions. CoRR **abs/2311.17620** (2023). https://doi.org/10.48550/arXiv.2311.17620
2. Berglund, M., Drewes, F., van der Merwe, B.: Analyzing catastrophic backtracking behavior in practical regular expression matching. In: Ésik, Z., Fülöp, Z. (eds.) Proceedings 14th International Conference on Automata and Formal Languages, AFL 2014, Szeged, Hungary, May 27-29, 2014. EPTCS, vol. 151, pp. 109–123 (2014)
3. Berglund, M., van der Merwe, B.: On the semantics of regular expression parsing in the wild. Theor. Comput. Sci. **679**, 69–82 (2017)
4. Brüggemann-Klein, A.: Regular expressions into finite automata. In: Simon, I. (ed.) LATIN 1992. LNCS, vol. 583, pp. 87–98. Springer, Heidelberg (1992). https://doi.org/10.1007/BFb0023820
5. Cox, R.: Regular expression matching: the virtual machine approach (2009). https://swtch.com/~rsc/regexp/regexp2.html. Accessed 19 Apr 2024
6. Cox, R.: Regular expression matching in the wild: A tour of RE2, an efficient, production regular expression implementation (2010). https://swtch.com/~rsc/regexp/regexp3.html. Accessed 30 Apr 2024
7. D'Antoni, L., Veanes, M.: The power of symbolic automata and transducers. In: Majumdar, R., Kunčak, V. (eds.) CAV 2017. LNCS, vol. 10426, pp. 47–67. Springer, Cham (2017). https://doi.org/10.1007/978-3-319-63387-9_3
8. Davis, J.C., IV, L.G.M., Coghlan, C.A., Servant, F., Lee, D.: Why aren't regular expressions a lingua franca? An empirical study on the re-use and portability of regular expressions. In: Dumas, M., Pfahl, D., Apel, S., Russo, A. (eds.) Proceedings of the ACM Joint Meeting on European Software Engineering Conference and Symposium on the Foundations of Software Engineering, ESEC/SIGSOFT FSE 2019, Tallinn, Estonia, August 26-30, 2019, pp. 443–454. ACM (2019)
9. Davis, J.C., Servant, F., Lee, D.: Using selective memoization to defeat regular expression denial of service (redos). In: 42nd IEEE Symposium on Security and Privacy, SP 2021, San Francisco, CA, USA, 24-27 May 2021, pp. 1–17. IEEE (2021)
10. Giammarresi, D., Ponty, J., Wood, D., Ziadi, D.: A characterization of Thompson digraphs. Discret. Appl. Math. **134**(1–3), 317–337 (2004)
11. Glushkov, V.M.: The abstract theory of automata. Russ. Math. Surv. **16**(5), 1–53 (1961)
12. van der Merwe, B., Mouton, J., van Litsenborgh, S., Berglund, M.: Memoized regular expressions. In: Maneth, S. (ed.) CIAA 2021. LNCS, vol. 12803, pp. 39–52. Springer, Cham (2021). https://doi.org/10.1007/978-3-030-79121-6_4
13. Sung, S., Cheon, H., Han, Y.: How to settle the ReDoS problem: back to the classical automata theory. In: Caron, P., Mignot, L. (eds.) CIAA 2022. LNCS, vol. 13266, pp. 34–49. Springer, Cham (2022)
14. Thompson, K.: Programming techniques: regular expression search algorithm. Commun. ACM **11**(6), 419–422 (1968)
15. Weideman, N., van der Merwe, B., Berglund, M., Watson, B.: Analyzing matching time behavior of backtracking regular expression matchers by using ambiguity of NFA. In: Han, Y.-S., Salomaa, K. (eds.) CIAA 2016. LNCS, vol. 9705, pp. 322–334. Springer, Cham (2016). https://doi.org/10.1007/978-3-319-40946-7_27

On the Complexity of Decision Problems for Parameterized Finite State Synchronous Transducers

Tianxiang Tang[ID] and Vladimir A. Zakharov[✉][ID]

Shenzhen MSU-BIT University, Shenghen, China
tangtx@smbu.edu.cn, zakh@cs.msu.su

Abstract. When designing and maintaining the information systems, there is a need to use parameterized automata, in which some components of the model that determine its behavior are replaced with variables. It has previously been shown that the complexity of many analysis and synthesis problems increases significantly when machines or programs have parameters. Nevertheless, it can be expected that for some classes of automata the complexity of certain important problems will remain low and the development of efficient algorithms will be possible. We present here the results of our study of some basic decision problems for finite state transducers parameterized on output data. When choosing this class of automata, we were guided by the intention of preserving the determinism of runs with uncertain values of some elements due to the presence of parameters in the model. We managed to show that for some decision problems their complexity remains low (NL-complete) even after parameters are added to the model, whereas the complexity of other problems that have efficient solutions for ordinary transducers increases significantly (from NL or P-completeness to NP-completeness and beyond) for their parameterized modifications.

Keywords: finite state transducer · parameterized model · decision problem · complexity class

Variables are everywhere in mathematics, and computer science is no exception. But the pragmatics of variables in automata theory is quite diverse. Most often, they are used to name arguments of functions and relations or as unknowns in constraint systems that specify words, languages, transductions, or automata.

There are also other uses for variables. They can denote special information stores that a machine is equipped with in order to increase its computing potency. The ability to operate with such variables is strictly limited so that an increase in the computational power of automata is not achieved at the expense of the undecidability inherent in universal computational models. Typical examples of such use of variables are the extentions of finite state machines developed in [5,9,10,14,15,17] to process languages over infinite alphabets. In these models a finite set of registers (variables) are assigned letters from the input word in the

© The Author(s), under exclusive license to Springer Nature Switzerland AG 2024
S. Z. Fazekas (Ed.): CIAA 2024, LNCS 15015, pp. 332–346, 2024.
https://doi.org/10.1007/978-3-031-71112-1_24

course of the run. In some of these models variables can be updated, in others transitions are supplied with guards that include such variables. In addition to the comparative analysis of the computational power and closedness properties of the proposed models, the authors of the above papers mainly studied the complexity of some basic decision problems. And it was found that even for deterministic versions of these models, a lot of problems that are NL-complete for ordinary finite state machines become intractable (NP-complete and even far beyond) for automata augmented with variables.

Variables in automata can also act as parameters. The difference between these two uses of variables—data stores and parameters—is quite significant. The values of the former can only be influenced by input data (words) and machine commands (transitions). They cannot be given or updated arbitrary. Manipulating the latter is completely beyond the power of an automaton; their values can only be changed by external means: as soon as the parameters are assigned values, the machine becomes completely ready for operation, and further the parameter values are not subject to change. The need to use parameters arises in many cases, of which it is enough to mention only two. When installing a computer program, the user selects certain options, thereby assigning values to system parameters, and expects to obtain the desirable behavior of such a specialized application. Then it is important to know what parameter values guarantee the required behavior of the system, how several instances can be combined into a single parameterized system, and how the number of such parameters can be reduced. Another example of the use of parameterization is in top-down system design when the final values of some components of the system are refined as the project evolves. In this case one might, perhaps, wonder whether two parameterized designs are compatible, is it possible for every choice of one set of parameters to assign certain values to some other set of parameters to guarantee a required behavior of the system. These and other similar questions motivate the study of decision and algorithmic problems for parameterized finite state machines.

Actually, any element of these machines (input and output data, states, transitions, time constraints, etc.) can be an object of parameterization. The most intensively studied were the parameterized models of timed automata introduced in [1] where numerical parameters are used in transition guards and state invariants. The increased expressive power of such models comes at the price of the undecidability of most important algorithmic problems in the general case: manipulations with timers allows one to simulate arithmetic operations. Currently, research in this direction is mainly focused on marking off decidable and undecidable cases when syntactic restrictions on automata are imposed (boundaries on the number of clocks and parameters, types of admissible constraints, etc.). However, as surveys [2,3] show, there are still many open questions there, and for decidable cases, the complexity estimates turn out to be quite high. Recent results confirm this state of art [11].

A thorough study of parameterized probabilistic automata was made in the remarkable graduate work [16]. The objects of parameterization, of course, were the probabilities of transitions firing in the automata. It was noticed that many algorithmic problems for parameterized probabilistic automata are closely related with well-known problems in linear algebra and linear programming. The reserach was focused mostly on the equivalence and bisimilarity checking problems. The author showed that both problems, which are known to be P-complete for ordinary probabilistic automata, migrate into complexity classes NP and PSpace when parameterization on the probability distributions on automata transitions is introduced.

The first (to the best of our knowledge) detailed study of finite state automata with parameterized inputs on transitions was undertaken recently in [6]. The authors of this paper just added to an alphabet of regular expressions free variables that can be interpreted as letters. Given an evaluation of parameters $\theta : \mathcal{X} \to \Sigma$, a parameterized formula e becomes a usual regular expression $e\theta$ which specifies a language $L(e\theta)$. The authors of [6] associated with every such formula e two languages $L_\square(e) = \bigcap_{\theta \in Eval} L(e\theta)$ and $L_\diamond(e) = \bigcup_{\theta \in Eval} L(e\theta)$ which correspond to universal and existential quantification over all possible evaluations of parameters. Clearly, both languages $L_\square(e)$ and $L_\diamond(e)$ are regular, but in [6] it was shown that parametrization provides a very powerful means for their succinct representation. The complexity of usual decision problems (membership, non-emptiness, universality, and containment) was studied for regular expressions thus parameterized in two variants of their semantics. Using fairly sophisticated reduction techniques, the authors of [6] were able to find that the complexity of these problems for parameterized regular expressions grows significantly (from NL and PSpace-completeness to NP and ExpSpace-completeness) (see Table 1). Of course, with so high complexity estimates, it is not easy to find an application area for this parameterized model that tolerates the computational hardness of the algorithmic problems.

Table 1. Complexity of decision problems for parameterized regular expressions

Problem\Semantics	$\diamond$-variant	$\square$-variant
Membership	NP-complete	coNP-complete
Non-emptiness	NL-complete	ExpSpace-complete
Universality	ExpSpace-complete	PSpace-complete
Containment	ExpSpace-complete	ExpSpace-complete

Although the authors of [6] have made considerable progress in the study of algorithmic problems for parameterized automata-type models, there are still several important questions remaining, without the solution of which the overall picture remains incomplete. Since regular expressions can be easily converted to finite state automata (FSAs) with small overhead, all complexity estimates presented in Table 1 are also applicable to nondeterministic parameterized FSAs. However, as is known, the complexity of decision problems is significantly different for deterministic and non-deterministic automata. One might wonder what happens to deterministic FSAs that are parameterized in the same way as was done with regular expressions in [6]? Unfortunately, when it comes to Rabin-Scott automata, input parameterization eliminates the syntactic difference between deterministic and non-deterministic FSAs. Therefore, in order to see what effect parameterization has on the complexity of decision problems for deterministic models, it is necessary to choose a different type of finite state machine.

To define the semantics of parameterized regular expressions, the authors of [6] used two languages $L_\square(e)$ and $L_\Diamond(e)$. This is quite reasonable when only unary properties that need to be checked are concerned. But in fact, every parameterized regular expression e specifies the entire family of regular languages $\mathcal{L}(e) = \{L(e\theta) : \theta \in Eval\}$, and when a binary property $R(e_1, e_2)$ is checked the relationships between regular languages from $\mathcal{L}(e_1)$ and $\mathcal{L}(e_2)$ can be of prime importance. Therefore, for parameterized automata it is necessary to refine the settings of some problems which involve binary relations.

Finally, for finite state machines there are algorithmic problems in which it is important to take into account not only the semantics of automata but the structural features of their program design. To formulate and study these problems, an algebraic model of parameterized regular expressions is no longer sufficient, and it is necessary to use computational models based on transition systems, grammars, or rewriting systems to be able to reason about the structure of programs and their computations

Taking these considerations into account, we decided to move from a one-dimensional model of Rabin-Scott automata to a two-dimensional model of finite state transducers, which makes it possible to combine parameterization over the outputs on transitions and deterministic behavior on the input data. Another advantage of the chosen model is that in the case of a unary input alphabet, finite state transducers turn out to be isomorphic to Rabin-Scott FSAs, and, therefore, all complexity results obtained in [6] are preserved for this model. The main goal of our study is to identify those cases in which the analysis of the behavior of finite state transducers preserves effective algorithmic solutions even after introducing data parameterization. We started with a simple model of synchronous finite state transducers to continue the research on data parameterized automata initiated in [6] with the intention of extending our results to broader classes of data-parameterized automata-based models of computation if our expectations will be confirmed for this class of finite state machines.

1 Preliminaries

A *word* over *alphabet* Σ is any finite sequence $w = a_1 a_2 \ldots a_k$ of letters in Σ. By $|w|$ we denote the length of w, and by $w[i]$ the i-th letter $a_i, 1 \leq i \leq |w|$, in the word w. Given a pair of words u and v, we write uv for their *concatenation*. A pair of words (u, w) is called *balanced* if $|u| = |w|$. The set of all words over an alphabet Σ is denoted by Σ^*. A *transduction* over alphabets Σ and Δ is any subset of $\Sigma^* \times \Delta^*$. A transduction T is called *balanced* if all its pairs of words are balanced.

A *synchronous finite state transducer* (briefly, FST) over an input alphabet Σ and an output alphabet Δ is a quadruple $\pi = \langle Q, q_0, F, \longrightarrow \rangle$, where Q is a finite set of *states*, q_0 is an *initial state*, $F \subseteq Q$ is a subset of *final states*, and $\longrightarrow$ is a finite *transition relation* of the type $Q \times \Sigma \times \Delta \times Q$. Quadruples (q, a, b, q') in $\longrightarrow$ are called *transitions* and depicted as $q \xrightarrow{a/b} q'$. A *run* of π on an input word $u = a_1 a_2 \ldots a_n$ is any finite sequence of transitions

$$q \xrightarrow{a_1/b_1} q_1 \xrightarrow{a_2/b_2} \cdots \xrightarrow{a_{n-1}/b_{n-1}} q_{n-1} \xrightarrow{a_n/b_n} q' . \tag{1}$$

The pair (u, w), where $w = b_1 b_2 \ldots b_n$, is called a *label* of a run (1). We will write $q \xrightarrow{u/w}_* q'$ to indicate that an FST π has a run labeled with (u, w) from a state q to a state q'. If $q = q_0$ and $q' \in F$ then (1) is a *complete* run. A *transduction relation realized by* an FST π is the set of labels of all complete runs of π, i.e. $TR(\pi) = \{(u, w) : q_0 \xrightarrow{u/w}_* q', q' \in F\}$. As can be seen, any transduction relation realized by a synchronous FST is balanced. By the *size* $|\pi|$ of an FST π we mean the number of transitions of π. An FST π is called *deterministic* if its transition relation $\longrightarrow$ is a function of the type $Q \times \Sigma \to \Delta \times Q$, i.e. for every input letter a and a state q there is at most one transition of the form $q \xrightarrow{a/b} q'$ in π.

Let $\mathcal{X}$ be an infinite set of variables disjoint with both alphabets Σ and Δ. Then a *parameterized synchronous finite state transducer* (briefly, PFST) is a quadruple $\Pi = \langle Q, q_0, F, \longrightarrow \rangle$, where Q, q_0, and F have the same meaning as in FST π, whereas $\longrightarrow$ is a transition relation of the type $Q \times \Sigma \times (\Delta \cup \mathcal{X}) \times Q$, i.e. some *parameterized transitions* in a PFST Π are labeled with variables (parameters) from $\mathcal{X}$ instead of letters (constants) from Δ. Runs of PFSTs are labeled with balanced pairs (u, α), where $u \in \Sigma^*$ and α is a sequence of output letters from Δ and variables from $\mathcal{X}$. A parametrization of PFST π is called *simple* if no variable from $\mathcal{X}$ occurs in two different transitions of Π.

When all variables in parameterized transitions of a PFST get output letters from Δ as their values, a PFST Π becomes an FST π. Such an evaluation of parameters is achieved by means of substitutions. Any function $\theta : \mathcal{X} \to \Delta$ is called a *ground substitution*. The set of all ground substitutions is denoted by *GSubst*. Applying a substitution θ to a PFST Π results in an FST $\Pi\theta$ in which every occurrence of any variable x in the transitions of Π is replaced by an output letter $\theta(x)$. In this case an FST $\pi = \Pi\theta$ is called an *instance* of Π.

Thus, each PFST Π generates a whole family of FSTs $\mathcal{F}(\Pi) = \{\Pi\theta : \theta \in GSubst\}$, and all typical decision problems for checking certain properties (like non-emptiness, functionality, minimality) or relations (like equivalence, bisimilarity) on FSTs can be quite naturally addressed to PFSTs as well. These decision problems can be specified by unary or binary predicates of the form $P(\pi)$ or $R(\pi', \pi'')$, and every such problem when referred to PFSTs can be reformulated in two ways:

- *existentially* ($\exists$-variant): given a PFST Π or a pair of PFSTs Π', Π'' check the predicates $\exists \pi \in \mathcal{F}(\Pi): P(\pi)$ or $\exists \pi' \in \mathcal{F}(\Pi'), \pi'' \in \mathcal{F}(\Pi''): R(\pi', \pi'')$;
- *universally* ($\forall$-variant): given a PFST Π or a pair of PFSTs Π', Π'' check the predicates $\forall \pi \in \mathcal{F}(\Pi): P(\pi)$ or $\forall \pi' \in \mathcal{F}(\Pi') \exists \pi'' \in \mathcal{F}(\Pi''): R(\pi', \pi'')$.

When a binary relation $R(\Pi', \Pi'')$ is considered it is assumed that PFSTs Π' and Π'' do not have common parameters from $\mathcal{X}$. Although our setting of $\forall$-decision problems possibly breaks the symmetry of certain relations $R(\pi', \pi'')$ (like, say, equivalence or bisimilarity), nevertheless, in some applications this variant is more meaningful than the symmetric one.

Clearly, when the output alphabet Δ is fixed the number of instances $\Pi\theta$ of every PFST Π does not exceed $|\Delta|^{|\Pi|}$. Therefore, we can immediately yield some rough estimates of the complexity of decision problems for PFSTs. If a decision problem for FSTs has time complexity $t(n)$ and space complexity $s(n)$ then the trivial upper bounds for the complexity of existential and universal counterparts of the problem are $2^{O(n)}t(n)$ and $O(n + s(n))$ respectively. Also if a decision problem for FSTs is PSpace-complete then both variants of this problem for PFSTs are PSpace-complete as well. The main objective of our research is to study in which cases for the well-known decision problems in automata theory these rough upper bounds can be significantly improved and when they are tight.

2   The Complexity of Decision Problems for Parameterized Transducers

We consider a series of standard decision problems for FSTs—membership, non-emptiness, functionality, equivalence, minimization, and bisimilarity,—and for each of these problems we will show the positions of its existential and universal analogues for deterministic and non-deterministic PFSTs in the hierarchy of complexity classes. In what follows it is assumed that $|\Delta| \geq 2$, since otherwise parametrization of FSTs is meaningless.

2.1   Membership

The general membership problem for FSTs is to check, given an FST π and a balanced pair of words $(u, w) \in \Sigma^* \times \Delta^*$, if $(u, w) \in TR(\pi)$. It is known that this problem is L-complete (under AC^0 reduction) for deterministic FSTs and NL-complete for non-deterministic FSTs [12, 13].

Proposition 1. *Both the $\exists$-membership and the $\forall$-membership problems are L-complete for deterministic PFSTs.*

Proof. It is enough to show that both problems are in L. If Π is a deterministic PFST then for every balanced pair (u, w) there is at most one run $q_0 \xrightarrow{u/\alpha}_* q'$ of Π on u such that for every $i, 1 \leq i \leq n$, either $\alpha[i] = w[i]$, or $\alpha[i] \in \mathcal{X}$. Clearly, $(u, w) \in TR(\Pi\theta)$ for every $\theta \in GSubst$ iff $q' \in F$ and $\alpha = w$, i.e. all transitions in this run are not parameterized. On the other hand, $(u, w) \in TR(\Pi\theta)$ for some $\theta \in GSubst$ iff 1) $q' \in F$, 2) $w[i] = \alpha[i]$ whenever $\alpha[i] \in \Delta$, and 3) $\alpha[i] = \alpha[j]$ implies $w[i] = w[j]$ whenever $\alpha[i]$ and $\alpha[j]$ are parameters. To check the latter in log-space, each time a parameterized transition appears in the run $q_0 \xrightarrow{u/\alpha}_* q'$, an auxiliary traversal of this run must be started until some transition with the same output parameter is reached.   □

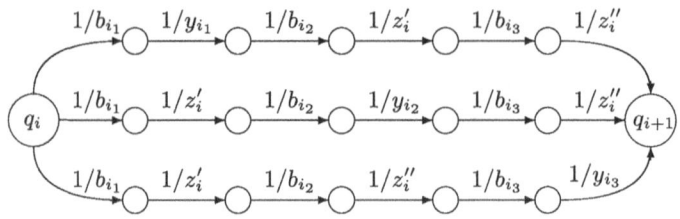

Fig. 1. A PFST Π_i for a clause $D_i = y_{i_1}^{\sigma_1} \vee y_{i_2}^{\sigma_2} \vee y_{i_3}^{\sigma_3}$.

Proposition 2. *The $\exists$-membership for nondeterministic PFSTs is* NP*-complete and the $\forall$-membership problem for nondeterministic PFSTs is* $\mathrm{co} - \mathrm{NP}$*-complete.*

Proof. Obviously, both variants of the membership problem belong to the corresponding complexity classes indicated above. Next we show how 3-SAT problem can be reduced to $\exists$-membership problem for PFSTs.

For every 3-CNF $\varphi = \bigwedge\limits_{i=1}^{n} D_i$ over a finite set of variables $\mathcal{Y}$ we define an output word w_φ of the size $|w_\varphi| = 6n$ and a PFST Π_φ of the size $|\Pi_\varphi| = 18n$ such that a balanced pair $(1^{6n}, w_\varphi)$ is in $TR(\Pi_\varphi\theta)$ iff $\varphi\theta \equiv 1$. For the sake of simplicity we will assume that $\mathcal{Y} = \{y_1, \dots, y_m\}$ and an output alphabet $\Delta = \{b_1, \dots, b_m, 0, 1\}$ has sufficiently many letters (in general case an encoding of variables from $\mathcal{Y}$ with words in the output alphabet Δ is required). In the PFST Π_φ the letters $b_1, \dots, b_m$ are used as "visiting cards" of the Boolean variables in φ, whereas the variables $y_1, \dots, y_m$ act as the output parameters.

For every clause $D_i = y_{i_1}^{\sigma_1} \vee y_{i_2}^{\sigma_2} \vee y_{i_3}^{\sigma_3}$ in CNF φ, where $\sigma_1, \sigma_2, \sigma_3 \in \{0, 1\}$, we introduce a clausal word $w_i = b_{i_1}\sigma_1 b_{i_2}\sigma_2 b_{i_3}\sigma_3$, and a clausal PFST Π_i as depicted on Fig. 1. As it can be seen, a clausal PFST Π_i has exactly 3 runs labeled with pairs of words $(1^6, b_{i_1}y_{i_1}b_{i_2}z_i'b_{i_3}z_i'')$, $(1^6, b_{i_1}z_i'b_{i_2}y_{i_2}b_{i_3}z_i'')$, and $(1^6, b_{i_1}z_i'b_{i_2}z_i''b_{i_3}y_{i_3})$, where z_i' and z_i'' are dummy variables that do not appear anywhere else.

Then the word $w_\varphi = u_1 u_2 \dots u_n$ and a PFST $\Pi_\varphi = \Pi_1; \Pi_2; \dots, \Pi_n$ are defined as a concatenation of clausal words and a sequential composition of clausal transducers respectively for all clauses of φ. It is easy to see that a pair $(1^{6n}, w_\varphi)$ is a label of some complete run of FST $\Pi_\varphi\theta$ iff a substitution θ satisfies every clause $D_i, 1 \le i \le n$, in φ.

As for the $\forall$-membership problem, its co-NP-completeness can be proved in the same way by means of dual constructions. $\qquad\qquad\square$

By making this design of w_φ and Π_φ a little more sophisticated, it can be shown that the membership problem remains NP-complete even for simple PFSTs.

2.2   Non-emptiness

An FST π is non-empty if $Tr(\pi) \neq \emptyset$. It is known (see [12,13]) that the non-emptiness problem is NL-complete for deterministic as well as for non-deterministic FSTs. Since an evaluation of parameters does not influence on the reachability of final states of PFSTs, both the existential and the universal variants of the non-emptiness problem for PFTSs have the same complexity. However, it is worth noticing that the non-emptiness problem for PFSTs may be formulated differently. If the $\forall$-variant of this problem is specified by a predicate $\exists (u,w) \in \Sigma^* \times \Delta^* \, \forall \pi \in \mathcal{F}(\Pi) : (u,w) \in TR(\pi)$ instead of $\forall \pi \in \mathcal{F}(\Pi) \, \exists (u,w) \in \Sigma^* \times \Delta^* : (u,w) \in TR(\pi)$ then, as it was shown in [6], the non-emptiness problem for PFSTs immediately becomes ExpSpace-complete.

2.3   Functionality

An FST π is called *functional* if its transduction relation $TR(\pi)$ is a function, i.e. $\forall (u_1, w_1), (u_2, w_2) \in TR(\pi) : (u_1 = u_2 \Rightarrow w_1 = w_2)$. Of course, functionality checking problem makes sense only for non-deterministic FSTs.

Proposition 3. *The functionality checking problem for FSTs as well as its existential and universal counterparts for PFSTs are NL-complete.*

Proof. Since NL = co-NL, it is more suitable to check that a given FST π is not functional. To this end one may guess, following the definition, a pair of complete runs $q_0 \xrightarrow{u/w'}_* q'$ and $q_0 \xrightarrow{u/w''}_* q''$ on the same input u of the length $|u| \leq 2|\pi|^2$, such that $w'[i] \neq w''[i]$ for some i. Clearly this non-deterministic search can be done in log-space. The $\forall$-functionality of PFSTs can be checked just in the same way by regarding different parameters in $\mathcal{X}$ as different letters.

To show that $\exists$-functionality of PFSTs is in NL one may rely on the following consideration which follows from the definition of functionality of FSTs: a PFST Π has no functional instances iff there exist m pairs $q_0 \xrightarrow{u_i/\alpha_i'}_* q'$, $q_0 \xrightarrow{u_i/\alpha_i''}_* q''$, $1 \leq i \leq m$, of complete runs of Π and m integers $k_1, \ldots, k_m$ such that

- m does not exceed the number of variables in Π,
- the length $|u_i|$ of each run is less than $2|\Pi|^2$,
- $k_i \leq |u_i|$ for every $i, 1 \leq i \leq m$;
- $\alpha'[k_1]$ and $\alpha''[k_m]$ are different output letters from Δ;
- $\alpha_i''[k_i] = \alpha_{i+1}'[k_{i+1}]$ for every $i, 1 \leq i < m$.

These pairs of runs and integers can be guessed and processed one by one in space logarithmic in the size of Π.   □

2.4   Equivalence

Two FSTs π' and π'' are *equivalent* if $TR(\pi') = TR(\pi'')$. It is well known that the equivalence checking problem is NL-complete for deterministic FSTs and it is PSpace-complete for non-deterministic FSTs [12]. And the situation is exactly the same with both variants of the equivalence checking problem for PFSTs.

Proposition 4. *Both variants of equivalence checking problem are* NL-*complete for deterministic PFSTs and* PSpace-*complete for non-deterministic PFSTs.*

Proof. We have already discussed the question of sufficient conditions for the PSpace-completeness of decision problems for PFSTs. Since the equivalence checking problem for non-deterministic FSTs is PSpace-complete it remains the same for non-deterministic PFSTs.

To certify that two deterministic PFSTs Π' and Π'' do not have equivalent instances π' and π'', one can first try to find such an input word u of the length $|u| \leq |\Pi'||\Pi''|$ that only one of the runs of π' and π'' on this word is complete. If such word does not exists, i.e. all transductions $TR(\Pi'\theta)$ and $TR(\Pi''\theta)$ have the same domain, one can next apply the technique that has been used in our study of the $\exists$-functionality problem: in fact we need to verify that the union of Π' and Π'' is a non-deterministic PFST which has no functional instances. The only modification to be made is to combine in every pair $q'_0 \xrightarrow{u_i/\alpha'_i}_* q'$, $q''_0 \xrightarrow{u_i/\alpha''_i}_* q''$, the runs of both Π' and Π'' and to alternate from pair to pair the order in which these PFSTs are considered.

To check that Π' is not $\forall$-equivalent to Π'' it is sufficient to find a pair of input words u_1, u_2 whose lengths do not exceed $2|\Pi'||\Pi''|$ and to show that they satisfy one of the following conditions:

1) only one of the runs $q'_0 \xrightarrow{u_1/\alpha'}_* q'$, $q''_0 \xrightarrow{u_1/\alpha''}_* q''$ is complete, whereas the other is not;

2) both runs $q'_0 \xrightarrow{u_1/\alpha'}_* q'$, $q''_0 \xrightarrow{u_1/\alpha''}_* q''$ are complete, and there exists some $i, 1 \leq i \leq |u_1|$, such that $\alpha'[i] \neq \alpha''[i]$ and $\alpha''[i] \in \Delta$;

1. the runs $q'_0 \xrightarrow{u_1/\alpha'}_* q'$, $q''_0 \xrightarrow{u_1/\alpha''}_* q''$, and $q'_0 \xrightarrow{u_2/\beta'}_* q'$, $q''_0 \xrightarrow{u_2/\beta''}_* q''$ are complete, and $\alpha'[i] \neq \beta'[j]$, $\alpha''[i] = \beta''[j]$ hold for some integers $i, 1 \leq i \leq |u_1|$ and $j, 1 \leq j \leq |u_2|$.

It is easy to see that all these checkings can be performed by a non-deterministic algorithm in logarithmic space. □

2.5   Minimization

The decision version of the minimization problem for FSTs is defined as follows: given an FST π and an integer k, check if there exists such an equivalent FST π' that $|\pi'| \leq k$. It is known that the minimization problem is NL-complete for deterministic FSTs (see [7]) and it is PSpace-complete for non-deterministic FSTs. Although the equivalence checking and minimization problems for deterministic finite automata are closely related and a solution to one of them gives a solution to the other with small overheads, this is not the case for PFSTs.

Proposition 5. *For deterministic PFSTs the* $\exists$-*minimization problem is* NP-*complete, and* $\forall$-*minimization problem is* co − NP-*complete.*

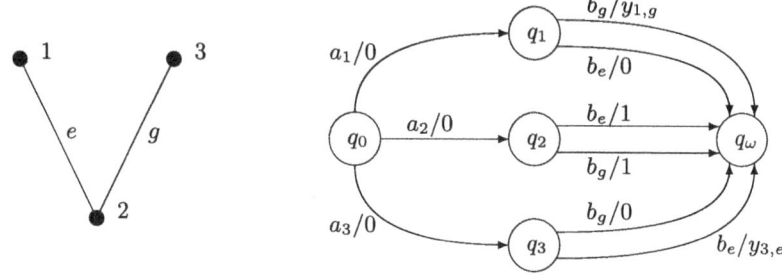

Fig. 2. An example of a graph G and the corresponding PFST Π_G.

Proof. We show how it is possible to reduce graph coloring problem to the ∃-minimization problem for deterministic PFSTs provided that quite a lot of input letters are available to indicate the nodes and edges of the graph. In a rigorous proof for a fixed input alphabet Σ, a plain encoding technique is required to compensate for insufficient number of letters.

For an arbitrary graph $G = (V, E)$ we build a deterministic PFST Π_G which operates over an input alphabet $\Sigma = \Sigma_V \cup \Sigma_E$ and an output alphabet $\Delta = \{0, 1\}$. Variables from the set $\mathcal{Y} = \{y_{v,e} : v \in V, e \in E\}$ are used as parameters. The letters in $\Sigma_V = \{a_v : v \in V\}$ are associated with nodes of G, and the letters in $\Sigma_E = \{b_e : e \in E\}$ are related with edges of E. A PFST Π_G has an initial state q_0, a final state q_ω, and $|V|$ distinguished states $q_v, v \in V$, that are in a one-to-one correspondence with the nodes of G. A transition function of Π_G is defined as follows:

– for every input letter $a_v \in \Sigma_V$ there is a transition $q_0 \xrightarrow{a_v/0} q_v$;
– for every input letter $b_e \in \Sigma_E$, where $e = (v_1, v_2) \in E$, there are two transitions $q_{v_1} \xrightarrow{b_e/0} q_\omega$, $q_{v_2} \xrightarrow{b_e/1} q_\omega$ labeled with output letters 0 and 1, and $|V| - 2$ transitions $q_v \xrightarrow{b_e/y_{v,e}} q_\omega$ for all other states $q_v, v \neq v_1, v \neq v_2$, labeled with fresh variables $y_{v,e} \in \mathcal{Y}$.

An example of such PFST is depicted on Fig. 2. A remarkable feature of Π_G is that it is acyclic and has simple parameterization. It should be noticed also that for every ground substitution $\theta \in GSubst$ it is true that

1) the labeling function $F_{G,\theta} : V \to 2^E$ such that

$$F_{G,\theta}(v) = \{e : q_v \xrightarrow{b_e/1} q_\omega \text{ is a transition in FST } \Pi_G\theta\},$$

is a correct coloring of nodes in G;
2) for every pair of nodes v_1, v_2 the states q_{v_1} and q_{v_2} are equivalent in FST $\Pi_G\theta$ iff $F_{G,\theta}(v_1) = F_{G,\theta}(v_2)$.

A simple parameterization of Π_G guarantees that any admissible correct coloring of nodes of G may be defined by means of an appropriate substitution θ. These considerations confirm the correctness of the described reduction.   □

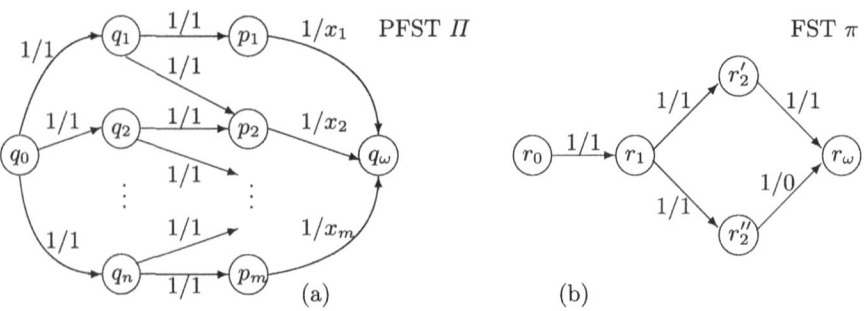

Fig. 3. A PFST Π which simulate the set splitting problem

2.6   Bisimilarity

A bisimulation relation $B \subseteq Q' \times Q''$ between two FSTs $\pi' = (Q', q_0', F', \longrightarrow)$ and $\pi'' = (Q'', q_0'', F'', \leadsto)$ is defined as usual: for every pair $(q', q'') \in B$

1) $q' \in F' \iff q'' \in F''$;

2) for every transition $q' \xrightarrow{a/b} p'$ in π' there exists a transition $q'' \overset{a/b}{\leadsto} p''$ in π'' such that $(p', p'') \in B$;

3) for every transition $q'' \overset{a/b}{\leadsto} p''$ in π'' there exists a transition $q' \xrightarrow{a/b} p'$ in π' such that $(p', p'') \in B$.

FSTs π' and π'' are *bisimilar* ($\pi' \sim \pi''$ in symbols) iff there exists such a bisim- ulation B between these FSTs that $(q_0', q_0'') \in B$. It is known (see [4]) that the bisimilarity checking problem for finite state automata (and, hence, for FSTs) is P-complete. But unlike the case of equivalence checking the complexity of bisimilarity checking is not preserved when moving from FSTs to PFSTs.

Proposition 6. *The $\exists$-bisimilarity checking problem for PFSTs is NP-complete, and $\forall$-bisimilarity checking problem for PFSTs is $\prod_2^p$-complete.*

Proof. We show that the set splitting problem which is known to be NP-complete (see [8]) is reducible to the $\exists$-bisimilarity checking problem for PFSTs. The set splitting problem is that of checking, given a family of finite sets $D_1, D_2, \ldots, D_n$, the existence of a partition of the union $U = \bigcup_{i=1}^{n} D_i$ into two subsets U_0 and U_1 such that $U_0 \cap D_i \neq \emptyset$ and $U_1 \cap D_i \neq \emptyset$ hold for every $D_i, 1 \leq i \leq n$.

Given a family of sets $D_1, D_2, \ldots, D_n$ such that $U = \bigcup_{i=1}^{n} D_i = \{e_1, \ldots, e_m\}$, define a simple acyclic PFST Π as depicted on Fig. 3(a). It operates over an input alphabet $\Sigma = \{1\}$ and an output alphabet $\Delta = \{0, 1\}$. In addition to the initial and the final states q_0 and q_ω, Π has n states $p_1, \ldots, p_n$ corresponding to the sets of this family, and m states $q_1, \ldots, q_m$ corresponding to the elements of U. The PFST Π has a transition $q_i \xrightarrow{1/1} p_j$ iff $e_j \in D_i$. All transitions $p_j \xrightarrow{1/x_j} q_\omega$ are labeled with pairwise different variables $x_j, 1 \leq j \leq m$.

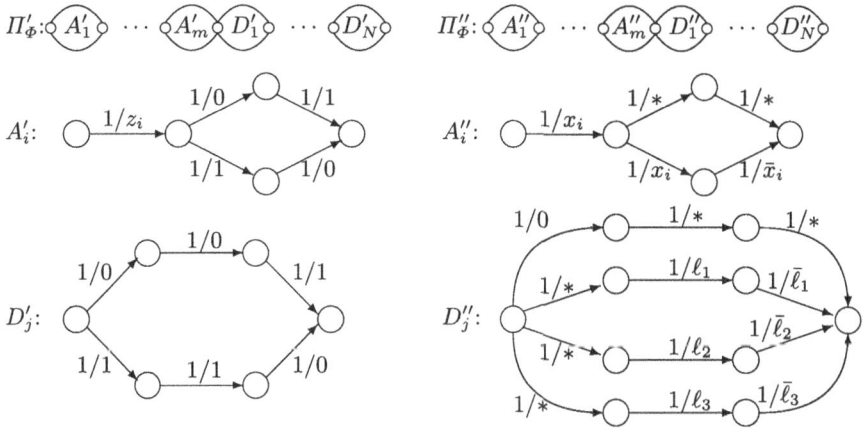

Fig. 4. PFSTs Π'_{Φ} and Π''_{Φ} corresponding to a QBF Φ

It is easy to see that for any ground substitution $\theta : \mathcal{X} \rightarrow \{0,1\}$ the FST $\Pi\theta$ is bisimilar to an FST π depicted on Fig. 3(b) iff θ provides a desirable partition of U: $U_0 = \{e_j : \theta(x_j) = 0\}$ and $U_1 = \{e_j : \theta(x_j) = 1\}$.

The belonging of $\forall$-bisimilarity checking problem for PFSTs to the complexity class $\prod_2^p$ obviously follows from the formulation of the problem and from the fact that bisimilarity checking for FSTs is P-complete.

Next we show that the satisfiability checking problem for quantified Boolean formulae with prefixes of the type $\forall\exists$ is reducible to the $\forall$-bisimilarity checking problem for PFSTs. Let $\Phi = \forall x_1 \ldots \forall x_m \exists x_{m+1} \ldots \exists x_{m+n}(D_1 \wedge D_2 \wedge \cdots \wedge D_N)$ be a $\forall\exists$-QBF, where every $D_i = \ell_1 \vee \ell_2 \vee \ell_3, 1 \leq i \leq N$, is a 3-literal clause. Consider a pair of PFSTs Π'_{Φ} and Π''_{Φ} depicted on Fig. 4. These PFSTs operate over an input alphabet $\Sigma = \{1\}$ and an output alphabet $\Delta = \{0,1\}$. Each of them is a concatenation of $m + N$ blocks.

PFST Π'_{Φ} has a set of m parameters $z_1, \ldots, z_m$. Using these parameters one can assign arbitrary values to those variables in Φ which are quantified universally. PFST Π''_{Φ} is supplied with $2(m+n)$ parameters $x_1, \ldots, x_{m+n}, \bar{x}_1, \ldots \bar{x}_{m+n}$ corresponding to all literals of Φ as well as with $2n + 5N$ dummy variables. Every dummy variable occurs in Π''_{Φ} only once, and, therefore, each of them is indicated on Fig. 4 by a sign $*$ common to all such variables. A notation $\bar{\ell}_j$ assumes that if ℓ_j is a negative literal $\bar{x}_k$ for some x_k then $\bar{\ell}_j = x_k$.

As can be seen from the composite structure of PFSTs Π'_{Φ} and Π''_{Φ}, the bisimilarity $\Pi'_{\Phi}\theta' \sim \Pi''_{\Phi}\theta''$ holds for some ground substitutions θ', θ'' iff all their components $A'_i\theta' \sim A''_i\theta''$, $1 \leq i \leq m$, and $D'_j\theta' \sim D''_j\theta''$, $1 \leq j \leq N$, are pairwise bisimilar. Bisimilarity of $A'_i\theta'$ and $A''_i\theta''$ holds iff $z_i\theta' = x_i\theta''$ and $x_i\theta'' = \neg\bar{x}_i\theta''$. Bisimilarity of $D'_j\theta'$ and $D''_j\theta''$ holds iff $\ell_k\theta'' = 1$ for some literal $\ell_k, k = 1, 2, 3$, in the clause $D_j = \ell_1 \vee \ell_2 \vee \ell_3$. So we come to the conclusion that PFSTs Π'_{Φ} and Π''_{Φ} are $\forall$-bisimilar iff for every evaluation of variables $x_1, \ldots, x_m$ there exists some evaluation of variables $x_{m+1}, \ldots, x_n$ for which every clause D_j in CNF $D_1 \wedge D_2 \wedge \cdots \wedge D_N$ takes value 1. □

3 Conclusion

The results obtained brought us to some conclusions about the complexity of analyzing the behavior of parameterized reactive systems modeled by finite state transducers and the possibility of constructing efficient decision algorithms.

1. When we add parameters to the model of finite state machine, the computational complexity of decision problems increases (as would be expected), but not as dramatically as was shown in [6] (see Table 2).

Table 2. Complexity of decision problems for PFSTs

	Deterministic PFSTs		Nondeterministic PFSTs	
Problems	∃-variant	∀-variant	∃-variant	∀-variant
Membership	L-comp.	L-comp.	NP-comp.	coNP-comp.
Non-emptiness	NL-comp.	NL-comp.	NL-comp.	NL-comp.
Functionality	—	—	NL-comp.	NL-comp.
Bisimilarity	—	—	NP-comp.	Π_2^p-comp.
Equivalence	NL-comp.	NL-comp.	PSpace-comp.	PSpace-comp.
Minimization	NP-comp.	coNP-comp.	PSpace-comp.	PSpace-comp.

This is partly due to the fact that the formulations of the existential and universal versions of the decision problems for parameterized transducers that we deal with are somewhat different from the similar problems studied in [6].

2. It is shown that for deterministic automata, adding parameters to the model does not lead to a significant change in the complexity of some problems. Thus, the proposed transition from regular expressions and finite state acceptors to a two-level model of finite state transducers turned out to be justified: it allows us to identify classes of parameterized automata for which some problems of behavior analysis remain to be tractable.

3. It is interesting to note that three problems—equivalence checking and minimization for deterministic machines, as well as bisimilarity checking for non-deterministic machines—which for ordinary finite state automata can be solved by similar set splitting algorithms in almost the same time $O(n \log n)$, for parameterized transducers fall into completely different complexity classes. Parameterization of automata emphasizes once again the diverse combinatorial nature of these problems.

In what follows we intend to continue our study of parameterized finite state machines in several directions. It is reasonable to extend the same concept of output data parameterization to asynchronous finite state string and tree transducers, real-time finite state machines and some other models that exhibit more complex behavior than synchronous automata, but at the same time admit to operate efficiently in the subclass of deterministic models.

For data parameterized automata, new problems also arise, such as, e.g., unification—computing the most general common instance of several parameterized models. The low computational complexity of the equivalence checking problems for deterministic PFSTs gives reason to believe that the unification problem for some classes of parameterized models can have an effective solution. Efficient unification algorithms are known to have many applications. Unification of parameterized finite state machines could be used, for example, to provide conformance between alternative drafts of reactive systems (controllers, drivers, etc.) at different stages of their development.

And finally, the question of how stable the class of proposed data parameterized automata is relative to natural composition operations deserves attention. A model of computation cannot be recognized perfect if it is not closed with respect to such operations, and from this point of view our model of PFSTs is imperfect. If we consider the sequential composition $\pi(\Pi)$ of a PFST Π and an FST π, where the latter takes the output stream of the former as its input, then an unfamiliar parameterized automaton is formed in which the parameters are not represented explicitly, but influence on system behavior in a hidden way by imposing some kind of entanglement on certain pairs of transitions. The effect of this kind is known in quantum physics as Einstein–Podolsky–Rosen paradox (see [18]) which reveals the incompleteness of quantum physics theory, and the study of its manifestation in models of data parameterized automata can also be one of the directions for further research of this model of computations.

The authors would like to thank the anonymous reviewers for their valuable comments and advice.

References

1. Alur R., Henzinger T.A., Vardi M.Y.: Parametric real-time reasoning. In: Proceedings of the 25-th Annual ACM Symposium on Theory of Computing (STOC), pp. 592–601 (1993)
2. André, É., Lime, D., Roux, O.H.: Decision problems for parametric timed automata. In: Ogata, K., Lawford, M., Liu, S. (eds.) ICFEM 2016. LNCS, vol. 10009, pp. 400–416. Springer, Cham (2016). https://doi.org/10.1007/978-3-319-47846-3_25
3. Andre, E.: What's decidable about parametric timed automata. Int. J. Softw. Tools Technol. Transf. **21**, 203–219 (2019)
4. Balcazar, J.L., Gabarro, J., Santha, M.: Deciding bisimilarity is P-complete. Formal Aspects Comput. **4**, 638–648 (1992)
5. Belkhir, W., Chevalier, Y., Rusinovitch, M.: Parametrized automata simulation and application to service composition. J. Symb. Comput. **69**, 40–60 (2015)
6. Barcelo, P., Reutter, J., Libkin, L.: Parameterized regular expressions and their languages. Theoret. Comput. Sci. **474**, 21–45 (2013)
7. Cho, S., Huynh, D.T.: The parallel complexity of finite-state automata problems. Inf. Comput. **97**, 1–22 (1992)

8. Garey, M., Johnson, D.: Computers and Intractability: A Guide to the Theory of NP-Completeness. Freeman, San Francisco (1979)

9. Grumberg O., Kupferman O., Sheinvald S.: Variable automata over infinite alphabets. In: Proceedings of the 4-th International Conference on Language and Automata Theory and Applications (LATA), pp. 561–572 (2010)

10. Grumberg, O., Kupferman, O., Sheinvald, S.: An automata-theoretic approach for reasoning about parameterized systems and specifications. In: Proceedings of the 11-th International Symposium on Automated Technology for Verification and Analysis (ATVA), pp. 397–411 (2013)

11. Goller, S., Hillair, M.: Reachability in two-parametric timed automata with one parameter is EXPSPACE-complete. In: Proceedings of the 38-th International Symposium on Theoretical Aspects of Computer Science (STACS), pp. 36:1–36:18 (2023)

12. Holzer, M., Kutrib, M.: Descriptional and computational complexity of finite automata - a survey. Inf. Comput. **209**, 456–470 (2011)

13. Jones, N.: Space-bounded reducibility among combinatorial problems. J. Comput. Syst. Sci. **11**, 68–85 (1975)

14. Kaminski, M., Francez, N.: Finite-memory automata. Theoret. Comput. Sci. **134**, 329–363 (1994)

15. Kaminski, M., Zeitlin, D.: Finite-memory automata with non-deterministic reassignment. Int. J. Found. Comput. Sci. **21**, 741–760 (2010)

16. Lenhardt, R.: Probabilistic automata with parameters. Master's Thesis, University of Oxford, p. 63 (2009)

17. Shemesh, Y., Francez, N.: Finite state unification automata and relational languages. Inf. Comput. **114**, 192–213 (1994)

18. Wiseman, H.M., Jones, S.J., Doherty, A.C.: Steering, entanglement, nonlocality, and the Einstein–Podolsky–Rosen paradox. Phys. Rev. Lett. **98** (2007)

Measuring Power of Commutative Group Languages

Takao Yuyama[1]([⊠]) and Ryoma Sin'ya[2]

[1] Research Institute for Mathematical Sciences, Kyoto University, Kyoto, Japan
yuyama@kurims.kyoto-u.ac.jp
[2] Akita University, Akita, Japan
ryoma@math.akita-u.ac.jp

Abstract. A language L is said to be $\mathcal{C}$-measurable, where $\mathcal{C}$ is a class of languages, if there is an infinite sequence of languages in $\mathcal{C}$ that "converges" to L. In this paper, we investigate the measuring powers of Gcom of the class of all languages recognised by finite commutative groups and its subclass named MOD. A language is in MOD if membership of a word in the language only depends on its length modulo some fixed integer. In particular, we show that, for a given regular language L, it is decidable whether L is Gcom-measurable (MOD-measurable, respectively) or not. Our results demonstrate that there is a huge gap between the expressive power of group languages and commutative group languages, even from a (very rough) measure theoretic point of view.

1 Introduction

The notion of $\mathcal{C}$-measurability for a class $\mathcal{C}$ of languages is introduced by [6] and it was used for classifying non-regular languages by using regular languages. A language L is said to be $\mathcal{C}$-measurable if there is an infinite sequence of languages in $\mathcal{C}$ that converges to L. Also, the $\mathcal{C}$-measurability can be defined by using so-called *Carathéodory extension* [7], a purely measure theoretic notion. Roughly speaking, L is $\mathcal{C}$-measurable means that it can be approximated by a language in $\mathcal{C}$ with arbitrary high precision: the notion of "precision" is formally defined by the *density* of formal languages. Hence that a language L is not $\mathcal{C}$-measurable ($\mathcal{C}$-*im*measurable) means that L has a complex shape so that it can not be approximated by languages in $\mathcal{C}$.

So far, the decidability and different characterisation of $\mathcal{C}$-measurability is systematically studied for some subclass $\mathcal{C}$ of star-free languages. For example, two simple and decidable characterisations of PT-measurable and AT-measurable languages are given in [8], where PT is the class of all piecewise testable languages and AT is the class of all alphabet testable languages (*i.e.*, languages definable by the first-order logic with only one variable). While the AT-measurability is strictly weaker than PT-measurability, it is interesting that the computational complexity of the AT-measurability is much higher than PT-measurability: deciding the AT-measurability of a given regular language L (represented by a deterministic automaton) is *PSPACE-complete* but deciding the PT-measurability of

© The Author(s), under exclusive license to Springer Nature Switzerland AG 2024
S. Z. Fazekas (Ed.): CIAA 2024, LNCS 15015, pp. 347–362, 2024.
https://doi.org/10.1007/978-3-031-71112-1_25

L can be done *in linear time* with respect to the number of states [10]. Further-more, in [8,9] it was shown that the GD-measurability and UPol-measurability are *equivalent*, where GD is the class of all generalised definite languages (*i.e.*, languages that can be defined by a finite Boolean combination of prefix and suffix tests of some bounded length) and UPol is the class of all unambiguous polyno-mials (*i.e.*, languages definable by the first-order logic with only two variables). Some properties of the measuring power of star-free languages is investigated in [7], but the decidability for regular languages is still unknown.

On the other hand, for the measuring power of *group languages*, nothing is yet known. A language is said to be a group language if its syntactic monoid is a finite group, equivalently, it can be recognised by a permutation automaton [11]. Although the definition of group languages is very simple from an algebraic point of view, this class dose not have a language theoretic nor logical different characterisation. To borrow the phrase used by Place–Zeitoun [4]: "This makes it difficult to get an intuitive grasp about group languages, which may explain why this class remains poorly understood."

This paper addresses the first development step for understanding the mea-suring power of group languages. In this paper, we investigate the measuring power of G the class of all group languages, Gcom the class of commutative group languages and MOD a subclass of Gcom (see Sect. 2 for the precise definition). The main results of this paper are three kinds.

(1) Every group language whose syntactic monoid is non-commutative is Gcom-*im*measurable (Theorem 1).
(2) A decidable characterisation of the Gcom-measurability for regular languages (Theorem 2).
(3) A simple decidable characterisation of the MOD-measurability (Theorem 3).

2   Preliminaries

This section provides the precise definitions of density, measurability, group lan-guages and its two subclasses. REG_A denotes the family of all regular languages over an alphabet A. For a word $w \in A^*$ and a letter $a \in A$, we write $|w|_a$ the number of occurrences of a in w. The complement of L over A is denoted by $L^c = A^* \backslash L$. We assume that the reader has a standard knowledge of algebraic language theory (*cf.* [2,3]).

The main targets of this paper are the following three subclasses of regular languages. The *group languages* G, the *commutative group languages* Gcom and the *modulo languages* MOD:

$$\mathsf{G}_A \stackrel{\text{def}}{=} \{\, L \subseteq A^* \mid L \text{ is recognised by a finite group} \,\}$$

$$\mathsf{Gcom}_A \stackrel{\text{def}}{=} \{\, L \subseteq A^* \mid L \text{ is recognised by a finite commutative group} \,\}$$

$$\mathsf{MOD}_A \stackrel{\text{def}}{=} \left\{\, L \subseteq A^* \,\middle|\, \begin{array}{l} L \text{ is recognised by a morphism } \eta \colon A^* \to G \text{ into a} \\ \text{finite group } G \text{ such that } \eta(a) = \eta(b) \text{ for all } a, b \in A \end{array} \right\}$$

A monoid M divides a monoid N if M is a monoid homomorphic image of a submonoid of N. It is well-known that a monoid M recognises a language L if and only if the syntactic monoid M_L of L divides M (cf. [2, Theorem 10.2.6]), hence L is in G_A (Gcom_A, respectively) if and only if M_L is a finite group (a finite commutative group, respectively). For $a \in A$, $q, r \in \mathbb{N}$ such that $r < q$, we let

$$L_{q,r}^a = \{\, w \in A^* \mid |w|_a \equiv r \mod q \,\} \text{ and } L_{q,r} = \{\, w \in A^* \mid |w| \equiv r \mod q \,\}.$$

It is also known that Gcom_A (MOD_A, respectively) is the smallest Boolean algebra containing all languages of $L_{q,r}^a$ ($L_{q,r}$, respectively) (cf. [3,4]). Hereafter, we only consider an alphabet A such that $\#(A) \geq 2$ (because $\mathsf{MOD}_A = \mathsf{Gcom}_A = \mathsf{G}_A$ if $\#(A) = 1$) and sometimes omit the subscript A for denoting these classes of languages.

2.1 Density and Measurability of Formal Language

For a set X, we denote by $\#(X)$ the cardinality of X. We denote by $\mathbb{N}$ the set of natural numbers including 0.

Definition 1 (cf [1]). *The* density $\delta_A(L)$ *of* $L \subseteq A^*$ *is defined as*

$$\delta_A(L) \stackrel{\text{def}}{=} \lim_{n \to \infty} \frac{1}{n} \sum_{k=0}^{n-1} \frac{\#(L \cap A^k)}{\#(A^k)}$$

if it exists, otherwise we write $\delta_A(L) = \bot$. *The language* L *is called* null *if* $\delta_A(L) = 0$, *and dually,* L *is called* co-null *if* $\delta_A(L) = 1$.

Example 1. It is known that every regular language has a rational density (cf. [1]) and it is computable. For each word w, the language $A^* w A^*$, the set of all words that contain w as a factor, is of density one (co-null). This fact follows from the so-called *infinite monkey theorem*: take any word w. A random word of length n contains w as a factor with probability tending to 1 as $n \to \infty$.

We list some basic properties of the density as follows.

Lemma 1. *Let* $K, L \subseteq A^*$ *with* $\delta_A(K) = \alpha, \delta_A(L) = \beta$. *Then we have:*
(1) $\delta_A(L \backslash K) = \beta - \alpha$ *if* $K \subseteq L$. *(2)* $\delta_A(K^c) = 1 - \alpha$. *(3)* $\delta_A(K \cup L) = \alpha + \beta$ *if* $K \cap L = \varnothing$.

Lemma 2. *Let* $\eta \colon A^* \to G$ *be a morphism onto a finite group. For any* $g \in G$, *we have* $\delta_A(\eta^{-1}(g)) = \#(G)^{-1}$.

For more detailed properties of δ_A, see Chapter 13 of [1].

The notion of "measurability" on formal languages is defined by a standard measure theoretic approach as follows.

Definition 2 ([6]). *Let $\mathcal{C}_A$ be a family of languages over A. For a language $L \subseteq A^*$, we define its $\mathcal{C}_A$-inner-density $\underline{\mu}_{\mathcal{C}_A}(L)$ and $\mathcal{C}_A$-outer-density $\overline{\mu}_{\mathcal{C}_A}(L)$ over A as*

$$\underline{\mu}_{\mathcal{C}_A}(L) \stackrel{\text{def}}{=} \sup\{\,\delta_A(K) \mid K \subseteq L, K \in \mathcal{C}_A, \delta_A(K) \neq \bot\,\} \quad and$$

$$\overline{\mu}_{\mathcal{C}_A}(L) \stackrel{\text{def}}{=} \inf\{\,\delta_A(K) \mid L \subseteq K, K \in \mathcal{C}_A, \delta_A(K) \neq \bot\,\}, \quad respectively.$$

A language L is said to be $\mathcal{C}_A$-measurable if $\underline{\mu}_{\mathcal{C}_A}(L) = \overline{\mu}_{\mathcal{C}_A}(L)$ holds. We say that an infinite sequence $(L_n)_n$ of languages over A converges to L from inner (from outer, respectively) if $L_n \subseteq L$ ($L_n \supseteq L$, respectively) for each n and $\lim_{n\to\infty} \delta_A(L_n) = \delta_A(L)$.

The following is an example of Gcom-measurable non-regular language.

Example 2 ([6]). The semi-Dyck language $D = \{\varepsilon, ab, aabb, abab, aaabbb, \ldots\}$ over $A = \{a, b\}$ is Gcom-measurable. For each $k \geq 2$, the language $L_k = \{w \in A^* \mid |w|_a = |w|_b \mod k\}$ is in Gcom, because $\eta^{-1}(0) = L_k$ holds for the morphism $\eta \colon A^* \to \mathbb{Z}/k\mathbb{Z}$ where $h(a) = 1$ and $h(b) = k - 1$. Obviously, $D \subseteq L_k$ holds and it follows from Lemma 2 that $\delta_A(L_k) = 1/k$ holds. Hence $\delta_A(L_k)$ tends to zero if k tends to infinity.

For a family $\mathcal{C}_A$ of languages over A, we denote by $\text{Ext}_A(\mathcal{C}_A)$ ($\text{RExt}_A(\mathcal{C}_A)$, respectively) the class of all $\mathcal{C}_A$-measurable languages ($\mathcal{C}_A$-measurable *regular* languages, respectively) over A.

Lemma 3 ([7]). *The operator Ext_A is a closure, i.e., it satisfies the following three properties for each $\mathcal{C} \subseteq \mathcal{D} \subseteq 2^{A^*}$: (extensive) $\mathcal{C} \subseteq \text{Ext}_A(\mathcal{C})$, (monotone) $\text{Ext}_A(\mathcal{C}) \subseteq \text{Ext}_A(\mathcal{D})$, and (idempotent) $\text{Ext}_A(\text{Ext}_A(\mathcal{C})) = \text{Ext}_A(\mathcal{C})$. Moreover, $\text{Ext}_A(\mathcal{C}_A)$ is closed under Boolean operations and quotients if $\mathcal{C}_A$ is closed under Boolean operations and quotients.*

2.2 Semilinear Sets and Some Background from Group Theory

Let $A = \{a_1, \ldots, a_d\}$. The *Parikh mapping* $\text{Pkh}_A \colon A^* \to \mathbb{N}^d$ is defined by $\text{Pkh}_A(w) \stackrel{\text{def}}{=} (|w|_{a_1}, \ldots, |w|_{a_d})$. This can be naturally extended to a map taking a language L over A: $\text{Pkh}_A(L) = \{\,\text{Pkh}_A(w) \mid w \in L\,\}$. A set $S \subseteq \mathbb{N}^d$ is called *linear* if S is of the form

$$S = \{\,\boldsymbol{c} + x_1\boldsymbol{p}_1 + \cdots + x_k\boldsymbol{p}_k \mid x_i \in \mathbb{N} \text{ for each } i\,\}$$

for some $k \in \mathbb{N}$ and some vectors $\boldsymbol{c}, \boldsymbol{p}_1, \ldots, \boldsymbol{p}_k \in \mathbb{N}^d$. In this case, we call $\boldsymbol{c}$ a constant vector and $\boldsymbol{p}_1, \ldots, \boldsymbol{p}_k$ are period vectors of (this representation of) S. A set $S \subseteq \mathbb{N}^d$ is called *semilinear* if it is a finite union of linear sets. It is well-known that (i) every regular language has a semilinear Parikh image, and (ii) the complement of a semilinear set is also semilinear (*cf.* [5]).

Definition 3 (span). *For each* $\mathbb{K} \in \{\mathbb{N}, \mathbb{Z}, \mathbb{Q}, \mathbb{R}\}$ *and* $\boldsymbol{p}_1, \ldots, \boldsymbol{p}_r \in \mathbb{K}^r$, *we define*

$$\mathrm{span}_{\mathbb{K}}(\boldsymbol{p}_1, \ldots, \boldsymbol{p}_r) \stackrel{\mathrm{def}}{=} \{\, x_1 \boldsymbol{p}_1 + \cdots + x_r \boldsymbol{p}_r \mid x_1, \ldots, x_r \in \mathbb{K} \,\}.$$

With this notation, a linear set $S \subseteq \mathbb{N}^d$ *is written as* $S = \boldsymbol{c} + \mathrm{span}_{\mathbb{N}}(\boldsymbol{p}_1, \ldots, \boldsymbol{p}_r)$.

A finitely generated commutative group M is called *free* if M is isomorphic to $\mathbb{Z}^r$ for some $r \in \mathbb{N}$. This r is called the *rank* of M, denoted by $\mathrm{rank}(M)$. It is known that every subgroup M of $\mathbb{Z}^r$ is also free of rank $\leq r$.

Proposition 1. *Let* $r \in \mathbb{N}$ *and* $M \subseteq \mathbb{Z}^r$ *be a subgroup such that* $\mathrm{rank}(M) = r$. *Then there exists* $N \in \mathbb{N}$ *such that* $N\mathbb{Z}^r \subseteq M$. *In particular, the index* $(\mathbb{Z}^r : M)$ *is finite.*

Proof. Let $\{\boldsymbol{e}_1, \ldots, \boldsymbol{e}_r\}$ be the standard basis of $\mathbb{Z}^r$. Since $\mathrm{rank}(M) = r$, there exists a free basis $\{\boldsymbol{p}_1, \ldots, \boldsymbol{p}_r\}$ of M such that $M = \mathrm{span}_{\mathbb{Z}}(\boldsymbol{p}_1, \ldots, \boldsymbol{p}_r)$. If we put $M_{\mathbb{Q}} = \mathrm{span}_{\mathbb{Q}}(\boldsymbol{p}_1, \ldots, \boldsymbol{p}_r)$, then $\dim_{\mathbb{Q}}(M_{\mathbb{Q}}) = r$, i.e., $M_{\mathbb{Q}} = \mathbb{Q}^r \supseteq \mathbb{Z}^r$. Therefore there exist $c_{i,j} \in \mathbb{Z}$ $(1 \leq i, j \leq r)$ and $N \in \mathbb{N}$ such that

$$\boldsymbol{e}_i = \frac{c_{i,1}}{N} \boldsymbol{p}_1 + \cdots + \frac{c_{i,r}}{N} \boldsymbol{p}_r$$

for each $i = 1, \ldots, r$. Thus $N\mathbb{Z}^r = \mathrm{span}_{\mathbb{Z}}(N\boldsymbol{e}_1, \ldots, N\boldsymbol{e}_r) \subseteq M$ and $(\mathbb{Z}^r : M) \leq (\mathbb{Z}^r : N\mathbb{Z}^r) = N^r < \infty$. $\qquad\square$

Definition 4 (rank and index of coset). *For every linear set* $\Lambda \subseteq \mathbb{Z}^r$, *there exist* $\boldsymbol{c} \in \mathbb{Z}^r$ *and* $\boldsymbol{p}_1, \ldots, \boldsymbol{p}_\mu \in \mathbb{Z}^r$ *such that* $\Lambda = \boldsymbol{c} + \mathrm{span}_{\mathbb{Z}}(\boldsymbol{p}_1, \ldots, \boldsymbol{p}_\mu)$. *Then* $M = \mathrm{span}_{\mathbb{Z}}(\boldsymbol{p}_1, \ldots, \boldsymbol{p}_\mu)$ *is a subgroup of* $\mathbb{Z}^r$, *and we write*

$$\mathrm{rank}(\Lambda) \stackrel{\mathrm{def}}{=} \mathrm{rank}(M), \qquad\qquad (\mathbb{Z}^r : \Lambda) \stackrel{\mathrm{def}}{=} (\mathbb{Z}^r : M).$$

Note that $\mathrm{rank}(\Lambda)$ *and* $(\mathbb{Z}^r : \Lambda)$ *do not depend on the choice of* $\boldsymbol{c}, \boldsymbol{p}_1, \ldots, \boldsymbol{p}_\mu$. *More generally, we write* $(G : gH) \stackrel{\mathrm{def}}{=} (G : H)$ *for a group* G, *a subgroup* H, *and an element* $g \in G$.

Lemma 4. *Let* G *be a group,* $H, K \subseteq G$ *be subgroups, and* $a, b \in G$. *If* $aH \cap bK \neq \varnothing$, *then* $aH \cap bK = x(H \cap K)$ *for any* $x \in aH \cap bK$.

Proof. If $x \in aH \cap bK$, then $xH = aH$ and $xK = bK$. Thus $aH \cap bK = xH \cap xK = x(H \cap K)$. $\qquad\square$

Thanks to Lemma 4, we can write $\mathrm{rank}(\Lambda_1 \cap \Lambda_2)$ consistently.

Proposition 2. *Let* $r \in \mathbb{N}$, $M \subseteq \mathbb{Z}^r$ *be a subgroup, and* $\pi_{n!} \colon \mathbb{Z}^r \to (\mathbb{Z}/n!\mathbb{Z})^r$ *be the natural surjection. Then*

$$\mathrm{rank}(M) < r \implies \lim_{n \to \infty} ((\mathbb{Z}/n!\mathbb{Z})^r : \pi_{n!}(M)) = \infty.$$

For the proof of Proposition 2, see the full version [12] of this paper.

For a group G, the *commutator* of two elements $x, y \in G$ is defined as $[x, y] \overset{\text{def}}{=}$ $xyx^{-1}y^{-1}$. Note that $[x, y]$ is the identity element if and only if x and y commute. The *commutator subgroup* of G, denoted by $[G, G]$, is the subgroup generated by all the commutators of G. Note that $[G, G]$ is a normal subgroup of G since $z[x, y]z^{-1} = [zxz^{-1}, zyz^{-1}]$ for all $x, y, z \in G$. The *abelianisation* of G is defined as the quotient group $G^{\mathrm{ab}} \overset{\text{def}}{=} G/[G, G]$.

3   (Counter)examples of **Gcom**-Measurable and **MOD**-Measurable Languages

To grasp an intuition of Gcom-measurability and MOD-measurability, first we examine concrete examples of Gcom-measurable and MOD-measurable languages.

Proposition 3. *Let* FIN *be the family of all finite and co-finite languages over* A, Com *be the all languages whose syntactic monoid is a finite commutative monoid, and* AT *be the Boolean algebra generated by the languages of the form* $A^* a A^*$.

(1) $\mathrm{Ext}_A(\mathsf{FIN}) \subsetneq \mathrm{Ext}_A(\mathsf{MOD})$.
(2) $\mathrm{Ext}_A(\mathsf{AT}) \subsetneq \mathrm{Ext}_A(\mathsf{Gcom})$.
(3) $\mathrm{Ext}_A(\mathsf{Gcom}) = \mathrm{Ext}_A(\mathsf{Com})$.

Proof. We first show the strictness of (1) and (2). Every language in $\mathrm{Ext}_A(\mathsf{FIN})$ or $\mathrm{Ext}_A(\mathsf{AT})$ is null or co-null since every language in FIN or AT is so. The language $L_{2,0}$ of even-length words belongs to MOD and has density $\delta_A(L_{2,0}) = 1/2$ by Lemma 2. Thus $L_{2,0} \in \mathrm{Ext}_A(\mathsf{MOD}) \backslash \mathrm{Ext}_A(\mathsf{FIN})$ and $L_{2,0} \in \mathrm{Ext}_A(\mathsf{Gcom}) \backslash \mathrm{Ext}_A(\mathsf{AT})$.

(1) Let $F \subseteq A^*$ be a non-empty finite language. Define $\ell_F = \{\, |w| \mid w \in F \,\} \subseteq \mathbb{N}$ and $N = \max \ell_F$. For each $k \geq 1$, the language

$$F_k = \{\, w \in A^* \mid (|w| \bmod N + k) \in \ell_F \,\}$$

clearly belongs to MOD. By construction, it is easy to see that $F \subseteq F_k$ and $\delta_A(F_k) = \#(\ell_F)/(N + k)$ holds. This means that the density of F_k tends to zero if k tends to infinity, *i.e.*, F_k converges to F from outer. Since MOD is closed under Boolean operations, we have $\mathrm{Ext}_A(\mathsf{FIN}) \subseteq \mathrm{Ext}_A(\mathsf{MOD})$ by Lemma 3.

(2) Let $a \in A$ and $B = A \backslash \{a\}$. First we show the Gcom-measurability of $B^* = (A^* a A^*)^c \in \mathsf{AT}$. The construction of approximations is similar with the proof of Item (1). For each $k \geq 2$, the language $L_{k,0}^a = \{\, w \in A^* \mid |w|_a \equiv 0 \mod k \,\}$ is a commutative group language as stated in Sect. 2. Clearly, $L_{k,0}^a$ satisfies $B^* \subseteq L_k$ and it is easy to see that $\lim_{k \to \infty} \delta_A(L_k) = 0$ holds. Also, Gcom is closed under Boolean operations hence we have $\mathrm{Ext}_A(\mathsf{AT}) \subseteq \mathrm{Ext}_A(\mathsf{Gcom})$ by Lemma 3.

(3) The inclusion $\mathrm{Ext}_A(\mathsf{Gcom}) \subseteq \mathrm{Ext}_A(\mathsf{Com})$ is clear because $\mathsf{Gcom} \subseteq \mathsf{Com}$ holds and by the monotonicity of Ext_A (Lemma 3). To show the reverse inclusion, it is enough to show that $\mathsf{Com} \subseteq \mathrm{Ext}_A(\mathsf{Gcom})$ holds thanks to the idempotency of Ext_A (Lemma 3). We use the following fact [3, Proposition 1.11]: Com is the Boolean algebra generated by the languages of the form $L(a,r) = \{\, w \in A^+ \,|\, |w|_a = r \,\}$ and $L_{q,r}^a$ where $a \in A$ and $0 \le r < q$. The Gcom-measurability of $L(a,r)$ can be shown by the same manner with the proof of Item (2). Hence we have $\mathrm{Ext}_A(\mathsf{Gcom}) = \mathrm{Ext}_A(\mathsf{Com})$ by Lemma 3. $\square$

If we want to show the $\mathcal{C}$-measurability of a given language, the tactics is rather clear: to create a convergent sequence of languages in $\mathcal{C}$. However, to show the $\mathcal{C}$-*im*measurability of a given language, there is no routine tactics and it is much harder in most cases. The following theorem gives infinitely many non-trivial examples of Gcom-immeasurable languages.

Theorem 1. *Every group language whose syntactic monoid is non-commutative is* Gcom-*immeasurable.*

Proof. Let $L \in \mathsf{G}_A$ and suppose that the syntactic monoid $G = M_L$ is a non-commutative group. Let $\alpha \colon A^* \to G$ be the canonical surjection. Since G is non-commutative, the commutator subgroup $[G, G]$ of G is nontrivial, i.e., $\{1\} \subsetneq [G, G] \subseteq G$. Hence the abelianisation $G^{\mathrm{ab}} = G/[G, G]$ of G satisfies $\#(G) > \#(G^{\mathrm{ab}})$. Let $\pi \colon G \to G^{\mathrm{ab}}$ be the natural surjection. Then L is not recognised by $\pi \circ \alpha$ (otherwise G *divides* G^{ab}, a contradiction). That is, there exist two words $u, v \in A^*$ such that $u \in L \not\ni v$ and $\pi \circ \alpha(u) = \pi \circ \alpha(v)$. Hence we have $\alpha(u)^{-1}\alpha(v) \in [G, G]$, i.e., there exist $x_1, y_1, \ldots, x_l, y_l \in G$ such that

$$\alpha(u)^{-1}\alpha(v) = [x_1, y_1]\cdots[x_l, y_l] = \prod_{i=1}^{l}[x_i, y_i]$$

(note that we do not have to consider the inverses of commutators since $[x, y]^{-1} = [y, x]$ in general). Since α is surjective, there exist $4l$ words $s_i, t_i, \bar{s}_i, \bar{t}_i \in A^*$ $(i = 1, \ldots, l)$ such that $\alpha(s_i) = x_i$, $\alpha(t_i) = y_i$, $\alpha(\bar{s}_i) = x_i^{-1}$, and $\alpha(\bar{t}_i) = y_i^{-1}$. Define two words $w, w' \in A^*$ by

$$w = \prod_{i=1}^{l} s_i t_i \bar{t}_i \bar{s}_i, \qquad\qquad w' = \prod_{i=1}^{l} s_i t_i \bar{s}_i \bar{t}_i.$$

Then we have

$$\alpha(w) = \prod_{i=1}^{l} x_i y_i y_i^{-1} x_i^{-1} = 1, \qquad \alpha(w') = \prod_{i=1}^{l}[x_i, y_i] = \alpha(u)^{-1}\alpha(v).$$

Note that w' is a rearrangement of w.

Consider two languages $\alpha^{-1}(\alpha(u))$ and A^*wA^* in REG_A. Their densities are $\delta_A(\alpha^{-1}(\alpha(u))) = \#(G)^{-1} > 0$ (by Lemma 2) and $\delta_A(A^*wA^*) = 1$ (by

Example 1). The intersection $I = \alpha^{-1}(\alpha(u)) \cap A^*wA^*$ also has density $\delta_A(I) = \#(G)^{-1} > 0$ by Lemma 1.

To prove that L is Gcom-immeasurable, it suffices to show that every $K, M \in$ Gcom$_A$ with $K \subseteq L \subseteq M$ satisfies $I \subseteq M \backslash K$ since I has positive density. Let $s \in I$. Since $s \in \alpha^{-1}(\alpha(u))$ and $u \in L$, we have $s \in L \subseteq M$. Since $s \in A^*wA^*$, there exist $t_1, t_2 \in A^*$ such that $s = t_1wt_2$. The rearrangement $s' = t_1t_2w'$ of s satisfies $\alpha(s') = \alpha(t_1)\alpha(w)\alpha(t_2)\alpha(w') = \alpha(s)\alpha(u)^{-1}\alpha(v) = \alpha(v)$, hence $s' \notin L \supseteq K$. Since $s' \notin K \in$ Gcom$_A$ and s' is a rearrangement of s, we have $s \notin K$. Thus $s \in M \backslash K$.   □

Corollary 1. *For any group language L whose syntactic monoid is non-commutative and for any* Gcom-*measurable language M, their symmetric difference $L \triangle M$ is an infinite set.*

Proof. Let $L \in \mathsf{G}_A$ and $M \in \mathrm{Ext}_A(\mathsf{Gcom}_A)$ such that M_L is non-commutative. Suppose contrarily that $L \triangle M$ is finite. Then $L \triangle M \in \mathrm{Ext}_A(\mathsf{Gcom}_A)$ and hence $(L \triangle M) \triangle M \in \mathrm{Ext}_A(\mathsf{Gcom}_A)$ since $\mathrm{Ext}_A(\mathsf{Gcom}_A)$ is closed under Boolean combination. Thus $L \in \mathrm{Ext}_A(\mathsf{Gcom}_A)$, which contradicts Theorem 1.   □

4   Decidable Characterisation of **Gcom**-Measurability

The goal of this section is to prove the following Theorem 2, which gives the decidability of Gcom-measurability for regular languages. The following definition extends the Parikh mapping into the set of integer vectors so that we can use group theoretic tools. Intuitively, the set of vectors $\mathbb{Z}\mathrm{Pkh}_A(L) \cap \mathbb{Z}\mathrm{Pkh}_A(L^c)$ defined in Theorem 2 is a "boundary" between L and its complement from the viewpoint of commutative group languages. Hence, the equivalence of Condition (1) and Condition (2) in Theorem 2 states that a regular language L is Gcom-measurable if and only if the boundary $\mathbb{Z}\mathrm{Pkh}_A(L) \cap \mathbb{Z}\mathrm{Pkh}_A(L^c)$ has a *smaller rank* (than $\#(A)$). In other words, L is Gcom-measurable if and only if the boundary is *negligible* from the viewpoint of commutative group languages.

Definition 5 ($\mathbb{Z}\mathrm{Pkh}_A$). *Let $L \in \mathrm{REG}_A$ and choose a representation of the semilinear set $\mathrm{Pkh}_A(L) \subseteq \mathbb{N}^{\#(A)}$:*

$$\mathrm{Pkh}_A(L) = \bigcup_i (\boldsymbol{c}_i + \mathrm{span}_{\mathbb{N}}(\boldsymbol{p}_{i,1}, \ldots, \boldsymbol{p}_{i,\mu(i)})). \tag{1}$$

Then we define a semilinear set $\mathbb{Z}\mathrm{Pkh}_A(L) \subseteq \mathbb{Z}^{\#(A)}$ as

$$\mathbb{Z}\mathrm{Pkh}_A(L) \overset{\mathrm{def}}{=} \bigcup_i (\boldsymbol{c}_i + \mathrm{span}_{\mathbb{Z}}(\boldsymbol{p}_{i,1}, \ldots, \boldsymbol{p}_{i,\mu(i)})).$$

Note that $\mathbb{Z}\mathrm{Pkh}_A(L)$ depends on the choice (1) of representation of semilinear set. Our results are, however, true for any choice of representation.

Theorem 2. *Let $L \in \mathrm{REG}_A$ and*

$$\mathbb{Z}\mathrm{Pkh}_A(L) = \bigcup_i \Lambda_i, \quad \Lambda_i = c_i + \mathrm{span}_{\mathbb{Z}}\left(p_{i,1}, \ldots, p_{i,\mu(i)}\right) \subseteq \mathbb{Z}^{\#(A)},$$

$$\mathbb{Z}\mathrm{Pkh}_A(L^c) = \bigcup_j \Lambda'_j, \quad \Lambda'_j = c'_j + \mathrm{span}_{\mathbb{Z}}\left(q_{j,1}, \ldots, q_{j,\nu(j)}\right) \subseteq \mathbb{Z}^{\#(A)},$$

$$\mathbb{Z}\mathrm{Pkh}_A(L) \cap \mathbb{Z}\mathrm{Pkh}_A(L^c) = \bigcup_k \Lambda''_k, \quad \Lambda''_k = c''_k + \mathrm{span}_{\mathbb{Z}}\left(r_{k,1}, \ldots, r_{k,\xi(k)}\right) \subseteq \mathbb{Z}^{\#(A)}.$$

Then the following are equivalent.

(1) L is Gcom-measurable.
(2) For each (i,j) with $\Lambda_i \cap \Lambda'_j \neq \varnothing$, $\mathrm{rank}(\Lambda_i) < \#(A)$ or $\mathrm{rank}(\Lambda'_j) < \#(A)$.
(3) For each k, $\mathrm{rank}(\Lambda''_k) < \#(A)$.
(4) For each (i,j) with $\Lambda_i \cap \Lambda'_j \neq \varnothing$,
  – $\dim(\mathrm{span}_{\mathbb{R}}\left(p_{i,1}, \ldots, p_{i,\mu(i)}\right)) < \#(A)$ or
  – $\dim(\mathrm{span}_{\mathbb{R}}\left(q_{j,1}, \ldots, q_{j,\nu(j)}\right)) < \#(A)$.
(5) For each k, $\dim(\mathrm{span}_{\mathbb{R}}\left(r_{k,1}, \ldots, r_{k,\xi(k)}\right)) < \#(A)$.

The intersection of two semilinear sets, which is again semilinear as stated in Sect. 2, and its rank can be effectively computable, thus we obtain the decidability of Gcom-measurability.

Corollary 2. *It is decidable whether a given regular language L is Gcom-measurable or not.*

Proving Theorem 2 involves several lemmata, propositions, and an approximation notion which we call *standard approximation*.

For each $n \in \mathbb{N}$, we write $\mathrm{Pkh}[n]_A \colon A^* \to (\mathbb{Z}/n\mathbb{Z})^{\#(A)}$ for the composition of $\mathrm{Pkh}_A \colon A^* \to \mathbb{N}^{\#(A)}$ and the natural surjection $\mathbb{N}^{\#(A)} \to (\mathbb{Z}/n\mathbb{Z})^{\#(A)}$.

Definition 6 (standard approximation). *Let $L \subseteq A^*$ be a (not necessarily regular) language. The* standard approximation *of L is the two sequences $(\overline{L}_n)_{n\in\mathbb{N}}, (\underline{L}_n)_{n\in\mathbb{N}}$ of regular languages defined as*

$$\overline{L}_n \stackrel{\text{def}}{=} \mathrm{Pkh}[n!]_A^{-1}(\mathrm{Pkh}[n!]_A(L)), \qquad \underline{L}_n \stackrel{\text{def}}{=} \mathrm{Pkh}[n!]_A^{-1}(\mathrm{Pkh}[n!]_A(L^c))^c.$$

It is easy to see that $\underline{L}_n \subseteq \underline{L}_{n+1} \subseteq L \subseteq \overline{L}_{n+1} \subseteq \overline{L}_n$ for each $n \in \mathbb{N}$. That is, $\overline{L}_n$ (resp. $\underline{L}_n$) approximates L from outer (resp. from inner).

Lemma 5. *For every $L \in \mathrm{Gcom}_A$, there exists some $d \in \mathbb{N}$ such that L is recognised by $\mathrm{Pkh}[d]_A \colon A^* \to (\mathbb{Z}/d\mathbb{Z})^{\#(A)}$.*

Proof. Let G be a finite commutative group and $\alpha \colon A^* \to G$ be a homomorphism recognising L. Define $d = \mathrm{lcm}\{\mathrm{ord}(\alpha(a)) \mid a \in A\}$, where $\mathrm{ord}(\alpha(a))$ denotes the *order* of $\alpha(a)$ in G. One can then define a well-defined group homomorphism

$\varphi\colon (\mathbb{Z}/d\mathbb{Z})^A \to G$ such that $\varphi((x_a)_{a\in A}) = \prod_{a\in A}\alpha(a)^{x_a}$ for each $(x_a)_{a\in A} \in (\mathbb{Z}/d\mathbb{Z})^A$. It is easy to see that $\alpha = \varphi \circ \mathrm{Pkh}[d]_A$ and

$$L \subseteq \mathrm{Pkh}[d]_A^{-1}(\mathrm{Pkh}[d]_A(L)) \subseteq \alpha^{-1}(\alpha(L)) = L,$$

thus $\mathrm{Pkh}[d]_A$ recognises L.   □

Lemma 6. *Let* $L \subseteq A^*$ *be a language and* $(\underline{L}_n \subseteq L \subseteq \overline{L}_n)_{n\in\mathbb{N}}$ *be the standard approximation of* L. *Then the following claims hold.*

(1) If $L \subseteq M \in \mathsf{Gcom}_A$, *then* $L \subseteq \overline{L}_n \subseteq M$ *for any sufficiently large* $n \in \mathbb{N}$.
(2) If $\mathsf{Gcom}_A \ni K \subseteq L$, *then* $K \subseteq \underline{L}_n \subseteq L$ *for any sufficiently large* $n \in \mathbb{N}$.

Proof.

(1) By Lemma 5, we may assume that M is recognised by $\mathrm{Pkh}[d]_A$ for some $d \in \mathbb{N}$. Let $n \in \mathbb{N}$ be sufficiently large so that d divides the factorial $n!$. Then the natural surjection $\rho_{n!}\colon (\mathbb{Z}/n!\mathbb{Z})^{\#(A)} \to (\mathbb{Z}/d\mathbb{Z})^{\#(A)}$ makes the diagram

$$
\begin{array}{ccc}
A^* & \xrightarrow{\mathrm{Pkh}[n!]_A} & (\mathbb{Z}/n!\mathbb{Z})^{\#(A)} \\
 & \searrow{\scriptstyle \mathrm{Pkh}[d]_A} & \downarrow{\scriptstyle \rho_{n!}} \\
 & & (\mathbb{Z}/d\mathbb{Z})^{\#(A)}
\end{array}
$$

commute. Thus

$$
\begin{aligned}
L &\subseteq \mathrm{Pkh}[n!]_A^{-1}(\mathrm{Pkh}[n!]_A(L)) = \overline{L}_n \\
&\subseteq \mathrm{Pkh}[n!]_A^{-1}(\rho_{n!}^{-1}(\rho_{n!}(\mathrm{Pkh}[n!]_A(L)))) \\
&= \mathrm{Pkh}[d]_A^{-1}(\mathrm{Pkh}[d]_A(L)) \subseteq \mathrm{Pkh}[d]_A^{-1}(\mathrm{Pkh}[d]_A(M)) = M.
\end{aligned}
$$

(2) Similarly, we may assume that K is recognised by $\mathrm{Pkh}[d]_A$ and

$$
\begin{aligned}
L = (L^c)^c &\supseteq \mathrm{Pkh}[n!]_A^{-1}(\mathrm{Pkh}[n!]_A(L^c))^c = \underline{L}_n \\
&\supseteq \mathrm{Pkh}[n!]_A^{-1}(\rho_{n!}^{-1}(\rho_{n!}(\mathrm{Pkh}[n!]_A(L^c))))^c = \mathrm{Pkh}[d]_A^{-1}(\mathrm{Pkh}[d]_A(L^c))^c \\
&\supseteq \mathrm{Pkh}[d]_A^{-1}(\mathrm{Pkh}[d]_A(K^c))^c
\end{aligned}
$$

since $\mathrm{Pkh}[d]_A$ is surjective and recognises K,

$$= \mathrm{Pkh}[d]_A^{-1}(\mathrm{Pkh}[d]_A(K)) = K.$$ □

Proposition 4. *For any language* $L \subseteq A^*$ *and its standard approximation* $(\underline{L}_n \subseteq L \subseteq \overline{L}_n)_{n\in\mathbb{N}}$, *the following are equivalent.*

(1) L is Gcom-*measurable.*
(2) $\lim_{n\to\infty} \delta_A(\overline{L}_n \backslash \underline{L}_n) = 0$.

Proof. The implication (2) $\Longrightarrow$ (1) is obvious. Conversely, assume that L is Gcom-measurable. Then there exist sequences $(K_n \subseteq L \subseteq M_n)_{n \in \mathbb{N}}$ such that $K_n, M_n \in \text{Gcom}_A$ and $\lim_{n \to \infty} \delta_A(M_n \backslash K_n) = 0$. By Lemma 6, there exists a non-decreasing sequence $(N(n))_{n \in \mathbb{N}}$ such that $K_n \subseteq \underline{L}_{N(n)} \subseteq L \subseteq \overline{L}_{N(n)} \subseteq M_n$ for each $n \in \mathbb{N}$. Thus $\lim_{n \to \infty} \delta_A(\overline{L}_n \backslash \underline{L}_n) = \lim_{n \to \infty} \delta_A(\overline{L}_{N(n)} \backslash \underline{L}_{N(n)}) \leq \lim_{n \to \infty} \delta_A(M_n \backslash K_n) = 0$. $\qquad \square$

Proposition 5. *For any language $L \subseteq A^*$ and its standard approximation $(\underline{L}_n \subseteq L \subseteq \overline{L}_n)_{n \in \mathbb{N}}$, we have*

$$\delta_A(\overline{L}_n \backslash \underline{L}_n) = \frac{\#(\text{Pkh}[n!]_A(L) \cap \text{Pkh}[n!]_A(L^c))}{(n!)^{\#(A)}}.$$

Proof. We have

$$\overline{L}_n \backslash \underline{L}_n = \text{Pkh}[n!]_A^{-1}(\text{Pkh}[n!]_A(L)) \backslash \text{Pkh}[n!]_A^{-1}(\text{Pkh}[n!]_A(L^c))^c$$
$$= \text{Pkh}[n!]_A^{-1}(\text{Pkh}[n!]_A(L) \cap \text{Pkh}[n!]_A(L^c))$$

and thus Lemma 2 completes the proof. $\qquad \square$

Proposition 6. *Let $L \in \text{REG}_A$ and*

$$\mathbb{Z}\text{Pkh}_A(L) = \bigcup_i \Lambda_i, \qquad\qquad \mathbb{Z}\text{Pkh}_A(L^c) = \bigcup_j \Lambda'_j,$$

where $\Lambda_i, \Lambda'_j \subseteq \mathbb{Z}^{\#(A)}$ are linear sets. Then the following are equivalent.

(1) $\displaystyle\lim_{n \to \infty} \frac{\#(\text{Pkh}[n!]_A(L) \cap \text{Pkh}[n!]_A(L^c))}{(n!)^{\#(A)}} = 0.$

(2) For any (i, j),

$$\lim_{n \to \infty} \frac{\#(\pi_{n!}(\Lambda_i) \cap \pi_{n!}(\Lambda'_j))}{(n!)^{\#(A)}} = 0,$$

where $\pi_{n!} \colon \mathbb{Z}^{\#(A)} \to (\mathbb{Z}/n!\mathbb{Z})^{\#(A)}$ is the natural surjection.

Proof. Note that

$$\#(\text{Pkh}[n!]_A(L) \cap \text{Pkh}[n!]_A(L^c)) = \#(\pi_{n!}(\mathbb{Z}\text{Pkh}_A(L)) \cap \pi_{n!}(\mathbb{Z}\text{Pkh}_A(L^c)))$$

$$= \#\left(\bigcup_{i,j} (\pi_{n!}(\Lambda_i) \cap \pi_{n!}(\Lambda'_j)) \right).$$

(1) $\Longrightarrow$ (2). Contrarily suppose that

$$\lim_{n \to \infty} \frac{\#(\pi_{n!}(\Lambda_{i_0}) \cap \pi_{n!}(\Lambda'_{j_0}))}{(n!)^{\#(A)}} > 0$$

for some (i_0, j_0) (note that the limit exists because of the monotonicity). Then

$$
\lim_{n\to\infty} \frac{\#(\mathrm{Pkh}[n!]_A(L) \cap \mathrm{Pkh}[n!]_A(L^c))}{(n!)^{\#(A)}} = \lim_{n\to\infty} \frac{\#\left(\bigcup_{i,j}(\pi_{n!}(\Lambda_i) \cap \pi_{n!}(\Lambda_j'))\right)}{(n!)^{\#(A)}}
$$

$$
\geq \lim_{n\to\infty} \frac{\#\left(\pi_{n!}(\Lambda_{i_0}) \cap \pi_{n!}(\Lambda_{j_0}')\right)}{(n!)^{\#(A)}} > 0.
$$

$(2) \implies (1)$. We have

$$
\lim_{n\to\infty} \frac{\#(\mathrm{Pkh}[n!]_A(L) \cap \mathrm{Pkh}[n!]_A(L^c))}{(n!)^{\#(A)}} = \lim_{n\to\infty} \frac{\#\left(\bigcup_{i,j}(\pi_{n!}(\Lambda_i) \cap \pi_{n!}(\Lambda_j'))\right)}{(n!)^{\#(A)}}
$$

$$
\leq \lim_{n\to\infty} \frac{\sum_{i,j} \#\left(\pi_{n!}(\Lambda_i) \cap \pi_{n!}(\Lambda_j')\right)}{(n!)^{\#(A)}} = \sum_{i,j} \lim_{n\to\infty} \frac{\#\left(\pi_{n!}(\Lambda_i) \cap \pi_{n!}(\Lambda_j')\right)}{(n!)^{\#(A)}}
$$

$$
= 0.
$$

$\square$

Proposition 7. *Let $\Lambda_1, \Lambda_2 \subseteq \mathbb{Z}^{\#(A)}$ be two linear sets such that $\Lambda_1 \cap \Lambda_2 \neq \varnothing$. Then the following are equivalent.*

(1) $\displaystyle\lim_{n\to\infty} \frac{\#(\pi_{n!}(\Lambda_1) \cap \pi_{n!}(\Lambda_2))}{(n!)^{\#(A)}} = 0$.
(2) $\mathrm{rank}(\Lambda_1) < \#(A)$ *or* $\mathrm{rank}(\Lambda_2) < \#(A)$.
(3) $\mathrm{rank}(\Lambda_1 \cap \Lambda_2) < \#(A)$.

Proof.

$(1) \implies (2)$. Contrarily suppose that $\mathrm{rank}(\Lambda_1) = \mathrm{rank}(\Lambda_2) = \#(A)$. Since the indices $(\mathbb{Z}^{\#(A)} : \Lambda_1)$ and $(\mathbb{Z}^{\#(A)} : \Lambda_2)$ are finite by Proposition 1, we have

$$
\lim_{n\to\infty} \frac{\#(\pi_{n!}(\Lambda_1) \cap \pi_{n!}(\Lambda_2))}{(n!)^{\#(A)}} = \lim_{n\to\infty} \frac{\#(\pi_{n!}(\Lambda_1) \cap \pi_{n!}(\Lambda_2))}{\#(\mathbb{Z}/n!\mathbb{Z})^{\#(A)}}
$$

$$
= \lim_{n\to\infty} \frac{1}{\left((\mathbb{Z}/n!\mathbb{Z})^{\#(A)} : \pi_{n!}(\Lambda_1) \cap \pi_{n!}(\Lambda_2)\right)}
$$

$$
\geq \lim_{n\to\infty} \frac{1}{\left((\mathbb{Z}/n!\mathbb{Z})^{\#(A)} : \pi_{n!}(\Lambda_1)\right)} \cdot \frac{1}{\left((\mathbb{Z}/n!\mathbb{Z})^{\#(A)} : \pi_{n!}(\Lambda_2)\right)}
$$

$$
= \lim_{n\to\infty} \frac{1}{\left(\mathbb{Z}^{\#(A)} : \Lambda_1 + n!\mathbb{Z}^{\#(A)}\right)} \cdot \frac{1}{\left(\mathbb{Z}^{\#(A)} : \Lambda_2 + n!\mathbb{Z}^{\#(A)}\right)}
$$

$$
\geq \frac{1}{\left(\mathbb{Z}^{\#(A)} : \Lambda_1\right)} \cdot \frac{1}{\left(\mathbb{Z}^{\#(A)} : \Lambda_2\right)} > 0.
$$

(2) $\implies$ (1). We may assume that $\mathrm{rank}(\Lambda_1) < \#(A)$. Then we have

$$\lim_{n\to\infty} \frac{\#(\pi_{n!}(\Lambda_1) \cap \pi_{n!}(\Lambda_2))}{(n!)^{\#(A)}} = \lim_{n\to\infty} \frac{1}{\left((\mathbb{Z}/n!\mathbb{Z})^{\#(A)} : \pi_{n!}(\Lambda_1) \cap \pi_{n!}(\Lambda_2)\right)}$$

$$\leq \lim_{n\to\infty} \frac{1}{\left((\mathbb{Z}/n!\mathbb{Z})^{\#(A)} : \pi_{n!}(\Lambda_1)\right)} = 0.$$

The last equality is exactly Proposition 2.

(2) $\implies$ (3). $\mathrm{rank}(\Lambda_1 \cap \Lambda_2) \leq \min\{\mathrm{rank}(\Lambda_1), \mathrm{rank}(\Lambda_2)\} < \#(A)$.

(3) $\implies$ (2). Contrarily suppose that $\mathrm{rank}(\Lambda_1) = \mathrm{rank}(\Lambda_2) = \#(A)$. Since Λ_1, Λ_2 are coset of $\mathbb{Z}^{\#(A)}$, there exist constant vectors $c_1, c_2 \in \mathbb{Z}^{\#(A)}$ and subgroups $M_1, M_2 \subseteq \mathbb{Z}^{\#(A)}$ such that $\Lambda_1 = c_1 + M_1$, $\Lambda_2 = c_2 + M_2$. Since $\Lambda_1 \cap \Lambda_2 \neq \varnothing$, there exists $c \in M_1 \cap M_2$ such that $\Lambda_1 \cap \Lambda_2 = c + (M_1 \cap M_2)$. Since $\mathrm{rank}(M_1) = \mathrm{rank}(\Lambda_1) = \#(A)$ and $\mathrm{rank}(M_2) = \mathrm{rank}(\Lambda_2) = \#(A)$, by Proposition 1, there exists $N_1, N_2 \in \mathbb{N}$ such that $N_1\mathbb{Z}^{\#(A)} \subseteq M_1$ and $N_2\mathbb{Z}^{\#(A)} \subseteq M_2$. Thus $N_1 N_2 \mathbb{Z}^{\#(A)} \subseteq M_1 \cap M_2$ and $\#(A) \geq \mathrm{rank}(\Lambda_1 \cap \Lambda_2) = \mathrm{rank}(M_1 \cap M_2) \geq \mathrm{rank}(N_1 N_2 \mathbb{Z}^{\#(A)}) = \#(A)$. $\qquad\square$

Proof (of Theorem 2). The equivalences (2) $\iff$ (4) and (3) $\iff$ (5) are clear. For the standard approximation $(\underline{L}_n \subseteq L \subseteq \overline{L}_n)_{n \in \mathbb{N}}$ of L,

$$(1) \iff \lim_{n\to\infty} \delta_A(\overline{L}_n \backslash \underline{L}_n) = 0 \qquad\qquad \text{(Proposition 4)}$$

$$\iff \lim_{n\to\infty} \frac{\#(\mathrm{Pkh}[n!]_A(L) \cap \mathrm{Pkh}[n!]_A(L^c))}{(n!)^{\#(A)}} = 0 \qquad \text{(Proposition 5)}$$

$$\iff \forall i,j \left[\lim_{n\to\infty} \frac{\#\left(\pi_{n!}(\Lambda_i) \cap \pi_{n!}(\Lambda'_j)\right)}{(n!)^{\#(A)}} = 0 \right] \qquad \text{(Proposition 6)}$$

$$\iff \forall i,j \begin{bmatrix} \Lambda_i \cap \Lambda'_j = \varnothing \ \vee \\ \mathrm{rank}(\Lambda_i) < \#(A) \vee \mathrm{rank}(\Lambda'_j) < \#(A) \end{bmatrix} \qquad \text{(Proposition 7)}$$

$$(\iff (2))$$

$$\iff \forall k \left[\mathrm{rank}(\Lambda''_k) < \#(A) \right] \qquad\qquad \text{(Proposition 7)}$$

$$(\iff (3)).$$

$$\square$$

5   Simple Characterisation of **MOD**-Measurability

The decidability of MOD-measurability for regular languages is essentially already given by Theorem 2. Moreover, we have a simple language theoretic characterisation of MOD-measurable languages as follows.

Theorem 3. *For any $L \in \mathrm{REG}_A$, the following are equivalent.*

(1) L *is* MOD-*measurable.*

(2) $\operatorname{len}_A(L) \cap \operatorname{len}_A(L^c) \subseteq \mathbb{N}$ *is a finite set, where* $\operatorname{len}_A \colon A^* \to \mathbb{N}$ *is the length function* $w \mapsto |w|$.

(3) *There exists a unique* $M \in \mathsf{MOD}_A$ *such that* $L \triangle M$ *is a finite set.*

Proof. The proof of the equivalence (1) $\Longleftrightarrow$ (2) is similar to that of Theorem 2. Indeed, it suffices to replace all the occurrences of Gcom, Pkh, and $\#(A)$ in Sect. 4 by MOD, len, and 1, respectively. (Note that $\mathbb{Z}\operatorname{len}_A(L) \cap \mathbb{Z}\operatorname{len}_A(L^c) \subseteq \mathbb{Z}$ is finite if and only if $\operatorname{len}_A(L) \cap \operatorname{len}_A(L^c) \subseteq \mathbb{N}$ is finite.)

(3) $\Longrightarrow$ (1). From the assumption, $L \triangle M$ and M are both MOD-measurable. The class $\mathrm{Ext}_A(\mathsf{MOD}_A)$ is closed under Boolean combination, hence $L = (L \triangle M) \triangle M \in \mathrm{Ext}_A(\mathsf{MOD}_A)$.

(2) $\Longrightarrow$ (3). Since $\operatorname{len}_A(L)$ is a semilinear set of $\mathbb{N}$, we may assume that

$$\operatorname{len}_A(L) = \bigcup_i (c_i + \operatorname{span}_{\mathbb{N}}(p_i)) \cup \bigcup_j \{d_j\} \quad (p_i > 0 \text{ for each } i).$$

Define $S = \bigcup_i (c_i' + \operatorname{span}_{\mathbb{N}}(p_i))$, where $c_i' = c_i \bmod p_i$. Then we have $\operatorname{len}_A^{-1}(S) \in \mathsf{MOD}_A$. Since $\operatorname{len}_A(L) \triangle S$ is a finite set by construction, the inverse image $\operatorname{len}_A^{-1}(\operatorname{len}_A(L) \triangle S) = \operatorname{len}_A^{-1}(\operatorname{len}_A(L)) \triangle \operatorname{len}_A^{-1}(S)$ is also finite. Since we have

$$L \triangle \operatorname{len}_A^{-1}(\operatorname{len}_A(L)) = \operatorname{len}_A^{-1}(\operatorname{len}_A(L)) \cap L^c \subseteq \operatorname{len}_A^{-1}(\operatorname{len}_A(L) \cap \operatorname{len}_A(L^c)),$$

the set $L \triangle \operatorname{len}_A^{-1}(\operatorname{len}_A(L))$ is finite by the assumption (2). Thus

$$L \triangle \operatorname{len}_A^{-1}(S) = (L \triangle \operatorname{len}_A^{-1}(\operatorname{len}_A(L))) \triangle (\operatorname{len}_A^{-1}(\operatorname{len}_A(L)) \triangle \operatorname{len}_A^{-1}(S))$$

is a finite set.

Suppose that there exist two languages $M_1, M_2 \in \mathsf{MOD}_A$ such that both $L \triangle M_1$ and $L \triangle M_2$ are finite sets. Then $M_1 \triangle M_2 = (L \triangle M_1) \triangle (L \triangle M_2)$ is also a finite set. Since MOD_A is closed under Boolean combination, $M_1 \triangle M_2 \in \mathsf{MOD}_A$. Thus $M_1 = M_2$ since the only finite set in MOD_A is $\varnothing$. $\square$

6   Conclusion and Future Work

As we described in Sect. 1, while the measuring power of subclasses of star-free languages are systematically studied so far, nothing is known for the measuring power of group languages before. This paper gave the first decidability results on this topic: the Gcom-measurability and the MOD-measurability for regular languages are both decidable thanks to Theorem 2. Also, Theorem 1 tells us that there is a huge gap between group and commutative group languages even from a (very rough) measure theoretic point of view. To clarify the computational complexity of these two measurability and the decidability of the G-measurability for regular languages are our important future work. Also, to give a purely algebraic characterisation of the G-measurability (the Gcom-measurability, respectively) is an interesting open problem for us. For the case of MOD-measurability, we have very simple language theoretic characterisation as stated in Theorem 3. We are interested whether it is possible to obtain a similar language theoretic characterisation of Gcom-measurability.

No different characterisation of REG-measurability is yet known, and only few examples of REG-*im*measurable languages are known (*cf.* [6]). The Krohn-Rhodes theorem states that, for every finite monoid M, there exists a sequence $G_1, \ldots, G_n$ of finite groups dividing M and a sequence $M_0, \ldots, M_n$ of aperiodic finite monoids such that M divides $M_0 \circ G_1 \circ M_1 \circ \cdots \circ G_n \circ M_n$ where $\circ$ is the wreath product operation (*cf.* [3]). Roughly speaking, this means that every regular language can be represented as some "combination" of group and star-free languages because a language is star-free if and only if its syntactic monoid is aperiodic. In this sense, we can consider the class of group languages and the class of star-free languages as two representative subclasses of regular languages. We hope that a deep understanding of the REG-measurable languages (or, group languages its self) might be obtained by a further study of the measuring power of group languages.

Acknowledgements. The authors thank to anonymous reviewers for many valuable comments. This work was supported by JSPS KAKENHI Grant Number JP20H05961 and JST ACT-X Grant Number JPMJAX210B.

References

1. Berstel, J., Perrin, D., Reutenauer, C.: Codes and Automata, Encyclopedia of Mathematics and its Applications, vol. 129. Cambridge University Press, Cambridge (2010)
2. Lawson, M.V.: Finite Automata. Chapman and Hall/CRC (2004)
3. Pin, J.E.: Mathematical foundations of automata theory (2022). https://www.irif.fr/~jep/PDF/MPRI/MPRI.pdf
4. Place, T., Zeitoun, M.: Group separation strikes back. In: 2023 38th Annual ACM/IEEE Symposium on Logic in Computer Science (LICS), pp. 1–13. IEEE Computer Society, Los Alamitos, CA, USA (2023)
5. Shallit, J.: A Second Course in Formal Languages and Automata Theory, 1st edn. Cambridge University Press, Cambridge (2008)
6. Sin'ya, R.: Asymptotic approximation by regular languages. In: Bureš, T., et al. (eds.) SOFSEM 2021. LNCS, vol. 12607, pp. 74–88. Springer, Cham (2021). https://doi.org/10.1007/978-3-030-67731-2_6
7. Sin'ya, R.: Carathéodory extensions of subclasses of regular languages. In: Moreira, N., Reis, R. (eds.) DLT 2021. LNCS, vol. 12811, pp. 355–367. Springer, Cham (2021). https://doi.org/10.1007/978-3-030-81508-0_29
8. Sin'ya, R.: Measuring power of locally testable languages. In: Diekert, V., Volkov, M. (eds.) DLT 2022. LNCS, vol. 13257, pp. 274–285. Springer, Cham (2022). https://doi.org/10.1007/978-3-031-05578-2_22
9. Sin'ya, R.: Measuring power of generalised definite languages. In: Nagy, B. (ed.) CIAA 2023. LNCS, vol. 14151, pp. 278–289. Springer, Cham (2023). https://doi.org/10.1007/978-3-031-40247-0_21

10. Sin'ya, R., Yamaguchi, Y., Nakamura, Y.: Regular languages that can be approximated by testing subword occurrences. Comput. Softw. **40**(2), 49–60 (2023). (written in Japanese)
11. Thierrin, G.: Permutation automata. Math. Syst. Theory **2**, 83–90 (1968)
12. Yuyama, T., Sin'ya, R.: Measuring power of commutative group languages (full version) (2024). http://t-yuyama.jp/pdfs/gcommeasure_full.pdf

Author Index

© The Editor(s) (if applicable) and The Author(s), under exclusive license
to Springer Nature Switzerland AG 2024
S. Z. Fazekas (Ed.): CIAA 2024, LNCS 15015, pp. 363–364, 2024.
https://doi.org/10.1007/978-3-031-71112-1

GPSR Compliance

The European Union's (EU) General Product Safety Regulation (GPSR) is a set of rules that requires consumer products to be safe and our obligations to ensure this.

If you have any concerns about our products, you can contact us on ProductSafety@springernature.com

In case Publisher is established outside the EU, the EU authorized representative is:

Springer Nature Customer Service Center GmbH
Europaplatz 3
69115 Heidelberg, Germany

The manufacturer's authorised representative in the EU is Springer
Nature Customer Service Centre GmbH, Europaplatz 3, 69115 Heidelberg,
Germany. If you have any concerns regarding our products, please
contact ProductSafety@springernature.com

Printed and bound by CPI Group (UK) Ltd, Croydon, CR0 4YY
24/04/2026
02096358-0013